Biosensors for Emerging and Re-emerging Infectious Diseases

Biosensors for Emerging and Re-emerging Infectious Diseases

Edited by

JAYASHANKAR DAS
Valnizen Healthcare PVT Limited, Mumbai, Maharashtra, India

SUSHMA DAVE
Department of Applied Science, JIET Group of Institutions, Jodhpur, Rajasthan, India

S. RADHAKRISHNAN
Department of Organic Materials and Fiber Engineering, Jeonbuk National University, Jeonju-si, Jeollabuk-do, South Korea

PADMAJA MOHANTY
Department of Botany, Bioinformatics and Climate change, University School of Sciences, Gujarat University, Ahmedabad, Gujrat, India

ELSEVIER

ACADEMIC PRESS
An imprint of Elsevier

Academic Press is an imprint of Elsevier
125 London Wall, London EC2Y 5AS, United Kingdom
525 B Street, Suite 1650, San Diego, CA 92101, United States
50 Hampshire Street, 5th Floor, Cambridge, MA 02139, United States
The Boulevard, Langford Lane, Kidlington, Oxford OX5 1GB, United Kingdom

ISBN: 978-0-323-88464-8

For Information on all Academic Press publications
visit our website at https://www.elsevier.com/books-and-journals

Publisher: Mara E. Conner
Acquisitions Editor: Carrie L. Bolger
Editorial Project Manager: Sara Greco
Production Project Manager: Surya Narayanan Jayachandran
Cover Designer: Christian J. Bilbow

Typeset by MPS Limited, Chennai, India

Contents

List of contributors

Amina Antonacci
Department of Chemical Sciences and Materials Technologies, Institute of Crystallography, National Research Council, Rome, Italy

Fabiana Arduini
Department of Chemical Science and Technologies, University of Rome Tor Vergata, Rome, Italy

Shahani Begum
Plant Biotechnology Laboratory, Post Graduate Department of Botany, Utkal University, Vani-Vihar, Bhubaneswar, Odisha, India

Stephen Rathinaraj Benjamin
Center for Research and Drug Development (NPDM), Federal University of Cearà, Fortaleza, Ceará, Brazil; Department of Physiology and Pharmacology, Faculty of Medicine, Federal University of Cearà, Fortaleza, Ceará, Brazil

Joydeep Chakraborty
Department of Microbiology, RK University, Rajkot, Gujarat, India

Seung-Cheol Chang
Department of Cogno-Mechatronics Engineering, College of Nanoscience and Nanotechnology, Pusan National University, Busan, Republic of Korea

Hardik Churi
Department of Biotechnology SDSM College, Mumbai, Maharashtra, India

Jayashankar Das
Valnizen Healthcare PVT Limited, Mumbai, Maharashtra, India

Kajari Das
Department of Biotechnology, College of Basic Science and Humanities, Odisha University of Agriculture and Technology, Bhubaneswar, Odisha, India

Trupti R. Das
CIPET: Institute of Petrochemicals Technology (IPT), Bhubaneswar, Odisha, India

Aditya Dave
IIT Madras (B.Sc. Data Science), Chennai, Tamil Nadu, India

Shivani Dave
Department of Computer Science, MBM Engineering College, Jodhpur, Rajasthan, India

Sushma Dave
Department of Applied Science, JIET Group of Institutions, Jodhpur, Rajasthan, India

Geanne Matos de Andrade
Center for Research and Drug Development (NPDM), Federal University of Cearà, Fortaleza, Ceará, Brazil; Department of Physiology and Pharmacology, Faculty of Medicine, Federal University of Cearà, Fortaleza, Ceará, Brazil

Tyciane de Souza Nascimento
Center for Research and Drug Development (NPDM), Federal University of Cearà, Fortaleza, Ceará, Brazil; Department of Clinical Medicine, Faculty of Medicine, Federal University of Cearà, Fortaleza, Ceará, Brazil

M. Dhanalakshmi
Research and Development Centre, Bharathiar University, Coimbatore, Tamil Nadu, India

Ayushman Gadnayak
Centre for Genomics and Molecular Therapeutics, IMS and SUMH, Siksha "O" Anusandhan University (Deemed to be), Bhubaneswar, Odisha, India

Loknath Garhnayak
Department of Prosthodontics, SCB Dental College & Hospital, Cuttack, Odisha, India

Mirna Garhnayak
Department of Prosthodontics, Institute of Dental Sciences, Siksha O Anusandhan (Deemed to be University), Bhubaneswar, Odisha, India

Faezeh Ghorbanizamani
Biochemistry Department, Faculty of Sciences, Ege University, Izmir, Turkey

Tuncay Goksel
Department of Pulmonary Medicine, Faculty of Medicine, Ege University, Izmir, Turkey

Swati Goswami
Department of Microbiology, RK University, Rajkot, Gujarat, India

Penny P. Govender
Department of Chemical Sciences, University of Johannesburg, Johannesburg, South Africa

Vivek A. Kamat
Electrical Engineering and Computing, Florida International University, Miami, FL, United States

Fahmida Khan
Department of Chemistry, National Institute of Technology, Raipur, Chhattisgarh, India

Vijay Kumar
Department of Microbiology, RK University, Rajkot, Gujarat, India

Aswini Kumar Kar
Department of Prosthodontics, Kalinga Institute of Dental Sciences, Bhubaneswar, Odisha, India

Abhijita Mahapatra
Department of Prosthodontics, Institute of Dental Sciences, Siksha O Anusandhan (Deemed to be University), Bhubaneswar, Odisha, India

Anjisha Maharshi
Department of Microbiology, Faculty of Science, Marwadi University, Rajkot, Gujarat, India

Remya Mohanraj
Houston Community College, Houston, TX, United States

Jatindra Nath Mohanty
Centre for Genomics and Molecular Therapeutics, IMS and SUMH, Siksha "O" Anusandhan University (Deemed to be), Bhubaneswar, Odisha, India

Padmaja Mohanty
Department of Botany, Bioinformatics and Climate change, University School of Sciences, Gujarat University, Ahmedabad, Gujarat, India

Abhijita Mohapatra
Department of Prosthodontics, Institute of Dental Sciences, Siksha O Anusandhan (Deemed to be University), Bhubaneswar, Odisha, India

Maheswata Moharana
Department of Chemistry, National Institute of Technology, Raipur, Chhattisgarh, India

Hichem Moulahoum
Biochemistry Department, Faculty of Sciences, Ege University, Izmir, Turkey

Ananya Nayak
Centre for Genomics and Molecular Therapeutics, IMS and SUMH, Siksha "O" Anusandhan University (Deemed to be), Bhubaneswar, Odisha, India

Reinaldo Barreto Oriá
Department of Morphology, Institute of Biomedicine, Laboratory of the Biology of Tissue Healing, Ontogeny and Nutrition, School of Medicine, Federal University of Cearà, Fortaleza, Ceará, Brazil

Sangram Panda
Department of Prosthodontics, Institute of Dental Sciences, Siksha O Anusandhan (Deemed to be University), Bhubaneswar, Odisha, India

Medha Pandya
Department of LifeSciences, Moharaja Krishnakumarsinhji Bhavnagar University, Bhavnagar, Gujarat, India

Rohan Panwar
Seth GS Medical College and KEM Hospital, Mumbai, Maharashtra, India

Santanu Patra
Institute of Advanced Materials, IAAM, Ulrika, Sweden; VBRI, New Delhi, Delhi, India

Subrat Kumar Pattanayak
Department of Chemistry, National Institute of Technology, Raipur, Chhattisgarh, India

Janak Paudyal
Department of Chemistry and Biochemistry, University of Colorado, Colorado Springs, CO, United States

S. Radhakrishnan
Department of Organic Materials and Fiber Engineering, Jeonbuk National University, Jeonju-si, Jeollabuk-do, South Korea

Shakti Rath
Central Research Laboratory, Institute of Dental Sciences, Siksha O Anusandhan (Deemed to be University), Bhubaneswar, Odisha, India

Cássia Rodrigues Roque
Department of Morphology, Institute of Biomedicine, Laboratory of the Biology of Tissue Healing, Ontogeny and Nutrition, School of Medicine, Federal University of Cearà, Fortaleza, Ceará, Brazil

Pragyan Roy
Department of Biotechnology, College of Basic Sciences and Humanities, Bhubaneswar, Odisha, India

Maheswata Sahoo
Centre for Genomics and Molecular Therapeutics, IMS and SUMH, Siksha "O" Anusandhan University (Deemed to be), Bhubaneswar, Odisha, India

Pradyumna Sahoo
Department of Prosthodontics, Institute of Dental Sciences, Siksha O Anusandhan (Deemed to be University), Bhubaneswar, Odisha, India

Swayamprabha Sahoo
Centre for Genomics and Molecular Therapeutics, IMS and SUMH, Siksha "O" Anusandhan University (Deemed to be), Bhubaneswar, Odisha, India

Priyanka Samal
Department of Clinical Hematology-Hemato-oncology and Stem Cell Transplant, IMS & SUM Hospital, Bhubaneswar, Odisha, India

Madhappan Santhamoorthy
Department of Biomaterials and Polymer Engineering, Yeongnam University, Gyeongsan, Republic of Korea

Viviana Scognamiglio
Department of Chemical Sciences and Materials Technologies, Institute of Crystallography, National Research Council, Rome, Italy

Sejal Shah
Department of Computer Science and Bioscience, Faculty of Engineering and Technology, Marwadi University, Rajkot, Gujarat, India

Richa Sharma
EaStCHEM, School of Chemistry, King's Buildings, University of Edinburgh, Edinburgh, United Kingdom

Sudheesh K. Shukla
Department of Chemical Sciences, University of Johannesburg, Johannesburg, South Africa; Department of Biomedical Engineering, School of Biological Engineering and Life Science, Shobhit Institute of Engineering and Technology (Deemed-to-be University), Meerut, Uttar Pradesh, India

Dinakaran Thirumalai
Department of Cogno-Mechatronics Engineering, College of Nanoscience and Nanotechnology, Pusan National University, Busan, Republic of Korea

Kokila Thirupathi
Department of Physics, Sri Moogambigai College of Arts and Science, Dharmapuri, Tamil Nadu, India

Suna Timur
Biochemistry Department, Faculty of Sciences, Ege University, Izmir, Turkey; Central Research Test and Analysis Laboratory Application and Research Center, Ege University, Izmir, Turkey

Kutsal Turhan
Department of Thoracic Surgery, Faculty of Medicine, Ege University, Izmir, Turkey

Figen Zihnioglu
Biochemistry Department, Faculty of Sciences, Ege University, Izmir, Turkey

CHAPTER 1

Biosensors: principle, fundamentals history, recent trends and applications

Trupti R. Das[1], Santanu Patra[2,3], Penny P. Govender[4] and Sudheesh K. Shukla[4,5]
[1]CIPET: Institute of Petrochemicals Technology (IPT), Bhubaneswar, Odisha, India
[2]Institute of Advanced Materials, IAAM, Ulrika, Sweden
[3]VBRI, New Delhi, Delhi, India
[4]Department of Chemical Sciences, University of Johannesburg, Johannesburg, South Africa
[5]Department of Biomedical Engineering, School of Biological Engineering and Life Science, Shobhit Institute of Engineering and Technology (Deemed-to-be University), Meerut, Uttar Pradesh, India

1.1 Introduction

Recent development in science and technology has changed the lifestyle and living status of human being. To lead a smooth life in this materialistic world, we are completely dependent on several types of appliances including computers, copy machines, mobile phones, microwave ovens, refrigerators, air conditioning, television remotes, smoke detectors, etc. Most of these devices explicitly depend on the principle of sensors [1]. In addition, several biomolecules are found in the environment as well as in human body. The altering of biomolecule concentrations in the human body causes adverse effects. The concentration of these biomolecules needs to be quantified through biosensors. A biosensor is a device which converts a biological response into an electrical signal. Cammann coined the term "biosensor" and the IUPAC introduced its definition [2].

1.1.1 History of the biosensor

A biosensor is a device which converts biological information to an electrical signal. The electrical signal is proportional to the concentration of the analyte (the biological substance to be quantified). So the analyte resulting from biological or chemical events are quantified by the biosensors. The timeline of biosensors started in 1962 with the development of an enzymatic sensor. Leland C. Clark is regarded as the father of the biosensor. His first experiment was measuring oxygenation of blood by

Biosensors for Emerging and Re-emerging Infectious Diseases
DOI: https://doi.org/10.1016/B978-0-323-88464-8.00010-5

reducing oxygen on the platinum electrode surface. His first biosensor failed owing to the adsorption of blood components on the electrode surface, which distorted the signal [2]. Latter, Clark used a cellophane wrapper of a cigarette packet on that sensor as it provided passage of low molecular weight substance i.e. oxygen. Hence the current response to the oxygen reductions can be recorded at the electrode surface. Currently, teflon is used as the membrane in place of cellophane wrapper. Clark quantified the glucose concentration by immobilizing the glucose oxidase on an amperometric oxygen electrode surface of semipermeable dialysis membrane [3–5]. To calibrate the sensor, Clark added an enzyme i.e. glucose oxidase to the solution. Later, the enzyme layer became an integrated part of the sensor. With this discovery, research from several disciplines including physics, chemistry, biology and material science have raised their interest in developing more reliable, sophisticated, durable and highly-sensitive biosensors. The device has achieved special attention in the field of medicines, agriculture, biotechnology as well with the military in bioterrorism detection and prevention.

1.1.2 Principle of biosensor

Fig. 1.1 shows the schematic representation of the biosensor, which follows three important steps including recognition of analytes, signalization and quantification. The device consists of several components such as analytes, bioreceptors, a transducer, electronics and a display. Analytes are the substances which are our main interest for detection and include: glucose, uric acid, dopamine, cysteine etc. Bioreceptors are the molecules that specifically identify the biomolecules or analytes. Generally, enzymes, deoxyribonucleic acid (DNA), cells, aptamers, and antibodies are some important examples of bioreceptors in a biosensing device. When the

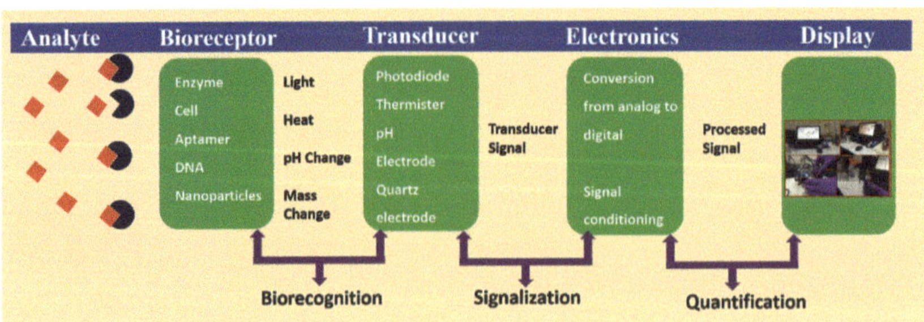

Figure 1.1 Schematic representation of the principle of biosensor.

bioreceptor interacts with the analytes, a signal in the form of light, heat, pH, charge or mass change is generated. This process is called bio-recognition. The transducer is a device which converts one form of signal to another form of energy. In the case of biosensors, the transducer converts a bio-recognition event into a measurable signal. The process of converting a mechanical event to a signal is called signalization. In most cases, the transducers produce optical and electrical signals which are dependent on the analyte-bioreceptor interaction. The transduced signals are prepared and processed through electronics and finally displayed. In the electronics, the amplification and conversion of signal from analog to digital is performed. The display unit of the biosensor quantifies the processed signal. Thus, finally in the display the interpretation curves related to the biomolecules are seen and from these the limit of detection and sensitivity of the biosensors are calculated. The display generates the output of the biosensor in the form of numeric, graphic, tabular or an image, depending on the requirements [5].

1.2 Recent trends and application

1.2.1 Nanosensor

Swift and dependable analytical tools are essential for environmental monitoring to achieve sample analysis with slight sample handling. Nanoparticle (NP)-based environmental sensors can detect heavy metals, toxins, and organic contaminants in water, air, and soil. They play a pivotal role in environmental monitoring. They can be helpful in both the detection and sensing of contaminants, and are utilized in developing new remediation technologies. In contrast with the conventional detection methods, NP sensors may have higher sensitivity, stability, and selectivity, plus they are inexpensive [6]. Gases like CO_2, NO_x, CO, CH_4, methanol, etc., are hazardous and toxic to the environment. These harmful agents are need to be detected and quantified depending on its toxicity. Several metal oxides thin films like zinc oxide, tin oxide, indium oxide etc. are being used for designing a conventional gas sensor.

1.2.1.1 Biosensor

Fields of magnetic NPs utilize a few kinds of iron oxides which generally incorporate Fe_3O_4 (magnetite), Fe_2O_3 (hematite, feebly ferromagnetic or antiferromagnetic), Fe_2O_3 (maghemite, ferrimagnetic), FeO (wustite, antiferromagnetic), out of which magnetite and maghemite, are the mainstream applicants as its biocompatibility have just demonstrated in bio-labeling and bio-separation [7]. Iron oxide NPs or magnetic NPs have a significant role in the development of electrochemical biosensors [8]. They have an extremely large surface area and great biocompatibility. Decontamination, concentration, separation and identification of molecules and explicit cells are mainly done by these magnetic NPs and they are specifically appropriate for integration in microfluidic gadgets [9,10]. Silica is frequently utilized as covering material above the surface of NPs. The inert behavior of silica, promotes the scattering of NPs. It possesses a high surface concentration while encouraging a broad assembly of surface reactions and the binding of bio particles (peptides, metals, polymers, antigen-antibodies, enzymes, proteins and nucleic acids) [11]. A novel glucose biosensor consisting of Prussian blue (PB) and glucose oxidase (GO_x) was recently arranged by Li et al. [12] with Fe_3O_4 NPs which attracted to the surface of strong paraffin-carbon paste electrode (CPE) by magnetic force. Amine-modified NPs were covalently conjugated to the glucose oxidase. Because of the high reactivity of hydrogen peroxide with the PB present in the glucose biosensor, the current created to decide the glucose concentration estimated the hydrogen peroxide framed during the enzymatic response with glucose. High affectability, selectivity, soundness and short reaction time were demonstrated by the biosensor. It displayed a straight reaction to glucose within the range from 5.0×10^{-7} to 8.0×10^{-5} M with a discovery breaking point of 1.0×10^{-7} M. The detection of glucose is not meddled by reductive substances in biological samples due to which biosensor was used to decide the glucose concentration in human blood samples. Depending on the immobilization of tyrosinase on the silica surface of modified magnetic $MgFe_2O_4$ NPs, a phenol biosensor was created by Yu and colleagues [13], which were additionally altered with amino groups on their surface. With the assistance of a permanent magnet, the subsequent magnetic bio-NPs were appended to the surface of a CPE. The immobilization matrix gave a satisfactory microenvironment for holding the tyrosinase movement. Phenol can be determined by direct reduction of the quinine species at -150 mV versus saturated calomel electrode (SCE). The species was created bio-catalytically. Within 20 s, a 95% consistent state current is achieved by phenol biosensor. Again it shows a high affectability of 54.2 A/mM, due to the high tyrosinase stacking in the immobilization matrix. The dependability and the utilization of the biosensor to phenol assurance in industrial wastewater were assessed with brilliant outcomes (sensitivity, selectivity, limit of detection).

A sensitive electrochemical DNA sensor dependent on a nonporous gold (NPAu) anode and multifunctional encoded AuNP was developed (Fig. 1.2) [14]. During the preparation of the NPAu electrode, a simple technique was observed, which disintegrated silver from an economically accessible 9-carat gold leaf (Ag/Au alloy, 50:50 wt.%) in nitric acid. Throughout scratching, silver atoms were specifically liquefied and deserted gold atoms amassed into the 3D porous structure. By an immobilizing catch test, DNA on the NPG electrode and hybridization with target DNA, the DNA biosensor was fabricated, which further hybridized with the correspondent DNA, that is, multifunctional encoded AuNP combined with $[Ru(NH_3)_6]^{3+}$. Two sorts of bio-barcode DNA are there in AuNP, one corresponding to the target DNA while other was not, decreasing the cross-response among the targets and reporter DNA on the same AuNP.

Reporter DNA is bounded by electrochemical signs of $[Ru\ (NH_3)_6]^{3+}$ through electrostatic connections and were estimated by chronocoulometry. The DNA biosensor can identify the DNA target in a range of $8.0 \times 10^{-17}–1.6 \times 10^{-12}$ M with a limit of detection as low as 28aM. This can be performed by exploiting the dual-amplification impacts of the AuNP electrode and multifunctional encoded AuNP. It has higher

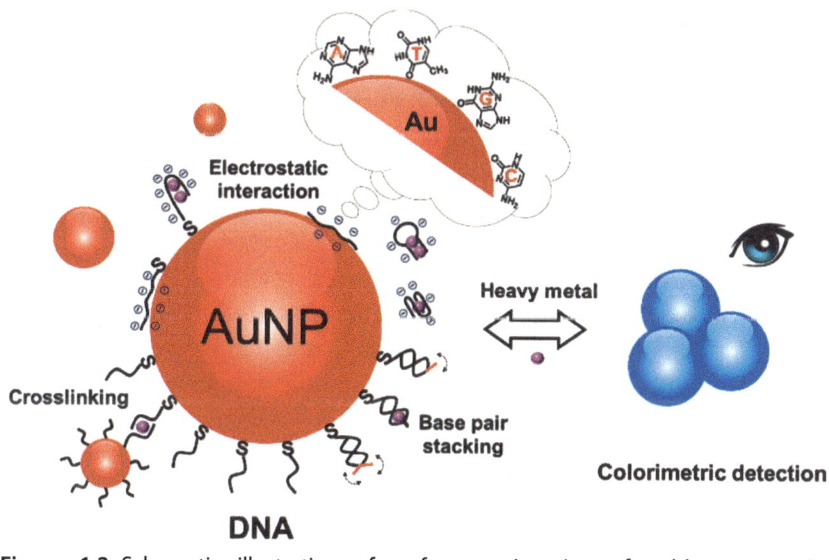

Figure 1.2 Schematic illustration of surface engineering of gold nanoparticles (AuNPs) with heavy metal specific-DNA. *This figure has been adopted from [15].*

selectivity for single crisscrossed DNA identification. By incubating the changed electrode in boiling water (90°C) for 1 minute, the NPG-based biosensor is recovered, by which hybridized DNA was evacuated by means of thermal denaturation. The AuNP-based biosensor nearly held its unique hybridization productivity after the recovery process was performed multiple times [15]. A schematic presentation for the utilization of gold nanoparticles for DNA sensing has been presented in Fig. 1.2.

Medintz et al. [16] utilized a related way to build up a FRET prototype based quantum dot (QD) biosensor capable of estimating paces of enzymatic processing. In their methodology, peptides were combined to have province explicit for QD connection, basic inflexibility, and protease acknowledgment properties and dye labeling. In this underlying state, fluorescence from the QD was fundamentally extinguished by the dye. As a suitable catalyst was added to solution a portion of the peptides were sliced bringing about a generous increment in fluorescence signal.

1.2.1.2 Nucleic acid biosensors

Nucleic acid biosensors are the progressively well-known devices that offer prevention, better diagnosis, and treatment of several human diseases [17–19]. To prevent and treat human disorders it is crucial to design reliable, resourceful, and low-priced tools. These tools can govern the genomic sequences, which have extensive potential applications comprising gene expression monitoring, viral and bacterial identification, forensic and genetic identification, clinical diagnostics, detection of bio-warfare and bioterrorism agents and pharmacogenomic research, drug discovery, etc. [6]. To achieve these breaks, experts have established a diversity of assays for DNA detection [20–23]. Either DNA hybridization or direct sequencing are standard analysis methods for the specific gene sequences. During DNA hybridization, a DNA probe categorizes the target gene sequence. It occurs by creating a double strand with its complementary nucleic acid.

Amid various conventional DNA detection methods, electrochemical biosensors signify a leading approach for rapid and complex resolutions for genetic disorders [24]. Due to their highly specific, fast, portable, and low-cost nature, electrochemical biosensors provide stirring opportunities for several decentralized clinical applications, varying from "alternative-site" testing emergency-room screening, home self-testing or bedside monitoring [24]. Such devices are useful in patient diagnosis by producing a simple, low-cost, yet a precise and subtle platform. The term electrochemical biosensor is applicable for a molecular sensing device jointly combining a biological recognition element to an electrode transducer. The electrochemical sensor alters the biological recognition event into a beneficial electrical signal. High performance, high speed, miniaturization, and low cost must be linked by future devices to utilize these opportunities and move DNA diagnostics out of the central laboratory [24,25]. The understanding of such robust devices needs groundbreaking efforts to expand new material design and novel fabrication processes. In this regard, experts have planned several conventional macromolecular material matrices for emerging electrochemical DNA biosensing devices. There are some safety issues in these electrochemical biosensing devices such as poor selectivity, sensitivity, and less solidity related to the radioisotope, fluorescent, and enzyme labels. Particularly, nanostructured materials are making way for new horizons for the application of electrochemical DNA biosensors. There is a recent review of the applications of nanostructured materials in electrochemical biosensors [25–30]. It has been proven that nanostructured materials are useful in the production of electrochemical DNA biosensing devices.

The enzymatic reaction on which the electrode surface occurs are also influenced by the nanostructured-modified electrodes, which involves improving the transducer's catalytic activity. The developed electrochemistry is due to the decreasing distance between the redox site of a protein and the electrode and because of the smaller size of NPs. The rate of electron transfer decreases with an increase in exponential distance between them, as the two are inversely proportional. Experts have investigated a range of nanostructured materials, comprising of nanocomb, nanotubes, nanofibers, nanobelt, nanorods, and nanowires, prepared from metals, semiconductors, carbon or polymeric species, for their capacity to augment the response of biosensors [31]. NPs are utilized in various ways, such as modifying electrode surfaces or modifying biological receptor molecules such as antibodies, enzymes, or oligonucleotides (ODNs). The nanostructured materials have distinguished achievements in comparison to other materials because of the ability for enhancing the feature for bioassays, enabling shrinking speed, decreasing the reagent and sample consumption, and streamlining the ability of different setups [26]. The usage of nanostructured materials consequently permits miniaturization of biosensors, microfluidic systems' growth, and enhancement in the bioassays' sensitivity.

1.2.1.3 Gas sensor

Gas sensors have been utilized in various applications. Having inexpensive sensors with great sensitivity, short response time, steadiness and reproducibility are vital for their applications. Meanwhile the response of this kind of device is brought about by the interface of the gas to be recognized with a functional layer that is detected by the existence of the gas. This creates materials with a bigger surface-to-volume proportion because upgrading the gas—solid interaction is of significant interest to develop the sensors' performance. In this sense, the utilization of nanomaterials is a method drawing much interest among researchers working in this domain of gas sensing. In 2016, diabetes biomarkers were detected by a vague impedance-metric chemical sensor array. The sensor is a blend of materials dependent on graphene oxide (GO) and it involves synchronous examination of water and gas analyte multivariate systems.

It was discovered that the electrical impedance of uncovered graphene either oxidized or after decrease indicated high explicitness for ammonia. The GO composite with cerium oxide and cyclodextrin could be used in the ammonia of the uncovered and reducible GO, for identifying the acetone [32].

In ordered to sense several gas molecule, nanostructured materials have achieved a special attention among others. These incorporate semiconductor metal oxide nanowires, nanoribbons, nanorods, carbon and further nanotubes, diamonds, graphene, and blends of these which make more fascinating nanostructures. A significant fact of nanomaterials in that gas sensor application is a result of high surface area to volume proportion. This perhaps ponders high affectability due to the incredible adsorption of gas species on the accessible molecular binding site and therefore frequently builds sensing ability [33]. In view of the materials, the accompanying gas sensors can be scrutinized for investigations and improvements. Semiconducting metal oxides are extensively used material in fabrication of nanostructured gas sensor. As the surface is visible to sensing gas, the electrical conductivity of semiconductor gets altered (Fig. 1.3). Owing to high surface activity and fashionable microstructure, the metal oxide thin film delivers extraordinary affectability. The sensing performance gets enhanced by the high adsorption

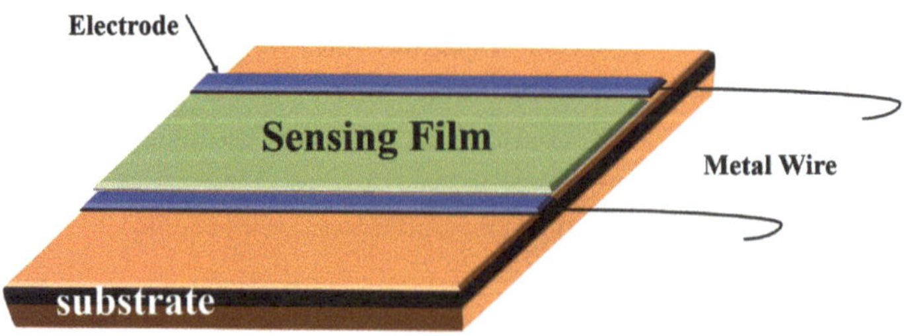

Figure 1.3 Schematic diagram of metal oxide thin film gas sensor.

of gas species and the resultant catalytic activity. This phenomena decreases the reaction time in contrast to the environmental microstructure gas sensor. The metal oxide gas sensors have the benefit of working in the lower-temperature range with high affectability.

Though, metal oxide nanosensors are costly as compared to bulk material gas sensors on account of the deposition of nanothin film using highly enlightened techniques and furthermore the utilization of progressive microscopic analysis for symbolizing the materials to improve its sensing attributes [34].

1.2.1.4 Importance of biosensor in Covid-19

On March 13, 2020 the pandemic was declared. SARS-CoV-2 has not only resulted in an unforeseen global public health crisis but also is becoming an unusual socio-economic burden [35]. In this situation, the World Health Organization (WHO) had urged the international community to do considerable diagnostic testing for fighting virus transmission and reduce the unobserved cases. Testing, together with risk management and the healthcare system is a necessity in all epidemics. Biosensing is the technique for determining several biological and biochemical agents using a biologically derived element. There are several available enzymatic and non-enzymatic biosensors for targeting the analyte (target molecules) in a highly precise way. Then one can transduce the binding event to an assessable signal directly. Impedance measurements or surface plasmon resonance, engaging signaling molecules i.e., labels comprising of fluorophores, enzymes, or electrochemically/optically active compounds are followed for accessing the signal [36,37].

Scientists initially developed polymerase chain reaction (PCR)-based tests around the globe, and they are possibly diagnosing COVID-19 through this gold standard method. PCR includes strengthening of minute traces of genetic material; i.e., the ribonucleic acid (RNA) of the virus, resulting in complex and highly specific detection (Fig. 1.4). Though, experts usually achieve PCR tests in centralized diagnostic services and the results may take from 4 hours to up to 3 days. Except for this, there are other limiting factors which can affect the PCR tests like sampling error, involved expert transportation and handling, requirement of extraction kits and lack of sensitivity in standard PCR methods. Sometimes even weeks after the full recovery of patients, PCR tests may report the existence of genetic material of the dead virus, thus, causing fabricated positive outcomes [38,39].

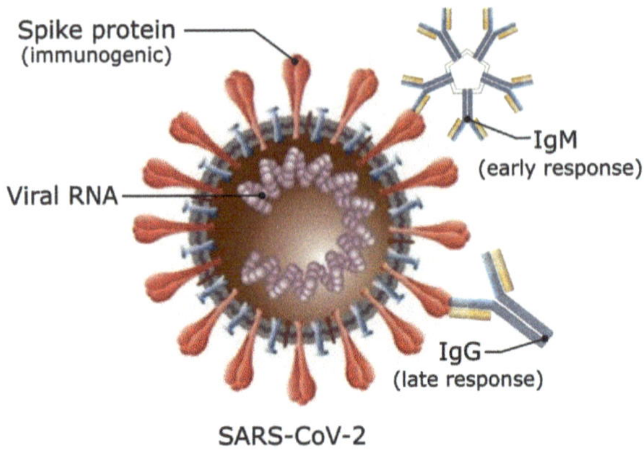

Figure 1.4 Schematic illustrating the structure of SARS-CoV-2 and related targeting sites (biomolecules) that can be used for the COVID-19 detection.

Through advancement, other biosensing technologies have provided a corresponding approach to standard PCR-based tests in terms of sensitivity, better diagnostics, and therapeutic decisions. Likewise, the biosensing research community is vigorously working to expand portability, time, and cost of PCR-based novel Coronavirus detection as well as to produce comprehensible PCR-based microfluidic devices with a unified sample preparation [40]. For the viral detection, other than the viral RNA, novel coronavirus exhibits immunogenic spike proteins; therefore, the immune system can produce immunoglobulin to activate an immune response besides the pathogen. Significantly, these immunoglobulins not only detect COVID-19, but also are valuable for its potential therapy. Immunoglobulin M (IgM) antibodies are made during the initial 4–10 days of the infectious disease, while immunoglobulin G (IgG) response is produced around 2 weeks later [41,42]. Substrate functionalized with recombinant viral antigen, can be more suitable to bind the immunoglobulin, which makes the detection of novel coronavirus (via IgM) or delivery an antiquity of past infections (via IgG). These antibodies exist in high numbers in the blood. Lateral flow assay (LFAs) are recognized as at-home pregnancy and ovulation tests. It provides apparatus free, user-friendly, and quick platform for optical on-site testing [43].

Additionally, as LFA employs cellulose-based materials, it's agreeable to mass production at an inexpensive price. Therefore, commercial LFAs simultaneously targeting the immunoglobulins have previously been established and accepted all over the globe. The drop of human serum, plasma or even whole blood, are used for detecting COVID-19, using the sensing device. It consumes less time i.e. 15 min for generating the result after diagnosis [43]. Other biosensing techniques using antibodies helps selectively capture the whole virus. The viral load in a saliva sample can be optically detected in 30 min through CONVAT project (It is funded by the H2020 Research and Innovation Programme of the European Commission). Then the result was followed with a PCR test on the same platform. Likewise, detections are now being conducted by utilizing specific antibodies rather than spike proteins. For this case, the graphene based field effect transistor, allowing portable devices are being engineered.

The following requirements need to be thoroughly satisfied by biosensing devices for testing infectious diseases:

1. Disposable, offering dependable large-scale manufacturing
2. Available and reasonable, permitting large-scale population screening
3. Easy-to-use by insignificantly prepared clients or even by the patients themselves
4. Equipment-free or operated with a reasonable and convenient read-out unit
5. Quick, empowering short sample-to-result times (within an hour)
6. Ready to work with low sample volumes and effectively accessible samples such as blood from a finger prick or non- invasive swab tests
7. Extremely selective and sensitive, conveying exact outcomes as per central laboratory findings
8. With an incorporated sample preparation
9. Effortlessly accessible and adaptable to recognize various targets using thc similar platform
10. Capable of multiplexing, concurrent detection of diverse analytes including controls [89ac].

1.2.1.5 Polymer-based nanosensor

For identifying several aromatic compounds, medication, ionic species, amine, organic fume and gases, organic polymer-based nanosensors have been specially created and designed [44]. The polymers behave as a thin film after depositing itself on a substrate in a layered fashion. In this type of sensor, the electrodes or field-impact transistors are used for screening the sensing current under a controlled electric field as shown in Fig. 1.5. The sensing area of the device and the thickness of the film influences the affectability of the polymer-based gas sensor, which is explicitly dependent on the expansion of metal oxides, carbon NPs, or filaments in the

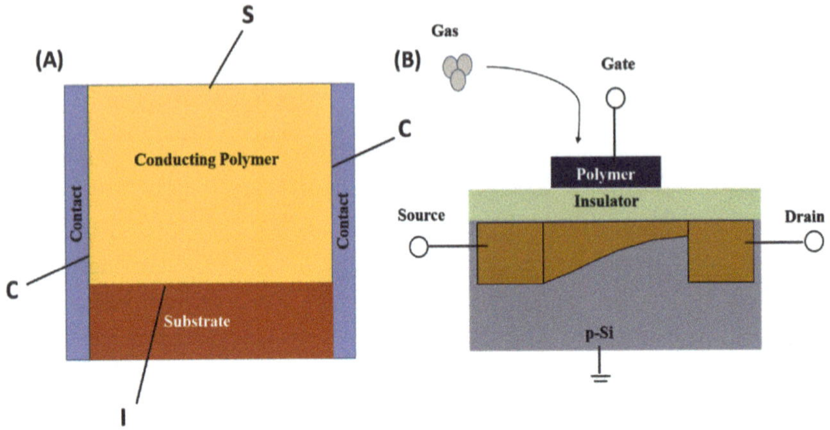

Figure 1.5 (A) Chemiresistor device structure: S-surface, I-interface with insulating substrate, C-contact (B) Conducting polymer active layer in field-effect transistor (FET). *FET*, Field-effect transistor.

polymer-based sensors. The deposition procedure of thin film is simple and cheap. The primary issue with a polymer-based sensor is its lower stability with temperature and time and absence of selectivity.

1.2.1.6 Metal and metal complex nanosensor

The adsorption of gases is enhanced by exposure to materials having a high surface to volume ratio. Such NPs, on interacting with the gas molecules, result in a quick variation of electronic properties with respect to the substrate. Such nanosensors are achieved by vaporizing metal precursor deposition techniques, which produce metal nanofilms and NPs. The exceptional characteristic of a metal complex nanosensor is the ability to make reversible interactions with different atoms. The metal complex enhances the reversible association of gas molecules with the device. The bond formation during the gas sensing method is widely influenced by the increase in temperature and the chemical environment. This kind of sensors, in light of the receptors, is intended to communicate with explicit gases to grow selectivity. Yet, development of metal complex sensing layers for nanosensors is usually costly.

1.2.1.7 Carbon-based nanosensor

Carbon-based materials are demonstrated to be very favorable amongst the nanostructured gas sensors because of high surface area, tremendous intrinsic electrical properties, and great thermal and chemical stability (Fig. 1.6) [45].

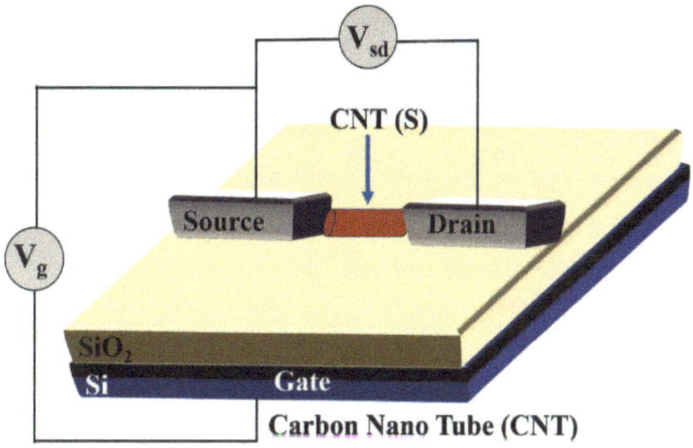

Figure 1.6 Schematic diagram of an FET device based on CNT(s). *FET*, Field-effect transistor.

The carbon allotrope, graphene, has been shown to be a remarkable material for several applications due to its fantastic electronic, optical, chemical, thermal, and mechanical properties. Graphene materials have exceptional sensing capability in gas sensing application, dependent on the two-dimensional structural property of pristine graphene (PG), GO and reduced GO (RGO) because of greater electron transport phenomena. In 2007, the charge density of graphene increased by the adsorbed gas molecules. The first graphene-based gas sensor was established by Schedin [44].

Differences in electrical conductivity arose owing to the adsorption of gas molecules. The adsorption process promotes the difference in electrical conductivity corresponding to the local carrier concentration of graphene [45–47]. The availability of surface functional groups on graphene surface and fascinating electrical properties, play a vital role in a sensing mechanism [48–51]. The adsorption of gas molecule on PG, influences charge career which offers a prominent enhancement in the electrical conductivity. Graphene oxide (GO) is unique compared to PG due to its physical and chemical properties. GO is a favorable material for gas sensing application due to its highly rich oxygen functional group. Prezioso et al. [52] by means of GO sensor device considered NO_2 gas sensing performance by varying the gas concentration at distinct operating temperatures to improve the sensing efficiency. Detection of low concentration of nitrogen dioxide, hydrogen, and other gases is done using GO-based gas sensors. The GO-based sensor has shown a higher and faster sensing response in contrast to RGO based devices. The RGO was synthesized from RGO. The RGO can be produced by reducing the oxygen containing functional groups in the GO sheets. In gas sensing application, RGO has a larger number of advantages than PG such as inexpensive to prepare, conceivable to amend structure, fine tuning of structure, etc. [53].

1.3 Commercialization

To this date, various biosensors for measuring the level of blood glucose, cholesterol along with uric acid, HIV, malaria, etc. have uncovered a remarkable commercial achievement, worldwide [54]. Additionally, some of the cancer and cardiac disease-based biosensors have also have excellent interest for commercial applications. Below, in Fig. 1.7, we have provided some the commercialized biosensors.

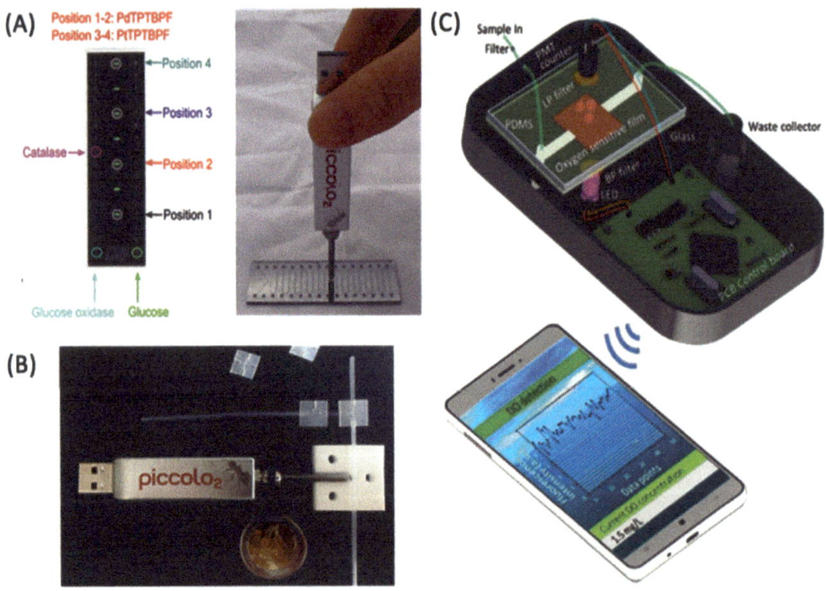

Figure 1.7 Commercial and portable optical readout systems for on-chip O2 monitoring. (A) A chip (left) designed for commercial optical readout devices (right). *Reproduced with permission [78], copyright 2016, Elsevier. (B) chip-based bacteria study with O2 measurement through the wall of tubing using a commercial readout device. Reproduced with permission [49], copyright 2016, American Chemical Society. (C) A portable handheld photodetector device connected to a mobile app for chip-based O2 monitoring [59].*

1.4 Conclusion and future scope

In conclusion, we have summarized the principles, history and various applications in the important fields. Biosensors extend multipurpose applications in the areas of technology, engineering, food safety monitoring, quality assurance, medicine, drug delivery, toxicology, separation devices, and disease identification and treatment. Glucometers using the approach of electrochemical detection of oxygen or hydrogen peroxide utilizing immobilized glucose oxidase electrode planted the breakthrough of biosensors. The latest innovations in biological methods and equipment containing fluorescence tag to nanomaterials have improved the sensitive limit of biosensors. Applications ot aptamers, peptide arrays, nucleotides or affibody molecules offer kits to create groundbreaking biosensors over the classical one. Multilingual methodologies offered a safer viewpoint for emerging sensitive and specific biosensors with high regenerative potentials [55]. With the implementation of nanomaterials in biosensors, all of us saw a precipitous development in biosensing technology in the current decade. The sensing approach has come to be more flexible, strong, and vibrant with the introduction of nanosized materials. The transduction process has been enhanced substantially by employing various nanomaterials like more rapidly detection, more sensitivity, less response time, and reproducibility [1]. Even though significant progress is has been done for utilization of nanostructured materials in biosensors applications, still there are certain drawbacks, which limit their uses for the next level. For example, the lack of selectivity continues to be a hindrance for most of the sensors, impeding its practice in nanobased sensing devices. Though, this obstacle can be surmounted by using certain artificial receptors like molecularly-imprinted polymers in the biosensing systems, this will enhance the selectivity and specificity in one side and the stability on the other side [56,57].

Innovative biosensors of next generations becoming a high demand for delivering fast and point-of-care analysis. The high requirement of diagnostic volume in the current pandemic due to the COVID-19 influenced the anthropology yet again limits not only the traditional methods but also the current commercial biosensing tools for their outreach to end-users. Thus, there is a serious demand for commercially mountable cost-effective prototypes to eliminate the potential prevalent outbreaks in the near future [58].

References

[1] V. Naresh, N. Lee, et al., A review on biosensors and recent development of nanostructured materials-enabled biosensors materials-enabled biosensors, Sensors 21 (2021) 1109. Available from: https://doi.org/10.3390/s21041109.

[2] R. Renneberg, D. Pfeiffer, F. Lisdat, G. Wilson, U. Wollenberger, F. Ligler, et al., Frieder Scheller and the short history of biosensors, Adv. Biochem. Eng./Biotechnol. 109 (2008) 1–18. Available from: https://doi.org/10.1007/10_2007_086.
[3] A. A. Karyakin. Biosensor of Environment, Health Security, (2009) 255–259.
[4] S.K. Mohanty, Elias, Biosensors: a tutorial review, Potentials, IEEE 25 (2006) 35–40. Available from: https://doi.org/10.1109/MP.2006.1649009.
[5] N. Bhalla, P. Jolly, N. Formisano, P. Estrela, Introduction to biosensors, Essays Biochem. 60 (2016) 1–8. Available from: https://doi.org/10.1042/EBC20150001.
[6] K. Iketania, R.D. Sunb, M. Tokib, K. Hirotaa, O. Yamaguchi, J. Phys. Chem. Solids 64 (2003) 507–513.
[7] L. Wang, W. Ma, L. Xu, W. Chen, Y. Zhu, C. Xu, et al., Mater. Sci. Eng. 70 (2010) 265–274. R.
[8] A.S. Teja, P.Y. Koh, Prog. Cryst. Growth Charact. Mater. 55 (2009) 22.
[9] I. Koh, L. Josephson, Sensors 9 (2009) 8130.
[10] X.M. Lin, A.C.S. Samia, J. Magnetism Magnetic Mater. 305 (2006) 100.
[11] S. Qadri, A. Ganoe, Y. Haik, J. Hazard. Mater. 169 (2009) 318.
[12] M. Alejandro-Arellano, et al., Pure Appl. Chem. 72 (2000) 257.
[13] J. Li, X. Wei, Y. Yuan, Sens. Actuators B 139 (2009) 400.
[14] Z. Liu, Y. Liu, H. Yang, Y. Yang, G. Shen, R. Yu, Anal. Chim. Acta 533 (2005) 3.
[15] Z. He, et al., Biosensors 10 (11) (2020) 167. Available from: https://doi.org/10.3390/bios10110167.
[16] I.L. Medintz, A.R. Clapp, F.M. Brunel, T. Tiefenbrunn, H.T. Uyeda, E.L. Chang, et al., Nat. Mater. 5 (2006) 581.
[17] K.J. Odenthal, J.J. Gooding, An introduction to electrochemical DNA biosensors, Analyst 132 (2007) 603–610.
[18] S. Kumar, A. Kumar, Recent advances in DNA biosensor, Sens. Transduc. J. 92 (2008) 122–133.
[19] A. Sassolas, B.D.L. Bouvier, L.J. Blum, DNA biosensors and microarrays, Chem. Rev. 108 (2008) 109–139.
[20] J. Wang, Electrochemical nucleic acid biosensors, Anal. Chim. Acta 469 (2002) 63–71.
[21] K. Kerman, M. Kobayashi, E. Tamiya, Recent trends in electrochemical DNA biosensor technology, Meas. Sci. Technol. 15 (2004) R1–R11.
[22] F. Valentini, G. Palleschi, Nanomaterials analytical chemistry, Anal. Lett. 41 (2008) 479–520.
[23] K. Kerman, M. Saito, S. Yamamura, Y. Takamura, E. Tamiya, Nanomaterial-based electrochemical biosensors for medical applications, Trends Anal. Chem. 27 (2008) 585.
[24] A. Erdem, Nanomaterial-based electrochemical DNA sensing strategies, Talanta 74 (2007) 318–325.
[25] L.G. Carrascosa, M. Moreno, M. Alvarez, L.M. Lechuga, Nanomechanical biosensors: a new sensing tool, Trends Anal. Chem. 25 (2006) 196–206.
[26] X. Luo, A. Morrin, A.J. Killard, M.R. Smyth, Application of nanoparticles in electrochemical sensors and biosensors, Electroanalysis 18 (2006) 319–326.
[27] A.S. Fauci, H.C. Lane, R.R. Redfield, Covid-19—navigating the uncharted, N. Engl. J. Med. 382 (2020) 1268–1269.
[28] A. Becker, Coronavirus: Rush. Dev. Rapid Tests (2020). DW.
[29] C. Dincer, R. Bruch, E. Costa-Rama, M.T. Fernandez-Abedul, A. Merkoçi, A. Manz, et al., Urban disposable sensors in diagnostics, food, and environmental monitoring, Adv. Mater. 31 (2019) 1806739.
[30] K. Bo-gyung, Tests in recovered patients found false positives, not reinfections, experts say, Korea Her (2020).
[31] J. An, X. Liao, T. Xiao, S. Qian, J. Yuan, H. Ye, et al., Clinical characteristics of recovered COVID-19 patients with re-detectable positive RNA test, Ann. Transl. Med. 8 (2020) 1084.

[32] E. Nunez-Bajo, M. Kasimatis, Y. Cotur, T. Asfour, A. Collins, U. Tanriverdi, et al., Ultra-Low-Cost Integrated Silicon-Based Transducer for On-Site, Genetic Detection of Pathogens, 2020.
[33] A. Grifoni, J. Sidney, Y. Zhang, R.H. Scheuermann, B. Peters, A. Sette, A sequence homology and bioinformatic approach can predict candidate targets for immune responses to SARS-CoV-2, Cell Host Microbe 27 (4) (2020) 671–680.
[34] L. Chen, J. Xiong, L. Bao, Y. Shi, Convalescent plasma as a potential therapy for COVID-19, Lancet Infect. Dis. (2020).
[35] C. Sheridan, Fast portable tests come online to curb coronavirus pandemic, Nat. Biotechnol. 38 (2020) 515–518.
[36] R. Bruch, G.A. Urban, C. Dincer, Unamplified gene sensing via Cas9 on grapheme, Nat. Biomed. Eng. 3 (2019) 419–420.
[37] C. Dincer, R. Bruch, A. Kling, P.S. Dittrich, G.A. Urban, Multiplexed point-of-care testing – xPOCT, Trends Biotechnol. 35 (2017) 728–742.
[38] T.C. Pearce, S.S. Schiffman, H.T. Nagle, J.W. Gardner, Handbook of machine olfaction: electronic nose technology, Br. Library (2006) 133–158.
[39] M.R. Cavallari, G.S. Braga, M.F.P. Silva, J.E.E. Izquierdo, L.G. Paterno, E.A.T. Dirani, et al., A hybrid electronic nose and tongue for the detection of ketones: improved sensor orthogonality using graphene oxide-based detectors, Sensors 14971 (2016) 1–8.
[40] R. Bogue, Nanomaterials for gas sensing: a review of recent research, Sens. Rev. 34 (1) (2014) 1–8.
[41] M.M. Arafat, B. Dinan, S.A. Akbar, A.S. Haseeb, Gas sensors based on one dimensional nanostructured metal-oxides: a review, Sensors 12 (6) (2012) 7207–7258.
[42] P.D. Harris, W.M. Arnold, M.K. Andrews, A.C. Partridge, Resistance characteristics of conducting polymer films used in gas sensors, Sens. Actuators B: Chem. 42 (3) (1997) 177–184.
[43] S. Hanna Varghese, R. Nair, B.G. Nair, T. Hanajiri, T. Maekawa, Y. Yoshida, et al., Sensors based on carbon nanotubes and their applications: a review, Curr. Nanosci. 6 (4) (2010) 331–346.
[44] T.R. Das, S. Patra, R. Madhuri, P.K. Sharma, Bismuth oxide decorated graphene oxide nanocomposites synthesized via sonochemical assisted hydrothermal method for adsorption of cationic organic dyes, J. Colloid Interface Sci. 509 (2018) 82–89.
[45] E. Roy, S. Patra, R. Madhuri, P.K. Sharma, Europium doped magnetic graphene oxide-MWCNT nanohybrid for estimation and removal of arsenate and arsenite from real water samples, Chem. Eng. J. 299 (2016) 244–254.
[46] S. Patra, E. Roy, R. Madhuri, P.K. Sharma, Fast and selective preconcentration of europium from wastewater and coal soil by graphene oxide/silane@Fe_3O_4 dendritic nanostructure, Environ. Sci. Technol. 49 (10) (2015) 6117–6126.
[47] E. Roy, S. Patra, R. Madhuri, P.K. Sharma, Introduction of selectivity and specificity to graphene using an inimitable combination of molecular imprinting and nanotechnology, Biosens. Bioelectron. 89 (2017) 234–248.
[48] S. Patra, E. Roy, R. Madhuri, P.K. Sharma, 2-Dimensional graphene as a route for emergence of additional dimension nanomaterials, Biosens. Bioelectron. 89 (2017) 234–248.
[49] S. Patra, E. Roy, R. Madhuri, P.K. Sharma, Nanocomposite of bimetallic nanodendrite and reduced graphene oxide as a novel platform for molecular imprinting technology, Anal. Chim. Acta 918 (2016) 77–88.
[50] S. Patra, E. Roy, R. Madhuri, P.K. Sharma, Graphene-Based Portable, Flexible, and Wearable Sensing Platforms: An Emerging Trend for Health Care and Biomedical Surveillance, 2017. Available from: https://doi.org/10.1016/B978-0-12-813349-1.00013-5.

[51] F. Schedin, A.K. Geim, S.V. Morozov, E.W. Hill, P. Blake, M.I. Katsnelson, et al., Detection of individual gas molecules adsorbed on grapheme, Nat. Mater. 6 (9) (2007) 652.
[52] S. Prezioso, F. Perrozzi, L. Giancaterini, C. Cantalini, E. Treossi, V. Palermo, et al., Graphene oxide as a practical solution to high sensitivity gas sensing, J. Phys. Chem. C 117 (20) (2013) 10683–10690.
[53] K. Mahato, P.K. Maurya, P. Chandra, Fundamentals and commercial aspects of nanobiosensors in point-of-care clinical diagnostics, 3 Biotech 8 (3) (2018) 149. Available from: https://doi.org/10.1007/s13205-018-1148-8.
[54] S. Vigneshvar, C.C. Sudhakumari, B. Senthilkumaran, H. Prakash, Recent advances in biosensor technology for potential applications – an overview. Front. Bioeng. Biotechnol. (2016). Available from: https://doi.org/10.3389/fbioe.2016.00011.
[55] S. Patra, E. Roy, R. Madhuri, P.K. Sharma, Nano-iniferter based imprinted sensor for ultra trace level detection of prostate-specific antigen in both men and women, Biosens. Bioelectron. 66 (2015) 1–10.
[56] E. Roy, S. Patra, R. Madhuri, P.K. Sharma, Gold nanoparticle mediated designing of non-hydrolytic sol–gel cross-linked metformin imprinted polymer network: a theoretical and experimental study, Talanta 120 (2014) 198–207.
[57] P. Bhattacherjee, et al., COVID-19 diagnostics: current approach, challenges, and technology adaptation, in: C.M. Hussain, S.K. Shukla (Eds.), Detection and Analysis of SARS Coronavirus: Advanced Biosensors for Pandemic Viruses and Related Pathogens, 2021. Available from: https://doi.org/10.1002/9783527832521.ch2.
[58] K. Hu, D. Lan, X. Li, S. Zhang, Anal. Chem. 80 (2008) 9124.

Further reading

M. Azimzadeh, P. Khashayar, M. Amereh, N. Tasnim, M. Hoorfar, M. Akbari, Microfluidic-based oxygen (O2) sensors for on-chip monitoring of cell, tissue and organ metabolism, Biosensors 12 (1) (2022) 6. Available from: https://doi.org/10.3390/bios12010006.

CHAPTER 2

Recent advances in the development of immunosensors for infectious diseases

Stephen Rathinaraj Benjamin[1,2], Tyciane de Souza Nascimento[1,3], Cássia Rodrigues Roque[4], Geanne Matos de Andrade[1,2] and Reinaldo Barreto Oriá[4]

[1]Center for Research and Drug Development (NPDM), Federal University of Cearà, Fortaleza, Ceará, Brazil
[2]Department of Physiology and Pharmacology, Faculty of Medicine, Federal University of Cearà, Fortaleza, Ceará, Brazil
[3]Department of Clinical Medicine, Faculty of Medicine, Federal University of Cearà, Fortaleza, Ceará, Brazil
[4]Department of Morphology, Institute of Biomedicine, Laboratory of the Biology of Tissue Healing, Ontogeny and Nutrition, School of Medicine, Federal University of Cearà, Fortaleza, Ceará, Brazil

2.1 Introduction

The emergence of infectious diseases in a global society is a cause of concern for the general population. Emerging viruses (e.g., Dengue virus, West Nile virus, Influenza A virus, etc.) have received a lot of attention in recent years, highlighting the need for more efficient diagnostic techniques in the medical field [1]. Viruses causes numerous illnesses lead to hundreds of thousands of deaths annually notably influenza, Ebola, the Middle East Respiratory Syndrome (MERS), HIV (human immunodeficiency virus), and the acquired immunodeficiency syndrome (AIDS) and more recently coronavirus disease (COVID-19). These are among the smallest identified infectious agents for humans and contain just a few genes typically embedded in a protein capsid and a lipid envelope [2]. The translation of these genes only takes place in the mechinery of the host cell, which is utilized by the intercellular parasite to manufacture millions of new viral particles that operate similarly and have severe effects. The early and accurate identification of viral illnesses in this situation is also significant for improving clinical results so that physicians may take appropriate measures. At present, the gold standard methods used to diagnose viral diseases include PCR (polymerase chain reaction) and ELISA (enzyme-linked immunosorbent assay). Both have sufficient detection limits and accuracy but need specialist reagents, staff, and equipment to be carried out with limited accessibility

Biosensors for Emerging and Re-emerging Infectious Diseases
DOI: https://doi.org/10.1016/B978-0-323-88464-8.00006-3

and late results. The development of novel and enhanced methods for the fast, easy, and onsite detection of low-cost viral infections is thus significant. Such technologies would enable decentralized findings to be achieved quickly, potentially significantly raising the number of tests that can be conducted. As a result, rapid, reliable, and repeatable analytical techniques are needed as a matter of urgency in order to be able to detect potential causative agents in a variety of matrices [3]. Today, researchers are working on fast and multi-point diagnostic procedures to offer an effective and accurate platform for detecting infectious viral diseases.

A robust biosensor system comprises highly distinct and sensitive components of biosignature, including nucleic acid, antibodies, glycoprotein, immunoglobulin, and enzyme, which are readily accessible for sensitive and selective quickly diagnosis. Detailed literature surveys revealed that most infectious disease virus sensors utilize one of three transducer modules, optical, piezoelectric, and electrochemical (EC). EC biosensors have received considerable interest among all kinds of biosensors, not only in research but also in the commercial sector. EC biosensors are favored over optical and mechanical due to its outstanding detectability, operational efficiency, and inexpensive cost production. In addition, EC biosensors enable experimental miniaturization with uncompromised analytical performance. Since these systems are primarily dependent on electrodes and electrical circuits, they may be incorporated easily for frequent monitoring into different electronic accessories.

In addition, EC biosensors perform better in the limit of detection and ranges, where the detection limits have been described to the subnanomolar (picomolar) level with more extensive linear ranges than other types. However, continuous research and design efforts have now breached the performance limits of this categorization. But additional advantages like minimization and integration of EC-based prototypes make EC biosensors the unique mix for the design of new commercial products [4]. EC immunosensors, on the other hand, are a kind of analytical technique that integrates EC sensors with immunoassays, and they provide innovation for traditional determination methods. Immunoassays are used to detect target analytes in a precise and sensitive process by taking advantage of the excellent specificity of the antigen (AG)—antibody (AB) interaction. As a result, EC immunosensors have gained growing interest in recent years due to their stable performance, ease of use, high accuracy, and acceptable efficacy in practical applications, among other characteristics.

An immunosensor (IC) has an AB as an EC detection element and a light-sensitive material transducer that may include many different sensors, such as fiber sensors, surface plasma resonance (SPR) sensors, etc. Based on the highly selective identification of AGs, immunosensors are extensively used in the monitoring of protein-related diseases as well as environmental, agricultural, processed food, and beverage applications. Many interfacial methods for selective biomarker detection have been designed and implemented. After specific foreign molecules, such as high molecular weight proteins (AGs) or chemical substances enter an organism, antibodies can recognize and selectively bind and eventually get rid of these molecules. The critical procedure between the AB and AG is a very sensitive immunoassay. Recently, immunosensors have been frequently used for detecting biological agents and hazardous substances in defense applications. Furthermore, EC immunosensors may be divided into two types: label-free (direct assay) and labeled (transient assay). The labeled immunosensor is used to label the analyte target prior to detection; it is also used to monitor the immunoassay response by measuring the labeled product produced by the labeling process. However, the nonspecific response of this kind of sensor is modest, and the detection procedure is complicated. It is possible to significantly simplify the preparation and operation processes using a label-free EC immunosensor that directly measures the physical and chemical changes that occur during the creation of AG−AB complexes. Label-free EC immunosensors have been created quickly in recent years due to these advances. The future study of EC immunosensors still has a significant potential for development. EC methods such as cyclic voltammetry (CV), differential pulse voltammetry (DPV), linear sweep voltammetry (LSV), square wave voltammetry (SWV), field-effect transistor (FET), electrochemcial impedance spectroscopy (EIS), amperometry, chronoamperometry (CA) have been used or versatile target analytes such as metal ion, biologic molecules have been detected. Most recently, a significant number of EC immunosensor detections have been used to develop highly quantitative and qualitative virus immunosensors. This chapter mainly focuses on and discusses the EC techniques for detecting various infectious viruses using biorecognition molecules.

2.2 Electrochemical immunosensors for infectious diseases

2.2.1 Influenza

Influenza A virus (IAVs) belongs to the *orthomyxoviridae* family and is spherical in form with a diameter of approximately 80−120 nm. Influenza A subtypes are further categorized as antigenic for hemagglutinin (HA) and neuraminidase (NA) surface glycoproteins. Avian influenza is an acute viral disease characterized by the avian influenza virus (AIV), resulting in increased economic consequences and significant public health concerns. Highly pathogenic avian influenza A H5N1 (HPAI H5N1) virus presents an important human and avian threat. Several biosensors were developed for the detection of AIV. Currently, colorimetric immunosensor [5,6], EC-SPR [7], chiro-immunosensor [8] methods can detect the H5N1 virus.

Jarocka et al. [9] proposed the biosensor based on the ternary metal chelate complex between DPM and the recombinant HA His-tagged (His6-H5 HA). The sensor allows the accurate detection of HA antibodies from the HPAI H5N1 virus. A particular interaction of His6-H5 HA and the relevant monoclonal AB was detected using Osteryoung SWV (OSWV). The proposed biosensor identified specific anti-H5 HA antibodies in hen serum samples. The biosensor showed a high detection of 2.4 pg/mL, a 7.2 pg/mL quantification limit, and a dynamic buffer range between 4.0 and 100.0 pg/mL. Zhang group [10] has previously reported on the premise of EC conversion of ECC (Electrochemical conversion) techniques. Here, the Sandwich immunosensor for fast detection of avian influenza H5N1 virus is required to generate PB (Prussian blue) for sensor development, using MNPs (Magnetic nanoparticls) as self-sacrificing labels. The ECC technique gives high sensitivity and detection limits up to 0.0022 hemagglutination units to biosensing technology.

Recently, the carbon nanomaterials (CNMs) used for nanoparticles (NPs) is receiving significant investigation, given their improved physical and outstanding synergistic characteristics. Furthermore, influenza virus DNA identification using the Au/MNPCNT sensing platform, modified with a DNA probe, and its detection limits were calculated at around 8.4 pM. In addition, the excellent selectivity of the Au/MNP-CNT system was identified against multiple conflicting DNA structures [11]. Based on the iterative SELEX technology, you can readily distinguish between nearly identical targets. This feature is important for various virus species or variants of the same family with high genome-like coefficients. Several researchers are now investigating various significant aptamer-based biosensors for the detection of viruses. Bhardwaj et al. [12] designed an EC biosensor without labels that uses a DNA aptamer specific to the H1N1 virus region of the HA stem. They utilized an indium thin oxide EC sensor for the detection technique and carried out measurements of CV and EIS, allowing them to identify the target virus at 3.7 PFU/mL.

In addition, the test has been validated by the use of various influenza A virus strains. Nevertheless, considering label-free electrical biosensors need simpler sensing protocols, portable devices have been developed. Singh et al. [13] have presented a new RGO (reduced graphene oxide)-modified EC immunosensor label-free and collaborated with the influenza A H1N1 microfluidics platform. All three microelectrodes were manufactured on the glass substratum, modified, and mAb. When lacking a linker or spacer, the amino groups on antibodies may establish a direct connection to quantities of carboxyl groups on the surface area of RGO. The microfluid label-free immunosensor showed an excellent linear range of $1-10^4$ PFU/mL and increased LOD (0.5 PFU/mL), which may be carried for clinical diagnosis by multi-analyst sensing devices. Label-free techniques do not include amplification steps that may restrict their sensitivity. Zhang et al. [14] developed a standard curve for detecting His-tag molecules on the surface area of redox-active monolayers produced on gold electrodes (GEs), which have attracted substantial attention among researchers in the immobilization of active bioanalytical molecules. This technique enables the stability and orientation of sensor immobilization elements for solid support with biological activity. Mikuła et al. [15] developed a highly sensitive EC biosensor with a redox monolayer to detect H1N1 anti-HA antibodies in the serum of the vaccinated mice. The biosensor was responsible for detecting 10^9 fold diluted serum sample antibodies. Several studies reporting on immunosensors for detecting various antibodies in buffer and actual samples have been done in order to evaluate their sensitivity with the specificity of the biosensor described in this study (Table 2.1).

Another approach is the detection of multiple viral AGs in the same sample using a multi EC immunosensor platform, which may be used in conjunction with other techniques. Recently, Han et al. [16] demonstrates that effectively, with excellent selectivity, a multi-detectable immunosensor can simultaneously and quantitatively recognize three subsets of influenza A viruses, H1N1, H5N1, and H7N9, via the amperometry method. According to this study, nanostructured ZnO NRs have an isoelectric point of about 9.5 and, when added to such proteins as H1N1, H5N1, and H7N9 antibodies, exhibit solid electrostatic interactions. The top inner surface of the nanoflow PDMS sensor area was produced with ZnO NRs hydrothermally. The suggested immunosensors were tested utilizing the following concentrations of H1N1, H5N1, and H7N9 AGs: 1 pg/mL, 10 pg/mL, 100 pg/mL,1 ng/mL, and 10 ng/mL. Table 2.1

Table 2.1 Electrochemical immunosensors for Influenza A viral disease detection.

Biosensor format/ transducer	Recognition matrix	Detection target	Linear range	Detection limit	Detection time	Samples	References
Electrochemical immunosensor CV, EIS	Au/1,6-hexanedithiol/ BSA/GCP/anti-M antibodies	Structural protein in the virion	10–100 pg/mL	20 pg/mL (80–100 virions/μL)	30 min	Throat swab	[97]
SPR biosensor	Gold nanoparticles on hydrophilic and hydrophobic substrates	Anti-HA	10 pg/mL to 10 μg/mL	50.5 pg/mL	–	Recombinant influenza virus A	[98]
Electrochemical biosensor	commercial Accu-Chek test strips	H1N1 influenza virus	10^3–10^8 PFU/sample	10^2 PFU/ sample	15 min	Nasal swab	[99]
Electrochemical impedance aptasensor	gold electrode virus samples (H1N1, B, and H7N9)	Inactivated influenza A virus subtype H1N1	9–900 ng/L	0.9 pg/uL	30 min	Inactivated H1N1 viruses	[100]
Electrochemical impedance sensor Linear sweep voltammetry	Au/MNP-CNT	Influenza virus DNA	1 pM to 10 nM	8.4 pM	–	Influenza virus DNA	[101]
Fluorescence assay, Voltammetry aptasensor, DPV	DNA aptamer	ssDNA	0–10^6 PFU/mL	3.7 PFU/mL	30 min	Mini-HA protein and H1N1 viruses	[102]
Upconversion luminescence resonance energy Biosensor	upconversion nanoparticles	H7 oligonucleotide	1 pM to 10 nM	7 pM	2 h	H7 oligonucleotide	[103]
Electrochemical DNA biosensor	DNA tetrahedral nanostructure	HA gene sequence	1 pM to 100 nM	100 fM	100 s	Throat swab	[104]
Electrochemical immunosensor LSV	Bifunctional magnetic nanobeads (bi-MBs)	H7N9 AIV	0.01–20 ng/mL	6.8 pg/mL	–	Inactivated H7N9 AIV	[105]

Electrochemical biosensor DPV	Isothermal exponential amplification (EXPAR) coupled with HCR	H7N9 virus DNA	50 fM to 100 pM	9.4 fM	150 min	Cell lysates	[106]
Electrochemical immunosensor LSV	PAb-AgNPs-G	AIV H7	1.6 pg/mL to 16 ng/mL	1.6 pg/mL	30 min	Viruses	[107]
Electrochemical immunosensor LSV	PAb-CNTs/MoSx	AIV H7	1–25 ng/mL	0.43 ng/mL	20 min	Real sample	[108]
IM-SPR biosensor	Antibody H7N9 SPR chip	H7N9 virus	2.3×10^2–2.3×10^5 copies/mL	402 copies/mL	10 min	H7N9 virus mixed with nasal mucosa	[109]
Electrochemical impedance aptasensor impedance analyzer	Gold nanoparticles/ H5N1-aptamer/ thiocyanuric acid	H5N1 virus	0.125–16 HAU/50 μL	H5N1 virus: 0.125 HAU/50 μL chicken swab: 1 HAU/ 50 mL	Less than 1 h	H5N1 virus and chicken swab	[110]
SPR biosensor Circular Dichroism study	Dual aptamer (IF10 aptamer and IF22 secondary aptamer)	H5N1 AIV	1×10^4–1×10^6 EID_{50}/mL	1000 EID_{50}/mL	–	H5N1-infected feces	[111]
FET biosensor extended gate	DNA aptamer	HA protein	10 pM to 10 nM	5.9 pM	–	Chicken serum	[112]
Eectrochemical biosensor EIS	Graphene–gold (Au) hybrid nanocomposite-SPCE	H9N2 NA	10^{-8} to 10^{-1} U/mL	10^{-8} U/mL	–	Diluent buffer	[113]

(*Continued*)

Table 2.1 (Continued)

Biosensor format/ transducer	Recognition matrix	Detection target	Linear range	Detection limit	Detection time	Samples	References
CV and Chronoamperometry	M2 antibody, iron MNPs-AuNPs-SPCE	H9N2 matrix Protein	8–128 HAU	16 HAU	160 s	Virus	[114]
Electrochemical biosensor CV, EIS	Shellac-derived TrGO/ ITO/glass	H1N1	0–10,000 PFU/mL	26.04 PFU/mL (PBS) 33.11 PFU/mL (saliva)	–	Buffer, saliva	[115]
Immunosensor OSWV	Dipyrromethene-Cu(II)-gold electrodes	His6-H5 HA	4.0–100.0 pg/mL	2.4 pg/mL	–	Hen sera	[116]
Microfludic electrochemical immunosensor Chrono ampero metry	TMP-HRP-ZnO nanorods-PDMS	Antigens of H1N1, H5N1, and H7N9	1–10 ng/mL (for all type)	1 pg/mL (for all type)	–	–	[117]
Impedimetric glycan-based biosensor, EIS	Gold electrode SAM	H3N2 influenza virus A	8 aM to 0.8 nM	5 aM	–	Lectins	[118]
Immunosensor DPV	Ab2/ALP/Pt-pZnO-hemin	Influenza virus A	0.001–60 ng/ mL	0.76 pg/mL		Serum	[119]
Electrochemical immunosensor EIS	Gold SPE 4,4′-Thiobisbenzenethiol SAM	Short and long Histagged HA (H5N1 viruses)	4.0–20.0 pg/mL (long) and 1–8 pg/mL (short)	0.6 pg/mL (long) and 0.9 pg/mL (short)		Serum	[120]
Electrochemical immunosensor CV, DPV	Pt/CeO_2/GO GCE	Influenza virus A	1.0×10^{-3}–1.0 ng/mL and 5.0–1.0×10^2 ng/mL	0.43 pg/mL		Serum	[121]

AIV, Avian influenza virus; *BSA*, bovine serum albumin; *CV*, cyclic voltammetry; *DPV*, differential pulse voltammetry; *EIS*, electrochemical impedance spectroscopy; *FET*, field-effect transistor; *HA*, hemagglutinin; *HCR*, hybridization chain reaction; *MNP*, magnetic nanoparticle; *OSWV*, Osteryoung Square Wave Voltammetry; *SPCE*, screen-printed carbon electrode; *SPR*, surface plasma resonance; *GCP*, Gold colloidal nanoparticles; *LSV*, Linear sweep voltammetry.

shows several analytical/ EC immunosensor diagnostic techniques for influenza virus detection.

2.2.2 Hepatitis

In recent years, viral hepatitis has become a severe health problem. Hepatitis (Family *Hepadnaviridae*) is an inflammation of the liver that may result in various health issues and can be deadly in certain cases. Hepatitis types A, B, C, D, and E are five distinct strains of the hepatitis virus. Even though they are all capable of causing liver disease, they vary in several critical aspects, including routes of transmission, the intensity of the disease, spatial distribution, and prevention techniques. Types B and C combined are the most common with symptoms of liver disease, carcinoma, and virus hepatitis-related mortality in hundreds of millions of individuals. Up to 325 million people live with hepatitis B and C, and most of these people have no access to diagnosis and treatment. All kinds of hepatitis viruses have the potential to cause acute hepatitis. Still, the hepatitis B virus (HBV), the hepatitis C virus (HCV), and the hepatitis D virus (HDV) are the most common viruses to cause chronic hepatitis, cirrhosis, and hepatocellular cancer. Despite this, approximately 96% of the mortality is caused by Hepatitis C, and Hepatitis B. Hepatitis B surface AG (HBsAg), hepatitis B surface AB (anti-HBs), hepatitis B e-AG (HBeAg), and total hepatitis B core AB (anti-HBc) are the standard five criteria used for HBV identification. These methods are unable to accurately depict the stage of viral replication that is important in the therapeutic intervention of the virus. Up until now, a series of analytical techniques have been used for the detection of HBV, such as ELISA or electrochemiluminescence immunoassay (ECLIA), recombinant immunoblot assay (RIBA), and ribonucleic acid (RNA) PCR technology, involves some limitations such as being relatively cost-ineffective, requiring large sample volumes lacking miniaturization for onsite analysis and they are time-consuming procedures.

Due to the high cost and technicality of PCR machinery, it is impractical for low-income nations to direct these tests routinely. On the other hand, EC methods are considered valuable and promising due to their robustness, cost-effectiveness, and miniaturization ability. EC immunosensors are developed using single electrode devices and examined for their target molecules, needing fast and reliable virus sensors for onsite detection.

HBV surface AG (HBsAg) is the significant diagnostic biomarker for hepatitis B infection-specific diagnosis. The HBsAg AG is a secreted covering protein that constantly enters the blood through viral liver cells. The early detection of hepatitis B is diagnosed based on HBsAg in serum. The precise and sensitive detection of low levels of HBsAg consequently represented an essential stage in identifying HBV infection [17]. Some recent studies have shown qualitative methods to detect HBsAg or HBeAg, such as EC DNA probes [18], Gold nanorods [19], ferrite magnetic NPs [20], chemiluminescence [21,22], Raman [23] differential pulse [24], EIS [25], and localized surface plasmon resonance (LSPR) [26].

Moreover, several reports have demonstrated HBsAg quantitatively based on EC techniques. Different NP-sensing devices for detecting low-concentration HBsAg in blood samples have been developed with varying modifications to nanomaterials and various subelectrodes. SPE is more efficient and straightforward to manage than glassy carbon and GEs. In order to identify and quantify HBsAg utilizing rGO/PB bimetallic Au@Pt nanocomposites, a new, ultrasensitive and label-free EC immunosensor was successfully produced. The rGO-TEPA/PB composite nanomaterials generated a wide surface and improved conductivity for increased signal production and enhanced biosensor amplification. The proposed immunosensor showed these benefits with a broad linear range, a low detection limit, excellent selectivity, reproductivity and stability [27].

EC immunosensors can be classified as label-free, sandwich (noncompetitive), and competitive immunosensors by immunoassay design. In addition, electrocatalyst labels provide promising alternatives to HRP labeling due to the current possibility of direct electron transfers to the surface of the transducer. Therefore, the immunoassay can be carried out without using redox mediators [28].

Since the EC reactions of the redox evaluation in solution are inherently diffusion-limited, the immunosensor can benefit from methods for immobilization on the transducer surface. Various electroactive substances, including PB [29], methylene blue [20], and ferrocene [30], were also employed as label-free immunosensors to detect viruses as redox mediators. Nevertheless, the redox material $[Fe(CN)_6]^{3-/4-}$ has been chosen for its excellent water solubility, well-known EC behavior, and cheap costs. Zhao et al. [31] constructed the composite of GO, ferrocene, and chitosan to construct an EC HBsAg immunosensor for surface B hepatitis. The composite was mixed with AuNPs and deposited as a film layer on a

GE as a composite. The presence of AuNPs enabled protein-binding sites and increased electron conductivity, which was critical for AB immobilization and utilized for AG HBsAg detection in the fact of AuNPs. The immunosensor demonstrated a broad linear range of 0.05 to 150 ng/mL and a LOD for HBsAg of 0.01 ng/mL using the EIS technique. Sandwich-type EC immunosensors have recently attracted various benefits, including easy operations, fast reaction, and signal amplification system coupling [28,32–37]. Pei et al. [35] developed another sandwich EC immunosensor for detecting HBsAg based on selective recognition of antibodies and AG with a required low detection limit (166 fg/mL) and broad linear range (0.0005–10 ng/mL).

A new graphene oxide-ferrocene-chitosan (GO-Fc-CS) matrix was recently designed to incorporate an EC immunosensor of GO-Fc-CS/Au-NPs film layer for the detection of HBsAg. The modified film of the layer exhibited enhanced electron capacitance. It also showed high reversible EC signals for current modifications, high biocompatibility, and excellent film potential to bind many antibodies. The immunosensor has a broad linear range of 0.05–150 ng/mL and a detection limit of 0.01 ng/mL for HbsAg [30].

Tan et al. [38] developed another label free immunosensor for hepatitis B detection. In this research, GCE was modified to enhance their peroxidase activities along with increased conductivity and surfaces in molybdenum disulfide nanosheets (MoS_2) with PtPd nanocubes (PtPd NCs) to maximize the sensitivity of the biosensor framework. The system has an LOD value of 10.2 fg/mL and a linear response ranging from 32 fg/mL to 100 ng/mL in a surface AG hepatitis B. In addition to its excellent sensitivity, the system offers speed, repeatability, and stability. A colorimetric immunosensor for detecting the surface AG of a HBV has also been developed using a sandwich-based immuno-assay technique with biotin and luminal gold NPs. A chemiluminescent immunosensor was constructed using gold-NPs with a 1.7–20 pg/mL detection range using hydrogen peroxide as the oxidant [22].

In general, label-free EC immunosensors offer many additional unique characteristics, including an absence of pollution and excellent sensitivity. However, although enzymes as signal amplification substrates yield positive results, they are time-consuming to prepare, store and reactivate. The development of nanoenzymes addresses these issues and their unique catalytic characteristics. Based on porous graphene nanozymes functionalized by the PdCu tripod, a label-free immunosensor has been designed to detect hepatitis B and the AG [39]. In addition, gold nanorods (AuNR) have recently received significant interest in research due to their remarkable electrical and optical characteristics. The EC DNA biosensor was also developed to detect HBV using highly conductive gold nanorods [40].

According to previous research, specific techniques may be used to determine two biomarkers simultaneously. Similarly, Liping and his colleagues reported simultaneous detection of two tumor markers (HBsAg and HBeAg) utilizing GS-SnO_2-BMNP (Graphene sheet hybridized by sea urchin–like bimetallic nanoparticles) modified electrodes Thi (Thionine) and Fc (Ferrocene)-functional. To improve the capacity of signal source molecules and, as a result, to raise signal intensity, GS-SnO2-BMNPs takes full use of the spacing effect. The linear concentration range of 0.01–100 ng/mL for HBsAg and 0.01–100 ng/mL for HBe-Ag was achieved with detection limits of 4.67 and 4.68 pg/mL, respectively [41]. Table 2.2 presents an overview of Hepatitis detection, including the detection technique, electrode, detection target, linear range, detection limit and response time, and the sample type.

2.2.3 Dengue

Dengue (DENV) is a primary mosquito-borne, single-positive nuclear RNA virus of the *Flaviviridae* (Flavivirus) family that transmits from infected Aedes female mosquitoes and produces dengue disease. The challenge of preventing the spread of infections and detecting viruses rapidly is a global health concern. The screening tests were a great effort to reduce the impact of DENV's high morbidity and death rates. Due to the scientific value of its recent recognition, the use of biosensors for sensing and measuring a disease with rapid detection is one of the best methods that are considered practical and simple with high sensitivity, precision, and specificity, as opposed to traditional technologies such as immunological and molecular approaches. Dengue includes four distinct species, including DENV-1, DENV-2, DENV-3, and DENV-4, which have been detected in tropical and subtropical areas but are also spreading to higher elevations globally. To increase diagnostic performance of electrical-based biosensors [42–44], EC [45–48], optical sensing methods [49–52] for dengue detection have been established [53]. Recently, several studies have shown that dengue nonstructural 1 (NS1) AB has become the most common biomarker for all serotypes. The significant benefit of NS1-based diagnostic methods, especially in the early clinical phase, is their improved circulating content in the blood. NS1-based

Table 2.2 Electrochemical immunosensors for Hepatitis B viral infectious disease detection.

Biosensor format/transducer	Recognition matrix	Detection target	Linear range	Detection limit	Detection time	Samples	References
Sandwich-type immunosensor EIS Amperometry	Au@Pd/MoS_2@MWCNTs	Hepatitis B e antigen	0.1–500 pg/mL	26 fg/mL	–	Serum	[122]
Label-free amperometric immunosensor EIS	GO/Fc-CS-Au NPs	Hepatitis B antigen	0.1–350 ng/mL	0.1 ng/mL	15 min	Human serum	[123]
Electrochemical immunosensor CV, EIS	GO-Fc-CS/Au NPs/GE	Hepatitis B antigen	0.05–150 ng/mL	0.01 ng/mL	25 min	Human serum	[124]
Label-free immunosensor CV, EIS	PdCu TPs/P-GN	Hepatitis B antigen	60 fg/mL to 100 ng/mL	20 fg/mL	40 min	Human serum	[125]
Electrochemical DNA biosensor, CV	1,6-hexanedithiol (HDT) and GNRs	Hepatitis B virus	1.0×10^{-12} to 10.0×10^{-6} mol/L	2.0×10^{-12} mol/L	–	Human serum	[126]
Electrochemical impedance immunosensor Nyquist impedance spectras	SPCE Anti-HBs-Anti-HBcGNP	HBsAg	5–3000 ng/mL	2.1 ng/mL	–	Human real	[127]
Sandwich-type immunosensor CV, EIS, DPV	Platinum nanowire inlaid globular SBA-15 (Pt NWs@g-SBA-15/Thi)	HBsAg	10 fg/mL to 100 ng/mL	3.3 fg/mL	–	Human serum	[128]

(Continued)

Table 2.2 (Continued)

Biosensor format/ transducer	Recognition matrix	Detection target	Linear range	Detection limit	Detection time	Samples	References
Sandwich-type electrochemical immunosensor CV, EIS	Palladium nanoparticles/ δ-manganese dioxide/ halloysite nanotubes (Pd/δ-MnO_2/HNTs)	HBsAg	0.001–20 ng/mL	0.3 pg/mL	–	Human serum	[129]
Sandwich-type electrochemical immunosensor CV, Amperometry	RhPt NDs/NH_2-GS	HBsAg	0.0005–10 ng/mL	166 fg/mL	–	Human serum	[130]
Label-free electrochemical immunosensor CV, EIS, DPV	rGO/PB-Au@Pt NPs	HBsAg	0.25 pg/mL to 400 ng/mL	0.08 pg/mL	–	Human serum	[131]
Electrochemical label-free immunosensor amperometry i–t	Ab1@Ni AuPt-NGs/ GCE	HBsAg	0.001–80 ng/mL	0.31 pg/mL	–	Human serum	[132]
Label-Free Immunosensor CV, EIS, DPV	PB/Fe_3O_4/GO nanocomposites and AuNPs	HBsAg	0.5 pg/mL to 200 ng/mL	0.166 pg/mL		Human serum	[133]
Electrochemical immunosensor DPV	(Fe_3O_4 – AuNPs) hemin-rGO-Au nanohybrid	HBAg	0.0001–0.3 ng/mL 0.0001^{-1}	60 fM 10 fM	–	Human blood serum	[134]

CV, Cyclic voltammetry; *DPV*, differential pulse voltammetry; *EIS*, electrochemical impedance spectroscopy; *GNPs*, graphene nanoparticles; *SPCE*, screen-printed carbon electrode; *GNR*, gold nanorods.

biosensors with diverse processes, comprising EC [54], quartz crystal microbalance [55], SPR [56] with various transduction materials, have been described. EC biosensors with screen-printed electrodes have been designed as suitable for point-of-care (POC) testing. In the literature, many SPE-based applications have been described as the detection of NS1. Nawaz et al. fabricated a development of NS1 AB immunosensors using a modified screen-printed carbon electrode combined with bovine serum albumin (BSA). The impedimetric immunosensor effectively identified dengue virus protein with a sensitivity of 0.3 ng/mL and a linear range of 1–200 ng/mL [57]. Recently, Siew et al. [58] developed the impedimetric immunosensor design based on a plant-derived antigenic glycoprotein envelope domain III (EDIII) to detect dengue virus (DENV) IgG antibodies. The graphene/TiO_2 nanocomposite immunosensor showed excellent sensitivity to DENV IgG over a broad linear working range (62.5–2000 ng/mL), with a detection limit value of 2.81 ng/mL. The same phenomenon was applied with graphene-coated screen-printed carbon electrode (SPCE) to detect DENV IgG antibodies [59].

Some of the published methods depend on the detection method of NS1 for dengue diagnosis (Table 2.3). Moreover, Awan and colleagues utilized Raman spectroscopy to analyze the surface of antibodies with silver NPs used in a sandwich EC immunosensor on a nanoprobe maintaining AB functionalized-silver-NPs (Ab-Ag NPs) for the detection of the dengue biomarker NS1 in human serum samples [60]. The sandwiched immunosensor enabled the development of a range of 3–300 ng/mL with a limit of detection of 0.5 ng/mL for NS1 detection. Carbon nanotubes (CNTs), namely single-wall carbon nanotubes (SWCNTs) and multiwall carbon nanotubes (MWCNTs), were increasingly attracted by sensors owing to their large surface area, electrical properties, solid thermal conductivity, and high mechanical strength. Single-walled carbon nanotubes coupled with antidengue NS1 monoclonal antibodies were used to construct a label-free chemi-resistive immunosensor to identify the DENV NS1 protein, as Wasik et al. [61]. However, Luna et al. [62] developed dengue virus immunosensors (DENV-1, DENV-2, DENV-3, and DENV-4) using gold NPs and self-assembled cysteine monolayers. The immunosensor potential to identify all four DENV subtypes tested was shown by EIS and CV measurements to demonstrate sensitivity and selectivity. Palomar et al. [63] employed MWCNTs and graphene NPs (GNPs) combined with a handmade GE to construct a nanostructured sensing platform in this research. In order to detect dengue toxin, this

Table 2.3 Electrochemical immunosensors for Dengue viral infectious disease detection.

Biosensor format/transducer	Recognition matrix	Detection target	Linearity	Limit of detection	Detection time	Sample	References
Immunosensor—DPV	Pt; NPAE/anti-DENV-2	DENV-2	$1-10^3$ pfu/mL	1.0 pfu/mL	1 h	Dengue virus	[135]
Immunosensor—EIS	NPA; NPAE/anti-DENV	DENV-2	1–900 pfu/mL	1.0 pfu/mL	–	Dengue virus	[136]
Immunosensor—EIS	Pt-AME, Pt-AME/$2H_2$	DENV-2	1–900 pfu/mL	0.230 pfu/mL (Denv-2) 0.710 pfu/mL (Denv-3)	40 min	Human serum	[137]
Immunosensor—EIS	Au; MUA/6COH/EDC/NHS/anti-NS1	NS1	0.01–1.00 g/mL	3 ng/mL	–	Human blood serum	[138]
Immunosensor—EIS	Au; Cysteamine/GA/IgY-NS1	NS1	0.1–10 μg/mL	9×10^{-8} g/mL	10 min	Commercial	[139]
Immunosensor—EIS	Au; Cysteine/EDC/NHS/MBA/AuNP/anti-DENV	Dengue viruses DENV-1, DENV-2, DENV-3, DENV-4.	1:10 to 1:100 dilution	1:100 dilution	20 min	Dengue virus	[140]
Immunosensor—EIS	ITO; SP/TMAP/PPD/diazonium/AuNP/EDC/NHS/IgG-NS1	NS1	5–4000 ng/mL	5 ng/mL	–	Human serum	[141]

Immunosensor—EIS	SPGE; MUA/EDC/NHS/IgG-NS1	NS1	1–25 ng/mL	0.5 ng/mL	15 min	Synthetic	[142]
Immunosensor-QCM QCM-D	GQC; PEE/CN/EDC/NHS/IgG-NS1	NS1	0.01–10 μg/mL	1×10^{-7} g/mL for QCM-D and 3.2×10^{-7} g/mL for QCM	–	Blood serum	[143]
Immunosensor—EIS	SPCE; BSA/EDC/NHS/Dengue NS1 antibody	NS1	1–200 ng/mL	0.3 ng/mL	1 h	Human serum	[144]
Immunosensor—EIS	GCE; MWCNT/poly (pyrrole-NHS) Dengue virus NS1	Anti-DENV-2 NS1	10,000–0.001 ng/mL	0.001 ng/mL	–	Bovine blood plasma	[145]
Immunosensor—EIS	SPCE; poly(4-APA)/recombinant NS1	Anti-NS1	–	1:1280 serum dilution	20 min	Human serum	[146]
Immunosensor – ECS	Au; PEG-thiol/11Fc/EDC/NS1	Anti-NS1 NS1	1–5000 ng/mL 1–1000 ng/mL	220 pg/mL 231 pg/mL	–	Human serum	[147]
Immunosensor—CV	SPGE; MUA/EDC/NHS/IgG-NS1	NS1	100 and 1000 ng/mL	50 ng/mL	Less than 30 min	Human serum	[148]
Immunosensor—DPV	PCGE; 4-CDS/EDC/NHS/anti-NS1	NS1	3–300 ng/mL	0.5 ng/mL	–	Human serum	[149]
Immunosensor—DPV	Au; MWCNT/AuNP/Cysteamine/EDC/NHS/anti-DENV2	DENV-2 NS1	1×10^{-12} and 1×10^{-6} g/mL	$3 \times 10 - 13$ g/mL	–	Human serum	[150]

(Continued)

Table 2.3 (Continued)

Biosensor format/ transducer	Recognition matrix	Detection target	Linearity	Limit of detection	Detection time	Sample	References
Immunosensor –EIS	SPCE; poly(4-ABA)/ recombinant NS1	Anti-NS1	1:40 to 1:1280 dilution	1:1280 serum dilution	30 min	Human serum	[151]
Impedimetric immunosensor, CV, EIS	Graphene/TiO_2	Dengue envelope domain III (EDIII)	62.5–2000 ng/mL	2.81 ng/mL	60 min	Mouse serum	[152]
Electrochemical capacitance spectroscopy EIS	Gold electrode	Anti-NS1	10–2000 ng/mL	3 ng/mL	–	Serum	[153]
Label-free electrochemical immunosensor, CV.	Polyaniline coated glassy carbon electrode	Anti-NS1	1–100 ng/mL	0.33 ng/mL	–	Blood serum	[154]

BSA, bovine serum albumin; *CV*, cyclic voltammetry; *DPV*, differential pulse voltammetry; *ECS*, Electrochemical capacitance spectroscopy; *EIS*, electrochemical impedance spectroscopy; *SPCE*, screen-printed carbon electrode.

electrode was conjugated by immobilizing dengue antibodies on GNPs through covalent bonding. The proposed immunosensor exhibited a wide linear range of 1×10^{-12} and 1×10^{-6} g/mL with a LOD value of 1×10^{-6} g/mL towards the recombinant viral dengue virus 2 NS1. Table 2.3 shows the analytical performance of EC immunosensors for detecting dengue viruses. Table 2.3 summarizes EC immunosensors for dengue detection, including the applied immobilization and technique, sample type, range, and LOD.

2.2.4 Zika virus

The Zika virus (ZIKV) is a flavivirus belonging to the *Flaviviridae* family, the same genus as the viruses that can cause dengue fever and yellow fever. It is transmitted via mosquitoes. ZIKV has evolved itself in people and can be sustained in huge populations via a mosquito-human-mosquito transmission cycle in which the presence of a nonhuman reservoir is not required. Although these symptoms produced by ZIKV and those caused by other flavivirus-infection disorders are similar, diagnosing specific ZIKV infections has proven to be difficult, and treatment decisions have been delayed. In addition, the lack of readily available efficient vaccines and medicines has increased the need for early-stage ZIKV detection, which is required for fast detection and infection progression monitoring to decide on treatment on time [64]. A new EC immunosensor for ZIKV was developed utilizing an intermicroelectrode microelectrode with a gold (IDE-Au) array function. The ZIKV-specific envelope protein AB (Zev-Abs) was immobilized on the functionalized IDE-Au chip using an immobilization method and an electrode width/gap of 10 μm. EIS and other experimental techniques were used to detect the electric signal of the IDE-Au immunosensing chip, and it was determined that the sensing chip is capable of explicitly detecting ZIKV protein. This sensitivity measured 12 kΩ/M and the LOD of 10 pM. In addition, the EC immunosensor identified the ZIKV-protein from 10 pM to 1 nM.

The IDE-Au immunosensing chip was coupled with a miniaturized potentiostat (MP) and thus created an easy-to-use POC application for detecting ZIKV early on in a POC research [65]. Another technique used was the printed circuit board (PCB) to detect the ZIKV. PCBs have the possibility of mass manufacturing, and three electrodes are attached. Aline et al. [66] presented an EC immunosensor based on ZnO nanostructures on the PCB immobilized with ZIKV-NS1 antibodies. CV evaluated the immunosensor's analytical responses. The biosensor is designed to enable quick detection in undiluted urine of the Zika virus without cross-reactive DENV-NS1 AG. The detection limit was found to be greatly below 1.00 pg/mL. The biosensor was quick, cheap, and easy to use; ideal for POC applications. A recent research report stated that a new EC biosensor constructed on graphene oxide (GO) and imprinted polymers (IPs) nanocomposite could detect both blood specimens and buffer samples for the virus known as ZIKV. This biosensor can be used to detect ZIKV because the biosensor uses an EC system that would be able to detect and quantify the ZIKV in both serum and buffer samples. The sensor could detect ZIKV in blood serum at a concentration of about 2.2×10^{-4} PFU/mL in PBS. Furthermore, the ZIKV LOD was quite close to the RT-PCR LOD [67].

The research claimed that a smartphone-integrated Point-of-Care (POCT) test was able to identify ZIKV in whole blood from such an infected person by utilizing NP-conjugated antibodies. The Au seeds were coated with platinum NPs and effectively produced Pt@Au core-shell NPs. This technique made the NPs evenly coat the glassy chip. ZIKV detection in endemic areas may require a more developed system, which can fulfill the current demands. In this research, chip-prepared vials were utilized to harvest the Zika virus. A serum spiked with a range of AGs, including recombinant vial and expressed, was added to a chip incubated at room temperature for 15 min. The optical analyzer-based findings were used to determine the results of the ZIKV sensor [68]. For the development of the immunosensor, CdS QDs were utilized as luminophores. This sandwich immunosensor for detecting ZIKV was developed based on its high Enhanced electrochemiluminescence (ECL) response. The proposed immunosensor has excellent linearity for ZIKV detection, ranging from 1.0 fg/mL to 1.0 ng/mL in human serum samples [69].

A label-free aptasensor for Zika virus detection in actual human serum samples was developed. Here the silsesquioxane-functionalized glassy carbon electrode has been modified to immobilize the aptamer sequences. CV, EIS, and Atomic Force Microscopy (AFM) experiments have been used to monitor the production stages of biosensing. The Zika virus EIS study in the 1.0 pM to 1 μM concentration range resulted in a low detection limit of 0.82 pM. In addition, the proposed impedimetric aptasensor showed excellent stability and acceptable sensitivity with a high detection selectivity of Zika virus in blood samples, which indicates its potential clinical applications in the early detection of Zika virus-related diseases [70]. Afsahi et al. [71] reported on Zika virus detection graphene-enabled immunosensors by covalently immobilizing specific mAbs. Field-effect biosensing via mAbs (monoclonal antibodies) combined with a graphene covalent bond makes the sensor capable of recognizing Zika in real-time and quantitatively. This technique was used in commercial graphene chips and incorporated into the electronic reading. The capacity change was linked to ZIKV AG concentration. The sensor proved capable of detecting a 450 pM ZIKV AG clinical significance level. This method was compassionate, cheap, selective, easy to detect, and cumulatively, the sensor can provide better quantitative information on nonclinical biophysical kinetic devices, making it adaptable for clinical research or detection. Nevertheless, the sensor was used for spiked serum samples. Table 2.4 shows the overview of several techniques for EC biosensing diagnosis for Zika virus identification.

2.2.5 Coronavirus disease

In late 2019, a new coronavirus known as severe acute respiratory syndrome coronavirus (SARS-CoV-2) was identified for the first time and has since spread across the globe [72]. Following that, the World Health Organization (WHO) declared the coronavirus disease 2019 (COVID-19) epidemic to be a worldwide pandemic. In the genus *β-coronavirus*, SARS-COV-2 (50–200 nm) is a subclass of viruses that are crown-like, enveloped, and positive-sense single-stranded RNA viruses [73]. There are at least four structural proteins in the SARS-COV-2 virus, which are designated as nucleocapsid (N), spike (S), small envelope (E), and membrane (M), in addition to numerous auxiliary proteins. The N protein is responsible for encoding the viral genome. The M and S glycoproteins are required for the binding of the virus. When RBD (Receptor binding domain) and S proteins interface with angiotensin-converting enzyme 2 (ACE-2) in host cells, the S glycoprotein is critical for connecting with the cells [74]. The S and N proteins are very useful AG markers for diagnosing COVID-19 disease. As a result, identifying RBD as a potential target of the COVID-19 virus should be the highest priority [75]. Magnetic systems are also increasing in popularity because of their adjustable features and directed methodology. They are also becoming a popular platform to build selective and sensitive biosensors for detecting SARS-CoV-2 viral protein. For the first time, femtomolar detection of SARS-CoV-2 viral protein using a toroidal

Table 2.4 Electrochemical immunosensors for Zika viral infectious disease detection.

Biosensor format/ transducer	Recognition matrix	Detection target	Linearity	Limit of detection	R detection time	Sa sample	References
Electrochemical Biosensor CV and EIS	IDE-GoxPol	ZIKV	5.2×10^{-20}–5.2×10^{-15} PFU/mL	1×10^{-8} PFU/mL	–	Serum	[155]
Label-free electrochemical DNA biosensor CV, EIS, DPV	Gold-PET electrode	ZIKV DNA RNA (NS5 protein)	54–340 nM	25.0 ± 1.7 nM	1.5 h	Serum	[156]
Label-free impedimetric DNA biosensor.	ox-GCE-[AuNPs-SiPy]	ZIKA DNA	1.0×10^{-12}–1.0×10^{-6} mol/L	0.82 pmol/L	50 min	Serum	[157]
Electrochemical Label-free	Graphene chips	Zika-Ag (NS1)	–	450 pmol/L	5 min	Serum	[158]
Immunosensor based on enzyme-free vial with a smartphone.	Platinum/gold core shell nanoparticle	ZIKV Vial immunosensor	1 pg/mL–1000 pg/mL	1 pg/mL	–	Urine, serum, and whole blood	[159]
Immunosensor—EIS	IDE-Au; Zev-Abs	ZIKV envelope protein	10 pM to 1 nM	10 pM	40 min	Real	[160]
Immunosensor—CV	Au; ZnO NRs/anti-ZIKV NS1	ZIKV NS1	0.1–100 ng/mL	1.00 pg/mL	–	Human urine	[161]
Biosensor is depending on recombinant forms of Zika- virus NS1 & the domain III of the "E" protein (EDIII). EIS, SWV	SPE-CNT	ZIKV-specific antibodies	8.7–5.8 mg/mL in the case of EDIII and 9.6–5.3 mg/mL for ΔNS1	53 fg/mL—EDIII, 17 fg/mL—ΔNS1	–	Serum and saliva	[162]

Immunosensor EIS	Graphene-coated screen-printed carbon electrode	Envelope glycoprotein domain III (EDIII) from plant	125–2000 ng/mL	22.5 ng/mL	–	Serum	[163]
LSPR-immunosensor Fluorescence enhancement	AuNPs-QDs	NS1	10–10^7 RNA copies/mL	NS1antigen 1.28 fg/mL (in DI water) 8.2 RNA copies/mL (in DI water) ~100 RNA copies/mL (inserum)	Human serum		[164]
Electrochemical genosensor EIS, DPV	GE/rGO/PTyr	Zika virus genomic RNA	1.72–1.72×10^{10} copies/mL	1.72 copies/mL	20 min	Serum	[165]
Electrochemical genosensor Chronoamperometry	SPGE sP-STV-MB/ oligonucleotide probe	Oligonucleotide targets	R1: 5–300 pmol/L R2: 5–400 pmol/L	R1: 0.7 pmol/L R2: 3 pmol/L	–	Specimens human	[166]

CV, Cyclic voltammetry; *DPV*, differential pulse voltammetry; *EIS*, electrochemical impedance spectroscopy; *SWV*, square wave voltammetry.

plasmonic sensor was recently reported by Ahmadivand et al. [76]. The terahertz (THz) generator was employed to generate a magneto-plasmonic polymer employing a nano-Au-modified magnetic platform in this work. A monoclonal AB subsequently detected the material against SARS-CoV-2 viral protein. Using toroidal electrodynamics, a plasmonic immunosensor can detect the spiked SARS-CoV-2 virus at 4.2 fM. This invention is still in its initial stages, and the authors have predicted that it will be used in POCT once it has been validated using actual samples. Fabiani and colleagues [77] used magnetic beads and an EC immunosensing method to identify SARS-CoV-2 in saliva samples. EC assay constructed using screen-printed electrodes employs magnetic beads to support immunological connections, which are used to associate AB spikes with protein capsids (such as the SARS-CoV-2 virus nucleocapsid protein) that have been spiked with specific concentrations of SARS-CoV-2 virus spike protein (19 ng/mL) or nucleocapsid protein (8 ng/mL). The sensor was able to identify SARS-CoV-2 viral protein and confirm its presence using saliva from COVID-19 patients infected with the virus [76]. Seo et al. [78] developed a FET-based biosensor system to detect SARS-CoV-2 utilizing spike protein antibodies as a sensor. The FET was covered with graphene sheets supported by immobilization by coupling the 1-pyrene-butyric acid N-hydroxysuccinimide ester (PBASE) antispike protein antibodies served as a sample linker. The FET performance was evaluated utilizing the COVID-19 patients' AG protein, cultivated virus, and nasal swab specimens. The FET biosensor detected a spike protein AG at 1 and 100 fM in phosphate buffering saline and universal transport medium (used for clinical suspension of nasopharyngeal swabs). The cultivated virus and nasal swab specimens revealed a LOD of 16 PFU/mL and 2.42/102 copies/mL, respectively.

The authors concluded establishing more effective and efficient nanoenabled sensing systems that can identify a COVID-19 using an EC SARS-CoV-2 virus detection method. A minimum threshold of viral detection is required to understand disease progression and evaluate the efficiency of prescribed therapy. Biosensing methods are highly supported by AI and IoMT for POC testing and disease analysis. Viral infection diagnosis and personalized treatment for overall intelligent healthcare are suggested for early-stage viral infection detection [79]. Mahari et al. [80] utilized a similar SARS-CoV-2 detection method. A new in-house biosensor was designed, eCovSens, whereas the electrode fluorine-doped tin oxide (FDTO) and AuNPs were used to detect SAR-CoV-2 spike protein. First, the AuNPs were immobilized by physisorption and electrostatic forces to produce AuNPs/SpAb via an antispike protein AB (SpAb). The device was manufactured using the FDTO coating of a glass electrode and the AuNPs/SpAb drop cast. The detection was achieved under optimal parameters from 1 fM to 1 μM, while with spike saliva samples; the authors were able to accomplish a LOD of 90 fM. The biosensors were also capable of detecting the spike protein within 10–30 s, enabling quick detection and potential clinical application. Liv et al. [81] developed a new technique for the voltammetric determination of SARS-CoV-2 spike protein using a more unique sensor combining bovine serum albumin, a SARS-CoV-2 spike AB, and a modified GCE (BSA/AB/f-GO/GCE) or a screen-printed electrode (BSA/AB/f-GO/SPE). BSA/AB/f-GO/SPE and BSA/AB/f-GO/GCE sensors were designed to detect 1 ag/mL of spike protein virus in 5 and 35 min synthetic, saliva, and oropharyngeal swab samples, and both sensors showed a dynamic response to the spike protein SARS-CoV-2 from 1 ag/mL to 10 fg/mL.

In order to identify SARS-CoV-2 spike antibodies, gold clusters covered by cysteamine and glutaraldehyde, including the spike protein SARS-CoV-2 AG and bovine serum albumin on a glassy carbon electrode, have been reported as an electric immunoassay platform [81]. The detection limits for SARS-CoV-2 spike AB were 0.01 ag/mL, both in the Synthetic Media and saliva or oropharyngeal. The linear Immuno-sensor response to the SARS-CoV-2 spike AB ranged from 0.1 to 1000 ag/mL. A label-free EC biosensor device was designed to simplify coronavirus detection as a promising analytical method for COVID-19 specific spike receptor-binding domain protein. They developed a new fabrication technique to produce gold NPs (GNPs), and then coated them with 11-mercaptoundecanoic acid (MUA). The GNPs served as an immobilization matrix material, with the attached sites of the MUA serving as attachment points for biomolecules. The designed RBD biosensor featured a twofold function, with special biorecognition and signal enhancement for RBD detection. This immunosensor based on sandwich impedance has a linear detection range of 0.002 to 100 pg/mL and a detection limit of 0.58 fg/mL [82]. Table 2.5 provides a synopsis of the EC diagnostic biosensing techniques for SARS-CoV-2 viral identification.

2.2.6 Human immunodeficiency virus

In most cases, the HIV is widely recognized to be incurable. HIV is present in two serotypes, HIV-1 and HIV-2, of which the first is the more prevalent

Table 2.5 Electrochemical immunosensors for COVID detection.

Biosensor format/ transducer	Recognition matrix	Detection target	Linear range	Detection limit	Detection time	Samples	References
Electrochemical immunosensor SWV	BSA/S-gene/GluAl/CysAm/ Au/GCE	SARS-CoV-2 spike antibody	0.1–1000 ag/mL	0.01 ag/mL	–	Saliva and nasopharyngeal sample	[167]
Immunosensor CV, DPV	Fluorine doped tin oxide (FTO) electrode with gold nanoparticle (AuNPs)	nCovid-19 spike antigen (nCovid-19Ag)	1 fM to 1 μM in standard buffer	10 fM (in-house built device) of nCovid-19 Ag	10–30 s	Spiked saliva samples	[168]
Label free Immunosensor CV, SWV	Carbon screen-printed electrodes with gold nanoparticles	SARS-CoV-2 nucleocapsid protein (N protein)	1.0 pg/mL to 100 ng/mL	0.4 pg/mL	15 min	Nasopharyngeal	[169]
Electrochemical immunosensor SWV	Carbon nanofiber	N antigen	1–1000 ng/mL	0.8 pg/mL	20 min	Nasopharyngeal sample	[170]
Field-rffect transistor-based biosensor CV, SWV	AuNPs on paper electrode	RNA of the N gene	585.4 copies/μL to 5.854 × 10^7 copies/μL	6.9 copies/μL	Less than 5 min	Nasopharyngeal sample	[171]
Biosensor CV	BSA/AB/f-GO/SPE	Spike protein	1 ag/mL and 10 fg/mL	1 ag/mL	Less than 5 min	Synthetic, saliva and oropharyngeal swab	[172]
Electrochemical immunosensor SWV	Graphene lattice; 1-pyrenebutyric acid *N*-hydroxysuccinimide ester (PBASE)	Spike subunit 1	260–1040 nM	260 nM	45 min	PBS buffer	[173]
Electrochemical immunosensor CV, DPV	Magnetic beads, alkaline phosphatase enzyme, antibody	S and N antigens (NAg)	0.04–10 μg/mL 0.01–0.6 μg/mL	19 and 8 ng/mL	30 min	Untreated saliva and nasopharyngeal sample	[174]

Electrochemical immunosensor CV, EIS	Cu_2O nanocubes-antibody-SPCE	Spike protein	0.25 fg/mL to 1 μg/mL	0.04 fg/mL	20 min	Saliva, artificial nasal, and UTM	[175]
FET biosensor DPV	Graphene sheets—pyrenebutyric acid *N*-hydroxysuccinimide ester (PBASE)	SARS-CoV-2 antigen protein, cultured SARS-CoV-2, or SARS-CoV-2 from clinical samples	Protein: 1 fg/mL to 100 pg/mL; SARS-CoV-2 in culture medium: $1.6 \times 10^1 - 1.6 \times 10^4$ PFU/mL; clinical samples: $1 \times 10^1 - 1 \times 10^5$ copies/mL	Protein: 1 fg/mL; SARS-CoV-2 in culture medium: 1.6 10^1 PFU/mL; clinical samples: 2.42 10^2 copies/mL	–	Culture medium; nasopharyngeal swabs	[176]
Field-effect transistor	Semiconducting (sc) SWCNT	S antigen (SAg) and N antigen	5.5 fg/mL to 5.5 pg/mL 16 fg/mL to 16 pg/mL	0.55 fg/mL for SAg and 0.016 fg/mL for NAg	–	Nasopharyngeal sample	[177]
FET	Redox dye-incorporated silica nanoparticles	N and S genes	1 to 1×10^9 copies/μL	1 copy/μL	Less than 2 h	RNA from Nasopharyngeal sample	[178]
FET	Graphene oxide paper-based electrode	SARS-CoV-2 IgG, IgM and S protein	1–1000 ng/mL	0.11 ng/mL	45 min	Serum	[179]
Isothermal RCA sandwich hybridization-Electrochemical biosensor	Molecular imprinted polymer	N antigen	2.22–111 fM	15 fM	15 min	Nasopharyngeal samples	[180]
Label-free detection-SWV	Magnetic nanobeads	N antigen	5 × diluted serum samples	230 pg/mL	50 min	Serum	[181]

BSA, Bovine serum albumin; *CV*, cyclic voltammetry; *DPV*, differential pulse voltammetry; *EIS*, electrochemical impedance spectroscopy; *FET*, field-effect transistor; *RCA*, rolling circle amplification; *SPCE*, screen-printed carbon electrode; *SWCNT*, single-walled carbon nanotube; *SWV*, square wave voltammetry; *UTM*, universal transport medium; *HCoV*, Human corona virus; *HFMD*, Hand, Foot, and Mouth Disease.

and virulent and infectious HIV-2 serotype. One of the HIV incidence reduction methods typically includes screening for HIV to start early treatment as feasible, such as antiretroviral therapy (ART). This approach reduces patients' viral load and reduces the risk of virus transmission. This therapy needs lifetime viral load monitoring, usually via a Nucleic Acid Amplification Test (NAAT) or CD4 + cell counts through flow cytometry [83]. These technologies are nevertheless complicated, costly, and need sufficient laboratory equipment. This generates the need for a cheaper and portable alternative test. In addition, HIV is known to have long-term latency, and carriers may remain asymptomatic for months to years before symptoms and potential development to acquired immunodeficiency virus (AIDS) [84]. During this period, the viral load is relatively low; consequently, a highly sensitive diagnostics test is required at this point to identify HIV.

HIV is usually tested using the ELISA system, identifying anti-HIV antibodies (IgG or IgM). The demand for simple, fast, sensitive, specific, and cheap HIV infection screening techniques have shifted to the HIV1-related capsid proteins p24, expressed in the viral cell core. EC immunosensors overcome this limitation using nanomaterials to improve sensitivity biosensors. A further attempt to identify the HIV-1 virus has resulted in discovering the capsid protein p24 in untreated human blood samples. The platform includes a single-walled carbon nanotube with screen-printed electrodes (SWCNT-SPCEs). Due to their superior characteristics as carbon nanotubes, such as effective immobilization of receptors, the development of the chemically diverse field, and substantial improvements in the electronic transmission process, SWCNT-SPCEs are the key to the detection process. The protein p24 was coupled with chitosan/glutaraldehyde (CS/GA) modified on the surface of SWCNT-SPCEs. The findings were very interesting, and there was a linear detection range of 10 pM to 1 nM, and the detection limit in a spiking human serum was equivalent to 2 pM [85].

Fang et al. [86] created an EC biosensor that detects p24 AG using a sandwich immunoreaction. An MWCNT, silica, and horseradish peroxidase matrix were used to immobilize a glassy carbon electrode (GCE). Anti-p24 antibodies were attached to the MWCNT/SiO_2HRP matrix using chitosan and glutaraldehyde (GA). Each of these anti-p24 antibodies was coupled to nanocarrier graphene oxide and thionine through HRP-modified anti-p24 antibodies. In the sandwich immunoreaction, both MWCNT/SiO_2/HRP and GO/TH antibodies bind to the AG, and the addition of hydrogen peroxide amplifies the signal by enzymatically oxidizing TH. The DPV technique detected p24 AG in the linear range of concentrations of 0.5−8500 pM, with a limit of detection of 0.15 pM from the amplified signal. Cerrutti et al. [87] designed a biosensor to detect anti-HIV p17 antibodies utilizing a layer-by-layer (LBL) technique to immobilize HIV p17 AG lignin (p17 is a matrix structural protein). Lignin/p17 LBL films were generated by varying immersion of a GE into lignin and p17 AG solutions. The authors could recognize the existence of antibodies from the 0.1 to 100 nM range using EIS. Fatin et al. [88] established a new FET-based biosensor that used MWCNT to immobilize the RNA aptamer as an HIV Tat protein sample. Tat AG is a regulatory protein that improves the effectiveness of the HIV transcription of double-stranded DNA (dsDNA), raising the number of transcripts produced. Tat protein was identified by voltage changes caused by a reduction in current flow at the concurrent door to bind to RNA aptamer. The authors could reach a detection limit of 600 p.m. Zheng et al. [89] have created an amperometric immunosensor sandwich for HIV p24 applied to gold-coated glassy carbon electrodes and showed linearity of 0.01 to 100 ng/mL. Most recently, Zhou and his colleagues presented the immunosensor electrochemiluminescence for measuring p24 using $Ru(bpy)_3^{2+}$ doped silica NP and gold-nano parts of graphene decorated and showed a linearity range of 10^9 to 10^5 mg/mL [90]. The EC immunosensors for HIV detection are summarized in Table 2.6.

2.2.7 Ebola

The Ebola virus (EBOV) is a pathogenic virus that invades most major organs and produces a human multisystem failure. The death rate for EBOV-infected disorders may rise to up to 90%. Due to its high incidence of mortality and its broad nonspecific early disease signs, quick, selective, and sensitive detection have significant early diagnostic consequences. Ebola is the cause of hemorrhagic lethality in hemorrhagic disease. However, during this epidemic, the diagnosis was based on the quantitative reverse transcription-polymerase reaction (qRT-PCR). Furthermore, EC detection provides a potential option with cheap cost, low power needs, sensitivity, and portable technology compatibility. Because of its interesting properties, biosensors constructed on reduced graphene oxide FETs (RGO-FET) are experiencing exponential growth in disease diagnosis as an advanced approach. These primarily electrical instruments, enhanced with particular receptor molecules, can detect

Table 2.6 Electrochemical immunosensors for HIV viral infectious disease detection.

Biosensor format/ transducer	Recognition matrix	Detection target	Linear range	Detection limit	Samples	References
Electrochemical sensor CV, DPV	GO/CNT/Silica electrode	p24 antigen	10^{-4}–2 ng/cm^3	0.083 pg/cm^3	Human serum	[182]
Sandwich-type enzyme immunosensor. DPV	HRP–HIV-p24-TH/GO	p24 antigen	0.5–8500 pM	0.15 pM	Human plasma	[183]
Electrochemical immunosensor DPV	p24 antigen/Chitosan/GA/ Single-walled CNT-SPE	Anti-HIV p24 antibody	10 pM to 1 nM	2 pM	Human serum	[184]
Electrochemical immunosensors based on FET	Antibody/amine-functionalized graphene-FET	p24 antigen	100 fM to 1 μM	100 fM	Real	[185]
Biosensor FET	RNA Aptamer/MWCNT-FET	HIV Tat antigen	0.2 nM to 1 μM	600 pM	Real	[186]
Label free liquid-ion gated FET	ssDNA probe/NiO/Liquid ion gate-FET	HIV-DNA	10^{-18} to 10^{-8} M	0.3 aM	Serum	[187]
Electrochemical DNA biosensor EIS	ssDNA/Graphene-Nafion/ GCE	Viral DNA	10^{-13} to 10^{-10} M	23 fM	–	[188]
Impedimetric DNA biosensor EIS	ssDNA/rGO/GCE	Viral genome	10^{-12} to 10^{-9} M	0.3 pM	–	[189]
Electrochemical biosensor DPV	Exonuclease III/ssDNA probe/Au Nanocluster/ GCE	Viral DNA	10^{-16} to 10^{-10} M	3×10^{-17}	Blood	[190]
DNA biosensor EIS	AuNPs/PABA/rGO/GCE	HIV DNA	0.1 fM to 10 nM	37 aM	Serum	[191]

CV, Cyclic voltammetry; *DPV*, differential pulse voltammetry; *EIS*, electrochemical impedance spectroscopy; *FET*, field-effect transistor.

targets and respond quickly due to their selective and sensitive sensor devices [91]. Even better, since it can be mass-manufactured utilizing microfabrication technology, the FET biosensor is among the most promising prospects for POC diagnostics.

An EC biosensor for human norovirus using new peptides has been reported. Eight peptides were evaluated as recognizable elements, which first separately confirmed their efficiency in viral sensing and found the improved binding performance of NoroBP peptide coated with the GE and detected a low detection limit of 1.7 copies/mL. The proposed sensor has been used effectively to detect norovirus in oysters [92]. Up to this point, many studies have been published in the literature on viral detection using FET biosensors. The authors [91] presented an immunoassay based on a FET to detect inactivated Ebola virus (EBOV). The equine AB was bound with reduced graphene oxide on the surface of the FET against EBOV glycoprotein (RGO). Antibodies were immobilized against EBOV in the modified FET, and a Dirac voltage shift assessed the EBOV response. The technology can detect EBAV across a range of 2.4×10^{-12}–1.2×10^{-7} g/mL with a detection limit of 2.4 pg/mL. Chen and colleagues [93] developed a FET biosensor based on rGO to detect the AG Ebola quickly. Anti-Ebola probes were attached to the surface of the rGO sheets and were used to capture the specific AG. The device exhibited outstanding semiconductor properties and could be utilized with real-time response to detecting EGP up to 1 ng/mL. The FET biosensor also exhibits excellent EGP selectivity, with insignificant responses from non-specific proteins. Several labels, including detection techniques, electrode modification, detection target, and samples, have been used to construct immunosensors for ebola viral detection, as shown in Table 2.7.

Table 2.8 demonstrates the analytical performance of recently published immunosensors to detect several viruses. Lee et al. presented a norovirus detection sensing platform utilizing multi-functional carbon nanotubes [11]. They manufactured the target DNA sensor using Au/MNP CNTs magnetically aligned to the Pt-integrated electrode (PtIDE). A carbon chain (C6) based spacer was used in the hybridization procedure. Nanomaterials can disperse well within the PtIDE electrode and improve the sensor performance under the magnetic field. This approach demonstrates improved sensing performance with various concentrations of target DNA (1 pM to 10 nM) and verified the LOD of about 8.8 pM, compared to previous nonaligned platforms. They also conducted influenza and ZKV viral DNA selectivity tests, but norovirus DNA showed a good response. Layqah and Eissa

Table 2.7 Electrochemical immunosensors for Ebola viral infectious disease detection.

Biosensor format/transducer	Recognition matrix	Detection target	Linear range	Detection limit	Detection time	Sample	References
FET	Antibody/AuNPs/rGO-FET	Viral glycoprotein	–	1 ng/mL	Within a few seconds	Diluted PBS buffer, human serum, and plasma	[192]
FET	Antibody/rGO-FET	Viral glycoprotein	2.4 pM to 12 μM	2.4 pM	–	Serum	[193]
Electrochemical DNA biosensor EIS and DPV	cDNA probe/Au-SPE	Viral cDNA	10–75 nM	4.7 nM	45 min	Real	[194]
Electrochemical genosensor SWV	HRP conjugated probe and Streptavidin-magnetic NPs-SPE	Viral cDNA	200 ymol to 200 zmol	33 cDNA copies	Below 2.5 h	Real	[195]

DPV, Differential pulse voltammetry; *EIS*, electrochemical impedance spectroscopy; *FET*, field-effect transistor; *SWV*, square wave voltammetry.

Table 2.8 Summary of the analytical performance of electrochemical immunosensors for different virus detection.

Virus	Biosensor format/ transducer	Recognition matrix	Detection target	Linearity	Limit of detection	Detection time	Sample	References
Rotavirus	Label-free biosensor peak wavelength value	3D PC structures	Rotavirus antigen	2.54 μg/mL to 1.27 mg/mL	6.35 μg/mL	–	PBS buffer	[196]
	FET biosensor	Anti-rotavirus IgG	Rotavirus	Pure rotavirus stock: $10-10^5$ PFU/mL Fecal sample: $10-10^4$ PFU/ mL	Pure rotavirus stock: 10^2 PFU/mL fecal sample: 10^3 PFU/mL	50 min	Pure rotavirus stock and fecal sample	[197]
Japanese encephalitis virus	Electrochemical biosensor EIS, CV	MRGO-FET	JEV positive serum	1–20 ng/mL	0.36 ng/mL	10 min	Human serum	[198]
	Electrochemical biosensor EIS, CV	CNP-SPCE	JEV antigen	0.1–20.0 ng/mL	2.60 ng/mL	20 min	–	[199]
Norovirus	Impedance electrochemical biosensor, EIS	Ag NPs-SPCE	Noro virus	10^1 to 10^5 copies/mL	2.47 copies/mL	30 min	Various food samples-Oyster	[200]
MERS-CoV	Electrochemical immunosensor EIS, CV	NoroBP-nonFoul-$(FlexL)_2$	Recombinant spike protein S1 MERS-CoV	0.001–100 ng/mL MERS, 0.01–10,000 ng/mL HCOV	0.4 (MERS) and 1.0 (HCOV) pg/mL	20 min	Spiked nasal samples	[201]

(Continued)

Table 2.8 (Continued)

Virus	Biosensor format/ transducer	Recognition matrix	Detection target	Linearity	Limit of detection	Detection time	Sample	References
Newcastle Disease Virus	Label-free electrochemical Immunosensor CV	Au-SPE	Inactivated NDV Newcastle disease virus	0.001–100 ng/mL	5 EID_{50}/mL	–	Chicken egg yolk antibodies	[202]
FMV	Label-free electrochemical Immunosensor DPV EIS	Array of carbon electrodes (DEP) modified with gold nanoparticles	FMV	10^2 EID50/mL–10^6 EID50/mL	100.95 EID50/ mL	–	Infected and healthy Fig leaf samples	[203]
Avian leukosis virus	Electrochemical immunosensor EIS, DPV	Au-SPE	Avian leukosis virus subgroup J (ALV-J)	0.1 nM to 1 μM	0.03 nM	–	Avian serum samples	[204]
Enterovirus 71	Colorimetric and electrochemical immunosensor Chronoamperometry. CV, SWV	Protein A/glutaraldehyde (PrA/GA) and self-assembly modification/*N*-succinimidyl active ester (SAM/NHS).	Human enterovirus 71 (EV71)	102.0–104.0 TCID50/mL	101.93 TCID50/ mL	2 h	Clinical positive samples of HFMD	[205]
CTV	Electrochemical immunosensor CV, EIS, DPV	Anti-FMV/MUA-MPA/Au electrode	CP of CTV	1 nM to 5 μM	0.27 nM	–	Real samples	[206]

Human papillomavirus	Electrochemical immunosensor EIS, SWV	Fc-AuNP-β-CD-Ab2	HPV-DNA type 16 human papillomavirus (HPV)	10–200 nM.	2.3 nM	–	Real DNA samples	[207]
	Electrochemical DNA sensor Sandwich hybridization DPV		HPV-DNA 16,18	0.1–1000 nM	150 pM (HPV types 16) 153 pM (HPV types 18)	-	Cell lines	[208]
	Interdigitated arrays SWV	ALV-J/BSA/Ab1/GR-PTCA/GCE	HPV-16-L1	10–50 nM	490 pM	–	Clinical samples	[209]
Chikungunya virus	Immunosensor—EIS	Dual-labeled (26) magnetic nanobeads	CHIKV-nsP3	25 ng/mL to 1 μg/mL	8×10^{-9} g/mL	–	Human serum	[210]
	Electrochemical DPV		Recombinant Chikungunya virus E1 protein (CHIK-VP)	10^{-14}–10^{-8} g/mL	3.27×10^{-14} g/mL	–	Real	[211]

BSA, Bovine serum albumin; *CP*, coat protein; *CTV*, citrus tristeza virus; *CV*, cyclic voltammetry; *DPV*, differential pulse voltammetry; *EIS*, electrochemical impedance spectroscopy; *FET*, field-effect transistor; *FMV*, fig mosaic virus; *MERS*, Middle East Respiratory Syndrome; *MRGO*, Micropatterned RGO; *PC*, photonic crystal; *SPCE*, screen-printed carbon electrode; *SWV*, square wave voltammetry.

have developed an EC immunosensor for MERS-CoV detection based on AuNPs-modified interdigitated carbon electrodes [94]. This method comprises an array of eight electrodes that may detect several CoV AGs in the same nasal fluid simultaneously using various combinations of electrodes. In addition to having excellent linearity (0.001–100 ng/mL), the interdigitated carbon electrode produced had a low LOD of 1 pg/mL and was easy to manufacture. Singhal et al. [95] reported developing the first EC biosensor for diagnosing the CHIGV by voltammetric signals based on molybdenum disulfide nanosheets (MoS2 NSs). The suggested genosensor showed the chikungunya viral DNA with a broad linear range of 0.1 nM to 100 μM. A simple, sensitive, and cheap technology has been created to detect CHIKV. The use of paper analytical devices (PADs) is another notable feature of the current biosensor technology. The same group of authors [95], for the first time, developed a simple paper-based nanoenabled EC biosensor for detecting the chikungunya virus. The researchers examined the use of the ultra-high charge-transfer capacity of gold NPs (Au) and the biocompatibility effect of magnetic NPs (Fe_3O_4) to construct a detecting method for CHIKV. According to the current sensor, the detection limit was 0.1 nM, and the linear range was from 0.1 nM to 100 μM. In the recent research, EIS-based immunosensors were used for label-free detection of one of the nonstructural proteins Chikungunya NsP3 (CHIKV-nsP3). On the Au surface, the respective Anti-CHIKV-nsP3 AB (Anti-CHIKV-nsP3Ab) to the target of the analyte was immobilized by a self-assembled monolayer (SAM) composed of 11-mercaptoundecanoic acid (11-MUA) and 6-mercapto-1-hexanol (6-MH). According to the methodology used in this research, the LOD and LOQ for the EIS biosensors were 8 and 17 ng/mL, respectively. The biosensor was highly specific CHIKV-nsP3 in a linear range from 25 ng/mL to 1 μg/mL [96].

2.3 Conclusion and future prospective

In recent decades, the sensors continues to be an active field in analytical chemistry research, as shown in the number of research studies until biosensors have generated considerable interest in detecting different diseases. The commercialization of biosensors and their uses for effects have been assessed, and earlier studies have shown the essential significance of biosensors in clinical trials. Nevertheless, we focused on the impact on disease-related analytics of immunosensors, a sub-class of biosensors and showed their distinctive features, discussing several new functionalized immunosensors. The emphasis is on electrode manufacturing, the immobilization methods, measuring instruments, and the relevant materials utilized as analytical tools for identifying diseases.

In addition to demonstrating benefits such as simplicity and cost-effectiveness, immunosensors have also been shown to display excellent sensitivity and selectivity operating under optimal circumstances. In some situations, low sample concentration is usually needed. In recent decades, the area of EC biosensors has seen a continuous progression toward greater sophistication. As a result of many improvements made to electrodes, signal amplifiers, and indicator labels, the detection limits have been reduced, and the linear response range has been expanded. Indeed, covering the electrode with different coatings and using sophisticated nanomaterials and appropriate indications may impact the sensor's overall performance. Although a low detection limit has been reached, the mini-potentiostat may react more slowly at greater concentrations in certain instances, even though the detection limit has been reduced.

On the other hand, a more significant number of clinical samples are required to process the samples. It is difficult to find literature methods that focus on decreasing sample quantities or the duration of analysis time. As a result, there is still a long way to go before EC immunosensors are extensively utilized in clinical laboratories not just in research laboratories, as they are now. Some major developing trends have a great potential for viral diagnostic technologies. Firstly, the device must incorporate wireless communication capabilities so that sophisticated machine learning and artificial intelligence (AI) approaches can be used to extract the required information from the results of the analyses. The results can be used to inform hospital consultant decision-making epidemiological models, quickly.

Furthermore, the use of biocompatible materials, which allows for the development of wearable and even implanted devices to be made possible, everything from the entire life cycle, comprising manufacturing process, the use of environmentally friendly, nontoxic materials, and the potential to recycle or repurpose the components after testing, must be carefully examined. Finally, it is necessary to utilize as few samples as possible, such as saliva, breath, tears, or urine. The field of molecular diagnostics continues to have a defined path for future research to improve sensitivity and stability.

Meanwhile, it is necessary to conduct analytical validation on more clinical samples obtained from patients. Additional factors must be considered as well as the type of the antigen (or enzyme), protein or other biomolecules (or the orientation of the immobilization). As a result, future studies will need to investigate the selection of the recognition element and optimize immobilization methods.

Declaration of competing interest

The authors declare that they have no known competing financial interests or personal relationships that could have influenced the work reported in this work.

Acknowledgments

The authors grateful to the Coordination for the Improvement of Higher Education Personnel (CAPES)-Brazil, PROJECT DATA: CORPORATE NAME: Reinaldo Barreto Oriá, Project Title: Risk factors, outcomes and impact of COVID-19 in patients with chronic diseases in Ceará, Process number: 88881.505364/2020–01 CAPES-EPIDEMICS Program for financial support.

References

[1] Y. Saylan, Ö. Erdem, S. Ünal, A. Denizli, An alternative medical diagnosis method: biosensors for virus detection, Biosensors 9 (2) (2019) 65. Available from: https://doi.org/10.3390/bios9020065.
[2] W.-S. Ryu, Virus structure, Molecular Virology of Human Pathogenic Viruses, Elsevier, 2017, pp. 21–29. Available from: https://doi.org/10.1016/B978-0-12-800838-6.00002-3.
[3] H.A.M. Faria, V. Zucolotto, Label-free electrochemical DNA biosensor for Zika virus identification, Biosens. Bioelectron. 131 (2019) 149–155. Available from: https://doi.org/10.1016/j.bios.2019.02.018.
[4] K. Mahato, P.K. Maurya, P. Chandra, Fundamentals and commercial aspects of nanobiosensors in point-of-care clinical diagnostics, 3 Biotech. 8 (3) (2018) 149. Available from: https://doi.org/10.1007/s13205-018-1148-8.
[5] S.R. Ahmed, J.C. Corredor, É. Nagy, S. Neethirajan, Amplified visual immunosensor integrated with nanozyme for ultrasensitive detection of avian influenza virus, Nanotheranostics 1 (3) (2017). Available from: https://doi.org/10.7150/ntno.20758.
[6] C. Lin, Y. Guo, M. Zhao, et al., Highly sensitive colorimetric immunosensor for influenza virus H5N1 based on enzyme-encapsulated liposome, Anal. Chim. Acta (2017) 963. Available from: https://doi.org/10.1016/j.aca.2017.01.031.
[7] A.H. Qatamin, J.H. Ghithan, M. Moreno, et al., Detection of influenza virus by electrochemical surface plasmon resonance under potential modulation, Appl. Opt. 58 (11) (2019). Available from: https://doi.org/10.1364/ao.58.002839.
[8] S.R. Ahmed, É. Nagy, S. Neethirajan, Self-assembled star-shaped chiroplasmonic gold nanoparticles for an ultrasensitive chiro-immunosensor for viruses, RSC Adv. 7 (65) (2017). Available from: https://doi.org/10.1039/c7ra07175b.
[9] U. Jarocka, R. Sawicka, A. Stachyra, et al., A biosensor based on electroactive dipyrromethene-Cu(II) layer deposited onto gold electrodes for the detection of antibodies

against avian influenza virus type H5N1 in hen sera, Anal. Bioanal. Chem. 407 (25) (2015) 7807–7814. Available from: https://doi.org/10.1007/s00216-015-8949-y.

[10] Q. Zhang, L. Lingyan, Q. Zhaohui, L. Chunyang, F. Yingchun, X. Qingji, et al., An impedance immunosensor based on low-cost microelectrodes and specific monoclonal antibodies for rapid detection of avian influenza virus H5N1 in chicken swabs, Acs.analchem (2017) 12145–12151. Available from: https://doi.org/10.1021/acs.analchem.7b02784.

[11] J. Lee, M. Morita, K. Takemura, E.Y. Park, A multi-functional gold/iron-oxide nanoparticle-CNT hybrid nanomaterial as virus DNA sensing platform, Biosens. Bioelectron. 102 (2018) 425–431. Available from: https://doi.org/10.1016/j.bios.2017.11.052.

[12] J. Bhardwaj, N. Chaudhary, H. Kim, J. Jang, Subtyping of influenza A H1N1 virus using a label-free electrochemical biosensor based on the DNA aptamer targeting the stem region of HA protein, Anal. Chim. Acta 1064 (2019) 94–103. Available from: https://doi.org/10.1016/j.aca.2019.03.005.

[13] R. Singh, S. Hong, J. Jang, Label-free detection of influenza viruses using a reduced graphene oxide-based electrochemical immunosensor integrated with a microfluidic platform, Sci. Rep. 7 (1) (2017) 42771. Available from: https://doi.org/10.1038/srep42771.

[14] H. Zhang, S. Zhang, N. Liu, Prevention and control of emergent infectious disease with high specific antigen sensor, Artif. Cells, Nanomed. Biotechnol. 45 (7) (2017). Available from: https://doi.org/10.3109/21691401.2016.1161638.

[15] E. Mikuła, C.E. Silva, E. Kopera, K. Zdanowski, J. Radecki, H. Radecka, Highly sensitive electrochemical biosensor based on redox – active monolayer for detection of anti-hemagglutinin antibodies against swine-origin influenza virus H1N1 in sera of vaccinated mice, BMC Vet. Res. 14 (1) (2018) 328. Available from: https://doi.org/10.1186/s12917-018-1668-9.

[16] J.-H. Han, D. Lee, C.H.C. Chew, T. Kim, J.J. Pak, A multi-virus detectable microfluidic electrochemical immunosensor for simultaneous detection of H1N1, H5N1, and H7N9 virus using ZnO nanorods for sensitivity enhancement, Sens. Actuators B Chem. 228 (2016) 36–42. Available from: https://doi.org/10.1016/j.snb.2015.07.068.

[17] B. Babamiri, R. Hallaj, A. Salimi, Ultrasensitive electrochemiluminescence immunosensor for determination of hepatitis B virus surface antigen using CdTe@CdS-PAMAM dendrimer as luminescent labels and Fe3O4 nanoparticles as magnetic beads, Sens. Actuators B Chem. 254 (2018) 551–560. Available from: https://doi.org/10.1016/j.snb.2017.07.016.

[18] X. Qian, S. Tan, Z. Li, Q. Qu, L. Li, L. Yang, A robust host-guest interaction controlled probe immobilization strategy for the ultrasensitive detection of HBV DNA using hollow HP5–Au/CoS nanobox as biosensing platform, Biosens. Bioelectron. 153 (2020) 112051. Available from: https://doi.org/10.1016/j.bios.2020.112051.

[19] M. Li, P. Wang, F. Pei, et al., Highly sensitive immunosensor for Hepatitis B surface antigen detection based on a novel signal amplification system of gold nanorods and mesoporous Au@Pd@Pt core-shell nanospheres, J. Electroanal. Chem. (2018) 809. Available from: https://doi.org/10.1016/j.jelechem.2017.12.044.

[20] X. Li, T. Liu, Y. Zhang, et al., A novel electrochemical immunosensor for hepatitis B surface antigen based on Fe3O4 nanoflowers and heterogeneous chain reaction signal amplification strategy, Talanta 221 (2021) 121459. Available from: https://doi.org/10.1016/j.talanta.2020.121459.

[21] S. Sabouri, H. Ghourchian, M. Shourian, M. Boutorabi, A gold nanoparticle-based immunosensor for the chemiluminescence detection of the hepatitis B surface antigen, Anal. Methods 6 (14) (2014) 5059–5066. Available from: https://doi.org/10.1039/c4ay00461b.

[22] M. Shourian, H. Ghourchian, M. Boutorabi, Ultra-sensitive immunosensor for detection of hepatitis B surface antigen using multi-functionalized gold nanoparticles, Anal. Chim. Acta 895 (2015) 1–11. Available from: https://doi.org/10.1016/j.aca.2015.07.013.
[23] F. Walters, S. Rozhko, D. Buckley, E.D. Ahmadi, M. Ali, Z. Tehrani, . . .O.J. Guy, Real-time detection of hepatitis B surface antigen using a hybrid graphene-gold nanoparticle biosensor, 2D Mater. 7 (2020) 024009. Available from: https://doi.org/10.1088/2053-1583/ab734c.
[24] J. Upan, P. Banet, P. Aubert, K. Ounnunkad, J. Jakmunee, Sequential injection-differential pulse voltammetric immunosensor for hepatitis B surface antigen using the modified screen-printed carbon electrode, Electrochim. Acta (2020) 136335. Available from: https://doi.org/10.1016/j.electacta.2020.136335.
[25] C. Akkapinyo, P. Khownarumit, D. Waraho-Zhmayev, R.P. Poo-arporn, Development of a multiplex immunochromatographic strip test and ultrasensitive electrochemical immunosensor for hepatitis B virus screening, Anal. Chim. Acta 1095 (2020) 162–171. Available from: https://doi.org/10.1016/j.aca.2019.10.016.
[26] J. Kim, S.Y. Oh, S. Shukla, et al., Heteroassembled gold nanoparticles with sandwich-immunoassay LSPR chip format for rapid and sensitive detection of hepatitis B virus surface antigen (HBsAg), Biosens. Bioelectron. 107 (2018) 118–122. Available from: https://doi.org/10.1016/j.bios.2018.02.019.
[27] S. Wei, H. Xiao, M. Gu, Z. Chen, L. Cao, Ultrasensitive label-free electrochemical immunosensor based on core-shell Au@PtNPs functionalized rGO-TEPA/PB nanocomposite for HBsAg detection, J. Electroanal. Chem. 890 (2021) 115216. Available from: https://doi.org/10.1016/j.jelechem.2021.115216.
[28] Z. Gao, Y. Li, X. Zhang, et al., Ultrasensitive electrochemical immunosensor for quantitative detection of HBeAg using Au@Pd/MoS2@MWCNTs nanocomposite as enzyme-mimetic labels, Biosens. Bioelectron. 102 (2018) 189–195. Available from: https://doi.org/10.1016/j.bios.2017.11.032.
[29] Wei S, Xiao H, Cao L, Chen Z. A label-free immunosensor based on graphene oxide/Fe3O4/prussian blue nanocomposites for the electrochemical determination of HBsAg. Biosensors. 2020;10(3):24. Doi: 10.3390/bios10030024
[30] F. Zhao, Y. Bai, R. Zeng, et al., An electrochemical immunosensor with graphene-oxide-ferrocene-based nanocomposites for hepatitis B surface antigen detection, Electroanalysis 30 (11) (2018) 2774–2780. Available from: https://doi.org/10.1002/elan.201800476.
[31] F. Zhao, L. Cao, Y. Liang, Z. Wu, Z. Chen, R. Zeng, Label-free amperometric immunosensor based on graphene oxide and ferrocene-chitosan nanocomposites for detection of hepatis B virus antigen, J. Biomed. Nanotechnol. 13 (10) (2017) 1300–1308. Available from: https://doi.org/10.1166/jbn.2017.2415.
[32] J. Yang, D. Li, J. Fu, F. Huang, Q. Wei, TiO2-CuCNFs based laccase biosensor for enhanced electrocatalysis in hydroquinone detection, J. Electroanal. Chem. 766 (2016) 16–23. Available from: https://doi.org/10.1016/j.jelechem.2016.01.030.
[33] P. Wang, M. Li, F. Pei, et al., An ultrasensitive sandwich-type electrochemical immunosensor based on the signal amplification system of double-deck gold film and thionine unite with platinum nanowire inlaid globular SBA-15 microsphere, Biosens. Bioelectron. 91 (2017) 424–430. Available from: https://doi.org/10.1016/j.bios.2016.12.057.
[34] Y. Li, L. Tian, L.L. Liu, et al., An ultrasensitive sandwich-type electrochemical immunosensor based on δ-MnO 2 and palladium nanoparticles covered natural halloysite nanotubes for the detection of hepatitis B surface antigen, N. J. Chem. 40 (1) (2016) 558–563. Available from: https://doi.org/10.1039/C5NJ01251A.

[35] F. Pei, P. Wang, E. Ma, et al., A sandwich-type electrochemical immunosensor based on RhPt NDs/NH2-GS and Au NPs/PPy NS for quantitative detection hepatitis B surface antigen, Bioelectrochemistry 126 (2019) 92–98. Available from: https://doi.org/10.1016/j.bioelechem.2018.11.008.

[36] F. Li, Y. Li, J. Feng, et al., Facile synthesis of MoS2@Cu2O-Pt nanohybrid as enzyme-mimetic label for the detection of the Hepatitis B surface antigen, Biosens. Bioelectron. 100 (2018) 512–518. Available from: https://doi.org/10.1016/j.bios.2017.09.048.

[37] Q. Yang, P. Wang, E. Ma, et al., A sandwich-type electrochemical immunosensor based on Au@Pd nanodendrite functionalized MoO2 nanosheet for highly sensitive detection of HBsAg, Bioelectrochemistry (2021) 138. Available from: https://doi.org/10.1016/j.bioelechem.2020.107713.

[38] Z. Tan, H. Dong, Q. Liu, et al., A label-free immunosensor based on PtPd NCs@MoS2 nanoenzymes for hepatitis B surface antigen detection, Biosens. Bioelectron. 142 (2019) 111556. Available from: https://doi.org/10.1016/j.bios.2019.111556.

[39] Z. Tan, L. Cao, X. He, et al., A label-free immunosensor for the sensitive detection of hepatitis B e antigen based on PdCu tripod functionalized porous graphene nanoenzymes, Bioelectrochemistry 133 (2020) 107461. Available from: https://doi.org/10.1016/j.bioelechem.2020.107461.

[40] Z. Shakoori, S. Salimian, S. Kharrazi, M. Adabi, R. Saber, Electrochemical DNA biosensor based on gold nanorods for detecting hepatitis B virus, Anal. Bioanal. Chem. 407 (2) (2015) 455–461. Available from: https://doi.org/10.1007/s00216-014-8303-9.

[41] L. Jiang, Y.Y. Li, Z. Xu, et al., Simultaneous electrochemical determination of two hepatitis B antigens using graphene-SnO2 hybridized with sea urchin–like bimetallic nanoparticles, Microchim. Acta 188 (4) (2021) 109. Available from: https://doi.org/10.1007/s00604-021-04763-8.

[42] X. Fang, O.K. Tan, M.S. Tse, E.E. Ooi, A label-free immunosensor for diagnosis of dengue infection with simple electrical measurements, Biosens. Bioelectron. 25 (5) (2010) 1137–1142. Available from: https://doi.org/10.1016/j.bios.2009.09.037.

[43] M.J. Huang, H. Xie, Q. Wan, L. Zhang, Y. Ning, G.-J. Zhang, Serotype-specific identification of dengue virus by silicon nanowire array biosensor, J. Nanosci. Nanotechnol. 13 (6) (2013) 3810–3817. Available from: https://doi.org/10.1166/jnn.2013.7220.

[44] D. Wasik, A. Mulchandani, M.V. Yates, Point-of-use nanobiosensor for detection of dengue virus NS1 antigen in adult *Aedes aegypti*: a potential tool for improved dengue surveillance, Anal. Chem. 90 (1) (2018) 679–684. Available from: https://doi.org/10.1021/acs.analchem.7b03407.

[45] K.Y.P.S.Y.P.S. Avelino, C.A.S.A.S. Andrade, C.P.P. de Melo, et al., Biosensor based on hybrid nanocomposite and CramoLL lectin for detection of dengue glycoproteins in real samples, Synth. Met. 194 (2014) 102–108. Available from: https://doi.org/10.1016/j.synthmet.2014.05.001.

[46] K. Navakul, C. Warakulwit, Pt Yenchitsomanus, A. Panya, P.A. Lieberzeit, C. Sangma, A novel method for dengue virus detection and antibody screening using a graphene-polymer based electrochemical biosensor, Nanomed. Nanotechnol., Biol. Med. 13 (2) (2017) 549–557. Available from: https://doi.org/10.1016/j.nano.2016.08.009.

[47] A.A. Odeh, Y. Al-Douri, C.H. Voon, et al., A needle-like Cu2CdSnS4 alloy nanostructure-based integrated electrochemical biosensor for detecting the DNA of Dengue serotype 2, Microchim. Acta 184 (7) (2017) 2211–2218. Available from: https://doi.org/10.1007/s00604-017-2249-5.

[48] G.A. Ortega, J.C. Zuaznabar-Gardona, E. Reguera, Electrochemical immunoassay for the detection of IgM antibodies using polydopamine particles loaded with PbS quantum dots as labels, Biosens. Bioelectron. 116 (2018) 30–36. Available from: https://doi.org/10.1016/j.bios.2018.05.046.
[49] N.T. Darwish, S.D. Sekaran, Y. Alias, S.M. Khor, Immunofluorescence–based biosensor for the determination of dengue virus NS1 in clinical samples, J. Pharm. Biomed. Anal. 149 (2018) 591–602. Available from: https://doi.org/10.1016/j.jpba.2017.11.064.
[50] E. Iswardy, T.-C. Tsai, I.-F. Cheng, T.-C. Ho, G.C. Perng, H.-C. Chang, A bead-based immunofluorescence-assay on a microfluidic dielectrophoresis platform for rapid dengue virus detection, Biosens. Bioelectron. 95 (2017) 174–180. Available from: https://doi.org/10.1016/j.bios.2017.04.011.
[51] Y. Mustapha Kamil, M.H. Abu Bakar, M.A. Mustapa, et al., Label-free dengue E protein detection using a functionalized tapered optical fiber sensor, Sens. Actuators B Chem. 257 (2018) 820–828. Available from: https://doi.org/10.1016/j.snb.2017.11.005.
[52] M. Sánchez-Purrà, M. Carré-Camps, H. de Puig, I. Bosch, L. Gehrke, K. Hamad-Schifferli, Surface-enhanced Raman spectroscopy-based sandwich immunoassays for multiplexed detection of Zika and dengue viral biomarkers, ACS Infect. Dis. 3 (10) (2017) 767–776. Available from: https://doi.org/10.1021/acsinfecdis.7b00110.
[53] O. Parkash, R. Shueb, Diagnosis of dengue infection using conventional and biosensor based techniques, Viruses 7 (10) (2015) 5410–5427. Available from: https://doi.org/10.3390/v7102877.
[54] O. Parkash, C. Yean, R. Shueb, Screen printed carbon electrode based electrochemical immunosensor for the detection of dengue NS1 antigen, Diagnostics 4 (4) (2014) 165–180. Available from: https://doi.org/10.3390/diagnostics4040165.
[55] Zainuddin A, Nordin A, Asri M, et al. Development of integrated electrochemical–quartzcrystal microbalance biosensor arrays: towards ultrasensitive, multiplexed and rapid point-of-care dengue detection, in: Proceedings of the Twelfth International Joint Conference on Biomedical Engineering Systems and Technologies, SCITEPRESS - Science and Technology Publications; 2019:220–227. doi:10.5220/0007523802200227.
[56] P. Jahanshahi, E. Zalnezhad, S.D. Sekaran, F.R.M. Adikan, Rapid immunoglobulin M-based dengue diagnostic test using surface plasmon resonance biosensor, Sci. Rep. 4 (1) (2015) 3851. Available from: https://doi.org/10.1038/srep03851.
[57] M.H. Nawaz, A. Hayat, G. Catanante, U. Latif, J.L. Marty, Development of a portable and disposable NS1 based electrochemical immunosensor for early diagnosis of dengue virus, Anal. Chim. Acta 1026 (2018) 1–7. Available from: https://doi.org/10.1016/j.aca.2018.04.032.
[58] Q.Y. Siew, E.L. Pang, H.S. Loh, M.T.T. Tan, Highly sensitive and specific graphene/TiO2 impedimetric immunosensor based on plant-derived tetravalent envelope glycoprotein domain III (EDIII) probe antigen for dengue diagnosis, Biosens. Bioelectron. (2021) 176. Available from: https://doi.org/10.1016/j.bios.2020.112895.
[59] Q.Y. Siew, S.H. Tan, E.L. Pang, H.-S.S. Loh, M.T.T.T. Tan, A graphene-based dengue immunosensor using plant-derived envelope glycoprotein domain III (EDIII) as the novel probe antigen, Analyst 146 (6) (2021) 2009–2018. Available from: https://doi.org/10.1039/D0AN02219E.
[60] M. Awan, S. Rauf, A. Abbas, et al., A sandwich electrochemical immunosensor based on antibody functionalized-silver nanoparticles (Ab-Ag NPs) for the detection of dengue biomarker protein NS1, J. Mol. Liq. 317 (2020) 114014. Available from: https://doi.org/10.1016/j.molliq.2020.114014.

[61] D. Wasik, A. Mulchandani, M. Yates, Salivary detection of dengue virus NS1 protein with a label-free immunosensor for early dengue diagnosis, Sensors 18 (8) (2018). Available from: https://doi.org/10.3390/s18082641.

[62] D.M.N. Luna, K.Y.P.S. Avelino, M.T. Cordeiro, C.A.S. Andrade, M.D.L. Oliveira, Electrochemical immunosensor for dengue virus serotypes based on 4-mercaptobenzoic acid modified gold nanoparticles on self-assembled cysteine monolayers, Sens. Actuators, B Chem. (2015) 220. Available from: https://doi.org/10.1016/j.snb.2015.05.067.

[63] Q. Palomar, X. Xu, C. Gondran, M. Holzinger, S. Cosnier, Z. Zhang, Voltammetric sensing of recombinant viral dengue virus 2 NS1 based on Au nanoparticle−decorated multiwalled carbon nanotube composites, Microchim. Acta 187 (6) (2020) 363. Available from: https://doi.org/10.1007/s00604-020-04339-y.

[64] S.S. Pawitwar, S. Dhar, S. Tiwari, et al., Overview on the current status of Zika virus pathogenesis and animal related research, J. Neuroimmune Pharmacol. 12 (3) (2017) 371−388. Available from: https://doi.org/10.1007/s11481-017-9743-8.

[65] A. Kaushik, A. Yndart, S. Kumar, et al., A sensitive electrochemical immunosensor for label-free detection of Zika-virus protein, Sci. Rep. 8 (1) (2018) 9700. Available from: https://doi.org/10.1038/s41598-018-28035-3.

[66] A.M. Faria, T. Mazon, Early diagnosis of Zika infection using a ZnO nanostructures-based rapid electrochemical biosensor, Talanta 203 (2019) 153−160. Available from: https://doi.org/10.1016/j.talanta.2019.04.080.

[67] C. Tancharoen, W. Sukjee, C. Thepparit, et al., Electrochemical biosensor based on surface imprinting for Zika virus detection in serum, ACS Sens. 4 (1) (2019) 69−75. Available from: https://doi.org/10.1021/acssensors.8b00885.

[68] Y.P. Hsu, N.S. Li, Y.T. Chen, H.H. Pang, K.C. Wei, H.W. Yang, A serological point-of-care test for Zika virus detection and infection surveillance using an enzyme-free vial immunosensor with a smartphone, Biosens. Bioelectron. (2020) 151. Available from: https://doi.org/10.1016/j.bios.2019.111960.

[69] H.-J. Zhang, J. Zhu, N. Bao, S.-N. Ding, Enhanced electrochemiluminescence of CdS quantum dots capped with mercaptopropionic acid activated by EDC for Zika virus detection, Analyst 146 (9) (2021) 2928−2935. Available from: https://doi.org/10.1039/D0AN02437F.

[70] M. Steinmetz, D. Lima, A.G. Viana, et al., A sensitive label-free impedimetric DNA biosensor based on silsesquioxane-functionalized gold nanoparticles for Zika Virus detection, Biosens. Bioelectron. (2019) 141. Available from: https://doi.org/10.1016/j.bios.2019.111351.

[71] S. Afsahi, M.B. Lerner, J.M. Goldstein, et al., Novel graphene-based biosensor for early detection of Zika virus infection, Biosens. Bioelectron. (2018) 100. Available from: https://doi.org/10.1016/j.bios.2017.08.051.

[72] F. Cui, H.S. Zhou, Diagnostic methods and potential portable biosensors for coronavirus disease 2019, Biosens. Bioelectron. 165 (2020) 112349. Available from: https://doi.org/10.1016/j.bios.2020.112349.

[73] T. Ji, Z. Liu, G. Wang, et al., Detection of COVID-19: a review of the current literature and future perspectives, Biosens. Bioelectron. 166 (2020) 112455. Available from: https://doi.org/10.1016/j.bios.2020.112455.

[74] E. Sheikhzadeh, S. Eissa, A. Ismail, M. Zourob, Diagnostic techniques for COVID-19 and new developments, Talanta 220 (2020) 121392. Available from: https://doi.org/10.1016/j.talanta.2020.121392.

[75] L. Chen, G. Zhang, L. Liu, Z. Li, Emerging biosensing technologies for improved diagnostics of COVID-19 and future pandemics, Talanta 225 (2021) 121986. Available from: https://doi.org/10.1016/j.talanta.2020.121986.

[76] A. Ahmadivand, B. Gerislioglu, Z. Ramezani, A. Kaushik, P. Manickam, S.A. Ghoreishi, Functionalized terahertz plasmonic metasensors: femtomolar-level

detection of SARS-CoV-2 spike proteins, Biosens. Bioelectron. 177 (2021) 112971. Available from: https://doi.org/10.1016/j.bios.2021.112971.

[77] L. Fabiani, M. Saroglia, G. Galatà, et al., Magnetic beads combined with carbon black-based screen-printed electrodes for COVID-19: a reliable and miniaturized electrochemical immunosensor for SARS-CoV-2 detection in saliva, Biosens. Bioelectron. 171 (2021) 112686. Available from: https://doi.org/10.1016/j.bios.2020.112686.

[78] G. Seo, G. Lee, M.J. Kim, et al., Rapid detection of COVID-19 causative virus (SARS-CoV-2) in human nasopharyngeal swab specimens using field-effect transistor-based biosensor, ACS Nano 14 (4) (2020) 5135−5142. Available from: https://doi.org/10.1021/acsnano.0c02823.

[79] A.K. Kaushik, J.S. Dhau, H. Gohel, et al., Electrochemical SARS-CoV-2 sensing at point-of-care and artificial intelligence for intelligent COVID-19 management, ACS Appl. Bio Mater. 3 (11) (2020) 7306−7325. Available from: https://doi.org/10.1021/acsabm.0c01004.

[80] Mahari S, Roberts A, Shahdeo D, Gandhi S. Ecovsens-ultrasensitive novel in-house built printed circuit board based electrochemical device for rapid detection of nCovid-19 antigen, a spike protein domain 1 of SARS-CoV-2. *bioRxiv*. January 2020:2020.04.24.059204. doi:10.1101/2020.04.24.059204.

[81] L. Liv, G. Çoban, N. Nakiboğlu, T. Kocagöz, A rapid, ultrasensitive voltammetric biosensor for determining SARS-CoV-2 spike protein in real samples, Biosens. Bioelectron. 192 (2021) 113497. Available from: https://doi.org/10.1016/J.BIOS.2021.113497.

[82] E.B. Aydın, M. Aydın, M.K. Sezgintürk, Highly selective and sensitive sandwich immunosensor platform modified with MUA-capped GNPs for detection of spike Receptor Binding Domain protein: A precious marker of COVID 19 infection, Sens. Actuators B Chem. 345 (2021) 130355. Available from: https://doi.org/10.1016/J.SNB.2021.130355.

[83] M.A. Lifson, M.O. Ozen, F. Inci, et al., Advances in biosensing strategies for HIV-1 detection, diagnosis, and therapeutic monitoring, Adv. Drug. Deliv. Rev. 103 (2016) 90−104. Available from: https://doi.org/10.1016/j.addr.2016.05.018.

[84] N.S. Padian, A. Buvé, J. Balkus, D. Serwadda, W. Cates, Biomedical interventions to prevent HIV infection: evidence, challenges, and way forward, Lancet 372 (9638) (2008) 585−599. Available from: https://doi.org/10.1016/S0140-6736(08)60885-5.

[85] M. Giannetto, M. Costantini, M. Mattarozzi, M. Careri, Innovative gold-free carbon nanotube/chitosan-based competitive immunosensor for determination of HIV-related p24 capsid protein in serum, RSC Adv. 7 (63) (2017) 39970−39976. Available from: https://doi.org/10.1039/C7RA07245G.

[86] Y.-S.S. Fang, X.-J.J. Huang, L.-S.S. Wang, J.-F.F. Wang, An enhanced sensitive electrochemical immunosensor based on efficient encapsulation of enzyme in silica matrix for the detection of human immunodeficiency virus p24, Biosens. Bioelectron. 64 (2015) 324−332. Available from: https://doi.org/10.1016/j.bios.2014.09.022.

[87] B.M. Cerrutti, M.L. Moraes, S.H. Pulcinelli, C.V. Santilli, Lignin as immobilization matrix for HIV p17 peptide used in immunosensing, Biosens. Bioelectron. 71 (2015) 420−426. Available from: https://doi.org/10.1016/j.bios.2015.04.054.

[88] M.F. Fatin, A. Rahim Ruslinda, S.C.B. Gopinath, M.K.M. Arshad, High-performance interactive analysis of split aptamer and HIV-1 Tat on multiwall carbon nanotube-modified field-effect transistor, Int. J. Biol. Macromol. 125 (2019) 414−422. Available from: https://doi.org/10.1016/j.ijbiomac.2018.12.066.

[89] L. Zheng, L. Jia, B. Li, et al., A sandwich HIV p24 amperometric immunosensor based on a direct gold electroplating-modified electrode, Molecules 17 (5) (2012) 5988−6000. Available from: https://doi.org/10.3390/molecules17055988.

[90] L. Zhou, J. Huang, B. Yu, Y. Liu, T. You, A novel electrochemiluminescence immunosensor for the analysis of HIV-1 p24 antigen based on P-RGO@Au@Ru-SiO2 composite, ACS Appl. Mater. Interfaces 7 (44) (2015) 24438–24445. Available from: https://doi.org/10.1021/acsami.5b08154.

[91] X. Jin, H. Zhang, Y.-T. Li, et al., A field effect transistor modified with reduced graphene oxide for immunodetection of Ebola virus, Microchim. Acta 186 (4) (2019) 223. Available from: https://doi.org/10.1007/s00604-019-3256-5.

[92] S.H. Baek, M.W. Kim, C.Y. Park, et al., Development of a rapid and sensitive electrochemical biosensor for detection of human norovirus via novel specific binding peptides, Biosens. Bioelectron. 123 (2019) 223–229. Available from: https://doi.org/10.1016/j.bios.2018.08.064.

[93] Y. Chen, R. Ren, H. Pu, et al., Field-effect transistor biosensor for rapid detection of Ebola antigen, Sci. Rep. 7 (1) (2017) 10974. Available from: https://doi.org/10.1038/s41598-017-11387-7.

[94] L.A. Layqah, S. Eissa, An electrochemical immunosensor for the corona virus associated with the Middle East respiratory syndrome using an array of gold nanoparticle-modified carbon electrodes, Microchim. Acta 186 (4) (2019) 224. Available from: https://doi.org/10.1007/s00604-019-3345-5.

[95] C. Singhal, M. Khanuja, N. Chaudhary, C.S. Pundir, J. Narang, Detection of chikungunya virus DNA using two-dimensional MoS2 nanosheets based disposable biosensor, Sci. Rep. 8 (1) (2018). Available from: https://doi.org/10.1038/s41598-018-25824-8.

[96] A. George, M.S. Amrutha, P. Srivastava, V.V.R. Sai, S. Sunil, R. Srinivasan, Label-free detection of chikungunya non-structural protein 3 using electrochemical impedance spectroscopy, J. Electrochem. Soc. 166 (14) (2019) B1356–B1363. Available from: https://doi.org/10.1149/2.1081914jes.

[97] D. Nidzworski, P. Pranszke, M. Grudniewska, E. Król, B. Gromadzka, Universal biosensor for detection of influenza virus, Biosens. Bioelectron. [Internet] 59 (2014) 239–242. Available from: https://linkinghub.elsevier.com/retrieve/pii/S0956566314002310.

[98] S.R. Ahmed, J. Kim, V.T. Tran, T. Suzuki, S. Neethirajan, J. Lee, et al., In situ self-assembly of gold nanoparticles on hydrophilic and hydrophobic substrates for influenza virus-sensing platform, Sci. Rep. [Internet] 7 (1) (2017) 44495. Available from: http://www.nature.com/articles/srep44495.

[99] X. Cui, A. Das, A.N. Dhawane, J. Sweeney, X. Zhang, V. Chivukula, et al., Highly specific and rapid glycan based amperometric detection of influenza viruses, Chem. Sci. [Internet] 8 (5) (2017) 3628–3634. Available from: http://xlink.rsc.org/?DOI = C6SC03720H.

[100] C. Bai, Z. Lu, H. Jiang, Z. Yang, X. Liu, H. Ding, et al., Aptamer selection and application in multivalent binding-based electrical impedance detection of inactivated H1N1 virus, Biosens. Bioelectron. [Internet] 110 (2018) 162–167. Available from: https://linkinghub.elsevier.com/retrieve/pii/S0956566318302203.

[101] J. Lee, M. Morita, K. Takemura, E.Y. Park, A multi-functional gold/iron-oxide nanoparticle-CNT hybrid nanomaterial as virus DNA sensing platform, Biosens. Bioelectron. [Internet] 102 (2018) 425–431. Available from: https://linkinghub.elsevier.com/retrieve/pii/S0956566317307819.

[102] J. Bhardwaj, N. Chaudhary, H. Kim, J. Jang, Subtyping of influenza A H1N1 virus using a label-free electrochemical biosensor based on the DNA aptamer targeting the stem region of HA protein, Anal. Chim. Acta [Internet] 1064 (2019) 94–103. Available from: https://linkinghub.elsevier.com/retrieve/pii/S0003267019302776.

[103] W.W. Ye, M.-K. Tsang, X. Liu, M. Yang, J. Hao, Upconversion luminescence resonance energy transfer (LRET)-based biosensor for rapid and ultrasensitive

detection of avian influenza virus H7 subtype, Small [Internet] 10 (12) (2014) 2390–2397. Available from: http://doi.wiley.com/10.1002/smll.201303766.
[104] S. Dong, R. Zhao, J. Zhu, X. Lu, Y. Li, S. Qiu, et al., Electrochemical DNA biosensor based on a tetrahedral nanostructure probe for the detection of avian influenza A (H7N9) virus, ACS Appl. Mater. Interfaces [Internet] 7 (16) (2015) 8834–8842. Available from: https://pubs.acs.org/doi/10.1021/acsami.5b01438.
[105] Z. Wu, C.-H. Zhou, J.-J. Chen, C. Xiong, Z. Chen, D.-W. Pang, et al., Bifunctional magnetic nanobeads for sensitive detection of avian influenza A (H7N9) virus based on immunomagnetic separation and enzyme-induced metallization, Biosens. Bioelectron. [Internet] 68 (2015) 586–592. Available from: https://linkinghub.elsevier.com/retrieve/pii/S0956566315000688.
[106] Y. Yu, Z. Chen, W. Jian, D. Sun, B. Zhang, X. Li, et al., Ultrasensitive electrochemical detection of avian influenza A (H7N9) virus DNA based on isothermal exponential amplification coupled with hybridization chain reaction of DNAzyme nanowires, Biosens. Bioelectron. [Internet] 64 (2015) 566–571. Available from: https://linkinghub.elsevier.com/retrieve/pii/S0956566314007672.
[107] J. Huang, Z. Xie, Z. Xie, S. Luo, L. Xie, L. Huang, et al., Silver nanoparticles coated graphene electrochemical sensor for the ultrasensitive analysis of avian influenza virus H7, Anal. Chim. Acta [Internet] 913 (2016) 121–127. Available from: https://linkinghub.elsevier.com/retrieve/pii/S0003267016301386.
[108] J. Tian, A high sensitive electrochemical avian influenza virus H7 biosensor based on CNTs/MoSx Aerogel, Int. J. Electrochem. Sci. [Internet] (2017) 2658–2668. Available from: http://www.electrochemsci.org/abstracts/vol12/120402658.pdf.
[109] Y.-F. Chang, W.-H. Wang, Y.-W. Hong, R.-Y. Yuan, K.-H. Chen, Y.-W. Huang, et al., Simple strategy for rapid and sensitive detection of avian influenza A H7N9 virus based on intensity-modulated SPR biosensor and new generated antibody, Anal. Chem. [Internet] 90 (3) (2018) 1861–1869. Available from: https://pubs.acs.org/doi/10.1021/acs.analchem.7b03934.
[110] S. Karash, R. Wang, L. Kelso, H. Lu, T.J. Huang, Y. Li, Rapid detection of avian influenza virus H5N1 in chicken tracheal samples using an impedance aptasensor with gold nanoparticles for signal amplification, J. Virol. Methods [Internet] 236 (2016) 147–156. Available from: https://linkinghub.elsevier.com/retrieve/pii/S0166093415300677.
[111] V.-T. Nguyen, H.B. Seo, B.C. Kim, S.K. Kim, C.-S. Song, M.B. Gu, Highly sensitive sandwich-type SPR based detection of whole H5Nx viruses using a pair of aptamers, Biosens. Bioelectron. [Internet] 86 (2016) 293–300. Available from: https://linkinghub.elsevier.com/retrieve/pii/S0956566316306030.
[112] J. Kwon, Y. Lee, T. Lee, J.-H. Ahn, Aptamer-based field-effect transistor for detection of avian influenza virus in chicken serum, Anal. Chem. [Internet] 92 (7) (2020) 5524–5531. Available from: https://pubs.acs.org/doi/abs/10.1021/acs.analchem.0c00348.
[113] Ü. Anik, Y. Tepeli, M. Sayhi, J. Nsiri, M.F. Diouani, Towards the electrochemical diagnostic of influenza virus: development of a graphene–Au hybrid nanocomposite modified influenza virus biosensor based on neuraminidase activity, Analyst [Internet] 143 (1) (2018) 150–156. Available from: http://xlink.rsc.org/?DOI = C7AN01537B.
[114] M. Sayhi, O. Ouerghi, K. Belgacem, M. Arbi, Y. Tepeli, A. Ghram, et al., Electrochemical detection of influenza virus H9N2 based on both immunomagnetic extraction and gold catalysis using an immobilization-free screen printed carbon microelectrode, Biosens. Bioelectron. [Internet] 107 (2018) 170–177. Available from: https://linkinghub.elsevier.com/retrieve/pii/S0956566318301064.

[115] S.R. Joshi, A. Sharma, G.-H. Kim, J. Jang, Low cost synthesis of reduced graphene oxide using biopolymer for influenza virus sensor, Mater. Sci. Eng.: C. [Internet] 108 (2020) 110465. Available from: https://linkinghub.elsevier.com/retrieve/pii/S0928493119332461.

[116] U. Jarocka, R. Sawicka, A. Stachyra, A. Góra-Sochacka, A. Sirko, W. Zagórski-Ostoja, et al., A biosensor based on electroactive dipyrromethene-Cu(II) layer deposited onto gold electrodes for the detection of antibodies against avian influenza virus type H5N1 in hen sera, Anal. Bioanal. Chem. [Internet] 407 (25) (2015) 7807–7814. Available from: http://link.springer.com/10.1007/s00216-015-8949-y.

[117] J.-H. Han, D. Lee, C.H.C. Chew, T. Kim, J.J. Pak, A multi-virus detectable microfluidic electrochemical immunosensor for simultaneous detection of H1N1, H5N1, and H7N9 virus using ZnO nanorods for sensitivity enhancement, Sens. Actuators B: Chem. [Internet] 228 (2016) 36–42. Available from: https://linkinghub.elsevier.com/retrieve/pii/S0925400515301258.

[118] A. Hushegyi, D. Pihíková, T. Bertok, V. Adam, R. Kizek, J. Tkac, Ultrasensitive detection of influenza viruses with a glycan-based impedimetric biosensor, Biosens. Bioelectron. [Internet] 79 (2016) 644–649. Available from: https://linkinghub.elsevier.com/retrieve/pii/S0956566315307533.

[119] Z.-H. Yang, Y. Zhuo, R. Yuan, Y.-Q. Chai, A nanohybrid of platinum nanoparticles-porous ZnO–hemin with electrocatalytic activity to construct an amplified immunosensor for detection of influenza, Biosens. Bioelectron. [Internet] 78 (2016) 321–327. Available from: https://linkinghub.elsevier.com/retrieve/pii/S0956566315305327.

[120] U. Jarocka, R. Sawicka, A. Góra-Sochacka, A. Sirko, W. Dehaen, J. Radecki, et al., An electrochemical immunosensor based on a 4,4′-thiobisbenzenethiol self-assembled monolayer for the detection of hemagglutinin from avian influenza virus H5N1, Sens. Actuators B: Chem. [Internet] 228 (2016) 25–30. Available from: https://linkinghub.elsevier.com/retrieve/pii/S0925400516300016.

[121] Z.-H. Yang, Y. Zhuo, R. Yuan, Y.-Q. Chai, An amplified electrochemical immunosensor based on in situ-produced 1-naphthol as electroactive substance and graphene oxide and Pt nanoparticles functionalized CeO2 nanocomposites as signal enhancer, Biosens. Bioelectron. [Internet] 69 (2015) 321–327. Available from: https://linkinghub.elsevier.com/retrieve/pii/S0956566315000366.

[122] Z. Gao, Y. Li, X. Zhang, J. Feng, L. Kong, P. Wang, et al., Ultrasensitive electrochemical immunosensor for quantitative detection of HBeAg using Au@Pd/MoS2@MWCNTs nanocomposite as enzyme-mimetic labels, Biosens. Bioelectron. [Internet] 102 (2018) 189–195. Available from: https://linkinghub.elsevier.com/retrieve/pii/S0956566317307625.

[123] F. Zhao, L. Cao, Y. Liang, Z. Wu, Z. Chen, R. Zeng, Label-free amperometric immunosensor based on graphene oxide and ferrocene-chitosan nanocomposites for detection of hepatis B virus antigen, J. Biomed. Nanotechnol. [Internet] 13 (10) (2017) 1300–1308. Available from: https://www.ingentaconnect.com/content/10.1166/jbn.2017.2415.

[124] F. Zhao, Y. Bai, R. Zeng, L. Cao, J. Zhu, G. Han, et al., An electrochemical immunosensor with graphene-oxide-ferrocene-based nanocomposites for hepatitis B surface antigen detection, Electroanalysis [Internet] 30 (11) (2018) 2774–2780. Available from: https://onlinelibrary.wiley.com/doi/10.1002/elan.201800476.

[125] Z. Tan, L. Cao, X. He, H. Dong, Q. Liu, P. Zhao, et al., A label-free immunosensor for the sensitive detection of hepatitis B e antigen based on PdCu tripod functionalized porous graphene nanoenzymes, Bioelectrochemistry [Internet] 133 (2020) 107461. Available from: https://linkinghub.elsevier.com/retrieve/pii/S1567539419307182.

[126] Z. Shakoori, S. Salimian, S. Kharrazi, M. Adabi, R. Saber, Electrochemical DNA biosensor based on gold nanorods for detecting hepatitis B virus, Anal. Bioanal. Chem. [Internet] 407 (2) (2015) 455–461. Available from: http://link.springer.com/10.1007/s00216-014-8303-9.

[127] C. Akkapinyo, P. Khownarumit, D. Waraho-Zhmayev, R.P. Poo-arporn, Development of a multiplex immunochromatographic strip test and ultrasensitive electrochemical immunosensor for hepatitis B virus screening, Anal. Chim. Acta [Internet] 1095 (2020) 162–171. Available from: https://linkinghub.elsevier.com/retrieve/pii/S0003267019312267.

[128] P. Wang, M. Li, F. Pei, Y. Li, Q. Liu, Y. Dong, et al., An ultrasensitive sandwich-type electrochemical immunosensor based on the signal amplification system of double-deck gold film and thionine unite with platinum nanowire inlaid globular SBA-15 microsphere, Biosens. Bioelectron. [Internet] 91 (2017) 424–430. Available from: https://linkinghub.elsevier.com/retrieve/pii/S0956566316313082.

[129] Y. Li, L. Tian, L.L. Liu, L.L. Liu, J. Li, Q. Wei, et al., An ultrasensitive sandwich-type electrochemical immunosensor based on δ-MnO 2 and palladium nanoparticles covered natural halloysite nanotubes for the detection of hepatitis B surface antigen, N. J. Chem. [Internet] 40 (1) (2016) 558–563. Available from: http://xlink.rsc.org/?DOI = C5NJ01251A.

[130] F. Pei, P. Wang, E. Ma, Q. Yang, H. Yu, C. Gao, et al., A sandwich-type electrochemical immunosensor based on RhPt NDs/NH2-GS and Au NPs/PPy NS for quantitative detection hepatitis B surface antigen, Bioelectrochemistry [Internet] 126 (2019) 92–98. Available from: https://linkinghub.elsevier.com/retrieve/pii/S1567539418303700.

[131] S. Wei, H. Xiao, M. Gu, Z. Chen, L. Cao, Ultrasensitive label-free electrochemical immunosensor based on core-shell Au@PtNPs functionalized rGO-TEPA/PB nanocomposite for HBsAg detection, J. Electroanalytical Chem. [Internet] 890 (2021) 115216. Available from: https://linkinghub.elsevier.com/retrieve/pii/S1572665721002423.

[132] L. Tian, L. Liu, Y. Li, X. Feng, Q. Wei, W. Cao, A novel label-free electrochemical immunosensor for the detection of hepatitis B surface antigen, Anal. Methods 8 (40) (2016) 7380–7386.

[133] S. Wei, H. Xiao, L. Cao, Z. Chen, A label-free immunosensor based on graphene oxide/Fe_3O_4/prussian blue nanocomposites for the electrochemical determination of HBsAg, Biosens. [Internet] 10 (3) (2020) 24. Available from: https://www.mdpi.com/2079-6374/10/3/24.

[134] N. Alizadeh, R. Hallaj, A. Salimi, Dual amplified electrochemical immunosensor for hepatitis B virus surface antigen detection using hemin/G-Quadruplex immobilized onto Fe_3O_4-AuNPs or (Hemin-Amino-rGO-Au) nanohybrid, Electroanalysis [Internet] 30 (3) (2018) 402–414. Available from: https://onlinelibrary.wiley.com/doi/10.1002/elan.201700727.

[135] M.S. Cheng, J.S. Ho, C.H. Tan, J.P.S. Wong, L.C. Ng, C.S. Toh, Development of an electrochemical membrane-based nanobiosensor for ultrasensitive detection of dengue virus, Anal. Chim. Acta 725 (2012) 74–80.

[136] B.T.T. Nguyen, A.E.K. Peh, C.Y.L. Chee, K. Fink, V.T.K. Chow, M.M.L. Ng, et al., Electrochemical impedance spectroscopy characterization of nanoporous alumina dengue virus biosensor, Bioelectrochemistry. 88 (2012) 15–21.

[137] A.E.K. Peh, S.F.Y. Li, Dengue virus detection using impedance measured across nanoporous alumina membrane, Biosens. Bioelectron. 42 (1) (2013) 391–396.

[138] J. Cecchetto, F.C. Carvalho, A. Santos, F.C.B. Fernandes, P.R. Bueno, An impedimetric biosensor to test neat serum for dengue diagnosis, Sens. Actuators B: Chem.

[Internet] (2015) [cited 2021 Jul 24];213:150–4. Available from: https://linkinghub.elsevier.com/retrieve/pii/S0925400515002427.

[139] A. Figueiredo, N.C.S. Vieira, J.F. dos Santos, B.C. Janegitz, S.M. Aoki, P.P. Junior, et al., Electrical detection of dengue biomarker using egg yolk immunoglobulin as the biological recognition element, Sci. Rep. [Internet] 5 (1) (2015) 7865. Available from: http://www.nature.com/articles/srep07865.

[140] D.M.N. Luna, K.Y.P.S. Avelino, M.T. Cordeiro, C.A.S. Andrade, M.D.L. Oliveira, Electrochemical immunosensor for dengue virus serotypes based on 4-mercaptobenzoic acid modified gold nanoparticles on self-assembled cysteine monolayers, Sens. Actuators, B: Chem. 220 (2015).

[141] N.T. Darwish, A.H. Alrawi, S.D. Sekaran, Y. Alias, S.M. Khor, Electrochemical immunosensor based on antibody-nanoparticle hybrid for specific detection of the dengue virus NS1 biomarker, J. Electrochem. Soc. [Internet] 163 (3) (2016) B19–B25. Available from: https://iopscience.iop.org/article/10.1149/2.0471603jes.

[142] P.D. Sinawang, V. Rai, R.E. Ionescu, R.S. Marks, Electrochemical lateral flow immunosensor for detection and quantification of dengue NS1 protein, Biosens. Bioelectron. [Internet] (2016) [cited 2021 Jul 18];77:400–8. Available from: https://linkinghub.elsevier.com/retrieve/pii/S0956566315304449.

[143] C.L. Pirich, R.A. de Freitas, R.M. Torresi, G.F. Picheth, M.R. Sierakowski, Piezoelectric immunochip coated with thin films of bacterial cellulose nanocrystals for dengue detection, Biosens. Bioelectron. 92 (2017).

[144] M.H. Nawaz, A. Hayat, G. Catanante, U. Latif, J.L. Marty, Development of a portable and disposable NS1 based electrochemical immunosensor for early diagnosis of dengue virus, Anal. Chim. Acta [Internet] (2018) [cited 2021 Jul 24];1026:1–7. Available from: https://linkinghub.elsevier.com/retrieve/pii/S0003267018305142.

[145] Q. Palomar, C. Gondran, R. Marks, S. Cosnier, M. Holzinger, Impedimetric quantification of anti-dengue antibodies using functional carbon nanotube deposits validated with blood plasma assays, Electrochim. Acta 274 (2018).

[146] T.C. Pimenta, C. Santos, C. da, R.L. Thomasini, L.F. Ferreira, Impedimetric immunosensor for dengue diagnosis using graphite screen-printed electrodes coated with poly(4-aminophenylacetic acid), Biomed. Microdev. [Internet] 20 (3) (2018) 78. Available from: http://link.springer.com/10.1007/s10544-018-0324-2.

[147] A. Santos, P.R. Bueno, J.J. Davis, A dual marker label free electrochemical assay for Flavivirus dengue diagnosis, Biosens. Bioelectron. 100 (2018).

[148] P.D. Sinawang, L. Fajs, K. Elouarzaki, J. Nugraha, R.S. Marks, TEMPO-based immuno-lateral flow quantitative detection of dengue NS1 protein, Sens. Actuators, B: Chem. 259 (2018).

[149] M. Awan, S. Rauf, A. Abbas, M.H. Nawaz, C. Yang, S.A. Shahid, et al., A sandwich electrochemical immunosensor based on antibody functionalized-silver nanoparticles (Ab-Ag NPs) for the detection of dengue biomarker protein NS1, J. Mol. Liq. 317 (2020) 114014.

[150] Q. Palomar, X. Xu, C. Gondran, M. Holzinger, S. Cosnier, Z. Zhang, Voltammetric sensing of recombinant viral dengue virus 2 NS1 based on Au nanoparticle–decorated multiwalled carbon nanotube composites, Microchim. Acta [Internet] 187 (6) (2020) 363. Available from: http://link.springer.com/10.1007/s00604-020-04339-y.

[151] C. da C Santos, P.C.M. Santos, K.L.S. Rocha, R.L. Thomasini, D.B. de Oliveira, D.L. Franco, et al., A new tool for dengue virus diagnosis: optimization and detection of anti-NS1 antibodies in serum samples by impedimetric transducers, Microchem. J. (2020) 154.

[152] Q.Y. Siew, E.L. Pang, H.S. Loh, M.T.T. Tan, Highly sensitive and specific graphene/TiO_2 impedimetric immunosensor based on plant-derived tetravalent

envelope glycoprotein domain III (EDIII) probe antigen for dengue diagnosis, Biosens. Bioelectron. 176 (2021).
[153] J. Cecchetto, F.C.B. Fernandes, R. Lopes, P.R. Bueno, The capacitive sensing of NS1 Flavivirus biomarker, Biosens. Bioelectron. [Internet] 87 (2017) 949–956. Available from: https://linkinghub.elsevier.com/retrieve/pii/S0956566316308533.
[154] R. Dutta, K. Thangapandi, S. Mondal, A. Nanda, S. Bose, S. Sanyal, et al., Polyaniline based electrochemical sensor for the detection of dengue virus infection, Avicenna J. Med. Biotechnol. [Internet] 12 (2) (2020) 77–84. Available from: http://www.ncbi.nlm.nih.gov/pubmed/32431791.
[155] C. Tancharoen, W. Sukjee, C. Thepparit, T. Jaimipuk, P. Auewarakul, A. Thitithanyanont, et al., Electrochemical biosensor based on surface imprinting for Zika virus detection in serum, ACS Sens. [Internet] 4 (1) (2019) 69–75. Available from: https://pubs.acs.org/doi/10.1021/acssensors.8b00885.
[156] H.A.M. Faria, V. Zucolotto, Label-free electrochemical DNA biosensor for Zika virus identification, Biosens. Bioelectron. [Internet] 131 (2019) 149–155. Available from: https://linkinghub.elsevier.com/retrieve/pii/S0956566319301204.
[157] M. Steinmetz, D. Lima, A.G. Viana, S.T. Fujiwara, C.A. Pessôa, R.M. Etto, et al., A sensitive label-free impedimetric DNA biosensor based on silsesquioxane-functionalized gold nanoparticles for Zika virus detection, Biosens. Bioelectron. 141 (2019).
[158] S. Afsahi, M.B. Lerner, J.M. Goldstein, J. Lee, X. Tang, D.A. Bagarozzi, et al., Novel graphene-based biosensor for early detection of Zika virus infection, Biosens. Bioelectron. (2018) 100.
[159] Y.P. Hsu, N.S. Li, Y.T. Chen, H.H. Pang, K.C. Wei, H.W. Yang, A serological point-of-care test for Zika virus detection and infection surveillance using an enzyme-free vial immunosensor with a smartphone, Biosens. Bioelectron. 151 (2020).
[160] A. Kaushik, A. Yndart, S. Kumar, R.D. Jayant, A. Vashist, A.N. Brown, et al., A sensitive electrochemical immunosensor for label-free detection of Zika-virus protein, Sci. Rep. [Internet] 8 (1) (2018) 9700. Available from: http://www.nature.com/articles/s41598-018-28035-3.
[161] A.M. Faria, T. Mazon, Early diagnosis of Zika infection using a ZnO nanostructures-based rapid electrochemical biosensor, Talanta 203 (2019) 153–160.
[162] G. Cabral-Miranda, A.R. Cardoso, L.C.S. Ferreira, M.G.F. Sales, M.F. Bachmann, Biosensor-based selective detection of Zika virus specific antibodies in infected individuals, Biosens. Bioelectron. [Internet] 113 (2018) 101–107. Available from: https://linkinghub.elsevier.com/retrieve/pii/S0956566318303270.
[163] Q.Y. Siew, S.H. Tan, E.L. Pang, H.-S. Loh, M.T.T. Tan, A graphene-based dengue immunosensor using plant-derived envelope glycoprotein domain III (EDIII) as the novel probe antigen, Analyst [Internet] 146 (6) (2021) 2009–2018. Available from: http://xlink.rsc.org/?DOI = D0AN02219E.
[164] K. Takemura, O. Adegoke, T. Suzuki, E.Y. Park, A localized surface plasmon resonance-amplified immunofluorescence biosensor for ultrasensitive and rapid detection of nonstructural protein 1 of Zika virus, PLoS ONE 14 (1) (2019).
[165] A.C.R. Moço, P.H. Guedes, J.M.R. Flauzino, H.S. da Silva, J.G. Vieira, A.C.H. Castro, et al., Electrochemical detection of Zika virus in biological samples: a step for diagnosis point-of-care, Electroanalysis 31 (8) (2019).
[166] D. Alzate, S. Cajigas, S. Robledo, C. Muskus, J. Orozco, Genosensors for differential detection of Zika virus, Talanta (2020) 210.
[167] L. Liv, Electrochemical immunosensor platform based on gold-clusters, cysteamine and glutaraldehyde modified electrode for diagnosing COVID-19, Microchem. J.

[Internet] 168 (2021) 106445. Available from: https://linkinghub.elsevier.com/retrieve/pii/S0026265X21005294.
[168] Mahari S, Roberts A, Shahdeo D, Gandhi S. Ecovsens-ultrasensitive novel in-house built printed circuit board based electrochemical device for rapid detection of nCovid-19 antigen, a spike protein domain 1 of SARS-CoV-2. bioRxiv [Internet]. 2020 Jan 1;2020.04.24.059204. Available from: http://biorxiv.org/content/early/2020/05/11/2020.04.24.059204.abstract
[169] S. Eissa, H.A. Alhadrami, M. Al-Mozaini, A.M. Hassan, M. Zourob, Voltammetric-based immunosensor for the detection of SARS-CoV-2 nucleocapsid antigen, Microchim. Acta [Internet] 188 (6) (2021) 199. Available from: https://doi.org/10.1007/s00604-021-04867-1.
[170] S. Eissa, M. Zourob, Development of a low-cost cotton-tipped electrochemical immunosensor for the detection of SARS-CoV-2, Anal. Chem. [Internet] 93 (3) (2021) 1826−1833. Available from: https://pubs.acs.org/doi/10.1021/acs.analchem.0c04719.
[171] M. Alafeef, K. Dighe, P. Moitra, D. Pan, Rapid, ultrasensitive, and quantitative detection of SARS-CoV-2 using antisense oligonucleotides directed electrochemical biosensor chip, ACS Nano [Internet] 14 (12) (2020) 17028−17045. Available from: https://pubs.acs.org/doi/10.1021/acsnano.0c06392.
[172] L. Liv, G. Çoban, N. Nakiboğlu, T. Kocagöz, A rapid, ultrasensitive voltammetric biosensor for determining SARS-CoV-2 spike protein in real samples, Biosens. Bioelectron. [Internet] (2021) [cited 2021 Aug 4];192:113497. Available from: https://linkinghub.elsevier.com/retrieve/pii/S0956566321005340.
[173] B. Mojsoska, S. Larsen, D.A. Olsen, J.S. Madsen, I. Brandslund, F.A. Alatraktchi, Rapid SARS-CoV-2 detection using electrochemical immunosensor, Sens. (Switz.) 21 (2) (2021) 1−11.
[174] L. Fabiani, M. Saroglia, G. Galatà, R. De Santis, S. Fillo, V. Luca, et al., Magnetic beads combined with carbon black-based screen-printed electrodes for COVID-19: a reliable and miniaturized electrochemical immunosensor for SARS-CoV-2 detection in saliva, Biosens. Bioelectron. 171 (2021) 112686.
[175] Z. Rahmati, M. Roushani, H. Hosseini, H. Choobin, Electrochemical immunosensor with Cu2O nanocube coating for detection of SARS-CoV-2 spike protein, Microchim. Acta [Internet] 188 (3) (2021) 105. Available from: http://link.springer.com/10.1007/s00604-021-04762-9.
[176] G. Seo, G. Lee, M.J. Kim, S.-H. Baek, M. Choi, K.B. Ku, et al., Rapid detection of COVID-19 causative virus (SARS-CoV-2) in human nasopharyngeal swab specimens using field-effect transistor-based biosensor, ACS Nano [Internet] 14 (4) (2020) 5135−5142. Available from: https://pubs.acs.org/doi/10.1021/acsnano.0c02823.
[177] W. Shao, M.R. Shurin, S.E. Wheeler, X. He, A. Star, Rapid detection of SARS-CoV-2 antigens using high-purity semiconducting single-walled carbon nanotube-based field-effect transistors, ACS Appl. Mater. Interfaces 13 (8) (2021).
[178] T. Chaibun, J. Puenpa, T. Ngamdee, N. Boonapatcharoen, P. Athamanolap, A.P. O'Mullane, et al., Rapid electrochemical detection of coronavirus SARS-CoV-2, Nat. Commun. [Internet] 12 (1) (2021) 802. Available from: http://www.nature.com/articles/s41467-021-21121-7.
[179] A. Yakoh, U. Pimpitak, S. Rengpipat, N. Hirankarn, O. Chailapakul, S. Chaiyo, Paper-based electrochemical biosensor for diagnosing COVID-19: detection of SARS-CoV-2 antibodies and antigen, Biosens. Bioelectron. 176 (2021) 112912.
[180] A. Raziq, A. Kidakova, R. Boroznjak, J. Reut, A. Öpik, V. Syritski, Development of a portable MIP-based electrochemical sensor for detection of SARS-CoV-2 antigen, Biosens. Bioelectron. [Internet] 178 (2021) 113029. Available from: https://linkinghub.elsevier.com/retrieve/pii/S0956566321000658.

[181] J. Li, P.B. Lillehoj, Microfluidic magneto immunosensor for rapid, high sensitivity measurements of SARS-CoV-2 nucleocapsid protein in serum, ACS Sens. [Internet] 6 (3) (2021) 1270–1278. Available from: https://pubs.acs.org/doi/10.1021/acssensors.0c02561.

[182] Y. Ma, X.-L. Shen, Q. Zeng, H.-S. Wang, L.-S. Wang, A multi-walled carbon nanotubes based molecularly imprinted polymers electrochemical sensor for the sensitive determination of HIV-p24, Talanta [Internet] 164 (2017) 121–127. Available from: https://linkinghub.elsevier.com/retrieve/pii/S0039914016309249.

[183] Y.-S. Fang, X.-J. Huang, L.-S. Wang, J.-F. Wang, An enhanced sensitive electrochemical immunosensor based on efficient encapsulation of enzyme in silica matrix for the detection of human immunodeficiency virus p24, Biosens. Bioelectron. [Internet] 64 (2015) 324–332. Available from: https://linkinghub.elsevier.com/retrieve/pii/S095656631400709X.

[184] M. Giannetto, M. Costantini, M. Mattarozzi, M. Careri, Innovative gold-free carbon nanotube/chitosan-based competitive immunosensor for determination of HIV-related p24 capsid protein in serum, RSC Advances [Internet] 7 (63) (2017) 39970–39976. Available from: http://xlink.rsc.org/?DOI = C7RA07245G.

[185] S. Islam, S. Shukla, V.K. Bajpai, Y.-K. Han, Y.S. Huh, A. Kumar, et al., A smart nanosensor for the detection of human immunodeficiency virus and associated cardiovascular and arthritis diseases using functionalized graphene-based transistors, Biosens. Bioelectron. [Internet] 126 (2019) 792–799. Available from: https://linkinghub.elsevier.com/retrieve/pii/S0956566318309382.

[186] M.F. Fatin, A. Rahim Ruslinda, S.C.B. Gopinath, M.K.M. Arshad, High-performance interactive analysis of split aptamer and HIV-1 Tat on multiwall carbon nanotube-modified field-effect transistor, Int. J. Biol. Macromol. [Internet] 125 (2019) 414–422. Available from: https://linkinghub.elsevier.com/retrieve/pii/S0141813018342491.

[187] S.M. Majd, A. Salimi, B. Astinchap, The development of radio frequency magnetron sputtered p-type nickel oxide thin film field-effect transistor device combined with nucleic acid probe for ultrasensitive label-free HIV-1 gene detection, Sens. Actuators B: Chem. [Internet] 266 (2018) 178–186. Available from: https://linkinghub.elsevier.com/retrieve/pii/S0925400518306099.

[188] Q. Gong, Y. Wang, H. Yang, A sensitive impedimetric DNA biosensor for the determination of the HIV gene based on graphene-Nafion composite film, Biosens. Bioelectron. [Internet] 89 (2017) 565–569. Available from: https://linkinghub.elsevier.com/retrieve/pii/S095656631630152X.

[189] Q. Gong, H. Yang, Y. Dong, W. Zhang, A sensitive impedimetric DNA biosensor for the determination of the HIV gene based on electrochemically reduced graphene oxide, Anal. Methods [Internet] 7 (6) (2015) 2554–2562. Available from: http://xlink.rsc.org/?DOI = C5AY00111K.

[190] Y. Wang, X. Bai, W. Wen, X. Zhang, S. Wang, Ultrasensitive electrochemical biosensor for HIV gene detection based on graphene stabilized gold nanoclusters with exonuclease amplification, ACS Appl. Mater. Interfaces [Internet] 7 (33) (2015) 18872–18879. Available from: https://pubs.acs.org/doi/10.1021/acsami.5b05857.

[191] M. Shamsipur, L. Samandari, A. Taherpour, (Arman), A. Pashabadi, Sub-femtomolar detection of HIV-1 gene using DNA immobilized on composite platform reinforced by a conductive polymer sandwiched between two nanostructured layers: a solid signal-amplification strategy, Analytica Chim. Acta [Internet] 1055 (2019) 7–16. Available from: https://linkinghub.elsevier.com/retrieve/pii/S000326701831448X.

[192] Y. Chen, R. Ren, H. Pu, X. Guo, J. Chang, G. Zhou, et al., Field-effect transistor biosensor for rapid detection of Ebola antigen, Sci. Rep. [Internet] 7 (1) (2017) 10974. Available from: http://www.nature.com/articles/s41598-017-11387-7.

[193] X. Jin, H. Zhang, Y.-T. Li, M.-M. Xiao, Z.-L. Zhang, D.-W. Pang, et al., A field effect transistor modified with reduced graphene oxide for immunodetection of Ebola virus, Microchim. Acta [Internet] 186 (4) (2019) 223. Available from: http://link.springer.com/10.1007/s00604-019-3256-5.

[194] H. Ilkhani, S. Farhad, A novel electrochemical DNA biosensor for Ebola virus detection, Anal. Biochem. [Internet] 557 (2018) 151−155. Available from: https://linkinghub.elsevier.com/retrieve/pii/S0003269718302604.

[195] S. Carinelli, M. Kühnemund, M. Nilsson, M.I. Pividori, Yoctomole electrochemical genosensing of Ebola virus cDNA by rolling circle and circle to circle amplification, Biosens. Bioelectron. [Internet] 93 (2017) 65−71. Available from: https://linkinghub.elsevier.com/retrieve/pii/S0956566316309836.

[196] B. Maeng, Y. Park, J. Park, Direct label-free detection of Rotavirus using a hydrogel based nanoporous photonic crystal, RSC Adv. [Internet] 6 (9) (2016) 7384−7390. Available from: http://xlink.rsc.org/?DOI = C5RA21665F.

[197] F. Liu, Y.H. Kim, D.S. Cheon, T.S. Seo, Micropatterned reduced graphene oxide based field-effect transistor for real-time virus detection, Sens. Actuators B: Chem. 186 (2013) 252−257.

[198] H.C. Lai, S.F. Chin, S.C. Pang, M.S. Henry Sum, D. Perera, Carbon nanoparticles based electrochemical biosensor strip for detection of Japanese encephalitis virus, J. Nanomaterials [Internet] 2017 (2017) 1−7. Available from: https://www.hindawi.com/journals/jnm/2017/3615707/.

[199] S.F. Chin, L.S. Lim, H.C. Lai, S.C. Pang, M.S.H. Sum, D. Perera, Electrochemical determination of Japanese encephalitis virus antigen using silver nanoparticles modified screen-printed carbon electrode, Nano Biomed. Eng. [Internet] 11 (4) (2019) 334−339. Available from: http://nanobe.org/Data/View/619?type = 100.

[200] S.H. Baek, M.W. Kim, C.Y. Park, C.-S. Choi, S.K. Kailasa, J.P. Park, et al., Development of a rapid and sensitive electrochemical biosensor for detection of human norovirus via novel specific binding peptides, Biosens. Bioelectron. [Internet] 123 (2019) 223−229. Available from: https://linkinghub.elsevier.com/retrieve/pii/S0956566318306754.

[201] L.A. Layqah, S. Eissa, An electrochemical immunosensor for the corona virus associated with the Middle East respiratory syndrome using an array of gold nanoparticle modified carbon electrodes, Microchimica Acta [Internet] 186 (4) (2019) 224. Available from: http://link.springer.com/10.1007/s00604-019-3345-5.

[202] L.T. Tran, T.Q. Tran, H.P. Ho, X.T. Chu, T.A. Mai, Simple label-free electro chemical immunosensor in a microchamber for detecting newcastle disease virus, J. Nanomater. [Internet] 2019 (2019) 1−9. Available from: https://www.hindawi.com/journals/jnm/2019/3835609/.

[203] H. Haji-Hashemi, M.R. Safarnejad, P. Norouzi, M. Ebrahimi, M. Shahmirzaie, M.R. Ganjali, Simple and effective label free electrochemical immunosensor for Fig mosaic virus detection, Anal. Biochem. [Internet] 566 (2019) 102−106. Available from: https://linkinghub.elsevier.com/retrieve/pii/S0003269718310716.

[204] S. Ning, M. Zhou, C. Liu, G.I.N. Waterhouse, J. Dong, S. Ai, Ultrasensitive electrochemical immunosensor for avian leukosis virus detection based on a β-cyclodextrin-nanogold-ferrocene host-guest label for signal amplification, Anal. Chim. Acta [Internet] 1062 (2019) 87−93. Available from: https://linkinghub.elsevier.com/retrieve/pii/S0003267019302302.

[205] Y.-H. Hou, J.-J. Wang, Y.-Z. Jiang, C. Lv, L. Xia, S.-L. Hong, et al., A colorimetric and electrochemical immunosensor for point-of-care detection of enterovirus 71, Biosens. Bioelectron. [Internet] 99 (2018) 186−192. Available from: https://linkinghub.elsevier.com/retrieve/pii/S0956566317304839.

[206] H. Haji-Hashemi, P. Norouzi, M.R. Safarnejad, M.R. Ganjali, Label-free electrochemical immunosensor for direct detection of Citrus tristeza virus using modified gold electrode, Sens. Actuators B: Chem. [Internet] 244 (2017) 211–216. Available from: https://linkinghub.elsevier.com/retrieve/pii/S0925400516321062.
[207] P. Teengam, W. Siangproh, A. Tuantranont, C.S. Henry, T. Vilaivan, O. Chailapakul, Electrochemical paper-based peptide nucleic acid biosensor for detecting human papillomavirus, Anal. Chim. Acta [Internet] 952 (2017) 32–40. Available from: https://linkinghub.elsevier.com/retrieve/pii/S0003267016314222.
[208] S. Jampasa, W. Siangproh, R. Laocharoensuk, P. Yanatatsaneejit, T. Vilaivan, O. Chailapakul, A new DNA sensor design for the simultaneous detection of HPV type 16 and 18 DNA, Sens. Actuators B: Chem. [Internet] 265 (2018) 514–521. Available from: https://linkinghub.elsevier.com/retrieve/pii/S0925400518305355.
[209] L.D. Tran, D.T. Nguyen, B.H. Nguyen, Q.P. Do, H. Le Nguyen, Development of interdigitated arrays coated with functional polyaniline/MWCNT for electrochemical biodetection: application for human papilloma virus, Talanta [Internet] 85 (3) (2011) 1560–1565. Available from: https://linkinghub.elsevier.com/retrieve/pii/S0039914011005339.
[210] A. George, M.S. Amrutha, P. Srivastava, V.V.R. Sai, S. Sunil, R. Srinivasan, Label-free detection of chikungunya non-structural protein 3 using electrochemical impedance spectroscopy, J. Electrochem. Soc. [Internet] 166 (14) (2019) B1356–B1363. Available from: https://iopscience.iop.org/article/10.1149/2.1081914jes.
[211] F. Nasrin, A.D. Chowdhury, A.B. Ganganboina, O.J. Achadu, F. Hossain, M. Yamazaki, et al., Fluorescent and electrochemical dual-mode detection of Chikungunya virus E1 protein using fluorophore-embedded and redox probe-encapsulated liposomes, Microchim. Acta 187 (12) (2020).

CHAPTER 3

Magnetic point-of-care biosensors for infectious disease diagnosis

Richha Sharma[1] and Vivek A. Kamat[2]
[1]EaStCHEM, School of Chemistry, King's Buildings, University of Edinburgh, Edinburgh, United Kingdom
[2]Electrical Engineering and Computing, Florida International University, Miami, FL, United States

3.1 Introduction

Infectious diseases, like respiratory infections and diarrheal diseases, are among the top ten leading causes of death globally [1]. The gravity of the situation worsens for low-income and middle-income countries where out of the top ten causes five and six, respectively, are communicable diseases. Besides the above mentioned conditions, lower income regions report serious infectious ailments like tuberculosis (TB), influenza, malaria and acquired immuno-deficiency syndrome. Moreover, regionally- generated pathogens are not limited to their source countries and often cause worldwide outbreaks. Timely diagnosis of infectious diseases is particularly crucial as it can prevent the spread of the infectious agent in the population and also avoid uninformed generic follow-up treatments, such as broad-spectrum antibiotic administration. Thus, development of low-cost but accurate diagnostic systems is a critical need to curb the rising mortality rates in these economically challenged regions, where, conventional diagnostics platforms [such as immunoassays, microscopy, microbial culturing and polymerase chain reactions (PCRs)], though accurate, often fail to meet the clinical management needs [2,3].

The prospect of diagnostics has improved in recent years with the invention of sophisticated biosensing techniques [4]. In brief, a biosensor involves a biospecific ligand that recognizes the analyte. This recognition/binding event is transduced to a readable signal. In most current biosensors, for transduction and amplification of transduced signals, optical and electrochemical tags are used. These give off the readable signals [5]. Magnetic tags are another kind that create a magnetically-read signal, such

Biosensors for Emerging and Re-emerging Infectious Diseases
DOI: https://doi.org/10.1016/B978-0-323-88464-8.00009-9

as magnetic resonance or resistance. In other cases, magnetic tags can be attached to either the analyte, biospecific ligand or the analyte-ligand complex to serve the purpose of separating the complex from the rest of the interfering matrix [6].

Point-of-care (POC) systems offer affordable, user-friendly, compactly-structured and rapid diagnosis arrangements [7]. They allow working with small volumes of samples, involve simple operational steps that can be performed by non-specialists, are miniaturized and designed to obtain test results on the spot, avoiding sample collection and transport, bulky analytical systems and delayed outcomes [8,9]. High sensitivity of assays are achieved by low sample volume systems in microfluidic channels [10–12], and capillary flow in cellulose and nitrocellulose [13,14]. The readout is simplified through innovative transduction methods [15,16], advanced data processing [17], web-based connectivity [18] as in smartphones [19,20] and e-textiles [21,22].

The success of POC systems centers on the sensitivity and the mode of transduction. Magnetic materials are extraordinary contributors to both these factors and hence find wide applications in general biosensors, and specifically field-applicable diagnostics [23–25]. For increasing sensitivity, magnetism helps in increasing the effective concentration of the target by its separation from other contaminants using a magnetic field. For achieving this, structures with magnetic properties (micro- and nano-beads in most cases) are attached with the specific biorecognition molecule. [26,27]. In the process of separation of biorecognition-target complex, often the magnetic particles (MPs) also generate their own signal, thereby, achieving the second objective of transduction. Uniqueness of these signals is the fact that the biological environment is largely non-magnetic. Hence, though optical and electrochemical transduction may suffer from noise, magnetic signals can efficiently avoid it [28].

3.2 Scope of the chapter

With an upsurge in thrust on disease diagnosis research from global and national health organizations, novel sensing modalities are being designed every day. This is evident from the steady incline in the number of research publications and reviews on the topic in the last few years [29–38]. Although literature compilations on magnetic biosensors are abundant, comprehensive reports on magnetic POC systems specific to infectious diseases, involving both magnetic separation and magnetic signaling techniques are scarce. It is important to keep pace with the plethora of original reports published in the last five years on the topic and summarize the significant achievements for easy reference by readers.

In addition to innovation in technology, the sensor industry also needs to comply by the guidelines laid out by World Health Organization (WHO) and other regulatory agencies to certify that the testing platform is accurate and acceptable. The WHO suggests that, besides the generic ASSURED (Affordable, Sensitive, Specific, User-Friendly, Robust and rapid, Equipment free, Deliverable) criteria, more stringent benchmarks need to be followed to certify a POC platform. The steps that need to be addressed are (1) defining the test's purpose (type of infection, qualitative/quantitative, end-users, required performance); (2) reviewing the market and checking product specifications (product shelf life, size, sample type, volume, reagents, number of steps, requirement of analysis experts and training); (3) reviewing the test's regulatory approval (whether approved by national agencies, complying to ISO, etc.); (4) evaluating diagnostic accuracy of the test under laboratory conditions; (5) evaluating diagnostic accuracy at end-user level; and (6) monitoring the performance in routine use (quality control, training and supervision) [39]. This chapter is therefore focused only on selected magnetic POC systems, explaining the theory behind their design and operations. It also provides an account of recent literature with the aim to encourage future POC research and development.

3.3 POC diagnostic systems

Infectious diseases continue to inflict health and economic burdens to resource-limited regions around the world. POC diagnostic methods have come to the rescue in the past decade by offering affordable yet accurate diagnosis. In traditional screening systems, a large number of samples are collected, transported, tested by experts with dedicated instruments in laboratories, during which clinicians are unable to prescribe specific treatments. POCs can be used in local, technically unequipped or less-equipped medical clinics. They are simple to use, portable so that a large number of test kits can be stored in the regional medical centers and give quick results.

Recent biomonitoring platforms developed for infectious agents often target specific biological markers like pathogen's nucleic acids, glycoproteins, enzymes and surface antigens or host's antibodies [40,41]. The transduction modes are optical (including visual signals), electrochemical and magnetic. Nanomaterials and microfluidics are commonly used [42,43]. Though many POC tools have been tested in laboratory setting with real samples, only a handful for infectious agents are in regular use in clinics with minimal laboratory capabilities. Some examples are lateral flow assays for HIV, Binax NOW for malaria, Xpert MTB/RIF for TB, and the recent MobiNAAT for *Chlamydia trachomatis* [44–49].

Magnetic methods have provided a strong boost to these compact diagnostic systems [50–52]. This can be attributed mainly to some remarkable properties of magnetic materials: (1) They can be manipulated by an external electrically-tunable magnetic field, hence remote-control is possible, (2) most such magnetic materials for example, magnetic beads and nanoparticles are easily synthesized, (3) the magnetic transduction can be easily integrated into standard integrated circuits, (4) biological samples do not have magnetic properties, so there is less background and, (5) magnetic probes, unlike optical ones, are more stable in different sample conditions [53–55] (Fig. 3.1).

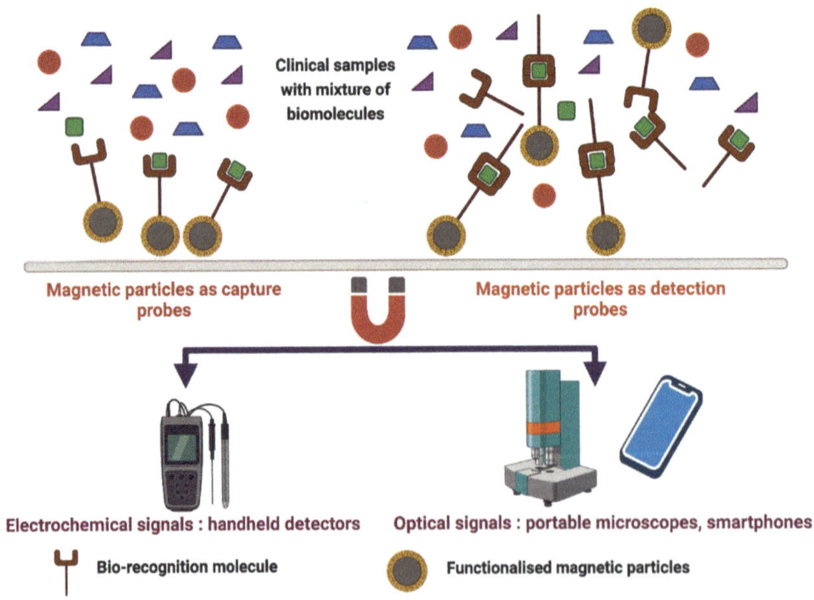

Figure 3.1 Application of magnetism in point of care diagnostic system for infectious diseases.

3.4 Design of magnetic materials for biosensing

The majority of biosensors use nano or micrometer sized MPs. MPs can be para-, superpara, ferro- or ferri-magnetic [56]. The size and material composition is tuned according to the end use. Increasing size of the particle increases its magnetic strength, however for many applications, a smaller size may be necessary. To be able to detect single molecules and bind to biological agents, the particle size should be comparable to the molecule size. This also allows a high surface area for binding and often introduces superparamagnetic properties (requiring very low strength magnetic field to induce change in magnetization) [57]. Polyol synthesis, thermal and hydrothermal processes, co-precipitation and microemulsion-based methods are used to synthesize nanomagnets of different shapes and sizes [58–62].

The MP can be homogeneously composed of, for example, single elements such as Fe, Ni or Co, two metallic elements like FePt or FeCo, iron compounds with non-metals such as magnetite (Fe_3O_4)/maghemite (γ-Fe_2O_3) complex and multimetal complexes, such as Fe/Zn oxide complexes, which can also be carbon doped [63–67]. The synthesis of such single core particles is straightforward and depends on reduction or decomposition. However, often to add novel properties, layers of different materials can be used to form a heterogeneous nanomagnet. These added properties can be: stability in external environment, biocompatibility, porosity, environment-sensitivity, solubility and enhanced optical, electro chemical and magnetic transduction. Examples are silica core-magnetic shell [68], silica core-MP embedded in gold shell [69], magnetic core with silica [70], styrene or polymer shell [71], magnetic core with gold [72] or silver shell [73], among many others.

As previously discussed, biosepecific agents need to be attached to the MP. The bare particles are functionalized with linkers to facilitate such attachment. The shells of silica or noble metals help, as they have an affinity for several ligands (as in gold–thiol bond). However, additional surface functionalization may be created to facilitate bio-attachment. MPs can be modified with hydroxyl groups (–OH), amino groups ($-NH_2$), carboxyl groups (–COOH), siloxane, carboxylates, phosphates, phosphonic acid and catechol. Longer polymers may also be attached such as polyethylene glycol (PEG), polyvinylpyrrolidone (PVP), starch, dextran, chitosan, poly (acrylic acid) (PAA) and polyethylenimine (PEI), poly(L-lysine), poly (dimethyl-aminoethyl acrylate), poly(tert-butyl acrylate) (PTBA), etc.

[74]. Caution should be exercised while selecting coating materials on the MPs as encapsulation may often lead to reduction in magnetic moments, weakening the magnetic strength of the MPs [57].

Often, this surface coverage itself has shown to bind to targets. For example, carboxy-dendrimer silane coated magnetic nanoparticles were found to concentrate HIV-1 from semen samples [75]. However, in most cases post-functionalization biospecific molecules need to be tethered to the MPs. Biomacromolecules such as nucleic acids, proteins (mostly antibodies), and lipids are quite common [76]. Shorter versions of these, for example, peptides and aptamers are also linked. A bio-ligand decorated MP now forms the key instrument in the biomonitoring assay.

3.5 Magnetic field based detection

3.5.1 Magnetic field for separation in POCs

Many efforts have been made to investigate target separation methods based on immunomagnetism [77]. The principle is that the magnetic beads are functionalized with antibodies that recognize the target. To separate out target pathogens from other contaminants in a sample running through flow channels, 3D-printed cylindrical magnetic coil channels [78], permanent magnets [79] and magnetized fluidized beds [80] have been used. The detection is often strengthened by using other nanoparticles that provide unique transduction approaches and high surface area [81].

3.5.1.1 Magnetic separation coupled with electrochemical detection

Electrochemical detection is often highly sensitive mainly due to highly advanced sensor-transducer systems that pick up small signals and also because background signals from biological signals are very low. In the last few years, scientists have tried to integrate electrochemical systems as closely with POC devices as possible.

Cortina et al. [82] developed rapid and uncomplicated electrochemical magnetic microbeads (MMBs)-based biosensor. In an unconventional approach, they attached the recombinant analyte (Brucella antigen and foot-and-mouth disease antigen) to the MMBs. The beads were then incubated with human serum samples. If infected, the antibodies generated and present in the serum would attach to the beads. Similar to set-up and molecules used in conventional ELISA, the secondary HRP-labeled antibodies attached to the specific bound primary antibodies, which then catalyzed a redox reaction. The bead complex was added on an electrode, and a redox current was measured. Disposable 8-channel electrochemical cells were used to concentrate the beads and readings were performed on these cartridges using a portable potentiostat. The assay displayed much better sensitivity and specificity over ELISA, and could efficiently detect antigens causing Chagas disease and brucellosis in humans and foot-and-mouth disease and bovine brucellosis in cattle. The assay could also be multiplexed and integrated with a smartphone read-out.

MMBs are often attached to the target-biospecific ligand complex after the biorecognition event for selection. Chand et al. [83] developed an miRNA detection system, where the signaling tag was $MoS_2-CuFe_2O_4$ nanocomposite sheets. These were functionalised with miRNA-specific thiolated molecular probe (a short oligonucleotide), which was in turn, tagged with biotin and ferrocene thiol. Only on target miRNA attachment would this oligonucleotide probe expose the biotin tag at its end. This mixture was incubated with streptavidin−MMBs. Hence MMBs only attached with successful recognition complexes. This entire sensing system was built on a microfluidic platform fabricated from silicon master mold by lithography. Dedicated chambers were used for thorough mixing of all individual complexes. Magnets were placed under electrodes that led to the capturing of recognition complexes, whereas unbound nanosheet-probes exited the system through outlets. The limit for detection of paratuberculosis (*Mycobacterium bovis*) miRNA was 0.48 pM with less than 5% error. Paratuberculosis is a proven fatal disease for dairy cattle and is considered a potential threat to humans.

Another fatal condition in humans is sepsis which affects all organ systems. An "integrated biosensor for sepsis" was developed by Min et al. in 2018 [84] to detect the biomarker Interleukin-3 that plays an important pathophysiological role in septic infections and in turn, organ failures (Fig. 3.2). The POC concept was similar to that developed by Cortina et al. [82] in a few ways and consisted of three main components: (1) a magnetic target enrichment system: MMBs were decorated by capture antibodies against Interleukin, and the binding was optimized. A magnetic pipette, having a sheathed magnet attached to a mechanical plunger, was designed. This helped in collection and release of the captured biomarkers, (2) signal generation and amplification system: the signal was generated in similar fashion to ELISA. The capture antibody-target-primary antibody-secondary antibody-HRP complex oxidized a substrate that could be reduced at an electrode. An 8-well sample detection strip was fabricated. These wells were preloaded with all the aforementioned

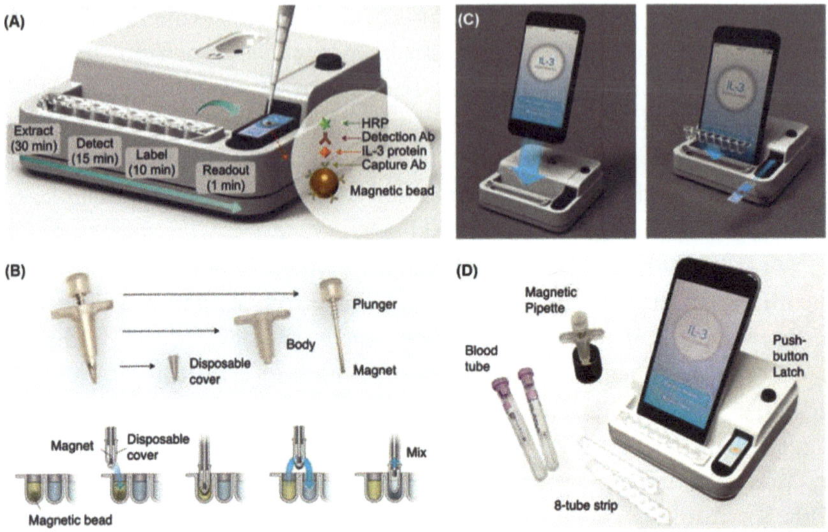

Figure 3.2 (reprinted) Integrated biosensor for sepsis. (A) Assay schematic. Interleukin-3 is captured on magnetic beads directly in plasma or whole blood and labeled antibodies. Electrical signal is generated through enzymatic reaction. The entire assay is performed using a portable device and complete within <1 h. (B) Magnetic kit for sample processing. A magnetic pipet was custom-built to facilitate the sample handling. The plunger mechanism allows for easy collection and release of magnetic beads. (C) The base station, housing measurement electronics, docks with a smartphone and a multiwell tray. An electrode slides in to the system for electrical connection. A smartphone app provides a user interface for the system control (via Bluetooth) and data upload to a cloud server. (D) Photo of the whole assay system. *Ab*, antibody; *HRP*, horseradish peroxidase (Fig. 3.1). *Reprinted with permission from J. Min, M. Nothing, B. Coble, et al. Integrated biosensor for rapid and point-of-care sepsis diagnosis, ACS Nano 12 (2018) 3378–3384. Available from: https://doi.org/10.1021/acsnano.7b08965. Copyright 2018 American Chemical Society.*

reagents (which can be stored in lyophilized form for long-term use). Blood was added to the first well, capturing was achieved using the magnetic pipette and the captured target was transferred to the second well for the enzyme action. No additional sample preparation was required. (3) A signal read-out system: the electrode was miniaturized and latched on to an electronic system with potentiostats, microcontroller and Bluetooth module. A smartphone loaded with a custom-designed app was docked to the system and functioned as a touch-screen interface for recording and storing data. A fascinating advantage of using the MMBs at this stage was the ability to concentrate the ELISA complex on top of the electrode by placing a small disk magnet underneath. Hence the amplification was twofold—one by the enzymatic reaction and second by enrichment of signal complex which contributed to 70% amplification. This POC system was an excellent display of a compact device as a standalone unit with ready kits and reagents, requiring minimum preparatory steps. Its detection time was below 1 h for native blood samples. Assay sensitivity was less than 10 pg/mL. The authors claimed that duration of assay was 5 times lesser and the sensitivity 10 times higher than ELISA.

3.5.1.2 Magnetic separation coupled with optical detection

Lateral flow assays (involving paper based analytical devices or PADs and fabricated microfluidic channels) have dominated sensor miniaturization research. Most flow systems are conjugated to an optical transduction approach. To circumvent elaborate instrumentation, the test is designed so that either a visual color is developed (through a microscale enzyme reaction or colored nanoparticles such as gold) or, in the case of luminescence, a portable optical device is used. Development of a visual signal elevates the convenience and applicable range of any bioanalytical method [85]. In this regard, gold nanoparticles as optical markers are extremely popular as signals can be quickly sensed due to strong plasmon signature of gold [86,87]. For example, Chen et al. [88] developed an immunoassay strip generating visual and chemiluminescent signals using gold nanoparticle-based lateral flow. The nanoparticles were conjugated with detection antibodies and horseradish peroxidase (HRP). The sample flows along the nitrocellulose strip, gets captured by capture antibodies on the sample line, and is sandwiched with gold nanoparticles while unbound nanoparticles bind to the control line. The naked eye could determine the red color of gold nanoparticles (for cancer biomarkers) and the strip can be viewed in a portable luminescence analyzer (for procalcitonin with detection limit 0.02 pg/mL). For POC tests, the manufacture of such portable devices has found its footing with enormous advancements in 3D printing of devices. Labs around the world have published open source designs for light-weight and small 3D printed microscopes, luminometers and fluorimeters. The following section will detail some of these marvelous optical/visual biomonitoring systems that engage MPs for sample enrichment.

Similar to the above work, Chen et al. [89] attempted bioluminescence based detection of antibody-antigen binding. Polystyrene beads and magnetic beads were separately labeled with detection and capture antibodies, respectively. In the presence of antigen, the MMB-polystyrene-antigen complex could be enriched in a vial. Thereafter, a competitive enzyme action was employed for bioluminescence generation. Alkaline phosphatase (already attached to polystyrene) could dephosphorylate adenine triphosphate. Adenine triphosphate is in turn a co-substrate for the bioluminescence reaction catalyzed by luciferase. Hence, increasing antigen concentration decreased the optical signal. Magnetic separation and use of two enzymes highly amplified the output which was read through a handheld luminometer. Femtomolar levels of procalcitonin, a disease biomarker for the invasion by pathogenic bacteria, fungi, and some parasites, could be quantified in less than 1 h.

As a fresh diversion from immunoassays, Dirkzwager and group [90] used aptamers (oligonucleotides or oligopeptides with recognition ability) [91] for target detection. The type of aptamer sensing is termed aptamer-tethered enzyme capture where the label is an enzyme. Once captured, the enzyme catalyzes a standard reaction generating a detectable product. The group could detect the malaria biomarker, *Plasmodium falciparum* lactate dehydrogenase, which converts lactate to pyruvate. The special cofactor used here for better results was 3-acetylpyridine adenine dinucleotide. This was coupled with a third conversion- nitrotetrazolium blue chloride to diformazan, a blue product. Magnetic beads were linked to aptamers by streptavidin-biotin chemistry. The assay was miniaturized using two approaches. The first (and more sensitive) one was a syringe-based test: a 3D printed syringe was fitted with an adapter, a paper membrane disk and a filter. The steps of magnetic bead-aptamer absorption on paper, drying of paper, sample withdrawal and reagent addition were achieved by syringe aspiration. The color developed on the paper was visible and was analyzed by ImageJ software. The second test set-up involved 3D prototype micro-wells. Magnetic bead-aptamer solution and sample were incubated in one well. Using a magnetic stick inserted in a sheath, beads were transferred from the first well to consecutive wash wells. Finally beads were deposited in a last well containing the color development reagent by removing the magnet stick from the sheath. Solutions could be removed for sample monitoring. The dehydrogenase enzyme could be detected down to ng/mL concentrations and the color development reagents required could be stabilized through desiccation and could withstand temperatures as high as 90°C. The disadvantage of the procedure was that blood samples needed to be lysed, processed and washed before the assay.

The same group later adapted the aforementioned detection principle and reagents to build a microfluidic assay [92]. The microfluidic device was composed of two separate transparent layers, which were printed using stereolithography. The top layer housed solution entry-ports while the bottom layer was etched with a network of chambers and channels. The assay was performed within a network of three distinct chambers connected by microfluidic stop valves, enabling reduced diffusion between chambers. MMBs acted as a solid mobile phase whose movement was controlled by an external magnet. Three separate chambers were there for incubation (magnet complex with lysed blood sample), wash (to separate non-specific blood components from the mobile phase) and color development (the MMB-aptamer-enzyme mobile phase incubated with development reagents). The strips were placed on the lit screen of a first-generation iPad, which emanated a homogenous white light. The iPad and sensors were covered by an opaque box and imaged by an iPhone 6, and then the image was further analyzed by ImageJ software.

Counting biomarker cells is a desirable tool to quantify infection. Most cell sorting data however needs to be accessed through fluorescence microscopes. Knowlton et al. [93] developed a portable multiplexed density-based cell sorting method via magnetic focusing. The sorted cells could then be imaged by brightfield, darkfield, or fluorescent imaging modes using a built-in smartphone camera. The 3D printed fluorescence microscope was fitted with different LEDs, lens and filters (optical components), sample tube, magnetic focusing chamber and two permanent magnets (enrichment cell-sorting components). Light produced by the LEDs is directed towards the sample from one side, passes through it, magnified by an aspheric lens and is focused on a smartphone camera. The lens is aligned to the phone camera and placed at a gap of 4 mm. The microcapillary sample tube (less than $1) was the only disposable element in the setting. The sample was stained with fluorescent dye, cells captured by antibodies-labeled MMBs and focused using the magnets in the capillary. The visual and ImageJ analyzed readings for ovarian cancer cell lines agreed well with those from haemocytometer. This $100 costing device finds clinical applications through cytometry and immunocytochemistry-based assays for infectious agents.

Similar to aptamer based capture methods, antibody recognition can also be visualized with the naked eye on PADs through color generation. Srisa-Art et al. [94] developed a very simple pathogenic bacteria paper sensor using two separate designs. One catered to a spot test, where wells were created on the paper using wax printing. The other was a chemometric distance-based design, where a channel is extended from the sample addition well. The color generation solution is wicked through this channel and the distance the color travels is directly proportional to the bacterial count. All reactions but the last assay reagent addition were done in Eppendorf tubes. Anti-Salmonella antibody coated MMBs were used to capture, separate and pre-concentrate bacteria from the sample matrix. Detection antibodies were tagged with biotin, which could attach streptavidin attached β-galactosidase enzyme. This mixture was dropped onto the wells on paper followed by addition of enzyme substrate chlorophenol red-β-D-galactopyranoside which on conversion produced a red color. This assay had a detection limit of 10^2 CFU/mL for *Salmonella typhimurium* in media, and 10^5 CFU/g and 10^3 CFU/mL, respectively, in Starling bird fecal samples and whole milk. Images could be visually seen, quantified visually in the chemometer method and through image analysis in the spot method.

Antibody capture followed by enzymatic color generation was given a new direction by Zheng et al. [95]. A microfluidic multi-chambered flow system was fabricated. The nano-MPs modified with the capture antibodies against *Escherichia coli* O157:H7 bound to the polystyrene microspheres modified with detection antibodies and catalase, if the pathogen was present. This recognition was achieved in the first mixing chamber. The MP–bacteria–polystyrene complex was captured in the second separation chamber by an external magnetic field. Thereafter hydrogen peroxide was injected in this chamber which was catalyzed and reduced by the catalases. In the third detection chamber, a mixture of the gold nanoparticles, HRP and the tyramine crosslinking agents was injected. If hydrogen peroxide was intact, it caused the aggregation of the AuNPs in presence of peroxidase enzyme and resulted in their color changes from blue to red. If, however, pathogen was present, the hydrogen peroxide is completely converted and none of it remains to form the crosslinking, leaving the gold nanoparticle solution red. The color change was detected using the Hue-Saturation-Lightness-based imaging app on an Android smartphone. *E. coli* O157:H7 in chicken samples was detected with a detection limit of 50 CFU/mL.

Alhogail and coworkers [96] developed a much simpler direct colorimetric PAD test. A gold chip was placed on the paper strip. One end of a specific protease substrate was linked to the gold while the other end was attached to nano-MPs. This coating made the strip appear black. On addition of *Pseudomonas aeruginosa* samples, the proteases present in it cleave the substrate, dislodging the particles from the gold surface. A magnet could be used to guide the MP-cleaved peptide off the gold chip, revealing the golden color of the sensor. A detection limit of 10^2 CFU/mL in could be visualized in less than 1 min.

3.5.1.3 Magnetic separation coupled with other transduction methods

3.5.1.3.1 Polymerase chain reaction

Quantification of nucleic acids could be an excellent way to accurately detect infectious agents, especially viruses which do not display most microbial activities/cell components. The most commonly used technique to produce a copious amount of nucleic acid for detection is amplification like PCR and loop-mediated isothermal amplification (LAMP). LAMP is more POC-friendly for direct screening of crude samples since a temperature cycler is not required and enzymes are more robust. Although PCR and LAMP techniques employ fluorescence signals as their readout, they cannot convincingly be grouped under optical detection as no special improvisation is done to develop or integrate the optical readout with biomonitoring. For portability however, the fluorescence signals are monitored using light emitting diodes (LEDs) and often recorded by smartphones.

Zika virus was detected in 2016 by Chan et al. [97] using PCR, since antibodies against it often cross-react with flaviviruses. The standard reverse transcription-PCR was converted to a portable, quick (less than 15 min) and low-cost molecular diagnostics system. It was also automatic, multiplexed for 12 viral urine samples and specific even in the presence of closely related dengue and chikungunya viruses. Similar efficiency was observed for a reverse transcription recombinase polymerase amplification reaction. Sample lysis and nucleic-acid extraction, involving MPs, was done by automated magnetic fields. For this, a 3D printer itself was used as a robotic device (rather than any printed structures). A thermal cycler was developed in-house for the PCR. The fluorescent signals were read by a smartphone.

Sample enrichment of viral nucleic acids requiring benchtop techniques was also successfully avoided by Shin et al. [98] in a quantitative PCR detector. They designed a thermoformed disposable cartridge coupled to a portable multiaxial magnetofluidic instrument which could perform real-time PCR assays. The user simply needed to inject the sample into a cartridge with MPs and load the cartridge into the instrument. Hepatitis C virus RNA load in blood serum could be quantified in 1 h. Multiple clinical samples were accurately tested with a limit of detection of 45 IU per 10 μL sample injection. This is very high advancement compared to the 60,000 IU/μL of conventional PCR.

This feat of fully automated detection, requiring only manual sample injection, was also achieved by Li et al. in 2019 [99] for the Hepatitis B virus. They designed a centrifugal microfluidic platform, where the disk was preloaded with all reagents. Several layers on the disk contained microchambers for the reagents, mixing, detection and disposal. A fully automated sequence of events—serum separation from blood, magnetic bead-based DNA extraction, collection of the nucleic acid by centrifugation, and real-time PCR was achieved by sequential release of the pre-stored liquid reagents guided by laser; 10^2 copies/mL of the DNA could be quantified in about 48 min.

A similar lab-on-a-disk method had already been utilized by Choi et al. [100] for malarial parasite DNA (Fig. 3.3). However, there were two major differences, (1) the disk was not subjected to centrifugation, which is an advantage in terms of energy requirements. The disk was

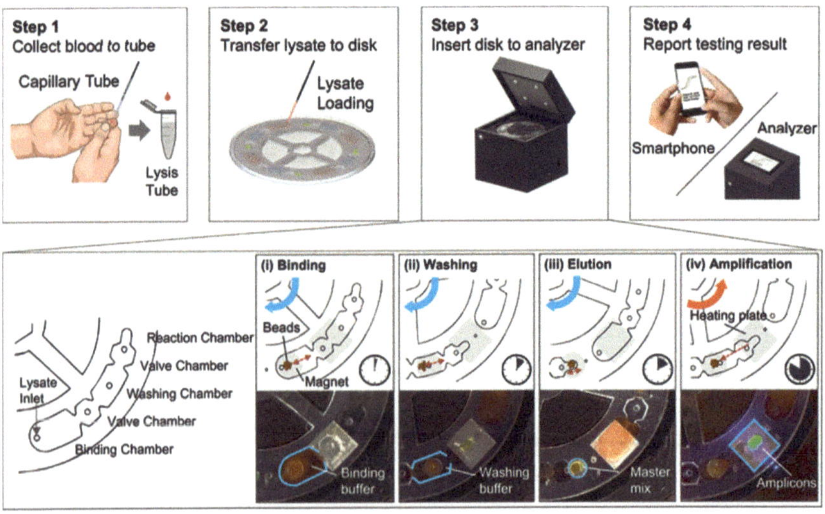

Figure 3.3 Workflow of the device. In step 1, whole blood is collected into the lysis buffer by the capillary tube. In step 2, the lysate is loaded into the binding chamber of the four testing units. In step 3, the disk is sealed and inserted into the analyzer, which then performs automated sample preparation and amplification. This automated process consists of four steps: binding (3 min), washing (4 min), elution (3 min), and amplification (40 min). During the amplification process, the real-time fluorescence signal from each testing unit was recorded and analyzed. Finally, in step 4, the testing results are reported (Fig. 3.2). *Reprinted with permission from G. Choi, T. Prince, J. Miao, L. Cui, W. Guan, Biosensors and bioelectronics sample-to-answer palm-sized nucleic acid testing device towards low-cost malaria mass screening, Biosens. Bioelectron. 115 (2018) 83–90. Available from: https://doi.org/10.1016/j.bios.2018.05.019.*

simply used for back-and-forth axial movement in order to mix and transport samples; (2) the amplification method chosen was LAMP that did not require thermal cycling programming. The magnetic collection was pH switchable, i.e. during binding pH 5, followed by transferring of nucleic acid-magnetic bead complex to the wash chamber by magnetic actuation, and finally in LAMP master mix chamber pH was raised to 8.8, which detached the nucleic acids that could be eluted for amplification. Each disk contains four parallel testing units and the turn-around time was 10 min. *P. falciparum* and *P. vivax* could be detected down to ~0.5 parasites/μL of whole blood. The disk could be fitted into a compact portable box, the data was stored in a non-volatile memory card and displayed on an LCD screen in real-time. Users also had the option to receive the data on a smartphone.

The use of LAMP technique to detect influenza virus (H1N1) was reported by Ma and group in two consecutive papers in 2019 [101,102]. Aptamer-conjugated MMBs were used for capture in a microfluidic chamber followed by low-temperature lysis of viruses, isothermal nucleic acid amplification and colorimetric detection using hydroxynaphthol blue dye. The actuation and mixing in the chambers were guided by hydrophobic valves, absorbent paper and external vibrations. A portable chamber to make the POC automated was designed. It comprised of two stepping motors, two sets of gears, two photo-interrupters, a punching-press, a microcontroller circuit, a thermal controller, a color sensor and a microfluidic chip that could connect readings to a smartphone, besides visual identification.

3.5.2 Magnetic field for signal generation in POCs

Several magnetic POC devices rely on a generation of signals from the sensors integrated in the circuit that detects a specific physical parameter and converts it to an electrical response. The output signal from the sensor is processed and used to provide a measurement. Magnetic sensors generating signal in POC devices are categorized as (1) Magnetic resonance (2) Magnetic resistive and (3) Magnetic relaxation sensors. In this section we will focus on these three types of sensors and their signal acquisition.

3.5.2.1 Magnetic resonance sensors

Magnetic resonance imaging sensors are devices that use radio frequency oscillations in magnetic fields to "excite" or "resonate" protons. As the protons resonate, "relax," and align again with the magnetic fields, they produce a magnetic resonance/relaxation signal that is detected by a scanner to form a computerized image [103]. Magnetic resonance sensors are noninvasive. With the rise of infectious disease cases worldwide, the use of such sensors for diagnosis has shown promising results. Resonance sensor techniques such as nuclear magnetic resonance (NMR) have proven to yield faster results in the detection of biological targets of infectious diseases than conventional methods [104].

Multiple studies have attempted to rapidly identify *Salmonella* in milk using nano-MPs and NMR sensors, the nanoparticles attached to antibodies would capture *Salmonella* and the sensor would then analyze the relaxation time of protons in water molecules with increased sensitivity of the NMR technique [105–107]. The study proved successful with detection of target in spiked milk samples up to 10^5 CFU/mL [106]. A similar method for *Salmonella* diagnosis was used in another study; however, superparamagnetic nanoparticles and time domain NMR (TD-NMR) biosensor was used. This study encountered the problem of agglutination of biological targets on the probe and created an alternative detection technique by using a streptavidin-biotin system to amplify the sensitivity of TD-NMR. In this method, the target size was larger and membrane filtration technology was able to solve the agglutination reaction. The study was successful in detecting *Salmonella* in less than 2 h and at levels as low as 10^4 CFU/mL [105]. A separate study observed the use of NMR in detection of *E. coli*. Similar MPs and filtration steps were employed but used a high-power radiofrequency transmitter and ultra-low noise sensing circuitry to be able to detect weaker NMR signals from hydrogen protons. With increased sensitivity the study yielded *E. coli* detection results as low as 76 CFU/mL in water and 92 CFU/mL in milk and produced results in 1 min (Fig. 3.4) [108]. Additionally, the use of an ultra-low frequency NMR relaxometry for the diagnosis of COVID-19 was used [109]. In a closed tube, magnetic graphene quantum dots were loaded with spike (S) protein COVID-19 antigen. This method proved to be faster than the reverse transcription-PCR (RT-PCR) test as it provided results in under 2 min, detected concentrations as low as 248 particles/mL, and the cost of a single test was $1.25. All studies acknowledged the high cost of NMR sensors and provided cost effective ways to manufacture them, however, one study focused on producing a compact and portable NMR sensor. The device was successful in detecting the relaxation of hydrogen protons in blood-based fungal infections and provided results in 3–5 h. It can be concluded that in view of the serious illness

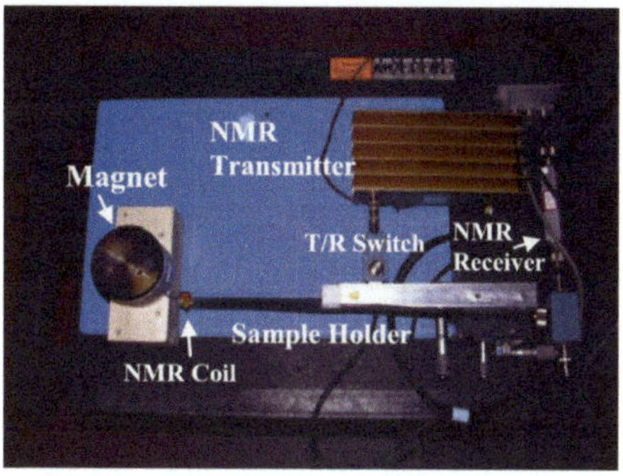

Figure 3.4 (Reprinted) System prototype of portable NMR. *NMR*, nuclear magnetic resonance (Fig. 3.1). *Reprinted with permission from Y. Luo, E.C. Alocilja, Portable nuclear magnetic resonance biosensor and assay for a highly sensitive and rapid detection of foodborne bacteria in complex matrices, J. Biol. Eng. 11(1) (2017) 1–8. Available from: https://doi.org/10.1186/s13036-017-0053-8.*

caused by bacterial infections, a rapid POC device using NMR sensors can be beneficial because they are fast, sensitive and have selective detection. Using nanoparticle systems paired with antibodies or aptamers, we can achieve high specificity and accuracy using magnetic resonance sensors. Further, NMR biosensors described above has tremendous potential as POC devices in biodefense and rapid diagnostic test systems during major pandemics [110,111].

3.5.2.2 Magnetic relaxation sensors

Recently POC devices implementing magnetic relaxation for detecting several different biomolecules to acquire fast, cost-efficient, and accurate measurements of cells, biomarkers and pathogens have gained significance. Magnetic relaxation sensors are devices that function in identifying and measuring the equilibrium of an analyte over time as the magnetic field is altered. The basic construction of a magnetic relaxation sensor, which makes this approach possible, is the utilization of magnetic nanoparticles, a modifier, and specific ligands. MPs are modified to bind to target-specific ligands. The ligands bind to the target molecules, and MPs attached to them generate the signals. In theory, magnetic relaxation works according to Néel's relaxation theory. Where K is the magnetic anisotropy energy density, V is the volume V, k_B is the Boltzmann constant, T the temperature and $\tau 0$ is a length of time [112].

$$\tau_N = \tau_0^{KV/k_B T}$$

The way in which the signal is measured can be compared to a compass in the way that the hydrogen matches the earth's natural spin on its axis with a north–south pole, making it possible for the hydrogen proton to act like a bar magnet. Thus, when hydrogen protons are placed in a strong magnetic field, the proton axes line up and create the magnetic vector that reflects the frequency of the hydrogen resonance portrayed by the MPs. The different receptors on the surface of the MPs then become amplified and cause the spin states to shift away from their regular thermal equilibrium values which makes the return of the magnetization to its equilibrium value to be known as the relaxation mechanism. At this point, it is known that the scattered and clumped MPs are going to be changed and once the analyte appears in the magnetic field, the visibility of the analyte in the magnetic field proceeds to affect the spin–spin relaxation time of the surrounding water molecules which generates the interpretative value of the signal [113]. The changes in signal are detected by magnetometers which can read the radio wave of the hydrogen switching. Magnetometers use the Hall effect to obtain and measure the signal. The Hall effect produces a difference of voltage across an electrical conductor, transverses to an electric current in the conductor and an applied magnetic field perpendicular to the current. In summary, magnetometers use current close to the testing site, and this current is affected by the magnetic fields that protons create. The current change is analyzed because the change of the magnetic field is proportional to the change voltage [114,115]. Recent trends of magnetic relaxation sensing allow for direct POC diagnoses of infectious agents like *Salmonella enterica, Listeria monocytogenes, P. aeruginosa*, SARS-CoV-2, and influenza virus, making it an excellent tool in the medical field [116]. An example of this is research conducted by Yue et al. [117]. The sensor was used to detect one of the major foodborne pathogen, *S. enterica*, sourced from seven milk samples in which there were different cell densities of the bacteria. The samples were mixed with variable amounts of antibody solution. With this mixture, the measurement of the spin–spin relaxation time is possible at the 59.095 magnetic field, detecting the bacteria in the milk samples. Another intriguing research is the 'One-Step Detection of Pathogens and Viruses' where magnetic relaxation switching, and magnetic separation are combined. This one-step detection is fast, with the entire process taking as little as 30 min [116]. The method detects the pathogens by adding MMBs and separating different sizes of them using a magnetic field. Then you can use the transverse relaxation time of the water molecules around the 30 nm MMBs as the signal readout of the immunoassay.

This same technique was used in detecting *P. aeruginosa* at a low concentration, by Jia et al. [106]. The detection depends on magnetic iron oxide nanoparticles, which covalently binds to the bacteria. The high sensitivity of the assay is because of magnetic amplification: a single MP can affect billions of surrounding water molecules and binding of MPs with bacteria leads to nanoparticle aggegration affecting nuclear spins. This interaction and binding allows for maximum recovery of 108% which means the sensor detects the bacteria in realistically complex food samples and drinking water.

3.5.2.3 Magnetoresistive sensors

Magnetoresistance property was discovered in 1857 when Lord Kevin observed a change in electrical resistance when placed in a magnetic field. The first magnetoresistive sensors were developed by IBM which were used to detect bits in the hard disk. They are now being implemented in detecting infectious pathogens and organisms. In general, magneto-resistive sensors are primarily used in electronic devices harnessing conversion of magnetic signal to electrical. To understand magnetoresistance, we need to study the Hunt element, which comprises a magneto-resistive film with a current I and applied magnetic field in vector M at an angle a to the current I (Fig. 3.5). When in a sensor the presence of magnetic field Hy will change the resistivity, which is probed by measuring current. In all the sensors, this principle is used to detect capture of target analyte using MPs. It presents variable resistance markers that depict the signal with extraordinarily low energy usage throughout the process. The traditional conventional frequency method was deemed to be complex and significantly more time consuming [118].

Magneto-resistive sensors are continually deemed promising in the research community due to the quick response time and efficacy presented when tested with biological samples. Presently, there are few methods for POC diagnosis of diseases using magneto-resistive sensors since they provide faster results, have a higher sensitivity, are easy to use

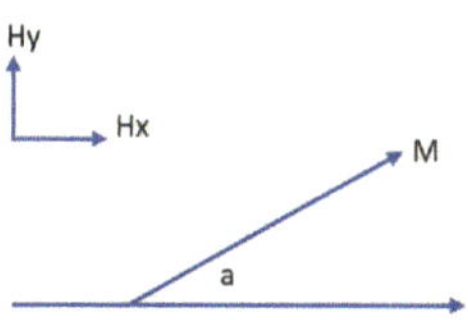

Figure 3.5 Geometry of the Hunt element.

and are inexpensive to manufacture. These sensors generate less noise and are not affected by external parameters such as change in temperature, nature of samples, or pH of the solution and are more stable than traditional sensors. In general, there are two major types of magneto-resistive sensors used in detection of biological analytes namely, magnetic tunnel junction (MTJ) sensors and giant magnetoresistance (GMR) sensors [119,120]. The GMRs are reported to have the lowest noise levels, but the downside is that these types have a small range. Recent trends demonstrate use of GMRs where regulatory activities in the body such as the reactions brought about by proteases can be monitored with this technology through employing a diversified protease assay. Herein magnetic resistance changes proportionately as MPs attach to the antibody-antigen interaction in the body, in real time [121]. Similarly, these devices have been applied in diagnosing different illnesses such as malaria and TB [122]. The use of a fabricated array of layers that include the base, pinned layer, barrier, and free layer make it possible to analyze the biological sample that has already been encased with the MPs or paramagnetic material. The biological samples and MPs are placed in the direction of magnetization within a magnetic field where magnetic resistance measurements can then be made and information pertaining to the data given off is demonstrated with an increased signal strength [123]. Such techniques have been recently used in microfluidic systems where Cunha et al. demonstrated multiplexed detection of *Enterococcus* and *Staphylococcus*. In this study, bacterial cells were labeled with MPs functionalized with RBP (receptor binding protein) and measured using magnetoresistive sensors. RBP-MP interactions are highly specific and resulted in the capture of more than 70% of the *Enterococcous* and *Staphylococcous*. The assay was able to detect both bacterial populations in low concentration (10 CFU/mL) in 2 h [124] (Fig. 3.6).

3.6 Special mentions

Adherence to reports of ready POC devices not involving dedicated instruments resulted in the exclusion of several promising magnetism-based proof-of-concept diagnostics from this critical review. However, some very innovative biosensing methods are worth mentioning as ideas translatable into POC devices. Opto-magnetic sensing has often been used to generate sensitive signals for target recognition. In most cases, MMBs are quantified by their Brownian motion using inherent

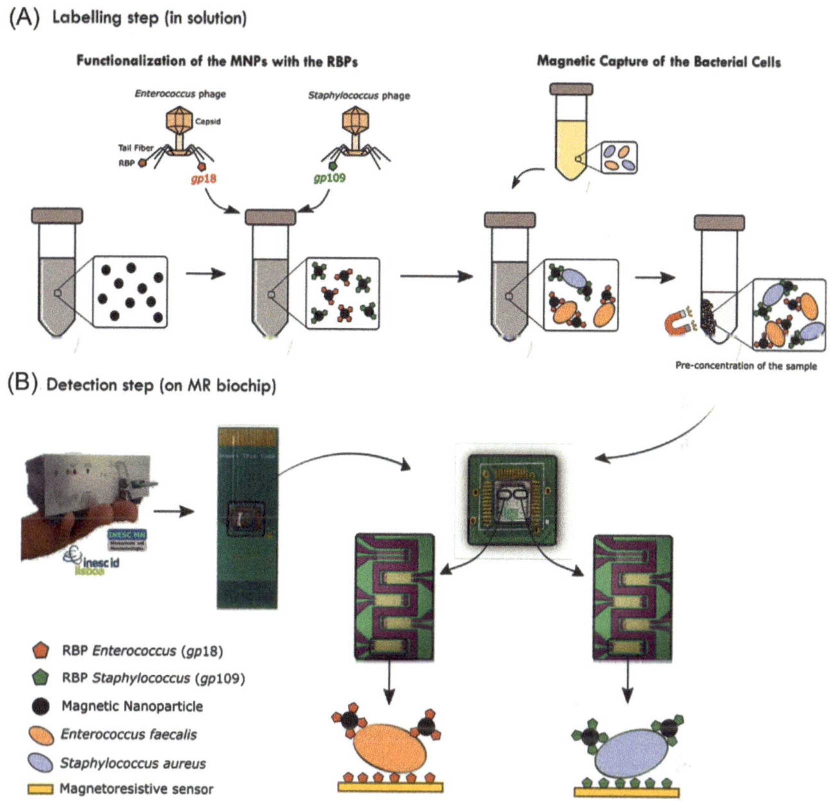

Figure 3.6 (reprinted) Schematic representation of the phage-based Mr lab-on-chip assay. (A) Labeling step based on the functionalization of the MPs with the RBPs (B) Detection step in the Mr platform, which is based on the biomolecular recognition strategy. The amplified image of the biochip consists of three groups of five spin-valve sensors (four active sensors covered with gold pads and a reference sensor) for detection. *RBP*, receptor binding protein (Fig. 3.1). *Reprinted with permission from A.P. Cunha, R. Henriques, S. Cardoso, P.P. Freitas, C.M. Carvalho, Rapid and multiplex detection of nosocomial pathogens on a phage-based magnetoresistive lab-on-chip platform, Biotechnol. Bioeng. Published online 2021. Available from: https://doi.org/10.1002/bit.27841.*

fluorescence. In one system, antibody and dye-conjugated iron oxide nanoparticles were used for bacteria recognition [125]. When MPs were mixed with bacterial colonies, they swarmed around the bacteria's outer membrane due to immunosensing. As a result of this clustering, the interaction between the MPs and their aqueous environment is inhibited, increasing the magnetic relaxation time. However to improve resolution, the fluorescence from the particles was also quantified. Tian et al. [126]

claimed that competitive MMB binding could improve detection limit by 20 times. They prepared avidin-coated magnetic beads to which biotinylated antibodies bound -if bacterial cells were present they would block the antibody sites creating a sterically-hindered cluster. A second set of streptavidin-coated MPs would not be able to bind and remain in solution, whereas, in the absence of bacteria, they would attach to the antibodies and remain clumped magnetic masses. The free streptavidin-coated MPs could be spectrophotometrically read by a novel blu-ray laser and visualized under a fluorescence microscope.

In recent years, biobarcoding has emerged as a brilliant sensing idea, where the target (mostly live cells or nucleic acids) is captured, and then the nucleic acids are hybridized with specific complementary strands that are nano-labeled. This allows for multiplexed detection as several complementary strands can be added to the detection mixture. The hybridized nucleic acids are then amplified and read as PCR output. In 2017, Cihalova et al. [127] developed a biobarcode assay for *Staphylococcus aureus*, methicillin-resistant *S. aureus* and *Klebsiella pneumoniae*, where MPs were surface functionalized to capture the species. The cells were then lysed and the DNA amplified. Specific single strand-linked gold nanoparticles were used to detect target DNA and the mixture was separated by magnetic field. Different colored quantum dots labeled with specific strands were used for generating fluorescent signatures.

Knowledge of fluid mechanics and dynamics has been brought into microfluidic devices to allow automated and thorough mixing. Petkovic et al. [128] used vacuum to draw fluid through the channels and a four/three-pin magnetic collection device, to concentrate particles. The mixing in microchambers was proven to be due to chaotic micromixing phenomenon. The detection protocol for one of the deadliest viruses, the Hendravirus, was as follows: collection of serum sample from infected patient, binding of the antibodies from sera to the antibody-specific sG protein conjugated to the MMBs, labeling of the MMB–sG–antibody conjugate with fluorescent protein R-Phycoerythrin, and separation of the magnetic complex to be read by an epifluorescence microscope. Du et al. [129] incorporated a key unique feature of using metered air bubbles to stir up the MMBs carrying the capture probes to effect rapid and efficient target capture. A pneumatic layer was introduced onto the microfluidic chip to guide the fluid layer mixing. Target Ebola virus RNA was hybridized to complementary fluorescent strands, if target was absent they bound to capture magnetic beads instead. These steps followed by washing and release of bound strands were carried out automatically on chip. Enrichment enabled to push detection limit down to 0.021 plaque forming U/mL. The washed supernatant was quantified using fluorimeter and spectrophotometer software.

Gold brings a strong optical signature to any detection. MPs have often been used to generate stronger signals. In one assay for C-reactive protein, MPs were used to enhance the surface plasmon resonance signals from gold nanoparticles. A microfluidic chip was designed by Liu et al. [130], based on sandwich immunocapture, where the detection antibody was labeled with gold nanoparticles. An array of pre fixed antibody modified by gold nanoparticles and MPs were used for rapid antigen capture. The microfluidic chip contained 3D optical-coupled cavities and was fabricated from a glass layer and polymethylmethacrylate bonded by a pressure sensitive adhesive. The self-drive flow was realized by capillary action and the adhesive hydrophilic membrane. This chip was analyzed by a surface plasmon resonance sensor that registered a pronounced signal due to the gold. The inflammation biomarker C-reactive protein was detected down to 0.5 ng/mL. Though all these assays achieve high sensitivity and appreciable miniaturization, they need either an installed instrument (fluorescence analyzer, surface plasmon resonance measurer, microscope etc.) or several manual steps to reach the final results. Hence, advancements in portable device design and automated work set-ups can push these excellent ideas towards successful stand-alone POC tests.

3.7 Future perspectives—overcoming the hurdles

The medical and scientific communities are engaged in a constant endeavor to fight against infectious diseases. Quickly obtaining test results near or at the point of patient care allows clinicians to arrive early at informed treatment plans. However, it is important to ensure that the POCs are sensitive, accurate, specific to the target and safe to use. Sensitivity requires specially designed biosensors that can pick up trace levels of the analyte, since we need to work with small sample volumes. This depends as much on the bioreceptor as on the transduction system. One challenge faced by magnetoresistance sensors is that the resistance signal is often influenced by the magnetic field of the ferromagnetic material. To manage this, an external electric field can be used to tune the magnetic properties—a useful approach for integration into portable formats. The transduction must also be able to avoid noise [57].

Every POC should exhibit sensitivity in a difficult and complex medium, such as blood, urine, food and water, proving their specificity. They should have short detection timelines. Additionally, there should be minimum false results, meaning the test should have the ability to respond to multiple confirmatory assays. The biggest challenge is however, to transform the concept into a physical test that does not require heavy dependence on laboratory or personnel. For instance, the very sensitive SQUID sensors with almost no sample preparation reactions require low temperature operation, achieved by liquid helium or nitrogen. This makes it almost impossible for a minimally trained staff to work with SQUID sensors. Such aspects of the ASSURED criteria have already been discussed before in this chapter with respect to literature on POCs. Some other major issues to address are the cost of the device (to be affordable by resource-constrained populations and remotely-located small establishments) and the toxicity of the test reagents (especially if used for a biomedical monitoring).

The combined cost of a POC test—taking into account the capital expenditure of a measuring instrument and consumable cost of strips/reagents, should be competitive to existing kits and heavy installed equipment. Several of the POC devices compiled in this chapter boast of lower prices, especially when these are manufactured in large scale. For example the optical well test by Dirkzwager et al. [90] for malaria was estimated to cost $0.36, whereas antibody based malaria kits could cost anywhere between $1 and $17 depending upon the country and the company. Even with inclusion of a custom-built reader, researchers were able to bring down the expense sharply. Knowlton et al. [93] claimed that the 3D printed fluorescence microscope to be fixed with the user's smartphone can be built within $100, and consumables cost under $1 per test. Similarly, the smartphone-based electrochemical sepsis diagnostic test by Min et al. [84] cost $50 for the reader and ~$5 for reagents per test, which gives this kit an advantage over ELISA that costs $11 or lateral flow strips costing $10–$20 per test. Materials such as paper and nitrocellulose contribute significantly to reduced costs.

Most MPs discussed here and used generally are composed of transition metals. Long-term storage may cause leaching, and biodegradation of MPs is likely to release free metal ions and generate reactive oxygen species. The reactive oxygen species and other components in MPs have been reported to cause inflammation, DNA damage and impaired mitochondrial function [131,132]. However, the required concentrations for such toxicity are often much higher (usually range of hundreds of mg/kg body weight) than the working concentrations in tests. Moreover, many MPs used for bioassays are rendered biocompatible due to suitable coating materials, such as, polymers, silica macromolecules and noble metals. It is important to remember that the operator will be working with potentially harmful clinical pathogens, thereby biological safety precautions need to be followed that will also be a barrier to any harm coming from test reagents.

3.8 Conclusions and future perspectives

Despite these significant hurdles, it is extremely encouraging to see that magnetic-based POCs are progressing with leaps. Patronage from government agencies and industry is also important for realistic implementation of many brilliantly developed diagnostic concepts. The biosensor market is reported to be currently valued at $25.5 billion (2021) and is projected to reach $36.7 billion by 2026. The majority of this will be contributed by nanosensors, with MPs occupying a considerable share. POCs and home diagnostics are the fastest growing applications of the biosensor market, as a result of the technological advancements in POC devices (many of which are discussed above), rising incidences of infectious diseases (as the COVID-19 pandemic has clearly indicated), and an increase in investments by key players. This is indirectly affecting ambulance and hospital costs in all countries. Also, the Asia-Pacific market for biosensors is supposed to grow at the highest rate among all continents, with the three foremost countries being China, Japan and India in that order [133]. Hence, it is safe to claim that POC for infectious diseases will soon rule the commercial arena and magnetic biosensor researchers have enormous opportunities to advance this lab-to-clinic revolution. It is reassuring to realize that this growth in economy and technology will ultimately lead to improvement of healthcare—literally and figuratively "bringing the diagnostics to the patient."

Acknowledgments

Richa Sharma thanks the Royal Society, United Kingdom, and the Science and Engineering Research Board (SERB, India) (grant number NIF/R1/192688) for a Newton International Fellowship.

References

[1] World Health Organization. The Top 10 Causes of Death. Published 2020. https://www.who.int/news-room/fact-sheets/detail/the-top-10-causes-of-death.

[2] K.V. Ragavan, S. Kumar, S. Swaraj, S. Neethirajan, Advances in biosensors and optical assays for diagnosis and detection of malaria, Biosens. Bioelectron. 105 (2018) 188–210. Available from: https://doi.org/10.1016/j.bios.2018.01.037.

[3] K.V. Ragavan, P. Egan, S. Neethirajan, Multi mimetic Graphene Palladium nanocomposite based colorimetric paper sensor for the detection of neurotransmitters, Sens. Actuators B: Chem. 273 (2018) 1385–1394. Available from: https://doi.org/10.1016/j.snb.2018.07.048.

[4] R. Sharma, K.S.M.S. Raghavarao, Nanoparticle-based aptasensors for food contaminant detection, Nanomater. Food Appl. (2019) 123–145. Available from: https://doi.org/10.1016/B978-0-12-814130-4.00006-3. Published online.

[5] M.S. Thakur, R. Ranjan, A.C. Vinayaka, K.S. Abhijith, R. Sharma, Nanoparticles and Biophotonics as Efficient Tools in Resonance Energy Transfer-Based Biosensing for Monitoring Food Toxins and Pesticides. Published online 2013:55–84. Available from: https://doi.org/10.1021/bk-2013-1143.ch004.

[6] R. Sharma, K.V. Ragavan, M.S. Thakur, K.S.M.S. Raghavarao, Recent advances in nanoparticle based aptasensors for food contaminants, Biosens. Bioelectron. 74 (2015) 612–627. Available from: https://doi.org/10.1016/j.bios.2015.07.017.

[7] M. Zarei, Infectious pathogens meet point-of-care diagnostics, Biosens. Bioelectron. 106 (2018) 193–203. Available from: https://doi.org/10.1016/j.bios.2018.02.007.

[8] C. Dincer, R. Bruch, A. Kling, P.S. Dittrich, G.A. Urban, Multiplexed point-of-care testing - xPOCT, Trends Biotechnol. 35 (8) (2017) 728–742. Available from: https://doi.org/10.1016/j.tibtech.2017.03.013.

[9] L. Bissonnette, M.G. Bergeron, Portable devices and mobile instruments for infectious diseases point-of-care testing, Expert. Rev. Mol. Diagn. 17 (5) (2017) 471–494. Available from: https://doi.org/10.1080/14737159.2017.1310619.

[10] D.I. Walsh 3rd, D.S. Kong, S.K. Murthy, P.A. Carr, Enabling microfluidics: from clean rooms to makerspaces, Trends Biotechnol. 35 (5) (2017) 383–392. Available from: https://doi.org/10.1016/j.tibtech.2017.01.001.

[11] Z. Li, Y. Bai, M. You, et al., Fully integrated microfluidic devices for qualitative, quantitative and digital nucleic acids testing at point of care, Biosens. Bioelectron. 177 (2021) 112952. Available from: https://doi.org/10.1016/j.bios.2020.112952.

[12] V. Kamat, S. Pandey, K. Paknikar, D. Bodas, A facile one-step method for cell lysis and DNA extraction of waterborne pathogens using a microchip, Biosens. Bioelectron. 99 (June 2017) (2018) 62–69. Available from: https://doi.org/10.1016/j.bios.2017.07.040.

[13] K.V. Ragavan, S.R. Ahmed, X. Weng, S. Neethirajan, Chitosan as a peroxidase mimic: paper based sensor for the detection of hydrogen peroxide, Sens. Actuators B Chem. 272 (2018) 8–13. Available from: https://doi.org/10.1016/j.snb.2018.05.142.

[14] K. Ratajczak, M. Stobiecka, High-performance modified cellulose paper-based biosensors for medical diagnostics and early cancer screening: a concise review, Carbohydr. Polym. 229 (2020) 115463. Available from: https://doi.org/10.1016/j.carbpol.2019.115463.

[15] V.X.T. Zhao, T.I. Wong, X.T. Zheng, Y.N. Tan, X. Zhou, Colorimetric biosensors for point-of-care virus detections, Mater. Sci. Energy Technol. 3 (2020) 237–249. Available from: https://doi.org/10.1016/j.mset.2019.10.002.

[16] H. Inan, M. Poyraz, F. Inci, et al., Photonic crystals: emerging biosensors and their promise for point-of-care applications, Chem. Soc. Rev. 46 (2) (2017) 366–388. Available from: https://doi.org/10.1039/C6CS00206D.

[17] Z.S. Ballard, H.-A. Joung, A. Goncharov, et al., Deep learning-enabled point-of-care sensing using multiplexed paper-based sensors, npj Digit. Med. 3 (1) (2020) 66. Available from: https://doi.org/10.1038/s41746-020-0274-y.

[18] T.K. Amukele, J. Street, K. Carroll, H. Miller, S.X. Zhang, Drone transport of microbes in blood and sputum laboratory specimens, J. Clin. Microbiol. 54 (10) (2016) 2622−2625. Available from: https://doi.org/10.1128/JCM.01204-16.

[19] A. Roda, E. Michelini, M. Zangheri, M. Di Fusco, D. Calabria, P. Simoni, Smartphone-based biosensors: a critical review and perspectives, TrAC. Trends Anal. Chem. 79 (2016) 317−325. Available from: https://doi.org/10.1016/j.trac.2015.10.019.

[20] I. Hussain, A.K. Bowden, Smartphone-based optical spectroscopic platforms for biomedical applications: a review [Invited], Biomed. Opt. Express. 12 (4) (2021) 1974−1998. Available from: https://doi.org/10.1364/BOE.416753.

[21] V. Lebedev, E. Laukhina, V. Laukhin, et al., Investigation of sensing capabilities of organic bi-layer thermistor in wearable e-textile and wireless sensing devices, Org. Electron. 42 (2017) 146−152. Available from: https://doi.org/10.1016/j.orgel.2016.12.034.

[22] G. Rajan, J.J. Morgan, C. Murphy, et al., Low operating voltage carbon−graphene hybrid E-textile for temperature sensing, ACS Appl. Mater. Interfaces 12 (26) (2020) 29861−29867. Available from: https://doi.org/10.1021/acsami.0c08397.

[23] J.C. Vidal, J.R. Bertolín, L. Bonel, L. Asturias, M.J. Arcos-Martínez, J.R. Castillo, Rapid determination of recent cocaine use with magnetic particles-based enzyme immunoassays in serum, saliva, and urine fluids, J. Pharm. Biomed. Anal. 125 (2016) 54−61. Available from: https://doi.org/10.1016/j.jpba.2016.03.004.

[24] T. Klein, W. Wang, L. Yu, et al., Development of a multiplexed giant magnetoresistive biosensor array prototype to quantify ovarian cancer biomarkers, Biosens. Bioelectron. 126 (2019) 301−307. Available from: https://doi.org/10.1016/j.bios.2018.10.046.

[25] S. Hash, M.P. Martinez-Viedma, F. Fung, et al., Nuclear magnetic resonance biosensor for rapid detection of Vibrio parahaemolyticus, Biomed. J. 42 (3) (2019) 187−192. Available from: https://doi.org/10.1016/j.bj.2019.01.009.

[26] V.D. Krishna, K. Wu, D. Su, M.C.J. Cheeran, J.-P. Wang, A. Perez, Nanotechnology: review of concepts and potential application of sensing platforms in food safety, Food Microbiol. 75 (2018) 47−54. Available from: https://doi.org/10.1016/j.fm.2018.01.025.

[27] M. Pastucha, Z. Farka, K. Lacina, Z. Mikušová, P. Skládal, Magnetic nanoparticles for smart electrochemical immunoassays: a review on recent developments, Mikrochim. Acta 186 (5) (2019) 312. Available from: https://doi.org/10.1007/s00604-019-3410-0.

[28] K. Wu, R. Saha, D. Su, et al. Magnetic Immunoassays: A Review of Virus and Pathogen Detection Before and Amidst the Coronavirus Disease-19 (COVID-19). Published online July 9, 2020. Accessed March 14, 2021. https://arxiv.org/abs/2007.04809.

[29] I. Gessner, J.W.U.U. Fries, V. Brune, S. Mathur, Magnetic nanoparticle-based amplification of microRNA detection in body fluids for early disease diagnosis, J. Mater. Chem. B 9 (1) (2021) 9−22. Available from: https://doi.org/10.1039/D0TB02165B.

[30] M. Aminul Islam, M. Ziaul Ahsan, Plausible approach for rapid detection of SARS-CoV-2 virus by magnetic nanoparticle based biosensors, Am. J. Nanosci. 6 (2) (2020) 6. Available from: https://doi.org/10.11648/j.ajn.20200602.11.

[31] S.K. Metkar, K. Girigoswami, Diagnostic biosensors in medicine − a review, Biocatal. Agric. Biotechnol. 17 (November 2018) (2019) 271−283. Available from: https://doi.org/10.1016/j.bcab.2018.11.029.

[32] A. Qureshi, J.H. Niazi, Biosensors for detecting viral and bacterial infections using host biomarkers: a review, Analyst 145 (24) (2020) 7825–7848. Available from: https://doi.org/10.1039/d0an00896f.

[33] S. Kim, M. Van Tieu, S.Y. Hwang, M. Lee, Magnetic Particles: Their Applications from Sample Preparations to Biosensing Platforms. Published online 2020:1–20.

[34] G. Lin, D. Makarov, G. Lin, O.G. Schmidt, Lab. a Chip 17 (11) (2017). Available from: https://doi.org/10.1039/c7lc00026j.

[35] E. Ng, C. Yao, T.O. Shultz, S. Ross-howe, S.X. Wang, Magneto-nanosensor smartphone platform for the detection of HIV and leukocytosis at point-of-care, Nanomed. Nanotechnol., Biol. Med. 16 (2019) 10–19. Available from: https://doi.org/10.1016/j.nano.2018.11.007.

[36] A. Romeo, T.S. Leung, S. Sánchez, Smart biosensors for multiplexed and fully integrated point-of-care diagnostics, Lab. Chip 16 (11) (2016) 1957–1961. Available from: https://doi.org/10.1039/C6LC90046A.

[37] L. Wang, J. Lin, Recent advances on magnetic nanobead based biosensors: From separation to detection, TrAC. Trends Anal. Chem. 128 (2020) 115915. Available from: https://doi.org/10.1016/j.trac.2020.115915.

[38] X. Wang, F. Li, Y. Guo, Recent trends in nanomaterial-based biosensors for point-of-care testing, Front. Chem. 8 (2020) 924. Available from: https://www.frontiersin.org/article/10.3389/fchem.2020.586702.

[39] C.S. Kosack, A. Page, P.R. Klatser, A Guide to Aid The Selection of Diagnostic Tests. 2017;(May):639–645.

[40] O. Pashchenko, T. Shelby, T. Banerjee, S. Santra, A comparison of optical, electrochemical, magnetic, and colorimetric point-of-care biosensors for infectious disease diagnosis, ACS Infect. Dis. 4 (8) (2018) 1162–1178. Available from: https://doi.org/10.1021/acsinfecdis.8b00023.

[41] R. Sharma, H. Rajagopalan, M. Klausen, M.V. Jeyalatha, M. Üçüncü, S. Venkateswaran, A.R. Anand, M. Bradley, Rapid detection of major Gram-positive pathogens in ocular specimens using a novel fluorescent vancomycin-based probe, Sens. Diagn. (2022). Available from: https://doi.org/10.1039/D2SD00061J.

[42] B. Heidt, W.F. Siqueira, K. Eersels, et al., Point of care diagnostics in resource-limited settings: a review of the present and future of PoC in its most needed environment, Biosens 10 (10) (2020). Available from: https://doi.org/10.3390/bios10100133.

[43] S. Yadav, N.N. Sharma, J. Akhtar, Nucleic acid analysis on paper substrates (NAAPs): an innovative tool for Point of Care (POC) infectious disease diagnosis, Analyst 146 (11) (2021) 3422–3439. Available from: https://doi.org/10.1039/D1AN00214G.

[44] K.P. Delaney, B.M. Branson, A. Uniyal, et al., Evaluation of the performance characteristics of 6 rapid HIV antibody tests, Clin. Infect. Dis. 52 (2) (2011) 257–263. Available from: https://doi.org/10.1093/cid/ciq068.

[45] M. Schito, T.F. Peter, S. Cavanaugh, et al., Opportunities and challenges for cost-efficient implementation of new point-of-care diagnostics for HIV and tuberculosis, J. Infect. Dis. 205 (2012) S169–S180. Available from: https://doi.org/10.1093/infdis/jis044. Suppl.

[46] M.A. Dimaio, I.T. Pereira, T.I. George, N. Banaei, Performance of BinaxNOW for diagnosis of malaria in a U.S. hospital, J. Clin. Microbiol. 50 (9) (2012) 2877–2880. Available from: https://doi.org/10.1128/JCM.01013-12.

[47] M.L. Wilson, Malaria rapid diagnostic tests, Clin. Infect. Dis. 54 (11) (2012) 1637–1641. Available from: https://doi.org/10.1093/cid/cis228.

[48] P.K. Drain, E.P. Hyle, F. Noubary, et al., Diagnostic point-of-care tests in resource-limited settings, Lancet Infect. Dis. 14 (3) (2014) 239–249. Available from: https://doi.org/10.1016/S1473-3099(13)70250-0.

[49] D.J. Shin, P. Athamanolap, L. Chen, et al., Mobile nucleic acid amplification testing (mobiNAAT) for Chlamydia trachomatis screening in hospital emergency department settings, Sci. Rep. 7 (1) (2017) 4495. Available from: https://doi.org/10.1038/s41598-017-04781-8.
[50] P. Balaraman, B. Balasubramanian, D. Kaliannan, et al., Phyco-synthesis of silver nanoparticles mediated from marine algae *Sargassum myriocystum* and its potential biological and environmental applications, Waste Biomass Valoriz. 11 (10) (2020) 5255–5271. Available from: https://doi.org/10.1007/s12649-020-01083-5.
[51] Y. Xianyu, Q. Wang, Y. Chen, Magnetic particles-enabled biosensors for point-of-care testing, TrAC. Trends Anal. Chem. 106 (2018) 213–224. Available from: https://doi.org/10.1016/j.trac.2018.07.010.
[52] S. Adem, S. Jain, M. Sveiven, X. Zhou, A.J. O'Donoghue, D.A. Hall, Giant magnetoresistive biosensors for real-time quantitative detection of protease activity, Sci. Rep. 10 (1) (2020) 7941. Available from: https://doi.org/10.1038/s41598-020-62910-2.
[53] R. Birla, Biosensors in Tissue and Organ Fabrication BT - Tissue Engineering for the Heart: A Case Study Based Approach. In: Birla R, (ed.) Springer International Publishing; 2016:31–57. Available from: https://doi.org/10.1007/978-3-319-41504-8_2.
[54] V. Nabaei, R. Chandrawati, H. Heidari, Magnetic biosensors: Modelling and simulation, Biosens Bioelectron, 103, 2018, pp. 69–86. Available from: https://doi.org/10.1016/j.bios.2017.12.023.
[55] I. Giouroudi, G. Kokkinis, Recent advances in magnetic microfluidic biosensors, Nanomater 7 (7) (2017) 1–10. Available from: https://doi.org/10.3390/nano7070171.
[56] Y.-T. Chen, A.G. Kolhatkar, O. Zenasni, S. Xu, T.R. Lee, Biosensing using magnetic particle detection techniques, Sensors 17 (10) (2017) 2300. Available from: https://doi.org/10.3390/s17102300.
[57] D.J. Denmark, X. Bustos-perez, A. Swain, M. Phan, S. Mohapatra, Readiness of magnetic nanobiosensors for point-of-care commercialization, J. Electron. Mater. 48 (8) (2019). Available from: https://doi.org/10.1007/s11664-019-07275-7.
[58] R. Hachani, M. Lowdell, M. Birchall, et al., Polyol synthesis, functionalisation, and biocompatibility studies of superparamagnetic iron oxide nanoparticles as potential MRI contrast agents, Nanoscale 8 (6) (2016) 3278–3287. Available from: https://doi.org/10.1039/C5NR03867G.
[59] M. Barbic, H. ElBidweihy, Effect of magnetic nanoparticle shape on flux amplification in inductive coil magnetic resonance detection, J. Appl. Phys. 120 (10) (2016) 104506. Available from: https://doi.org/10.1063/1.4962451.
[60] T. Zargar, A. Kermanpur, Effects of hydrothermal process parameters on the physical, magnetic and thermal properties of Zn0.3Fe2.7O4 nanoparticles for magnetic hyperthermia applications, Ceram. Int. 43 (7) (2017) 5794–5804. Available from: https://doi.org/10.1016/j.ceramint.2017.01.127.
[61] L.B. de Mello, L.C. Varanda, F.A. Sigoli, I.O. Mazali, Co-precipitation synthesis of (Zn-Mn)-co-doped magnetite nanoparticles and their application in magnetic hyperthermia, J. Alloy. Compd. 779 (2019) 698–705. Available from: https://doi.org/10.1016/j.jallcom.2018.11.280.
[62] I. Bibi, H. Maqbool, S. Iqbal, et al., La1-xGdxCr1-yNiyO3 perovskite nanoparticles synthesis by micro-emulsion route: dielectric, magnetic and photocatalytic properties evaluation, Ceram. Int. 47 (4) (2021) 5822–5831. Available from: https://doi.org/10.1016/j.ceramint.2020.11.033.
[63] C.S. Tiwary, S. Kashyap, K. Biswas, K. Chattopadhyay, Synthesis of pure iron magnetic nanoparticles in large quantity, J. Phys. D. Appl. Phys 46 (38) (2013) 385001. Available from: https://doi.org/10.1088/0022-3727/46/38/385001.

[64] T. Saito, H. Tsuruta, A. Watanabe, T. Ishimine, T. Ueno, Pure-iron/iron-based-alloy hybrid soft magnetic powder cores compacted at ultra-high pressure, AIP Adv. 8 (4) (2017) 47708. Available from: https://doi.org/10.1063/1.4993687.
[65] B. Heer, A. Bandyopadhyay, Compositionally graded magnetic-nonmagnetic bimetallic structure using laser engineered net shaping, Mater. Lett. 216 (2018) 16–19. Available from: https://doi.org/10.1016/j.matlet.2017.12.129.
[66] S.K. Avugadda, S. Wickramasinghe, D. Niculaes, et al., Uncovering the magnetic particle imaging and magnetic resonance imaging features of iron oxide nanocube clusters, Nanomater. 11 (1) (2021). Available from: https://doi.org/10.3390/nano11010062.
[67] S. Rezania, M.A. Kamboh, S.S. Arian, et al., Conversion of waste frying oil into biodiesel using recoverable nanocatalyst based on magnetic graphene oxide supported ternary mixed metal oxide nanoparticles, Bioresour. Technol. 323 (2021) 124561. Available from: https://doi.org/10.1016/j.biortech.2020.124561.
[68] A.R. Iliasov, T.R. Nizamov, V.A. Naumenko, et al., Non-magnetic shell coating of magnetic nanoparticles as key factor of toxicity for cancer cells in a low frequency alternating magnetic field, Colloids Surf. B Biointerfaces 206 (2021) 111931. Available from: https://doi.org/10.1016/j.colsurfb.2021.111931.
[69] L. Wang, G. Jang, D.K. Ban, et al., Multifunctional stimuli responsive polymer-gated iron and gold-embedded silica nano golf balls: nanoshuttles for targeted on-demand theranostics, Bone Res. 5 (1) (2017) 17051. Available from: https://doi.org/10.1038/boneres.2017.51.
[70] M. Asgari, T. Miri, M. Soleymani, A. Barati, A novel method for in situ encapsulation of curcumin in magnetite-silica core-shell nanocomposites: a multifunctional platform for controlled drug delivery and magnetic hyperthermia therapy, J. Mol. Liq. 324 (2021) 114731. Available from: https://doi.org/10.1016/j.molliq.2020.114731.
[71] K. Kamari, A. Taheri, Preparation and evaluation of magnetic core–shell mesoporous molecularly imprinted polymers for selective adsorption of amitriptyline in biological samples, J. Taiwan. Inst. Chem. Eng. 86 (2018) 230–239. Available from: https://doi.org/10.1016/j.jtice.2018.02.031.
[72] A. Tomitaka, S. Ota, K. Nishimoto, H. Arami, Y. Takemura, M. Nair, Dynamic magnetic characterization and magnetic particle imaging enhancement of magnetic-gold core–shell nanoparticles, Nanoscale 11 (13) (2019) 6489–6496. Available from: https://doi.org/10.1039/C9NR00242A.
[73] V. Haribabu, K. Girigoswami, A. Girigoswami, Magneto-silver core–shell nanohybrids for theragnosis, Nano-Structures & Nano-Objects 25 (2021) 100636. Available from: https://doi.org/10.1016/j.nanoso.2020.100636.
[74] R. Keçili, F. Ghorbani-Bidkorbeh, İ. Dolak, G. Canpolat, M. Karabörk, C.M. Hussain, Functionalized magnetic nanoparticles as powerful sorbents and stationary phases for the extraction and chromatographic applications, TrAC. Trends Anal. Chem. (2021) 116380. Available from: https://doi.org/10.1016/j.trac.2021.116380. Published online.
[75] A. Barrios-gumiel, D. Sepúlveda-crespo, J.L. Jiménez, et al., Dendronized magnetic nanoparticles for HIV-1 capture and rapid diagnostic, Colloids Surf. B: Biointerfaces 181 (May) (2019) 360–368. Available from: https://doi.org/10.1016/j.colsurfb.2019.05.050.
[76] C.-B. Kim, E.-G. Lim, S.W. Shin, H.J. Krause, H. Hong, Magnetic immunoassay platform based on the planar frequency mixing magnetic technique, Biosens. Bioelectron. 83 (2016) 293–299. Available from: https://doi.org/10.1016/j.bios.2016.04.076.
[77] C.L. Pirich, R.A. de Freitas, R.M. Torresi, G.F. Picheth, M.R. Sierakowski, Piezoelectric immunochip coated with thin films of bacterial cellulose nanocrystals

for dengue detection, Biosens. Bioelectron. 92 (2017) 47–53. Available from: https://doi.org/10.1016/j.bios.2017.01.068.
[78] K. Luo, K.-B. Jeong, S.-M. You, D.-H. Lee, J.-Y. Jung, Y.-R. Kim, Surface-engineered starch magnetic microparticles for highly effective separation of a broad range of bacteria, ACS Sustain. Chem. Eng. 6 (10) (2018) 13524–13531. Available from: https://doi.org/10.1021/acssuschemeng.8b03611.
[79] L. Armbrecht, C. Dincer, A. Kling, J. Horak, J. Kieninger, G. Urban, Self-assembled magnetic bead chains for sensitivity enhancement of microfluidic electrochemical biosensor platforms, Lab. Chip 15 (22) (2015) 4314–4321. Available from: https://doi.org/10.1039/C5LC00796H.
[80] I. Pereiro, A. Bendali, S. Tabnaoui, et al., A new microfluidic approach for the one-step capture, amplification and label-free quantification of bacteria from raw samples, Chem. Sci. 8 (2) (2017) 1329–1336. Available from: https://doi.org/10.1039/C6SC03880H.
[81] R. Sharma, K.V. Ragavan, K.S.M.S. Raghavarao, M.S. Thakur, Nano-aptamer Based Quantitative Detection of Chloramphenicol, Biotechnol. Biochem. Eng. (2016) 187–195. Available from: https://doi.org/10.1007/978-981-10-1920-3_21. Published online.
[82] M.E. Cortina, L.J. Melli, M. Roberti, et al., Electrochemical magnetic microbeads-based biosensor for point-of-care serodiagnosis of infectious diseases, Biosens. Bioelectron. 80 (2016) 24–33. Available from: https://doi.org/10.1016/j.bios.2016.01.021.
[83] R. Chand, S. Ramalingam, S. Neethirajan, A 2D Transition-Metal Dichalcogenide MoS2 based Novel Nanocomposite and Nanocarrier for Multiplex miRNA Detection. Published online 2018:8217–8225. Available from: https://doi.org/10.1039/c8nr00697k.
[84] J. Min, M. Nothing, B. Coble, et al. Integrated biosensor for rapid and point-of-care sepsis diagnosis, ACS Nano 12 (2018) 3378–3384. Available from: https://doi.org/10.1021/acsnano.7b08965.
[85] R. Sharma, H. Krishna, K. Raghavarao, Metal ion-enhanced quantification of chlor-amphenicol in milk using imipramine hydrochloride as Diazo-coupling agent, Food Anal. Methods 13 (2020) 2321–2329. Available from: https://doi.org/10.1007/s12161-020-01837-w/Published.
[86] K.S. Abhijith, R. Sharma, R. Ranjan, M.S. Thakur, Facile synthesis of gold–silver alloy nanoparticles for application in metal enhanced bioluminescence, Photochem. Photobiol. Sci. 13 (7) (2014) 986–991. Available from: https://doi.org/10.1039/C4PP00046C.
[87] R. Sharma, K.V. Ragavan, K.S. Abhijith, A. Akanksha, M.S. Thakur, Synergistic catalysis by gold nanoparticles and metal ions for enhanced chemiluminescence, RSC Adv. 5 (40) (2015) 31434–31438. Available from: https://doi.org/10.1039/C5RA01078K.
[88] Y. Chen, J. Sun, Y. Xianyu, et al. Flow Test for Multiplex and Ultrasensitive Detection. 2016;(Cl):15205–15212. Available from: https://doi.org/10.1039/c6nr04017a.
[89] Y. Chen, Y. Xianyu, J. Wu, et al. Double-Enzymes-Mediated Bioluminescent Sensor for Quantitative and Ultrasensitive Point-of-Care Testing. Published online 2017. Available from: https://doi.org/10.1021/acs.analchem.7b00239.
[90] R.M. Dirkzwager, S. Liang, J.A. Tanner, Development of Aptamer-Based Point-of-Care Diagnostic Devices for Malaria Using Three-Dimensional Printing Rapid Prototyping. Published online 2016:1–7. Available from: https://doi.org/10.1021/acssensors.5b00175.
[91] R. Sharma, U.S. Akshath, P. Bhatt, K. Raghavarao, Fluorescent aptaswitch for chloramphenicol detection – quantification enabled by immobilization of aptamer,

Sens. Actuators B Chem. 290 (2019) 110–117. Available from: https://doi.org/10.1016/j.snb.2019.03.093.
[92] L.A. Fraser, A.B. Kinghorn, R.M. Dirkzwager, et al., A portable micro fluidic Aptamer-Tethered Enzyme Capture (APTEC) biosensor for malaria diagnosis, Biosens. Bioelectron. 100 (September 2017) (2018) 591–596. Available from: https://doi.org/10.1016/j.bios.2017.10.001.
[93] S. Knowlton, A. Joshi, P. Syrrist, Lab on a Chip. Published online 2017:2839–2851. Available from: https://doi.org/10.1039/c7lc00706j.
[94] M. Srisa-art, K.E. Boehle, B.J. Geiss, C.S. Henry, Highly Sensitive Detection of Salmonella typhimurium Using a Colorimetric Paper-Based Analytical Device Coupled with Immunomagnetic Separation. Published online 2018. Available from: https://doi.org/10.1021/acs.analchem.7b04628.
[95] L. Zheng, G. Cai, S. Wang, M. Liao, Y. Li, J. Lin, A micro fluidic colorimetric biosensor for rapid detection of *Escherichia coli* O157: H7 using gold nanoparticle aggregation and smart phone imaging, Biosens. Bioelectron. 124–125 (October 2018) (2019) 143–149. Available from: https://doi.org/10.1016/j.bios.2018.10.006.
[96] S. Alhogail, G.A.R.Y. Suaifan, F.J. Bikker, et al. Rapid Colorimetric Detection of *Pseudomonas aeruginosa* in Clinical Isolates Using a Magnetic Nanoparticle Biosensor. Published online 2019:21684-21688. Available from: https://doi.org/10.1021/acsomega.9b02080.
[97] K. Chan, S.C. Weaver, W. Pui-yan, S. Lie, E. Wang, G. Mathilde, Rapid, affordable and portable medium-throughput molecular device for Zika virus, Nat. Publ. Gr. (December)(2016) 1–12. Available from: https://doi.org/10.1038/srep38223.
[98] D.J. Shin, A.Y. Trick, Y. Hsieh, D.L. Thomas, T. Wang, Sample-to-answer droplet magnetofluidic platform for point- of-care hepatitis C viral load quantitation, Sci. Rep. (May)(2018) 1–12. Available from: https://doi.org/10.1038/s41598-018-28124-3.
[99] L. Li, B. Miao, Z. Li, Z. Sun, N. Peng, Sample-to-Answer Hepatitis B Virus DNA Detection from Whole Blood on a Centrifugal Micro Fluidic Platform with Double Rotation Axes. Published online 2019. Available from: https://doi.org/10.1021/acssensors.9b01270.
[100] G. Choi, T. Prince, J. Miao, L. Cui, W. Guan, Sample-to-answer palm-sized nucleic acid testing device towards low-cost malaria mass screening, Biosens. Bioelectron. 115 (May) (2018) 83–90. Available from: https://doi.org/10.1016/j.bios.2018.05.019.
[101] Y. Ma, K. Li, Y. Chen, et al., Lab. a Chip 19 (22) (2019). Available from: https://doi.org/10.1039/c9lc00797k.
[102] Y. Ma, Y. Chen, G. Lee, An integrated self-driven micro fl uidic device for rapid detection of the influenza A (H1N1) virus by reverse transcription loop-mediated isothermal amplification, Sens. Actuators B: Chem. 296(June) (2019). Available from: https://doi.org/10.1016/j.snb.2019.126647.
[103] E. Terreno, D.D. Castelli, A. Viale, S. Aime, Challenges for molecular magnetic resonance imaging, Chem. Rev. 110 (5) (2010) 3019–3042. Available from: https://doi.org/10.1021/cr100025t.
[104] J.M. Chan, P.M. Valencia, L. Zhang, R. Langer, O.C. Farokhzad, Polymeric nanoparticles for drug delivery, Methods Mol. Biol. 624 (2010) 163–175. Available from: https://doi.org/10.1007/978-1-60761-609-2_11.
[105] L. Jin, T. Li, B. Wu, et al., Rapid detection of Salmonella in milk by nuclear magnetic resonance based on membrane filtration superparamagnetic nanobiosensor, Food Control. (2020) 110. Available from: https://doi.org/10.1016/j.foodcont.2019.107011.

[106] L. Jin, T. Li, T. Yang, et al., NMR rapid detection of Salmonella in milk based on ultra-small iron oxide nanobiosensor, Int. Dairy. J. 110 (2020) 104807. Available from: https://doi.org/10.1016/j.idairyj.2020.104807.

[107] D. Zou, L. Jin, B. Wu, et al., Rapid detection of Salmonella in milk by biofunctionalised magnetic nanoparticle cluster sensor based on nuclear magnetic resonance, Int. Dairy. J. 91 (2019) 82–88. Available from: https://doi.org/10.1016/j.idairyj.2018.11.011.

[108] Y. Luo, E.C. Alocilja, Portable nuclear magnetic resonance biosensor and assay for a highly sensitive and rapid detection of foodborne bacteria in complex matrices, J. Biol. Eng. 11 (1) (2017) 1–8. Available from: https://doi.org/10.1186/s13036-017-0053-8.

[109] Y. Li, P. Ma, Q. Tao, et al., Magnetic graphene quantum dots facilitate closed-tube one-step detection of SARS-CoV-2 with ultra-low field NMR relaxometry, Sens. Actuators, B Chem. 337 (January) (2021) 129786. Available from: https://doi.org/10.1016/j.snb.2021.129786.

[110] V.I. Chizhik, M.S. Tagirov, Applied magnetic resonance under conditions of a pandemic, Appl. Magn. Reson. 51 (12) (2020) 1621–1624. Available from: https://doi.org/10.1007/s00723-020-01290-2.

[111] K. Wu, M.C.J. Cheeran, J.P. Wang, et al., Magnetic-nanosensor-based virus and pathogen detection strategies before and during covid-19, ACS Appl. Nano Mater. 3 (10) (2020) 9560–9580. Available from: https://doi.org/10.1021/acsanm.0c02048.

[112] P.P. Liu, K. Skucha, Y. Duan, et al., Magnetic relaxation detector for microbead labels, IEEE J. Solid-State Circuits 47 (4) (2012) 1056–1064. Available from: https://doi.org/10.1109/JSSC.2012.2185339.

[113] J.L. Pool, Magnetic resonance imaging, Biomed. Instrum. Technol. 36 (5) (2002) 341–346. Available from: https://doi.org/10.2345/0899-8205(2002)36[341:TFOMRI]2.0.CO;2.

[114] S.C. McBain, H.H.P. Yiu, J. Dobson, Magnetic nanoparticles for gene and drug delivery, Int. J. Nanomed. 3 (2) (2008) 169–180. Available from: https://doi.org/10.2147/IJN.S1608.

[115] M.A. Khan, J. Sun, B. Li, A. Przybysz, J. Kosel, Magnetic sensors—a review and recent technologies, Eng. Res. Express. 3 (2) (2021) 022005. Available from: https://doi.org/10.1088/2631-8695/ac0838.

[116] Y. Chen, Y. Xianyu, Y. Wang, et al., One-step detection of pathogens and viruses: Combining magnetic relaxation switching and magnetic separation, ACS Nano 9 (3) (2015) 3184–3191. Available from: https://doi.org/10.1021/acsnano.5b00240.

[117] Y. Li, L. Wu, Z. Wang, K. Tu, L. Pan, Y. Chen, A magnetic relaxation DNA biosensor for rapid detection of *Listeria monocytogenes* using phosphatase-mediated Mn(VII)/Mn(II) conversion, Food Control. 125 (January) (2021) 107959. Available from: https://doi.org/10.1016/j.foodcont.2021.107959.

[118] K. Kim, D.A. Hall, C. Yao, et al., Magnetoresistive Biosensors with On-Chip Pulsed Excitation and Magnetic Correlated Double Sampling. 2018;(July):1–10. Available from: https://doi.org/10.1038/s41598-018-34720-0.

[119] C.C. Lee, Y.S. Yen, C.H. Lai, Alignment-free sensing module for absolute and incremental lines in linear positioning system based on tunneling-magnetoresistance sensors, Sensors 21 (12) (2021). Available from: https://doi.org/10.3390/s21124137.

[120] D. Su, K. Wu, R. Saha, C. Peng, J.P. Wang, Advances in magnetoresistive biosensors, Micromachines 11 (1) (2020). Available from: https://doi.org/10.3390/mi11010034.

[121] S. Adem, S. Jain, M. Sveiven, X. Zhou, A.J. O'Donoghue, D.A. Hall, Giant magnetoresistive biosensors for real-time quantitative detection of protease activity, Sci.

Rep. 10 (1) (2020) 1−10. Available from: https://doi.org/10.1038/s41598-020-62910-2.

[122] S. Gupta, P. Bhatter, V. Kakkar, Point-of-care detection of tuberculosis using magnetoresistive biosensing chip, Tuberculosis 127 (June 2020) (2021) 102055. Available from: https://doi.org/10.1016/j.tube.2021.102055.

[123] Y. Li, S. Zuo, V. Nabaei, et al. Tunnel Magnetoresistance Sensor for Point-of-Care and Rapid Label-free Malaria Diagnosis. (c):2−3.

[124] A.P. Cunha, R. Henriques, S. Cardoso, P.P. Freitas, C.M. Carvalho, Rapid and multiplex detection of nosocomial pathogens on a phage-based magnetoresistive lab-on-chip platform, Biotechnol. Bioeng. (2021). Available from: https://doi.org/10.1002/bit.27841. Published online.

[125] T. Banerjee, S. Sulthana, T. Shelby, et al., HHS Public Access. 2019;2 (10):667−673. Available from: https://doi.org/10.1021/acsinfecdis.6b00108. Multiparametric.

[126] B. Tian, R.S. Bejhed, P. Svedlindh, M. Strömberg, Blu-ray optomagnetic measurement based competitive immunoassay for Salmonella detection, Biosens. Bioelectron. 77 (2016) 32−39. Available from: https://doi.org/10.1016/j.bios.2015.08.070.

[127] K. Cihalova, D. Hegerova, A. Maria, et al., Antibody-free detection of infectious bacteria using quantum dots-based barcode assay, J. Pharm. Biomed. Anal. 134 (2017) 325−332. Available from: https://doi.org/10.1016/j.jpba.2016.10.025.

[128] K. Petkovic, G. Metcalfe, H. Chen, et al., Rapid detection of Hendra virus antibodies: an integrated device with nanoparticle assay and, Lab Chip (2017) 169−177. Available from: https://doi.org/10.1039/c6lc01263a.

[129] K. Du, H. Cai, M. Park, et al., Multiplexed efficient on-chip sample preparation and sensitive amplification-free detection of Ebola virus, Biosens. Bioelectron.91 (December 2016) (2017) 489−496. Available from: https://doi.org/10.1016/j.bios.2016.12.071.

[130] C. Liu, N. Xu e, H. Cai, J. Sun, Z. Qi, P. Zhao, Nanoparticles Enhanced Self-Driven Microfludic Biosensor, 1−9.

[131] N. Singh, G.J.S. Jenkins, R. Asadi, S.H. Doak, Potential toxicity of superparamagnetic iron oxide nanoparticles (SPION), Nano Rev. (2010) 1. Available from: https://doi.org/10.3402/nano.v1i0.5358.

[132] H. Markides, M. Rotherham, A.J. El Haj, Biocompatibility and toxicity of magnetic nanoparticles in regenerative medicine. Li X, (ed.), J. Nanomater. 2012 (2012) 614094. Available from: https://doi.org/10.1155/2012/614094.

[133] Markets and Markets, Biosensors Market with COVID-19 Impact by Type, Product (Wearable, Non-Wearable), Technology, Application (POC, Home Diagnostics, Research Lab, Environmental Monitoring, Food & Beverages, Biodefense) and Region - Global Forecast to 2026; 2021. 5314267.

CHAPTER 4

Advancements in electrochemical biosensors development for infectious disease diagnosis

Dinakaran Thirumalai and Seung-Cheol Chang
Department of Cogno-Mechatronics Engineering, College of Nanoscience and Nanotechnology, Pusan National University, Busan, Republic of Korea

4.1 Overview

The World Health Organization warned in its 2007 report that infectious diseases are occurring at an unprecedented rate. Since the 1970s, nearly 40 infectious diseases have been identified, including severe acute respiratory syndrome (SARS), Middle East respiratory syndrome coronavirus (MERS-CoV), Ebola, Chikungunya, avian influenza, swine flu, Zika virus, and more recently, the novel coronavirus (SARS-CoV-2) [1].

Biosensors have emerged over the past 25 years to complement the polymerase chain reaction (PCR) and enzyme-linked immunosorbent assay (ELISA) for infectious disease diagnosis. The biosensor relies on the biorecognition component and the direct integration of the sensor component, providing an additional PCR and ELISA platform for infectious disease identification and quantification. According to the International Union of Pure and Applied Chemistry (IUPAC), a biosensor must include an element of biorecognition in direct spatial contact with a transduction element. In addition, biosensors must provide quantitative or semi-quantitative analytical information and measurements without additional processing steps or reagents. Although the biosensor must be a standalone integrated device, measurement methods can vary from droplet format to continuous flow format requiring a suitable fluid handling system. Biosensors detect infectious disease-causing pathogens sensitively and selectively in a variety of real-time environments without the need for sample preparation. For example, biosensors have made it possible to detect a wide variety of pathogens in a variety of matrices and media, including food, biological fluids, and object surfaces. In addition to the protocol without sample preparation, the biosensor is compatible with the label-free protocol.

Biosensors for Emerging and Re-emerging Infectious Diseases
DOI: https://doi.org/10.1016/B978-0-323-88464-8.00001-4

Although different types of transducers have been investigated for pathogen biosensitivity, including mechanical and optical sensors such as cantilever biosensors or surface plasmon resonance biosensors, electrochemical biosensors have been widely used to detect pathogens. Electrochemical biosensors for pathogen detection use conductive and semiconducting materials as transformers, commonly referred to as electrodes. The chemical energy associated with the binding between the target pathogen and the biometric element immobilized on the electrode is converted into electrical energy using an electrochemical method consisting of an electrode and an electrolyte containing the pathogen. Today, electrochemical biosensors provide pathogen detection in a variety of matrices without sample preparation and with on-site detection of surface pathogens, rapid detection of pathogens through a low-cost platform, detection of multiple pathogens in functional matrices, wireless pathogen detection and data collection formats. As a result, electrochemical biosensors have been extensively investigated for food and water safety, medical diagnostics, environmental monitoring, and pathogen detection for biological threats [2].

4.2 Electrochemical biosensors for infectious disease diagnosis

Currently, electrochemical biosensors are being studied for possible point-of-care testing (POCT) of infectious diseases. This is one of the most studied and advanced methods due to its potential for miniaturization, portability, and cost-effectiveness [3]. Electrochemical biosensors are classified based on measurable parameters such as potential (potentiometric biosensors), current (amperometric biosensor), voltage difference (voltammetric biosensor), and impedance (impedimetric biosensor).

4.2.1 Potentiometric biosensors

The working principle of the potentiometric biosensor is that the potential difference between the working electrode and the reference electrode directly depends on the concentration of the analyte being studied at zero current (Fig. 4.1). Ion-selective electrodes, gas-sensitive electrodes, and field-effect transistors (FETs) are commonly used in transducers of potentiometric biosensors, which are selected depending on the type of material being inspected. Since the preparation of biosensors for POCT devices is the main goal of researchers, potentiometric biosensors based on FETs

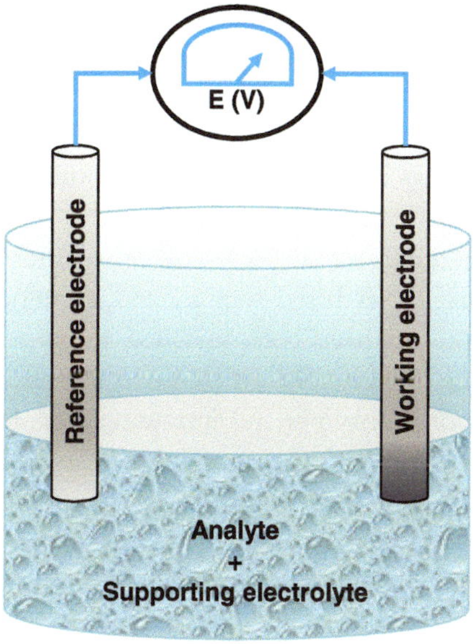

Figure 4.1 Schematic representation of general potentiometric biosensor.

have recently received a lot of attention, which may be due to the possibility of built-in miniaturization of FETs [4].

Recently, FET biosensors for infectious diseases are being developed. The most interesting and important in this scenario is the FET-based immune sensor to detect the virus that causes coronavirus disease (COVID-19). Seo et al. developed a SARS-CoV-2 viral detection platform with a graphene-based FET biosensing device functionalized with the anti-SARS CoV-2 spike antibody [5]. Shao et al. developed a SARS-CoV-2 antigen (Ag) FET nano-biosensor by employing a high-purity semiconducting single-walled carbon nanotube (SWCNT) functionalized with a specific antibody to access the presence of two SARS-CoV-2 structural proteins: spike protein (S antigen, SAg) and nucleocapsid protein (N antigen, NAg) [6]. Table 4.1 provides a detailed summary of the literature discussed in the potentiometric biosensors used for the detection of infectious diseases.

4.2.2 Amperometric biosensors

The literature clearly shows that the era of electrochemical biosensors began with current and potentiometric transducer platforms. The amperometric glucose biosensor itself was the starting point for developing a cost-effective, reliable, fast, and convenient POCT device and has since revolutionized medical diagnostics. The electro-oxidation/reduction of reaction time of the electroactive particles at the optimum potential is monitored by amperometry (Fig. 4.2). The generated current is proportional to the concentration of the electroactive particles, and the excellent detection selectivity provided by this potentiostatic method has been widely used in chemical sensor development [10].

In recent years, there has been a lot of research on amperometric enzyme-based sensors, DNA-based sensors, and immunosensors for clinical analysis. Several biosensors have been included above to detect biomarkers for infectious diseases. Interestingly, DNA-based sensors are currently at the forefront of conventional enzymes and immunosensors for infectious disease detection, which may be due to the surprising stability found in nucleic acids. Recently, a reagent-free electrochemical biosensor platform for detecting SARS-CoV-2 S protein within 5 min in saliva samples has been reported [11]. They used a DNA linker and ferrocene redox reagent to move specific antibodies to SARS-CoV-2 to create a sensor chip. Two opposing forces acting on the sensor probe, hydrodynamic and induced electric fields were responsible for the diagnostic process. Amperometry has shown that S protein levels are up to 1.0 pg/mL in clinical trials. Teengam et al. described the label-free strategy [12], screen-printed graphene electrodes were modified with gold and β-cyclodextrin (β-CD) nanoparticles onto anti-hepatitis B virus (HBV) monoclonal antibodies. The HBV concentration was determined by measuring the current $[Fe(CN)_6]^{3-/4-}$. Electrochemical measurements were

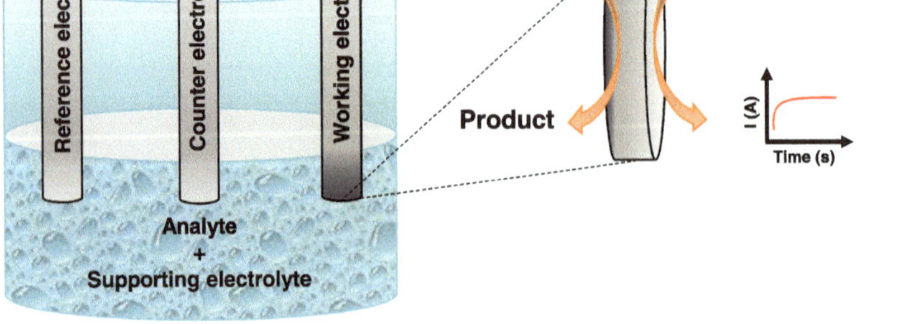

Figure 4.2 Schematic representation of general amperometric biosensor.

Table 4.1 A summary of potentiometric biosensors used for the detection of infectious diseases.

Transducer	Target	Biosensor format	Linear range	Limit of detection	References
Potentiometry	Influenza virus	Glycan based FET biosensor	$10^{0.5}$–$10^{8.5}$ $TCID_{50}$/mL	$10^{0.5}$ $TCID_{50}$/mL	Hideshima et al. [7]
Potentiometry	SARS-CoV-2	Immunosensor	–	Culture medium-1.6 PFU/mL and Clinical samples-242 copies/mL	Seo et al. [5]
Potentiometry	Plasmodium falciparum glutamate dehydrogenase (Malaria biomarker)	Aptamer based FET biosensor	10×10^{-15}–10×10^{-9} M	Spiked buffer—16.7×10^{-15} M and serum samples—48.6×10^{-15} M	Singh et al. [8]
Potentiometry	Avian influenza virus	Aptamer based FET biosensor	10×10^{-15}–10×10^{-9} M	5.9×10^{-15} M	Kwon et al., [9]
Potentiometry	SARS-CoV-2	Immunosensor	5.5 fg/mL to 5.5 pg/mL for SAg and 16 fg/mL to 16 pg/mL for NAg	0.55 fg/mL for SAg and 0.016 fg/mL for NAg	Shao et al. [6]

FET, Field-effect transistor.

performed using an NFC-based short form potentiostat. Measurements can be made by simply placing a smartphone on the NFC antenna, and data can be processed and transmitted with the developed application. Here we have summarized the recently used amperometric biosensors (see Table 4.2) for the detection of infectious diseases.

4.2.3 Voltammetric biosensors

Voltammetric biosensors exist within the scientific community because they can provide information about biological systems by converting them into electrical signals. These belong to an electroanalytical detection method; the generated current is monitored between the working electrode and the counter electrode by changing the potential between the working electrode and the reference electrode. In voltammetry, heterogeneous electron transfer occurs at the electrode-electrolyte interface, where electroactive particles arrive by mass transfer in the bulk solution. The measured current is the result of an electroactive process for oxidation or reduction of particles occurring on the surface of the working electrode.

Table 4.2 A summary of amperometric biosensors used for the detection of infectious diseases.

Transducer	Target	Biosensor format	Linear range	Limit of detection	References
Amperometric	Ebola virus	DNA sensor	–	–	Ciftci et al. [13]
Amperometric	Influenza A virus	Immunosensor	$1–10^4$ PFU/mL	0.5 PFU/mL	Singh et al. [14]
Amperometric	Avian Influenza A virus	Immunosensor	–	16 HAU	Sayhi et al. [15]
Amperometric	Influenza A virus	Glycan based	–	–	Cui et al. [16]
Amperometric	Dengue virus	DNA sensor	–	17 nM	Odeh et al. [17]
Amperometric	SARS-CoV-2	DNA sensor	–	1.0 pg/mL	Yousefi et al. [11]
Amperometric	Hepatitis B virus	Immunosensor	$10-200$ μg/mL	0.17 μg/mL	Teengam et al. [12]

Voltammetric techniques are classified into cyclic voltammetry (CV), square wave voltammetry (SWV), differential pulse voltammetry (DPV), and anodic stripping voltammetry (ASV), according to the waveform of the applied potential. In a voltammetric biosensor, the biological recognition process between the recognition layer and the analyte causes a current response through the redox process or labeling of the analyte (Fig. 4.3). The actual reaction is usually observed as a peak depending on the concentration of the electroactive particles. To overcome certain limitations of bare electrodes, the researchers focused on developing a modified coating on the electrode surface. These modifications include a wide range of materials such as nanomaterials, polymer films, metal composites, etc., and these transducers are commonly referred to as chemically modified electrodes. While advantages such as higher sensitivity and shorter analysis times make voltammetric biosensors better, there are problems associated with insufficient selectivity (compared to conventional methods) and requirements aimed at redox capabilities in many possible complex systems than other electroanalytical techniques [4].

Identification of infectious pathogens has received a lot of attention because it is important to public health. A variety of pathogens were detected with the voltammetric biosensor method, and a simple and complex sensor platform was used, including specific nanostructures, carbon nanomaterials, and polymers. In addition to layers of biomolecular modifications such as aptamers, DNA, proteins, or antibodies. Torrente-Rodríguez et al. [18] demonstrated a

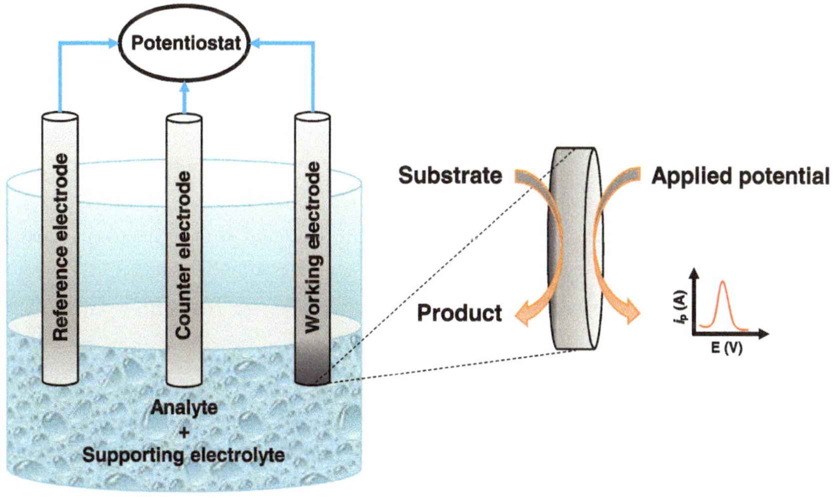

Figure 4.3 Schematic representation of general voltammetric biosensor.

wireless electrochemical POCT multiplex for SARS-CoV-2 detection called RapidPlex. The sensor can detect CRP with viral antigen antibodies, nucleocapsid proteins, immunoglobulin M (IgM) and G (IgG) antibodies, and laser-induced graphene electrodes. The pool of biomarkers allowed us to identify viral infections, immune responses, and disease severity. The electrochemical sensor was connected to a microcontroller-based potentiostat with a built-in Bluetooth module for easy data transfer to a smartphone. The sensor was placed on clinical blood and saliva samples that were positive and negative for COVID-19. This groundbreaking work is showing promising results in the field of high-volume POCT testing for COVID-19 diagnosis, infection control, and disease monitoring. Srisomwat et al., [19] developed a 3D paper-based microfluidic origami device for label-free HBV detection. The origami design enables multi-step operation in a single device with minimal sample handling, preventing contamination and reducing exposure to pathogens. Interestingly, the authors used the inherent polyol chemistry to immobilize a highly specific and selective pyrrolidinyl peptide nucleic acid (acpcPNA) probe for HBV DNA, avoiding the tedious modification steps required by the classical method. With the addition of the $[Fe(CN)_6]^{3-/4-}$ probe, the decrease in the electrochemical signal obtained from DPV is related to HBV. Pop-up DNA probes have been successfully applied to detect HBV DNA isolated from plasmid structures, and more recent voltammetric biosensing platforms for virus detection are presented in Table 4.3.

4.2.4 Impedimetric biosensors

Electrochemical impedance spectroscopy (EIS) can be used effectively for label-free detection, making it a powerful tool in biosensor applications. The emergence of biological recognition processes always entails some changes in the various physical and chemical properties of the electrode-electrolyte interface, and the EIS method uses changes in charge transfer resistance (Rct) or the ability of the interface to identify biochemical property changes that occur on the sensor surface. A general schematic for the fabrication of an impedimetric biosensor is shown in Fig. 4.4.

In general, impedance can be viewed as the resistance to the flow of current in an electrical circuit. The information obtained by measuring impedance consists mainly of resistive and capacitive parts with Faraday and non-Faraday sensors. In a Faraday biosensor, an electrode containing a bioreceptor is immersed in an electrochemical redox probe solution and the

Table 4.3 A summary of voltammetric biosensors used for the detection of infectious diseases.

Transducer	Target	Biosensor format	Linear range	Limit of detection	References
CV, DPV	Dengue virus	DNA sensor	1×10^{-6}–100×10^{-6} M	4.3×10^{-5} M	Singhal et al. [20]
DPV	Ebola virus	DNA sensor	–	4.7×10^{-9} M	Ilkhani and Farhad [21]
DPV	Influenza A and B	DNA sensor	–	5×10^{-9} M for Influenza A and 21×10^{-9} M for Influenza B	Subak and Ozkan-Ariksoysal [22]
DPV	Dengue virus	DNA sensor	1×10^{-18}–1×10^{-6} M	120×10^{-21} M	Tripathy et al. [23]
SWV	Zika virus	DNA sensor	84×10^{-12}–1.4×10^{-9} M	25.4×10^{-12} M	da Fonseca Alves et al. [24]
CV	Hantavirus	Immunosensor	5.0 ng/mL to 1.0 μg/mL	0.14 ng/mL	Martins et al. [25]
DPV	Norovirus	DNA sensor	100×10^{-12}–3.5×10^{-9} M	100×10^{-12} M	Chand and Neethirajan [26]
CV	Avian Influenza virus	Immunosensor	0.0025–0.16 HAU/6 μL	0.0022 HAU/6 μL	Zhang et al. [27]
SWV	MERS-CoV	Immunosensor	0.001–100 ng/mL	1.04 pg/mL	Layqah and Eissa [28]
CV	Hanta virus	Immunosensor	30–240 μg/mL	22 μg/mL	Martins et al. [29]
SWV	SARS-CoV-2	Immunosensor	1–1000 ng/mL	0.8 pg/mL	Eissa and Zourob [30]
SWV	SARS-CoV-2	Immunosensor	1–1000 ng/mL	1 ng/mL	Yakoh et al. [31]

(Continued)

Table 4.3 (Continued)

Transducer	Target	Biosensor format	Linear range	Limit of detection	References
DPV	SARS-CoV-2	Immunosensor	–	–	Torrente-Rodríguez et al. [18]
DPV	Hepatitis B virus	DNA sensor	0.05–100 nM	1.45 pM	Srisomwat et al. [19]
ASV	Hepatitis B virus	DNA sensor	10 pM to 2.0 μM	7.23 pM	Srisomwat et al. [32]
DPV	SARS-CoV-2	DNA sensor	10^{-17}–10^{-12} M	200 copies/mL	Zhao et al. [33]

CV, Cyclic voltammetry; DPV, differential pulse voltammetry; SWV, square wave voltammetry.

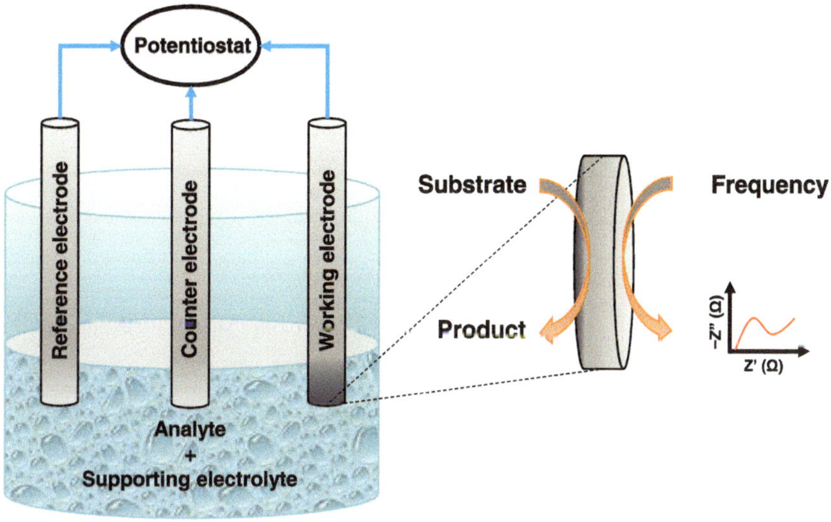

Figure 4.4 Schematic representation of general impedimetric biosensor.

resistance of the electrode to electron transfer is measured, but for a non-Faraday sensor, the redox probe is not required because it measures the capacity of the interface. (i.e., accumulate system charge when voltage is applied). Impedance can be represented by the Nyquist plot (for Faraday processes) and Bode plot (for capacitive or non-Faraday processes). This section covers impedimetric biosensors based on the Faraday process.

Impedimetric biosensors for infectious diseases are currently undergoing active research. As the bioreceptor immobilized on the electrode surface captures the target analyte, there is a difference in the Rct value of the redox probe that can be related to the concentration of the target analyte [4].

A summary of impedimetric biosensors used for the detection of infectious diseases has been shown in Table 4.4. For example, Rashed et al. developed a label-free impedimetric biosensor to detect SARS-CoV-2 antibodies in serum samples within 5 min. They prepared a solution from the well and then extracted it from the receptor-binding domain (RBD) of SARS-CoV-2. Capacitive measurements were performed according to the possible interactions between the SARS-CoV-2 and SARS-CoV-2 RBD antibodies. It can also detect more than 1000 samples at a time [40]. However, the biosensor's limit of detection (LOD) has not yet been determined and additional sampling is required.

Table 4.4 A summary of impedimetric biosensors used for the detection of infectious diseases.

Transducer	Target	Biosensor format	Linear range	Limit of detection	References
Impedimetric	SARS-CoV-2	DNA sensor	1.0×10^{-12}–1.0×10^{-6} M	0.82 pmol/L	Steinmetz et al. [34]
Impedimetric	Dengue virus	Immunosensor	10^{-12}–10^{-5} g/mL	10^{-12} g/mL	Palomar et al. [35]
Impedimetric	Influenza A virus	Immunosensor	1×10^{-8}–1×10^{-1} U/mL	1×10^{-8} U/mL	Anik et al. [36]
Impedimetric	Dengue virus	Immunosensor	1–2×10^{3} PFU/mL	0.12 pfu/mL	Navakul et al. [37]
Impedimetric	Norovirus	Immunosensor	0–10^{5} copies/mL	1.7 copies/mL	Baek et al. [38]
Impedimetric	Zika virus	DNA sensor	54–340×10^{-9} M	25×10^{-9} M	Faria and Zucolotto [39]
Impedimetric	SARS-CoV-2	Immunosensor	–	–	Rashed et al. [40]

4.3 Summary

In the current SARS-CoV-2 pandemic scenario, millions of people around the world are at risk for severe respiratory disease. The most in-demand tools are for quick, economical, and early diagnosis. Electrochemical biosensors are very specific, sensitive, and precise. There are also promising platforms that can be used for biosensors due to their availability, cost-effectiveness, and miniaturization. However, integration with other devices, such as microfluidics, can further improve sensitivity and target multiple analytes at the same time. As a result, time is reduced, and test costs are optimized Electrochemical biosensors can detect trace assay levels and detect femtomolar/attomolar concentrations within seconds, even in clinical microliter samples. The production and use of biosensors is much easier compared to other techniques and does not require expensive and complex instruments and reagents. Sometimes these devices can be reused by rinsing with acid or other reagents without interfering with their function. In current technological developments, electrochemical platforms could open new possibilities for the real-time detection of viral infections.

Acknowledgments

Dinakaran Thirumalai expresses his thanks to the educational research center for personalized healthcare based on Cogno-Mechatronics at Pusan national university and for its kind financial support. Dinakaran Thirumalai and Seung-Cheol Chang are grateful for the financial support of the project titled "Omics based on fishery disease control, technology development, and industrialization (20150242)," funded by the Ministry of Oceans and Fisheries, Korea.

References

[1] J. Stephenson, Coronavirus outbreak—an evolving global health emergency, JAMA Health Forum 1 (2) (2020) e200114.

[2] E. Cesewski, B.N. Johnson, Electrochemical biosensors for pathogen detection, Biosens. Bioelectron. 159 (2020) 112214.

[3] O. Pashchenko, T. Shelby, T. Banerjee, S. Santra, A comparison of optical, electrochemical, magnetic, and colorimetric point-of-care biosensors for infectious disease diagnosis, ACS Infect. Dis. 4 (8) (2018) 1162–1178.

[4] S. Menon, M.R. Mathew, S. Sam, K. Keerthi, K.G. Kumar, Recent advances and challenges in electrochemical biosensors for emerging and re-emerging infectious diseases, J. Electroanal. Chem. 878 (2020) 114596.

[5] G. Seo, G. Lee, M.J. Kim, S.-H. Baek, M. Choi, K.B. Ku, et al., Rapid detection of COVID-19 causative virus (SARS-CoV-2) in human nasopharyngeal swab specimens using field-effect transistor-based biosensor, ACS Nano 14 (4) (2020) 5135–5142.

[6] W. Shao, M.R. Shurin, S.E. Wheeler, X. He, A. Star, Rapid detection of SARS-CoV-2 antigens using high-purity semiconducting single-walled carbon nanotube-based field-effect transistors, ACS Appl. Mater. Interfaces (2021).
[7] S. Hideshima, H. Hayashi, H. Hinou, S. Nambuya, S. Kuroiwa, T. Nakanishi, et al., Glycan-immobilized dual-channel field effect transistor biosensor for the rapid identification of pandemic influenza viral particles, Sci. Rep. 9 (1) (2019) 11616.
[8] N.K. Singh, P.D. Thungon, P. Estrela, P. Goswami, Development of an aptamer-based field effect transistor biosensor for quantitative detection of Plasmodium falciparum glutamate dehydrogenase in serum samples, Biosens. Bioelectron. 123 (2019) 30−35.
[9] J. Kwon, Y. Lee, T. Lee, J.-H. Ahn, Aptamer-based field-effect transistor for detection of avian influenza virus in chicken serum, Anal. Chem. 92 (7) (2020) 5524−5531.
[10] S. Borgmann, A. Schulte, S. Neugebauer, W. Schuhmann, Amperometric biosensors, in: C. Richard, D.M.K. Alkire, L. Jacek (Eds.), Advances in Electrochemical Science and Engineering, Wiley-VCH Verlag GmbH & Co. KGaA, Weinheim, Germany, 2011, pp. 1−83.
[11] H. Yousefi, A. Mahmud, D. Chang, J. Das, S. Gomis, J.B. Chen, et al., Detection of SARS-CoV-2 viral particles using direct, reagent-free electrochemical sensing, J. Am. Chem. Soc. 143 (4) (2021) 1722−1727.
[12] P. Teengam, W. Siangproh, S. Tontisirin, A. Jiraseree-amornkun, N. Chuaypen, P. Tangkijvanich, et al., NFC-enabling smartphone-based portable amperometric immunosensor for hepatitis B virus detection, Sens. Actuators, B 326 (2021) 128825.
[13] S. Ciftci, R. Cánovas, F. Neumann, T. Paulraj, M. Nilsson, G.A. Crespo, et al., The sweet detection of rolling circle amplification: glucose-based electrochemical genosensor for the detection of viral nucleic acid, Biosens. Bioelectron. 151 (2020) 112002.
[14] R. Singh, S. Hong, J. Jang, Label-free detection of influenza viruses using a reduced graphene oxide-based electrochemical immunosensor integrated with a microfluidic platform, Sci. Rep. 7 (2017) 42771.
[15] M. Sayhi, O. Ouerghi, K. Belgacem, M. Arbi, Y. Tepeli, A. Ghram, et al., Electrochemical detection of influenza virus H9N2 based on both immunomagnetic extraction and gold catalysis using an immobilization-free screen printed carbon microelectrode, Biosens. Bioelectron. 107 (2018) 170−177.
[16] X. Cui, A. Das, A.N. Dhawane, J. Sweeney, X. Zhang, V. Chivukula, et al., Highly specific and rapid glycan based amperometric detection of influenza viruses, Chem. Sci. 8 (5) (2017) 3628−3634.
[17] A.A. Odeh, Y. Al-Douri, C. Voon, R.M. Ayub, S.C. Gopinath, R.A. Odeh, et al., A needle-like Cu 2 CdSnS 4 alloy nanostructure-based integrated electrochemical biosensor for detecting the DNA of Dengue serotype 2, Microchim. Acta 184 (7) (2017) 2211−2218.
[18] R.M. Torrente-Rodríguez, H. Lukas, J. Tu, J. Min, Y. Yang, C. Xu, et al., SARS-CoV-2 RapidPlex: a graphene-based multiplexed telemedicine platform for rapid and low-cost COVID-19 diagnosis and monitoring, Matter 3 (6) (2020) 1981−1998.
[19] C. Srisomwat, P. Teengam, N. Chuaypen, P. Tangkijvanich, T. Vilaivan, O. Chailapakul, Pop-up paper electrochemical device for label-free hepatitis B virus DNA detection, Sens. Actuators, B 316 (2020) 128077.
[20] C. Singhal, C. Pundir, J. Narang, A genosensor for detection of consensus DNA sequence of Dengue virus using ZnO/Pt-Pd nanocomposites, Biosens. Bioelectron. 97 (2017) 75−82.
[21] H. Ilkhani, S. Farhad, A novel electrochemical DNA biosensor for Ebola virus detection, Anal. Biochem. 557 (2018) 151−155.

[22] H. Subak, D. Ozkan-Ariksoysal, Label-free electrochemical biosensor for the detection of Influenza genes and the solution of guanine-based displaying problem of DNA hybridization, Sens. Actuators, B 263 (2018) 196–207.
[23] S. Tripathy, S.R.K. Vanjari, V. Singh, S. Swaminathan, S.G. Singh, Electrospun manganese (III) oxide nanofiber based electrochemical DNA-nanobiosensor for zeptomolar detection of dengue consensus primer, Biosens. Bioelectron. 90 (2017) 378–387.
[24] R. da Fonseca Alves, D.L. Franco, M.T. Cordeiro, E.M. de Oliveira Junior, R.A.F. Dutra, M.D.P.T. Sotomayor, Novel electrochemical genosensor for Zika virus based on a poly-(3-amino-4-hydroxybenzoic acid)-modified pencil carbon graphite electrode, Sens. Actuators, B 296 (2019) 126681.
[25] G. Martins, J.L. Gogola, F.R. Caetano, C. Kalinke, T.R. Jorge, C.N. Santos, et al., Quick electrochemical immunoassay for hantavirus detection based on biochar platform, Talanta 204 (2019) 163–171.
[26] R. Chand, S. Neethirajan, Microfluidic platform integrated with graphene-gold nano-composite aptasensor for one-step detection of norovirus, Biosens. Bioelectron. 98 (2017) 47–53.
[27] Q. Zhang, L. Li, Z. Qiao, C. Lei, Y. Fu, Q. Xie, et al., Electrochemical conversion of Fe_3O_4 magnetic nanoparticles to electroactive Prussian blue analogues for self-sacrificial label biosensing of avian influenza virus H5N1, Anal. Chem. 89 (22) (2017) 12145–12151.
[28] L.A. Layqah, S. Eissa, An electrochemical immunosensor for the corona virus associated with the Middle East respiratory syndrome using an array of gold nanoparticle-modified carbon electrodes, Microchim. Acta 186 (4) (2019) 224.
[29] G. Martins, J.L. Gogola, L.H. Budni, B.C., Janegitz, L.H. Marcolino-Junior, M.F. Bergamini, 3D-printed electrode as a new platform for electrochemical immunosensors for virus detection, Anal. Chim. Acta 1147 (2021) 30–37.
[30] S. Eissa, M. Zourob, Development of a low-cost cotton-tipped electrochemical immunosensor for the detection of SARS-CoV-2, Anal. Chem. 93 (3) (2021) 1826–1833.
[31] A. Yakoh, U. Pimpitak, S. Rengpipat, N. Hirankarn, O. Chailapakul, S. Chaiyo, Paper-based electrochemical biosensor for diagnosing COVID-19: detection of SARS-CoV-2 antibodies and antigen, Biosens. Bioelectron. (2021) 112912.
[32] C. Srisomwat, A. Yakoh, N. Chuaypen, P. Tangkijvanich, T. Vilaivan, O. Chailapakul, Amplification-free DNA sensor for the one-step detection of the hepatitis B virus using an automated paper-based lateral flow electrochemical device, Anal. Chem. 93 (5) (2021) 2879–2887.
[33] H. Zhao, F. Liu, W. Xie, T.-C. Zhou, J. OuYang, L. Jin, et al., Ultrasensitive supersandwich-type electrochemical sensor for SARS-CoV-2 from the infected COVID-19 patients using a smartphone, Sens. Actuators, B 327 (2021) 128899.
[34] M. Steinmetz, D. Lima, A.G. Viana, S.T. Fujiwara, C.A. Pessôa, R.M. Etto, et al., A sensitive label-free impedimetric DNA biosensor based on silsesquioxane-functionalized gold nanoparticles for Zika Virus detection, Biosens. Bioelectron. 141 (2019) 111351.
[35] Q. Palomar, C. Gondran, R. Marks, S. Cosnier, M. Holzinger, Impedimetric quantification of anti-dengue antibodies using functional carbon nanotube deposits validated with blood plasma assays, Electrochim. Acta 274 (2018) 84–90.
[36] Ü. Anik, Y. Tepeli, M. Sayhi, J. Nsiri, M.F. Diouani, Towards the electrochemical diagnostic of influenza virus: development of a graphene–Au hybrid nanocomposite modified influenza virus biosensor based on neuraminidase activity, Analyst 143 (1) (2018) 150–156.

[37] K. Navakul, C. Warakulwit, P.-t Yenchitsomanus, A. Panya, P.A. Lieberzeit, C. Sangma, A novel method for dengue virus detection and antibody screening using a graphene-polymer based electrochemical biosensor, Nanomed. Nanotechnol. Biol. Med. 13 (2) (2017) 549–557.
[38] S.H. Baek, M.W. Kim, C.Y. Park, C.-S. Choi, S.K. Kailasa, J.P. Park, et al., Development of a rapid and sensitive electrochemical biosensor for detection of human norovirus via novel specific binding peptides, Biosens. Bioelectron. 123 (2019) 223–229.
[39] H.A.M. Faria, V. Zucolotto, Label-free electrochemical DNA biosensor for zika virus identification, Biosens. Bioelectron. 131 (2019) 149–155.
[40] M.Z. Rashed, J.A. Kopechek, M.C. Priddy, K.T. Hamorsky, K.E. Palmer, N. Mittal, et al., Rapid detection of SARS-CoV-2 antibodies using electrochemical impedance-based detector, Biosens. Bioelectron. 171 (2021) 112709.

CHAPTER 5

Plant-based biosensors in disease diagnosis

Remya Mohanraj
Houston Community College, Houston, TX, United States

5.1 Introduction

Biosensors are devices that aid in measuring the concentration of a particular analyte. They are designed to generate a digital electronic signal in proportion to the concentration of a specific biochemical or a set of biochemicals among numerous interfering species. A typical design of a biosensor includes both the biological components and transducers [1]. In 1962, a biosensor was developed to monitor blood gas levels during surgery [2].

Research on biosensors is moving at a very fast pace due to the fact that biosensors are easy, rapid, low-cost, highly sensitive, and contribute to advances in medicine such as in individualized medicine and the precise detection of disease-specific markers [3]. Biosensors are fast and precise as compared to conventional methods that take longer to provide results. They rely on simple instrumentation that does not require highly trained professionals to operate [4]. Their immense potential in medical diagnostics, has lead to the development of biosensor technologies that could potentially outshine conventional in vitro diagnostics for disease detection and health monitoring [5]. Various types of biosensors encompassing enzyme-based, tissue-based, immunosensors, DNA biosensors, thermal and piezoelectric biosensors not only find applications in many fields such as food, medicine, marine sector, etc., but also offer better stability and sensitivity in comparison to traditional methods [6].

A biosensor consists of bioreceptors such as enzymes, cells, aptamers, DNA and antibodies that are able to specifically interact with an analyte that needs to be detected. The next component is a transducer that converts the physical or chemical change from the bio-recognition event into a measurable signal, proportional to the amount of analyte–bioreceptor interactions. The electronics part processes the transduced signal by amplification and conversion of signals from analog into digital format and

Biosensors for Emerging and Re-emerging Infectious Diseases
DOI: https://doi.org/10.1016/B978-0-323-88464-8.00011-7

prepares it for display by the display unit. The display unit displays the signal in a numeric, graphic, tabular or an image form depending on user requirements [7]. A schematic diagram of a biosensor is given below:

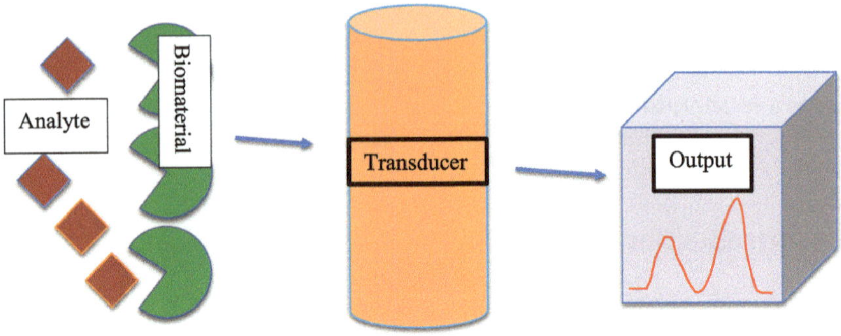

Biosensors are classified by their biorecognition molecule. For example, biosensors that catalyze specific reactions are classified as enzyme-based. Antibody-based biosensors depend on interaction with corresponding antigens; oligonucleotides based ones recognize complementary sequences; respiration, growth, fluorescence or bioluminescence of cells and whole organisms can help detect the presence of certain chemicals; tissues, as enzyme sources; and synthetic bioreceptors [8,9].

Biosensors are also classified according to the transducer used such as electrochemical transducers (amperometric, conductometric and potentiometric); optical transducers [absorbance-based colorimetric (UV–vis)]; emission based (bioluminescence, fluorescence); heat-sensitive transducers, thermistor-based biosensors; and mass-sensitive transducers (piezoelectric like quartz crystal microbalance) [9].

The bioelements (enzyme, antibody, DNA, microorganism, etc.) recognizes the specific analytes and regulates the specificity and sensitivity of the device. Various techniques could be used for immobilization of the bioelements. These techniques include physical adsorption, cross-linking, encapsulation, and entrapment [10]. https://www.news-medical.net/health/What-are-Biosensors.aspx

Biosensors could play a significant role in the detection and monitoring of different types of diseases. Biomolecules such as nucleic acids, proteins, and cells that are associated with diseases could be detected using biosensors [11].

With the COVID-19 pandemic that is currently prevalent world wide, rapid and accurate clinical diagnosis is becoming more important. In addition, it has become apparent that these techniques should be easily available and simple to use. Development of biosensors to suit the above requirements would go a long way in the management of new and emerging diseases.

The key advantages offered by suitable biosensors are their selectivity, rapid response, low time-consumption during analysis, easy handling, and lower price for analysis. However, challenges like instability of biological components and insufficient reproducibility are likely to occur while designing biosensors for commercial applications [12].

This review explores the potential and evidence for using plant-based components as bioelements in biosensors for clinical and other diagnostic purposes.

5.2 Plant-based biosensors

Plants harbor an array of phytochemicals, many of which could specifically interact with analytes. Phytochemical is a broad term that encompasses a wide variety of compounds that occur naturally in plants [13]. Phytochemicals have been classified into six major categories: carbohydrates, lipids, phenolics, terpenoids, alkaloids, and other nitrogen-containing compounds [13–15].

These phytochemicals offer a variety of optical and electrochemical features that enable their development as biosensors for clinical/health diagnostics [16]. In addition, biocatalytic sensors that use plant tissue materials along with electrochemical elements could be an alternative to biosensors based on isolated enzymes. Such plant-based sensors not only possess attractive analytical properties but could also be constructed with simplicity at a low cost with reduced cofactor requirements [17].

This has led to the design of biosensors using various plant-derived materials. It is worth noting that in response to the assortment of environmental changes they encounter, plants generate different types of intracellular and intercellular electrical signals mainly as action and variation potentials. These bioelectrochemical signals resemble nerve impulses [18,19].

Some of the key advantages of plant-based biosensors include high stability and level of activity, long lifetime, high reproducibility, low cost, avoidance of enzyme extraction and purification steps, cofactors for the enzyme present within the same tissue, simplicity in manufacturing and the diversity of plants which provide the opportunity to design a wide range of biosensors [8,12].

Plant-based biosensors are constructed by retaining a thin tissue slice on the surface of the electrode with a dialysis membrane or by entrapping the tissue into a carbon paste matrix. In biosensors based on membranes, the membrane prevents enzyme leakage away from the electrode surface but permits and modifies substrate diffusion into the plant tissue. This prolongs the response time as compared to electrodes based on carbon paste that have to reduce the response time to seconds. A majority of plant tissue-based biosensors offer electrochemical detection. In addition, higher sensitivity and faster response time could be achieved using optical techniques like chemiluminescence or fluorescence [8,12].

A basic design for a plant-based biosensor is presented in Fig. 5.1.

5.3 Potential plant tissues for biosensors

One of the early plant-tissue based biosensors, utilized a tissue slice from a yellow squash and a carbon dioxide electrode and was developed by Kuriyama and Rechnitz [20]. Glutamate decarboxylase from the yellow squash tissue catalyzes the conversion of glutaminic acid to carbon dioxide which in turn, is detected by the carbon dioxide electrode generating a response proportional to the concentration of glutaminic acid in the sample [20]. Following this significant breakthrough in plant-based biosensors, various plant-tissue-based biosensors using tissue materials from different plant parts have been developed [21].

In an attempt to emphasize the use of plant tissues in the development of biosensors, a brief summary of various plant tissues that have been used in biosensors as reviewed by Singh and Verma [9] has been provided below.

Sidwell and Rechnitz [21] used banana tissue as a source of polyphenol oxidase for biosensor development. They placed a thin slice of banana pulp tissue onto the surface of a Clark-type oxygen electrode to measure oxygen consumption required during the conversion of dopamine into

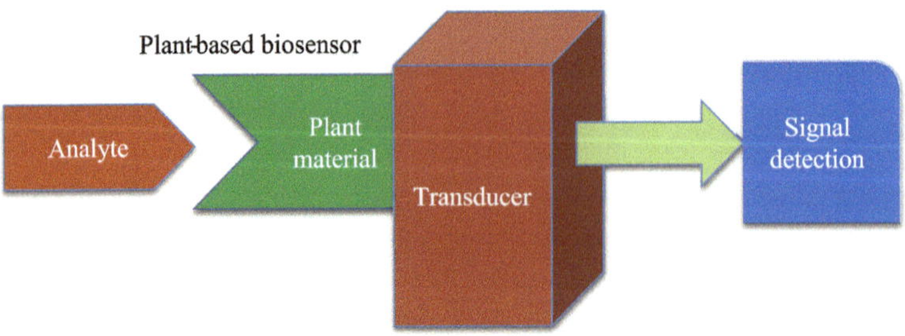

Figure 5.1 Plant-based biosensor.

1,2-benzoquinone. Wang and Lin [22] and Eggins et al. [23] demonstrated the use of banana pulp tissue. Cummings et al. [24] entrapped apple tissues in carbon paste, while Lima et al. [25] entrapped tissue from *Cocus nucifera* in the capillary of a glass reactor and demonstrated it's use as a biocatalyst to analyze catechol. An amperometric biosensor for paracetamol detection was developed by Fatibello-Filho et al. [26] using a mixture of graphite/vaseline with avocado tissue as the source of polyphenol oxidase. Felix et al. [27] showed the potential of biosensors based on palm tree fibers to detect epinephrine through the catalytic conversion of epinephrine to epinephrine quinone by polyphenol oxidase.

Horseradish tissue was immobilized onto the surface of a carbon disk electrode and used as a source of peroxidase by Wang et al. [28]. Vieira and Fatibello-Filho [29] designed a biosensor based on *Ipomea batatas* tissue (as a source of peroxidase) mixture with paraffin/graphite, using the peroxidase for the detection of hydroquinone. The crude extract from *Solanum gilo* was immobilized in chitosan biopolymer by de Oliveira et al. [30] for the development of a biosensor to detect hydroquinone and rutin. Oungpipat et al. [31] entrapped ground tissue of *Asparagus officinalis* along with ferrocene in a carbon paste matrix to develop a glass capillary-based oxide biosensor. Kozan et al. [32] reported the use of coconut fibers for the design of a biosensor for the detection of peroxide. Campanella et al. [33], Campanella et al. [34] used potato tissue as a potential source of acid phosphatase and other glucose oxidase enzymes for the development of a bi-enzymatic biosensor to detect phosphate.

Sezgintürk and Dinçkaya [35] demonstrated that sulfite oxidase from *Malva vulgaris* can be used in the development of a biosensor for the detection of sulfite. The oxygen utilized during the oxidation of sulfur containing amino acids by sulfite oxidase was monitored by an amperometric transducer. Sezgintürk and Dinçkaya [36] immobilized *Spinacia oleracea* tissue, which is a source of oxalate oxidase on a Clark-type oxygen electrode with albumin. Qin et al. [37] used soya bean tissue (source of urease) for the development of a biosensor for urea detection. Here, the soya bean tissue was entrapped in a miniaturized glass column and the flow system was designed to detect urea in urine samples. Botrè et al. [38] used a tissue homogenate from *Cucumis sativus* (a potential source of ascorbic acid) in the development of biosensors to detect reduced glutathione. Horie and Rechnitz [39] used mandarin peel tissue as a source of pectin esterase and alcohol oxidase enzymes and reported the development of a hybrid biosensor for the detection of pectin. The authors had suggested that biosensors based on plant tissues could accept the absorption of substances having a high molecular weight.

Feng et al. [40] developed a biosensor based on a ligand-binding domain (LBD) linked to a fluorescent protein or a transcriptional activator that can be readily detected. LBD is destabilized by mutation. Fusion to a suitable reporter protein makes it stable and it accumulates only in cells containing the target molecule generating a signal. The authors used this method to develop biosensors for the steroids digoxin and progesterone. Naturally-occurring LBDs can be engineered to be conditionally stable [41—43], so that they can be converted into biosensors for target ligands. The authors indicated that designed LBDs could also be used, where natural binding proteins do not exist or lack sufficient specificity or bioorthogonality.

Genetically-encoded biosensors that have the potential to interact with a molecule of interest were originally developed with fusion proteins for the detection of calcium ions. Since then, the technology has lead to the development of a diverse array of biosensors that have enhanced our comprehension of molecules which exert profound influence on plant physiology and development [44].

5.4 Callus culture for identical plant cells

Callus culture refers to the culture of dedifferentiated plant cells under in vitro conditions on media that generally contain relatively high auxin concentrations or a combination of auxin and cytokinin. Where the desired metabolites are present in leaves, in vitro cultures could be established from leaves and used for the extraction of compounds [45]. A variety of useful secondary metabolites could be produced using plant cell and tissue cultures [46].

In vitro cultured tobacco callus tissue, a source of peroxidase, was used to construct an amperometric biosensor for the detection of hydrogen peroxide by Navaratne and Rechnitz [47]. The tissue was entrapped in a carbon paste matrix along with ferrocene. The authors demonstrated better sensitivities from in vitro derived tissues, indicating the advantages of this approach. In addition, the sensor exhibited a extended lifetime [47].

Though in vitro cultures of plant tissue offer the potential to generate uniform tissues, in vitro regenerated plants show considerable variability. This leads to the isolation of rare somaclonal variants which are a result of an existing genetic variability or changes during the in vitro culture conditions [48—50]. Cells in culture are amenable to genetic modification by various techniques and they could be induced to produce the desired secondary metabolite. In addition, callus cultures could be established from thin slices of explants containing only a few cells. Callus culture could potentially increase the yield of active constituents in cultured plant cells and pave the way for the production of various classes of therapeutic compounds from a variety of plants by standardization of culture conditions [51].

Also, plant cell suspension cultures in bioreactors under standardized culture conditions could be used in the production of secondary metabolites on a large scale and in a season independent manner. Cell suspension culture could be initiated by transferring friable calli into liquid medium and is maintained under suitable conditions of aeration, agitation, light, temperature and other physical parameters [52].

Bioreactors could be accommodated in small spaces and culture conditions could be precisely controlled. Biotic and abiotic elicitors could be used to enhance the production of the necessary secondary metabolite. The main challenge usually is in the standardization of culture conditions necessary for the establishment and maintenance of the cell suspension culture under sterile conditions. Selection of the appropriate type of bioreactor is also critical. A variety of secondary metabolites have been produced by cell suspension culture. We successfully reported the production of 4-ipomeanol from cell suspension cultures of *I. batatas* [53].

5.5 Plant-based biosensors for disease diagnosis

Enzyme-based biosensors, that use enzymes as recognition elements and combine enzyme specificity with the unique advantages of biosensors, have found extensive application in various fields. Such biosensors, because of their high sensitivity and specificity, offer huge potential for future bioanalysis and contain biocatalysts or immobilized materials. The biocatalytic transformations are caused by optical and electronic electrodes [54]. The most commonly used biological components are microorganisms and enzymes like tyrosinase, peroxidase, laccase, glucose dehydrogenase, cellobiose dehydrogenase, etc. [55]. A review of existing studies indicate that these enzymes could be found in plants.

Tyrosinase belonging to phenol oxidases is quite prevalent in nature and can be found in animals, plants, fungi and bacteria. They aid in the biosynthesis of melanin pigments from tyrosine, and also from other phenolic compounds [56]. They are found in fruits and vegetables where they are responsible for enzymatic browning reactions [57].

Peroxidases are found in animals, plants and microorganisms and they catalyze the redox reaction between hydrogen peroxide and some reductors. They aid in numerous cellular processes including development and stress responses in plants. In addition to the above, they help regulate hormonal and cell wall metabolism and antioxidant defense [58].

Laccase from plants has been reported from the Japanese lacquer tree *Toxicodendron verniciflum* (*Rhus vernicifera*) [59]. These enzymes have been associated with biosynthesis and polymerization of lignin [60,61], elongation [62–64] and in stress response [65–67].

Glucose-6-phosphate dehydrogenase, a key enzyme of the plant PPP pathway is involved in regulating the metabolic rates of many physiological processes [68].

In addition to the above, plants possess dehydrogenase enzymes that could be used in amperometric biosensors. Alcohol dehydrogenase belongs to the superfamily of medium-chain dehydrogenases/reductases encompassing different enzyme activities, such as alcohol, sorbitol, xylitol, threonine dehydrogenase and quinone reductase [69]. They convert acetaldehyde into ethanol via fermentation under conditions of low oxygen concentrations or low temperatures [70]. Glutamate dehydrogenase catalyzes deamination of glutamate in higher plants [71]. Lactate dehydrogenase is present in all green land plants ranging from flowering plants to mosses [72].

Various human diseases are caused due to infection by pathogens including viruses, bacteria and fungi. The majority of biosensors that detect pathogens causing infectious diseases are based on the principle of electrochemical reaction. These types of biosensors are cost effective, independent of solution turbidity, have a low power requirement, have high sensitivity, and are of simple instrumentation [73].

A wide variety of human diseases are characterized by abnormal metabolic states that disrupt normal physiology leading to severe tissue dysfunction [74]. During infection, changes in host metabolism occur in the cellular, tissue, organ and physiological levels [75–77]. A study to identify metabolite changes in blood to predict the onset of tuberculosis (TB) was performed by Weiner et al. [78]. The authors had identified changes in serum or plasma metabolic compounds in individuals who progressed to clinical Tuberculosis (TB) and generated a trans-African metabolic biosignature for TB. This could pave the path for timely interventions so that disease progression and transmission could be prevented [78].

Xiao et al. [79] performed metabolomics and cytokine/chemokine profiling on serum samples from healthy controls and mild to severe COVID-19 patients. This showed correlations between metabolites and proinflammatory cytokines/chemokines implying a potential regulatory crosstalk between arginine, tryptophan, purine metabolism and hyperinflammation. Chen et al. [80] suggested that the COVID-19 infection affected the cell signal, nucleic acid metabolism and amino acid metabolism networks. The metabolomic profiles of COVID-19 patients were found to be distinctive from the disease control group. The authors indicated cysteine sulfinic acid, phosphocholine and 3-sulfo-L-alanine as the top three biomarkers that can be used to differentiate COVID-19 patients from disease control patients.

Kimhofer et al. [81] characterized the metabolic effects of SARS-CoV-2 infection on human blood plasma using Nuclear Magnetic Resonance (NMR) spectroscopy and liquid chromatography—mass spectrometry and identified markers of inflammation including elevated α-1-acid glycoprotein and an increased kynurenine/tryptophan ratio. They also reported abnormal lipoprotein, glucose, and amino acid signatures consistent with diabetes and coronary artery disease in addition to multiple amino acid markers for liver dysfunction.

Multiude of host biochemical pathways and molecular mechanisms that assist the various defense systems used by the body to respond to microbial invasions are influenced by infection. The magnitude of these changes depend on the severity and duration of the illness. Losses of body constituents and nutrients are the most visible metabolic effects of infectious diseases [82].

Changes in metabolite levels that occur during an infection could be monitored by plant-based biosensors. Various plant components and phytochemicals could help indicate the precise level of disease-specific markers in a sample because of their potential to interact specifically with the target analyte.

5.6 Future prospects

Keeping the above background and supporting evidences in mind, it is proposed that biosensors using plant-based compounds could be used to detect changing levels of metabolites during infectious diseases. By virtue of their ability to interact with various compounds and produce a physical or a chemical change, these phytochemicals offer a tremendous potential to be used as biorecognition elements in biosensors. The various phytochemicals present in plants could be isolated and purified and could be used for more precise detection of the analyte. In addition, biosensors can be genetically modified to detect a variety of molecules.

Another advancement to consider is the use of nanoparticles that could act as carriers of phytochemicals. Nanoparticles could improve solubility, stability and absorption of phytochemicals, permeation and retention in target tissues, improve bioavailability, prevent premature degradation in the body, exhibit high differential uptake efficiency in the target cells and prolong their circulation time [83].

A thorough study on the mechanism of action of various phytochemicals and the types of changes they can generate following their interactions with biomarkers specific to the infection is warranted. This would enable harnessing the immense potential of the multitude of phytochemicals present in the plant kingdom.

5.7 Conclusion

Design of newer biosensors for the detection of various infectious diseases is definitely a current need. Biosensors derived from plant-based components offer the potential for the design of biosensors that are cost effective, accurate, simple and fast. This surpasses the advantages offered by other biorecognition elements.

References

[1] H. Muguruma, Biosensors: enzyme immobilization chemistry, in: K. Wandelt (Ed.), Encyclopedia of Interfacial Chemistry, Elsevier, 2018, pp. 64−71.
[2] A. Michelmore, Thin film growth on biomaterial surfaces, in: H.J. Griesser (Ed.), Thin Film Coatings for Biomaterials and Biomedical Applications, Woodhead Publishing, 2018, pp. 29−47.
[3] A. Kawamura, T. Miyata, Biosensors, in: M. Ebara (Ed.), Biomaterials Nanoarchitectonics, William Andrew Publishing, 2016, pp. 157−176.
[4] A. Kumar, M. Malinee, A. Dhiman, A. Kumar, T.K. Sharma, Aptamer technology for the detection of foodborne pathogens and toxins, in: Inamuddin, R. Khan, A. Mohammad, A.M. Asiri (Eds.), Advanced Biosensors for Health Care Applications, Elsevier, 2019, pp. 45−69. 2019.
[5] M. Malinee, A. Kumar, A. Dhiman, T.K. Sharma, Aptamer-mediated nanobiosensing for health monitoring, in: Inamuddin, R. Khan, A. Mohammad, A.M. Asiri (Eds.), Advanced Biosensors for Health Care Applications, Elsevier, 2019, pp. 227−248.
[6] P. Mehrotra, Biosensors and their applications − a review, J. Oral Biol. Craniofac. Res. 6 (2) (2016) 153−159.
[7] N. Bhalla, P. Jolly, N. Formisano, P. Estrela, Introduction to biosensors, Essays Biochem. 60 (1) (2016) 1−8.
[8] M. Campàs, R. Carpentier, R. Rouillon, Plant tissue-and photosynthesis-based biosensors, Biotechnol. Adv. 26 (2008) 370−378.
[9] A.K. Singh, N. Verma, Plants and plant-derived materials used for biosensor development, in: M. Yadav, V. Kumar, N. Sehrawat (Eds.), Industrial Biotechnology Plant Systems, Resources and Products, De Gruyter, 2019.
[10] B.D. Malhotra, A. Chaubey, Biosensors for clinical diagnostics industry, Sens. Actuators B Chem. 91 (2003) 117−127.

[11] L. Castillo-Henríquez, M. Brenes-Acuña, A. Castro-Rojas, R. Cordero-Salmerón, M. Lopretti-Correa, J.R. Vega-Baudrit, Biosensors for the detection of bacterial and viral clinical pathogens, Sensors 20 (2020) 6926.
[12] R. Monošík, M. Stred'anský, E. Šturdík, Application of electrochemical biosensors in clinical diagnosis, J. Clin. Lab. Anal. 26 (1) (2012) 22–34.
[13] Y. Huang, D. Xiao, B.M. Burton-Freeman, I. Edirisinghe, Chemical changes of bioactive phytochemicals during thermal processing, Reference Module in Food Science, Elsevier, 2016.
[14] R. Campos-Vega, B.D. Oomah, Chemistry and Classification of Phytochemicals. Handbook of Plant Food Phytochemicals: Sources, Stability and Extraction, 2013, pp. 5–48.
[15] J.B. Harborne, H. Baxter, Phytochemical dictionary, A Handbook of Bioactive Compounds From Plants, Taylor & Francis, London, 1993, 1993.
[16] T. Naghdi, S. Faham, T. Mahmoudi, N. Pourreza, R. Ghavami, H. Golmohammadi, Phytochemicals toward green (bio)sensing, ACS Sens. 5 (12) (2020) 3770–3805.
[17] J.S. Sidwell, G.A. Rechnitz, Progress and challenges for biosensors using plant tissue materials, Biosensors 2 (1986) 221–233.
[18] A. Goldsworthy, The evolution of plant action potentials, J. Theor. Biol. 103 (1983) 645–648.
[19] A.G. Volkov, D.R. Ranatunga, Plants as environmental biosensors, Plant. Signal. Behav. 1 (3) (2006) 105–115.
[20] S. Kuriyama, G.A. Rechnitz, Plant tissue-based bioselective membrane electrode for glutamate, Anal. Chim. Acta 131 (1981) 91–96.
[21] J.S. Sidwell, G.A. Rechnitz, "Bananatrode" – an electrochemical biosensor for dopamine, Biotech. Lett. 7 (1985) 419–422.
[22] J. Wang, M.S. Lin, Mixed plant tissue-carbon paste bioelectrode, Anal. Chem. 60 (1988) 1545–1548.
[23] B.R. Eggins, C. Hickey, S.A. Toft, D.M. Zhou, Determination of flavanols in beers with tissue biosensors, Anal. Chim. Acta 347 (1997) 281–288.
[24] E.A. Cummings, P. Mailley, S. Linquette-Mailley, B.R. Eggins, E.T. McAdams, S. McFadden, Amperometric carbon paste biosensor based on plant tissue for the determination of total flavanol content in beers, Analyst 123 (1998) 1975–1980.
[25] A.W.O. Lima, V.B. Nascimento, J.J. Pedrotti, L. Angnes, Coconut-based plant tissue reactor for biosensing of catechol in flow injection analysis, Anal. Chim. Acta 31 (354) (1997) 325.
[26] O. Fatibello-Filho, K. Omuro Lupetti, I.C. Vieira, Chronoamperometric determination of paracetamol using an avocado tissue (*Persea americana*) biosensor, Talanta 55 (2001) 685–692.
[27] F.S. Felix, M. Yamashita, L. Angnes, Epinephrine quantification in pharmacological formulations utilizing plant tissue biosensors, Biosens. Bioelectron. 21 (2006) 2283–2289.
[28] J. Wang, N. Naser, H.S. Kwon, M.Y. Cho, Tissue bioelectrode for organic-phase enzymatic assays, Anal. Chim. Acta 264 (1992) 7–12.
[29] I.C. Vieira, O. Fatibello-Filho, Biosensor based on paraffin/graphite modified with sweet potato tissue for the determination of hydroquinone in cosmetic cream in organic phase, Talanta 52 (2000) 681–689.
[30] I.R.W.Z. de Oliveira, S.C. Fernandes, I.C. Vieira, Development of a biosensor based on gilo peroxidase immobilized on chitosan chemically cross linked with epichlorohydrin for determination of rutin, J. Pharm. Biomed. 41 (2006) 366–372.
[31] W. Oungpipat, P.W. Alexander, P. Southwell-Keely, A reagentless amperometric biosensor for hydrogen peroxide determination based on asparagus tissue and ferrocene mediation, Anal. Chim. Acta 309 (1995) 35–45.

[32] J.V.B. Kozan, R.P. Silva, S.H.P. Serrano, A.W.O. Lima, L. Angnes, Biosensing hydrogen peroxide utilizing carbon paste electrodes containing peroxidases naturally immobilized on coconut (*Cocus nucifera* L.) fibers, Anal. Chim. Acta 591 (2007) 200–207.
[33] L. Campanella, M. Cordatore, F. Mazzei, M. Tomassetti, Determination of inorganic phosphate in drug formulations and biological fluids using a plant tissue electrode, J. Pharm. Biomed. Anal. 8 (1990) 711–716.
[34] L. Campanella, M. Cordatore, F. Mazzei, M. Tomassetti, G. Volpe, Phosphate determination in foodstuffs using a plant tissue electrode, Food Chem. 44 (1992) 291–297.
[35] M.K. Sezgintürk, E. Dinçkaya, Direct determination of sulfite in food samples by a biosensor based on plant tissue homogenate, Talanta 65 (2005) 998–1002.
[36] M.K. Sezgintürk, E. Dinçkaya, A novel amperometric biosensor based on spinach (*Spinacia oleracea*) tissue homogenate for urinary oxalate determination, Talanta 59 (2003) 545–551.
[37] W. Qin, Z. Zhang, Y. Peng, Plant tissue-based chemiluminescence flow biosensor for urea, Anal. Chim. Acta 407 (2000) 81–86.
[38] F. Botrè, C. Botrè, G. Lorenti, F. Mazzei, F. Porcelli, G. Scibona, Plant tissue biosensors for the determination of biogenic diamines and of their amino acid precursors: effect of carbonic anhydrase, Sens. Actuators B Chem. 15 (1993) 135–140.
[39] H. Horie, G.A. Rechnitz, Hybrid tissue/enzyme biosensor for pectin, Anal. Chim. Acta 306 (1995) 123–127.
[40] J. Feng, B.W. Jester, C.E. Tinberg, D.J. Mandell, M.S. Antunes, R. Chari, et al., A general strategy to construct small molecule biosensors in eukaryotes, eLife 4 (2015) e10606. Available from: https://doi.org/10.7554/eLife.10606. 2015.
[41] M. Iwamoto, T. Björklund, C. Lundberg, D. Kirik, T.J. Wandless, A general chemical method to regulate protein stability in the mammalian central nervous system, Chem. Biol. 17 (2010) 981–988.
[42] L.A. Banaszynski, L.-C. Chen, L.A. Maynard-Smith, A.G.L. Ooi, T.J. Wandless, A rapid, reversible, and tunable method to regulate protein function in living cells using synthetic small molecules, Cell 126 (2006) 995–1004.
[43] Y. Miyazaki, H. Imoto, L.-C. Chen, T.J. Wandless, Destabilizing domains derived from the human estrogen receptor, J. Am. Chem. Soc. 134 (2012) 3942–3945.
[44] A. Walia, R. Waadt, A.M. Jones, Genetically encoded biosensors in plants: pathways to discovery, J. Annu. Rev. Plant. Biol. (2018) 497–524.
[45] P. Srivastava, M. Singh, G. Devi, R. Chaturvedi, Herbal medicine and biotechnology for the benefit of human health, in: A.S. Verma, A. Singh (Eds.), Animal Biotechnology, Academic Press, 2014, pp. 563–575.
[46] N. Vijayasree, P. Udayasri, K.Y. Aswani, B.B. Ravi, K.Y. Phani, V.M. Vijay, Advancements in the production of secondary metabolites, J. Nat. Prod. 3 (2010) 112–123.
[47] A. Navaratne, G.A. Rechnitz, Improved plant tissue-based biosensor using in vitro cultured tobacco callus tissue, Anal. Chim. Acta 257 (1992) 59–66.
[48] C.E. Flick, D.A. Evans, W.R. Sharp, Organogenesis, in: D.A. Evans, W.R. Sharp, P.V. Amrnitato, Y. Yamada (Eds.), Handbook of Plant Cell Culture, vol. 1, MacMillan, New York, 1983, pp. 13–18. 1.
[49] P.J. Larkin, W.R. Scowcraft, Somaclonal variation – a novel source of variability from cell culture for plant improvement, Theor. Appl. Genet. 60 (1981) 197–214.
[50] V. Raghavan, Embryogenesis in Angiosperms, Cambridge University Press, Cambridge, 1986.
[51] E.D. Benjamin, G.A. Ishaku, F.A. Peingurta, A.S. Afolabi, Callus culture for the production of therapeutic compounds, Am. J. Plant. Biol. 4 (2019) 76–84.

[52] S. Chattopadhyay, S. Farkya, A.K. Srivastava, V.S. Bisaria, Plant tissue culture: current status and opportunities, Biotechnol. Bioprocess. Eng. 7 (2002) 138−149.

[53] R. Mohanraj, S. Subha, Optimization of process parameters for bioproduction, isolation and purification of 4-ipomeanol from cell suspension cultures of *Ipomoea batatas* (L.) Lam, Indian J. Exp. Biol. 55 (3) (2017) 191−196.

[54] Y.-C. Zhu, L.-P. Mei, Y.-F. Ruan, N. Zhang, W.-W. Zhao, J.-J. Xu, et al., Enzyme-based biosensors and their applications, in: R.S. Singh, R.R. Singhania, A. Pandey, C. Larroche (Eds.), Biomass, Biofuels, Biochemicals, Advances in Enzyme Technology, Elsevier, 2019, pp. 201−223.

[55] F. Karim, A.N.M. Fakhruddin, Recent advances in the development of biosensor for phenol: a review, Rev. Env. Sci. Biotechnol. 11 (2012) 261−274.

[56] H. Claus, H. Decker, Bacterial tyrosinases, Syst. Appl. Microbiol. 29 (2006) (2006) 3−14.

[57] J. Buchert, E. Selinheimo, K. Kruus, M.-L. Mattinen, R. Lantto, K. Autio, Using cross-linking enzymes to improve textural and other properties of food, in: R. Rastall (Ed.), Woodhead Publishing Series in Food Science, Technology and Nutrition. Novel Enzyme Technology for Food Applications, Woodhead Publishing, 2007, pp. 101−139.

[58] H. Jouili, H. Bouazizi, E. El Ferjani, Plant peroxidases: biomarkers of metallic stress, Acta Physiol. Plant 33 (2011) 2075.

[59] H. Yoshida, Chemistry of lacquer (urushi) part 1, J. Chem. Soc. 43 (1883) 472−486.

[60] Y. Tobimatsu, M. Schuetz, Lignin polymerization: how do plants manage the chemistry so well? Curr. Opin. Biotechnol. 56 (2019) 75−81.

[61] Q. Zhao, J. Nakashima, F. Chen, Y.B. Yin, C.X. Fu, J.F. Yun, et al., Laccase is necessary and nonredundant with peroxidase for lignin polymerization during vascular development in *Arabidopsis*, Plant. Cell 25 (2013) 3976−3987.

[62] V.K. Balasubramanian, K.M. Rai, S.W. Thu, M.M. Hu, V. Mendu, Genome-wide identification of multifunctional laccase gene family in cotton (*Gossypium* spp.); expression and biochemical analysis during fiber development, Sci. Rep. 6 (2016) 34309.

[63] M. Liang, E. Davis, D. Gardner, X. Cai, Y. Wu, Involvement of AtLAC15 in lignin synthesis in seeds and in root elongation of *Arabidopsis*, Planta 224 (2006) 1185−1196.

[64] D. Nigam, P. Kavita, R.K. Tripathi, A. Ranjan, R. Goel, M. Asif, et al., Transcriptome dynamics during fibre development in contrasting genotypes of *Gossypium hirsutum* L, Plant. Biotechnol. J. 12 (2014) 204−218.

[65] X. Cai, E.J. Davis, J. Ballif, M. Liang, E. Bushman, V. Haroldsen, et al., Mutant identification and characterization of the laccase gene family in *Arabidopsis*, J. Exp. Bot. 57 (2006) 2563−2569.

[66] H.Y. Cho, C. Lee, S.G. Hwang, Y.C. Park, H.L. Lim, C.S. Jang, Overexpression of the OsChI1 gene, encoding a putative laccase precursor, increases tolerance to drought and salinity stress in transgenic *Arabidopsis*, Gene 552 (2014) 98−105.

[67] Q. Liu, L. Luo, X. Wang, Z. Shen, L. Zheng, Comprehensive analysis of rice laccase gene (OsLAC) family and ectopic expression of OsLAC10 enhances tolerance to copper stress in *Arabidopsis*, Int. J. Mol. Sci. 18 (2017) 209.

[68] N.J. Kruger, A. von Schaewen, The oxidative pentose phosphate pathway: structure and organisation, Curr. Opin. Plant Biol. 6 (3) (2003) 236−246.

[69] B. Persson, T. Bergman, W.M. Keung, U. Waldenström, B. Holmquist, B.L. Vallee, et al., Basic features of class-I alcohol dehydrogenase: variable and constant segments coordinated by inter-class and intra-class variability. Conclusions from characterization of the alligator enzyme, Eur. J. Biochem. 15 (1993) 49−56.

[70] R. Dolferus, M. Jacobs, W.J. Peacock, E.S. Dennis, Differential interactions of promoter elements in stress responses of the Arabidopsis Adh gene, Plant. Physiol. 105 (4) (1994) 1075−1087.

[71] T. Tercé-Laforgue, M. Bedu, C. Dargel-Grafin, F. Dubois, Y. Gibon, F.M. Restivo, et al., Resolving the role of plant glutamate dehydrogenase: II. Physiological characterization of plants overexpressing the two enzyme subunits individually or simultaneously, Plant. Cell Physiol. 54 (2013) 1635–1647.
[72] P. O'Carra, P. Mulcahy, Lactate dehydrogenase in plants: distribution and function, Phytochemistry 42 (1996) 581–587.
[73] B. Srinivasan, S. Tung, Development and applications of portable biosensors, J. Lab. Autom. 20 (4) (2015) 365–389.
[74] R.J. DeBerardinis, C.B. Thompson, Cellular metabolism and disease: what do metabolic outliers teach us? Cell 148 (6) (2012) 1132–1144.
[75] J.S. Ayres, Immunometabolism of infections, Nat. Rev. Immunol. 20 (2020) 79–80.
[76] J.S. Ayres, The biology of physiological health, Cell 181 (2020) 250–269.
[77] K. Troha, J.S. Ayres, Metabolic adaptations to infections at the organismal level, Trends Immunol. 41 (2020) 113–125.
[78] J. Weiner, J. Maertzdorf, J. Sutherland, et al., Metabolite changes in blood predict the onset of tuberculosis, Nat. Commun. 9 (2018) 5208.
[79] N. Xiao, M. Nie, H. Pang, et al., Integrated cytokine and metabolite analysis reveals immunometabolic reprogramming in COVID-19 patients with therapeutic implications, Nat. Commun. 12 (2021) 1618.
[80] X. Chen, M. Gu, T. Li, Y. Sun, Metabolite reanalysis revealed potential biomarkers for COVID-19: a potential link with immune response, Future Microbiol. 16 (8) (2021) 577–588.
[81] T. Kimhofer, S. Lodge, L. Whiley, N. Gray, R.L. Loo, N.G. Lawler, et al., Integrative modeling of quantitative plasma lipoprotein, metabolic, and amino acid data reveals a multiorgan pathological signature of SARS-CoV-2 infection, J. Proteome Res. 19 (11) (2020) 4442–4454.
[82] W.R. Beisel, Metabolic effects of infection, Prog. Food Nutr. Sci. 8 (1–2) (1984) 43–75.
[83] S. Wang, S. Rui, S. Nie, M. Sun, J. Zhang, D. Wu, et al., Application of nanotechnology in improving bioavailability and bioactivity of diet-derived phytochemicals, J. Nutr. Biochem. 25 (4) (2014) 363–376.

CHAPTER 6

Development of optical biosensors for the diagnosis of pathogens

Pragyan Roy
Department of Biotechnology, College of Basic Sciences and Humanities, Bhubaneswar, Odisha, India

6.1 Introduction

Currently, there is a growing need to monitor and control the spread of infectious diseases. Infectious diseases can be of three types; the pathogens that were less virulent in the past but currently have created havoc (SARS-CoV-2), those whose occurrence and geographic range has increased in the present (foot-and-mouth disease), and the diseases which might be a threat in the near future (avian influenza) [1]. Regular disease surveillance and fast diagnosis are the only methods by which the spread of the disease can be checked. This can only be achieved by an adequate and prompt detection system that can check the spread of disease. Due to climate change, demographic shifts, greater global movement of human and animals, a disease outbreak is capable of completely paralyzing the economic and health sector. A rapid detection of viruses, bacteria or antigens could completely control the spread and also contribute towards the eradication of disease. The major hurdle is in the collection of samples from remote areas, transportation to the laboratories and the time taken in the analysis of the result. For communicable diseases, early warning signals are the most important for disease vigilance, as it may only take a few hours to spread amongst the human or animal population. Thus there is a need for the development of rapid and sensitive detection tools that can be used in the field without involving much mechanization. One such established tool is a biosensor capable of providing information on the disease outbreak, monitoring food and water qualities.

Another very challenging scenario, even for the most experienced doctor is to distinguish between bacterial and viral diseases. The symptoms for bacterial and viral diseases are more or less similar, so

Biosensors for Emerging and Re-emerging Infectious Diseases
DOI: https://doi.org/10.1016/B978-0-323-88464-8.00005-1

symptomatic analysis is difficult. To prevent the spread of disease and reduce mortality, empirical antibiotics are administered. Currently C-reactive protein and procalcitonin (PCT) are used as biomarkers to differentiate between bacterial and viral infections but they are not very sensitive as they are produced in low quanities. Earlier, viruses and pathogens were detected by the reverse transcription polymerase chain reaction (RT-PCR) [2]. This classical method of identification requires isolation, culturing and biochemical tests followed by immunological methods like the enzyme-linked immunosorbent assay (ELISA) [3]. All these processes are time consuming and require two to three days to give proper results. Therefore newer approaches have to be considered that have the facility for field testing and can detect pathogens in a much shorter timeframe. Today, biosensors and nanobiosensors are widely used for the identification of viral and bacterial pathogens. A specific molecular biosensor is a device that works on certain biochemical reactions between target analytes like enzymes, antigens, tissues, whole cells, organelles and their receptors. Early detection of disease is possible and its rapid spread can be checked. These devices are intelligent, fast, practical, efficient and have been used in the detection of Ebola virus, human immunodeficiency virus (HIV), Coranavirus 2, as well as bacteria like *Salmonella* spp. and *Escherichia coli* [4,5].

6.2 Infectious agents: bacteria and viruses

Pathogens are microorganisms that cause disease and they can be either bacteria or viruses. Bacteria release toxins and viruses destroy cells. To control bacterial infections, either antibiotics or antitoxins are given. Antibiotics are able to control non-resistant bacteria but not viruses. The efficient control of any disease depends upon timely identification of the causative agent and appropriate drugs to check its spread. Moreover, the correct dose of antibiotics will help to solve the various complications that arise due to antibiotic resistance. The analyses of bacteria and viruses uses different body fluids, so more sensitive and advanced methods have to be used to quantify the pathogen load.

Unambiguous identification of Gram positive or Gram negative bacteria is a challenge because bacteria as low as 1–10 CFU/mL of blood is capable of causing life threatening disease [6]. If RNA is used as the target then it should be present as 10^3 or 10^4 copy number which is actually a subfemtomolar concentration as compared to the large amount of non-target RNA. Thus for rapid diagnosis, molecular biology methods are preferred after bacterial culture. Urinary tract infection is caused by *Proteus mirabilis, E. coli, Pseudomonas aeruginosa, Staphylococcus saprophyticus* and *Klebsiella pneumonia.* Methods for accurate and rapid detection of the above pathogens are very important to avoid antibiotic resistance and to administer the best line of antibiotics. Similarly for tuberculosis, identification of pathogen from sputum is very difficult due to the viscosity of the sample and a high bacterial load.

Viral infection in the host can also be detected by nucleic acid sequences of the target virus. If the concentration of bloodborne viruses is 10^6 IU/mL of blood, then it is equivalent to femtomolar concentration of viral RNA [7]. For instance, dengue viral RNA can be detected in biological samples (blood, urine or saliva) if its concentration is in the range of 10^2–10^6 per mL [8]. *Flavivirus influenza* loads are 10^6–10^9 in aspirate samples and 10^3 in nasal samples [9]. In the case of zika viruses, 10^2–10^5 copies/mL are detected in plasma or urine.

6.2.1 Bacterial markers

C-reactive protein and PCT are the biomarkers present at elevated levels during bacterial infection. Apart from CRP and PCT, interleukin6 (IL-6) and interleukin 8 (IL-8) are frequently evaluated biomarkers for bacterial infection [10]. Further, the concentration of serum IL-6 is more in the case of Gram negative bacteria as compared to Gram positive bacteria. So for better sensitivity and specificity, physicians prefer to use a combination of all these biomarkers to differentiate between bacterial and viral infections [11]. Adrenomedulin, a peptide secreted during bacterial community-acquired pneumonia is absent during pneumonia due to non bacterial infections like fungus or virus. Similarly, presepsin is another biomarker that is secreted during bacterial phagocytosis [12].

6.2.2 Viral biomarkers

TRAIL (Tumor necrosis factor-related apoptosis-inducing ligand), IP-10 (Interferon gamma γ-induced protein-10), MxA (Myxovirus resistance potein A) are the prominent markers of viral infection. TRAIL is an apoptosis inducer, IP-10 is produced during inflammation, and angiogenesis are present in much higher concentration in viral infections as compared to bacterial infections. MxA controls virus multiplication in cells and is produced in patients with synctial and rotavirus infections and even in children with respiratory rhinovirus infection. Pentraxin 3 (PTX3), an acute-phase protein, plays a vital role in complement activation and pathogen recognition.

6.3 Traditional methods for pathogen detection

The conventional methods of pathogen detection involve identification of biochemical or microbiological constituents. Basically, there are three methods followed for pathogen detection; immunological methods, counting of colony and culturing methods and molecular methods. In immunological detection methods, antigen-antibody immunoprecipitation reactions are used for the detection of Gram negative bacteria *Salmonella* and *E. coli* [13]. Immunological detection methods include: Enzyme Immunoassay (EIA), Enzyme-linked fluorescent Assay (ELFA), ELISA and Flow injection Immunoassay [14]. The antibodies produced during viral infection are used to detect viral pathogens. Apart from the above mentioned immunodiagnostic methods, viral pathogens are also detected by hemagglutinationn inhibition assay and western blotting (Fig. 6.1).

The culturing techniques give the best results due to their high sensitivity and selectivity. Modern cell culture methods are used to detect

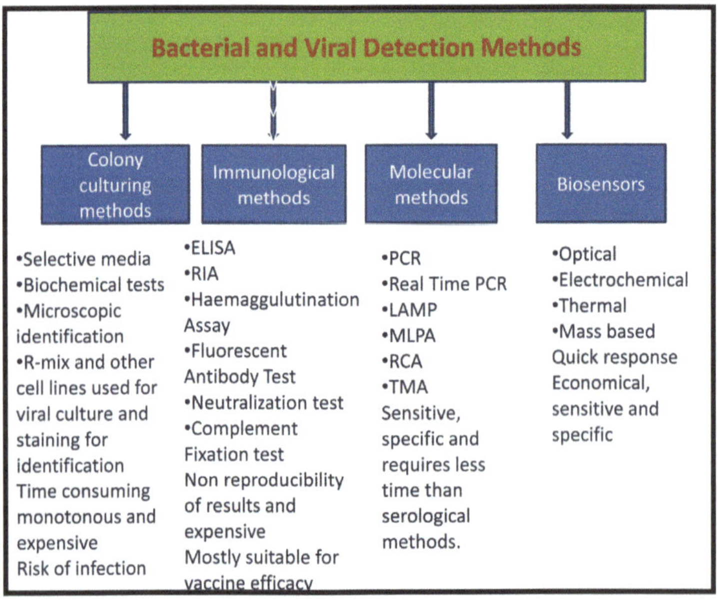

Figure 6.1 Various methods used for the detection of pathogens.

different viruses by inoculating samples in different cell lines. The most commonly used cell lines for viral diagnosis include: primary rabbit kidney cells, human foreskin fibroblasts, rhesus monkey kidney cells, MRC-5, Hep-2, and A549. With the use of specific monoclonal antibodies, respiratory viruses could be identified in 3 h by DFA staining. Currently, R-Mix Shell vials and tube culture is also done for virus isolation and identification. Conventional methods for viral or bacterial detection include virus or microorganism growth and isolation from culture. In the case of bacteria, selective media is used while for viruses, tissue culture methods are used. Though these methods are effective and precise, they are also expensive, time consuming (2–10 days) and labor-intensive.

Development of a valid diagnostic assay for early and accurate pathogen detection and identification is a challenge for the researchers. Therefore biosensors are the attractive alternative for efficient detection of pathogens due to their simplicity, preciseness and capability for real time analysis. In the last three decades, advancements in the field of biotechnology have completely modernized biosensors for viral and bacterial detection. Some biosensors have rapid pathogen detection system and can identify pathogens even in a very low quantity of samples.

6.4 Biosensors

A biosensor is a device that measures living organisms or chemicals present in living organisms. It is a machine composed of a transducer and a bioreceptor. The biological part is broadly composed of biomolecules like antibodies, nucleic acids, enzymes or tissues, microorganisms, organelles, etc. This immobilized signal-detecting element, actually recognizes the target biomarker specific to a particular pathogen. Due to its physiological properties, bioreceptors are highly specific and sensitive and they carefully avoid interferences from other organisms if present in the tested sample. This biochemical interaction between the biomarker and the bioreceptor is converted into a measurable signal by the transducer (physicochemical detector). Signal recording and display is then used for pathogen detection. The biological material for bioreceptors is chosen depending upon storage, specificity, and environmental stability. Biomolecules like antibodies, enzymes, receptors, microorganisms, cells and tissues of higher plants and animals all act as biological-sensing molecules. These biomolecules have become an integral part of biosensor technology.

Different types of bioreceptors play an important role in pathogen identification and functioning of the biosensors.

6.4.1 Biosensing components or bioreceptors

The bioreceptors can be either a catalytic type or an affinity type. Enzymes, cells, tissues or organelles belong to the catalytic type while antibodies, receptors and nucleic acids are of the analytical type. The enzymes are either directly immobilized on the surface of the transducer or attached to the surface by physical adsorption or covalent bonding. The enzymes used are oxidase type, that selectively react and produce H_2O_2. Since the enzymes are not used up in the reaction, these biosensors can be repeatedly used. Microbial cells as biosensing components are cheaper than enzymes and antibodies and can acclimatize to all conditions. Microbial biosensors are mostly used in the environmental field for detection of harmful bacteria, pesticides, etc. Cell organelles like the lysozyme, chloroplast and mitochondria are also used as bioreceptors. The cells have high sensitivity to physical parameters so they are used to monitor toxicity, stress condition and the effects of drugs. Antibody sensors are also known as immunosensors and are attached to the transducers through covalent bonding. They are fast with analysis and detection of pathogens but have limitations like irreversible reaction and strong bonding. In nucleic acid biosensors, oligonucleotides integrate with the signal transducer. Oligonucleotide probes detect DNA/RNA fragments through the code of complementarity (Adenine pairs with Thymine and Guanine pairs with Cytosine).

6.4.2 Biotransducer component

Biosensors are broadly classified into different groups based on signal generation and conduction such as acoustic, optical, thermometric, electrochemical, piezoelectric and magnetic. In order to propose a suitable biosensor for pathogen detection, it is important to choose the right recognition element as well as the transducer that converts the biochemical response into a readable physical signal. The interaction between analyte and recognition element can be recorded by various sensing techniques. The various sensing techniques include piezoelectrical, calorimetrical, surface plasmon resonance (SPR), Surface Enhanced Raman Spectroscopy (SERS), Impedance Spectroscopy, DNA microarray, Atomic Force Microscopy (AFM), and Scanning Probe Microscopy

(SPM). Out of the above mentioned methods, electrochemical and optical are the most preferred. Depending on the detection principle, electrochemical biosensors are categorized as potentiometric, amperometric, voltametric and conductometric. The electrochemical biosensor measures variation in currents and voltages produced due to the molecular reaction between analyte and recognition element and is directly proportional to the concentration of the analyte to be analyzed. Potentiometric biosensors have a reference electrode, a selective electrode and a device that measures potential difference between the two. The selective electrodes are made up of conducting materials, which quantify the analyte immersed in a solution by recording the changes in the potential. Amperometric biosensors record changes in current flows produced due to the electrochemical reaction at a constant potential, where the current strength is directly proportional to the concentration of the analyte oxidized and reduced at the electrode surface. These sensors are used for quantifying aminoglycoside antibiotics, anticancer drugs and bronchodialators. Unlike in potentiometric biosensors, in amperometric biosensors, measured species are consumed. In a conductometric biosensor, a change in conduction is recorded upon the utilization or production of ionic species produced during metabolic process. They are used to monitor metabolic changes in living systems. Among the electrochemical biosensors, the most versatile one is a voltametric biosensor, as it combines both current and potential differences giving a reasonably good response. Piezoelectric biosensors have an oscillating piezoelectric crystal resonating at a frequency controlled by an electric current. When the analyte comes in to contact with the detection material, it alters the vibration frequency and hence the current being recorded which is directly proportional to the concentration of analyte. These biosensors are used to measure immunoglobulins, insulin and particulate pollutants from industrial waste. Optical sensors are the most preferred class of biosensors as they are cost effective, rapid, and allow for real-time, label-free identification of various analytes. Further, their small size makes them highly sensitive and specific. Currently used biosensors (listed below) are the product of interdisciplinary areas such as chemistry, nanotechnology, microelectrical mechanical systems, molecular biology and micro-electronics.

1. Resonant biosensors
2. Optical biosensors
3. Thermal Biosensors
4. Electrochemical biosensors

5. Bioluminescence based biosensors
6. Nucleic acid based biosensors
7. Nanobiosensors (Fig. 6.2)

6.5 Optical biosensors

Optical biosensors are optical devices designed to detect changes in the properties of light like absorption, luminescence, fluorescence, refractive index, fluorescence or light scattering due to the interaction between an analyte and a receptor. The signal that is produced is proportional to the analyte concentration measured by bioreceptors like enzymes, antigens, antibodies, receptors, complete tissues and cells. Optical biosensors can sense the interaction of the microorganism with the analyte and the optical signal emitted is proportional to the concentration of the target compound. These biosensors have catalytic or affinity receptors making them the most versatile and precise biosensors, be easily suited for rapid, multiparameter detection. When compared to other biosensors, they facilitate, direct, real-time, and label free detection of various chemical and biological substances. Due to tremendous research and development, new optical biosensors are based on advanced concepts with interdisciplinary approaches including molecular biology, biotechnology,

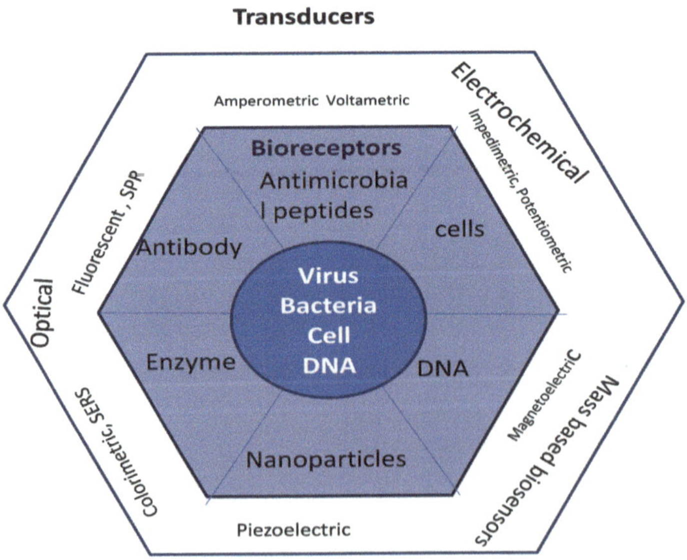

Figure 6.2 Different types of biosensors and their components.

nanotechnology, chemistry, microelectromechanical systems and microelectronics [15]. Optical biosensors have tremendous use in the health care industry, drug discovery, environmental fields and the biotechnology industry.

Leland C. Clark, Jr is known as the "father of biosensors" who developed first biosensor for oxygen detection. Later in 1962, Clark's amperometric enzyme electrode was used in the biosensor for the detection of glucose. Lubbers and Opitz described the first fiber optic biosensor in 1975. Since then, tremendous development in the field of optical biosensor research has promoted many new innovations using techniques such as anisotrophy, luminous energy transfer, quenching efficiency, nonradiative, etc. (Fig. 6.3).

6.6 Classification of optical biosensors

There are various ways of classifying optical biosensors such as based on the recognition method, that is either probe-based or reaction-based. In probing biosensors, the interaction between the analyte and recognition element is affinity-based and can be measured by various methods. While

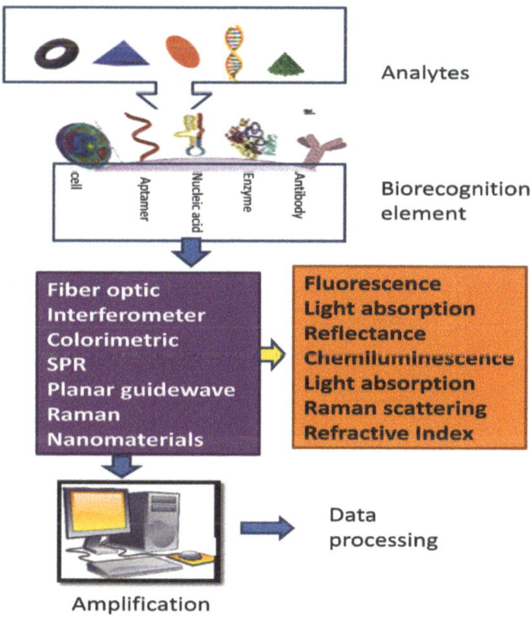

Figure 6.3 Different types of optical biosensors and their components.

in reacting biosensors, the optical responses are due to chemical reactions that can be recorded by numerous methods. Optical biosensors are sometimes grouped as Bio-optrodes and evanescent field based ones. The word optrode is the fusion of two words optical and electrode. Optrode are optical fiber based biosensors that are made up of a source of light, immobilized bioreceptor, an optical fiber used in light transmission, and a detector. When the analyte reacts with the recognition element on the surface of the fiber, biocatalysis occurs resulting in variation of the optical signal. This change is optically measured by dyes, bio- or chemiluminescents and fluorescent molecules. Optical fiber based biosensors are used for microorganism detection. Evanescent field-based biosensors are designed on electromagnetic waveguides which transmit light due to multiple internal reflections thereby creating an evanescent field that can be modified by the receptor. Evanescent field based biosensors are widely used for identifying chemical and biological substances because of their high degree of sensitivity and selectivity. Fiber-optic devices use bio-optrode biosensors while evanescent field-based biosensors use SPR-based, total internal reflection fluorescence (TIRF), surface-enhanced Raman scattering (SERS), optical waveguide interferometer and elipsometric. Sometimes optical biosensing can be classified either as label-free or label-based. In label-free mode, the signal is generated when the analyte reacts with the transducer. In label-based sensing, a label is used and the optical signal is generated by a colorimetric, luminescent or fluorescent method.

6.6.1 Fluoresence-based biosensors

Fluorescent biosensors detect the concentration, location and activity of biomolecules by absorbing the electromagnetic radiation released by the fluorophores, dyes, or labeled biomolecules, converting the energy to fluorescent emission. Fluorescent biosensors consist of an excitation light source (LED or lasers), fluorophore molecules that label the target molecules, and a detector that records the fluorescence intensity. The fluorophore molecules are small molecules like quantum dots (QD), protein used to bind lipids, nucleic acids and other proteins (antigens, and outer membrane proteins in the cell membrane of pathogens). The excitation signal is recognized by several techniques like FLM (fluorescence lifetime imaging), FRET (forster resonance energy transfer), FCS (fluorescence correlation spectroscopy), FI (changes in fluorescence intensity) (Fig. 6.4).

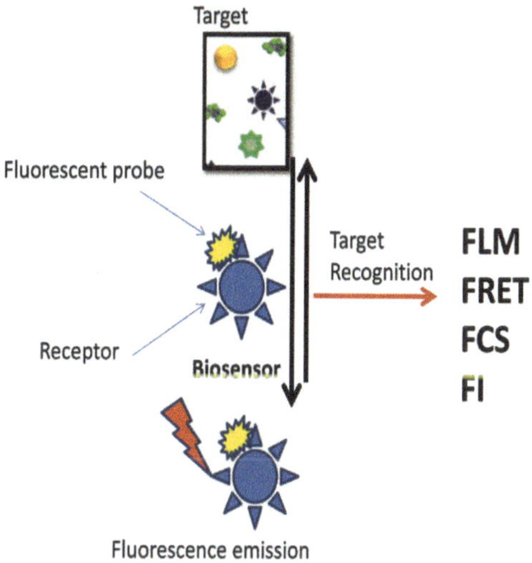

Figure 6.4 Conceptual illustration of a fluorescence biosensor.

6.6.2 Surface enhanced Raman scattering-based biosensors

SERS is a biosensing technique where surface sensitive methods are used to enhance Raman scattering of molecules adsorbed to the surface of transducers. When the biomolecule is close to nanoparticles or nano coated metal surfaces, the vibration spectra of the molecule increases several folds. This increase in SERS as compared to Raman scattering is either due to an electromagnetic or chemical mechanism. Since SERS is based on vibrational spectroscopy, it can be used to identify unknown target molecules as it gives information about its structure. It is also used to analyze samples in aqueous medium like food samples, as water does not interfere in SERS signals and it can penetrate deeply. The wavelength shift is attributed to the oscillation of conduction electrons in metal nanoparticles resulting in the excitation of localized SPR. The chemical mechanism involves several different transition processes that transfer charge between energy levels of analyte molecules. SERS active surface coated on the tip of the optical fiber is used to detect cancer protein in a very small sample volume (10 nL).

The accuracy of label-based SERS devices depends on the binding of the label to the target analyte, while in the case of label-free SERS devices, the signal depends upon either the analyte or the element-analyte complex. For small molecules, the Raman vibration bonds of the analyte when it is adsorbed on the reporter surface, are used for detection. However, sensing a small concentration of target analyte is a challenge if the target analyte is hydrophobic, as SERS surface receptors (Ag and Au) are hydrophilic in nature. So most of the biorecognition element used in SERS receptors are antibodies, aptamers, molecular imprinted polymers, etc. The first report of the use of antibodies as a biorecognition element in SERS devices was for the detection of Thyroid Stimulating Hormone [16].

6.6.3 Surface plasmon resonance based-biosensors

SPR method was used for the first time in biosensors by Liedberg et al. [17]. Since then, tremendous research has been done in this field to design new SPR-based biosensors and to study the interaction between the target analyte and bio-recognition element. The first commercial SPR-based biosensor was innovated by Pharmacia Biosensor AB and was later named as Biacore. SPR-based biosensors are highly sensitive, versatile, and the most preferred and advanced label-free biosensor. The past decade has seen massive research on SPR-based biosensors resulting in a label-free biosensing platform for characterization and quantification of a real time bio-molecular interaction. SPR occurs due to the total reflection of the incident light between the metallic interface and dielectric. Since the metal absorbs energy of the incident light, there is a decrease in reflected light intensity. Resonance angle is dependent on the refractive index of the medium and changes in the refractive index result in the shift of the resonance angle defined as SPR. When biomolecules are adsorbed or interact with the already immobilized molecules, if there is an increase in surface concentration, there is a shift in resonance angle, as seen in the Fig. 6.5. The incident light is totally internally reflected from the prism.

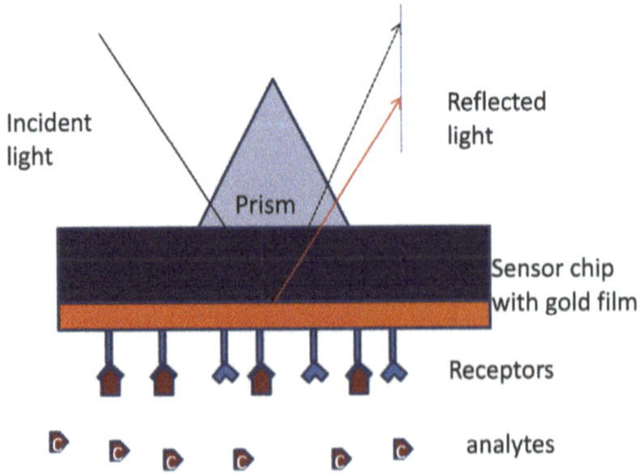

Figure 6.5 The principle of SPR. *SPR*, Surface plasmon resonance.

Interaction between the evanescent waves outside the prism and plasma waves on the surface of the metal (gold or silver) induces plasmon resonance.

There are four methods to excite SPR: prism coupling, waveguide coupling, fiber optic coupling and grating coupling. Prism couplers are the best sensors but are bulky and difficult to integrate. Waveguide couplers are better alternatives to prism couplers, as they are easy to integrate to other electrical and chemical components. The performance of the SPR-based biosensors depends upon the chemical interactions, functional kinetics of the bio-molecules and the arrangement of the optical, electrical and structural components of the instrument [18]. It is equally important that there is a robust bio-molecule coating on the sensor surface that is stable and regenerable. Further, the coating and immobilization of the biomolecule should not degrade the biological activity of the recognition element, as well as it should ensure that there is no non-specific binding. Antibody-based SPR sensors are called SPR immunosensors and are used for the detection of ochratoxin A (OTA) immobilized anti-OTA monoclonal antibody, benzoylecgonine (BZE), and estradiol detection. Aptamer-based SPR sensors are competitive assay types and can detect target analytes up to 0.005 ng/mL.

6.6.3.1 Localized surface plasmon resonance technology

Within nanosphere lithographic techniques, metal nanoparticles can be used to make a highly dispersed array on the surface. The resulting nanoparticle array gives a wavelength-specific LSPR peak, thus forming an ideal LSPR substrate. Biomolecules like biotin-streptavidin, aptamers, and antigen-antibody have been extensively used in LSPR. Basically, LSPR is used to detect biomolecular interactions and is sometimes used in the detection of bacterial colonies. Both SPR and LSPR are refractive index-based technologies. Refractive index of the medium depends upon analyte concentration and the association and dissociation of the target molecules. Several factors like the metal nanoparticle used, its dimension, shape and inter-particle distance controls the reflected color change and absorption peak shift. By controlling sensor fabrication sensitivity, the LSPR can be controlled.

6.6.4 Colorimetric biosensors

Colorimetric biosensors are used to identify specific compounds in the analytes through a color change that can be observed by the naked eye.

These sensors do not require any high-end instruments and are portable so they can be used as POC diagnostics. Several materials are used for the fabrication, like carbon nanotubes, noble metals and metal oxide nanoparticles (Fe_3O_4 and CeO_2) [19]. Gold nanoparticles are used because they produce color change when the analytes are aggregated or dispersed in the solution state. Metal oxide NPs and carbon nanotubes can produce a color change due to their peroxidase activity. However, the major challenge with gold nanoparticles is their intrinsic property to aggregate in the presence of impurities or high ionic strength. In one method, multicolor gold nanoparticles were conjugated with virus-specific antibodies and were used to distinguish between dengue, yellow fever and Ebola viruses [14]. Polydiacetylene (PDA) bilayer were used for the first time to detect H1N1 (Human influenza virus) colorimetrically at very low limits, so they were used for POC testing methods. In 1993, PDA bilayer was used for the first time for detection and since then tremendous modifications have been done like the functionalization of PDA to increase the colorimetric detection limit to 10^5 PFU or 0.2 HAU, as used in many antigen test kits.

6.6.5 Other optical biosensors

6.6.5.1 Evanescent wave fluorescence biosensors

In these biosensors, the biological element is recognized and is bounded within the evanescent field. The behavior of light within the optical fiber or waveguide results in evanescent waves. Confined light is totally internally reflected at the interface between the waveguide fiber and the surrounding medium having lower a refractive index. This results in an electromagnetic field and an evanescent wave is released from the interface. Since release of an evanescent wave is a surface phenomenon, its excitation detects biomolecules that are labeled with fluorescent tags. These waveguide-based biosensors are used for the detection of HIV, Syphhills and Hepatitis C [20].

6.6.5.2 Optical waveguide interferometric biosensors

These biosensors are based on evanescent waves and optical phase difference sensing methods. The near surface region is probed with the evanescent field and any change in the refractive index was compared with the reference field. The alteration in the refractive index is proportional to the concentration of the analyte. This technique is also called resonant waveguide grating (RWG). It is used for studying cellular contents, cellular responses and cellular processes and was also used for the detection of avian influenza virus [21].

6.6.5.3 Reflectometric interference spectroscopy biosensors

These optical biosensors are based upon label-free detection of biomolecules. In this method, white light interference at thin layers is studied. Change in the polarized light is due to the variation in thickness and refractive index of the adsorbed protein layer. This biosensor is used for the detection of circulating tumor cells within a concentration range of 1000–100,000 cells [22].

6.7 Biosensors and biomarkers to detect pathogens

Infectious diseases are caused by several classes of pathogenic microorganisms like viruses, fungi, bacteria, protozoa and parasites. Amongst the infectious diseases, bacterial disease accounts for one-third of human global mortality. Some of the bacteria responsible for high death rate include *Mycobacterium tuberculosis, Streptococcus pneumoniae, Staphylococcus aureus, Vibrio cholerae, Salmonella typhi, E. coli, Helicobacter pylori*. Multiple microorganisms are the cause of diarrheal diseases in children. Microbial toxins like Staphylococcal enterotoxin from *S. aureus*, botulinum toxin from *Clostridium botulinum*, emetic enterotoxins and diarrheagenic toxin from *Bacillus cereus*, mycotoxins from toxigenic molds, seafood toxins from algae can cause severe illness. Presently, culture-based methods, biochemical tests, and serological assays are used for the detection of viable microbial cells in food. These processes are labor-intensive and time-consuming. Therefore, researchers are using biosensor-based methods for faster detection of pathogens. Nanobiosensors are used for point of contact analysis of pathogens as they are very small, portable and give accurate results in a short time. A biomarker is an indicator of a disease or physical state of living tissues under infection. They can be molecular or phenotypic in nature. A biosensor is a device used to detect pathogens, DNA analysis, blood glucose measurements, contamination control and effect of drugs, so biomarkers can be used to detect the pathogens, trace them and then they can be used to deliver the drugs to the target tissues. Innovations in biosensor technology have resulted in the commercialization of biosensors and in the detection of biomolecules and biomarkers. Commercial biosensors have a simple construction, are easy to operate and are for point of care (POC) testing.

Label-free antibody-based SPR are used to detect bacteria and SPR imaging to monitor bacterial growth. An aptamer can be used instead of an antibody to bind to the target. Bacteriophages can be used to capture *E. coli* from drinking water. With a light-scattering sensor such as BARDOT (bacterial rapid detection using optical scattering technology), sensitivity can be improved by using immunomagnetic beads to capture target bacteria or by using selective media for bacterial growth and then laser identification. Biosensors offer the specificity and sensitivity of biological systems integrated into small, low-cost devices. Thanks to their high analytical performance, biosensors represent a powerful alternative to conventional methods in pathogen detection. Biosensors may integrate specific nucleic acid sequences, proteins (like antibodies or enzymes), and even whole cells or cell elements as recognition elements for a specific pathogen. The sensitive detection is enabled due to the transduction of molecular recognition of the target pathogen to a measurable signal that is processed and displayed. In this chapter, the different strategies of optical biosensors used to detect some important bacterial pathogens are discussed.

6.7.1 Bacterial pathogen

6.7.1.1 Vibrio cholerae

V. cholerae is a Gram negative facultative anaerobe that causes cholera. People are infected by it from consuming contaminated food and water and it spreads very fast due to its secretory nature. Thus it is very important to check the spread of the disease by identification of the early to avoid an epidemic. Ultrasensitive DNA-based fluorescence assays are used for the detection of *V. cholerae*. Virulence of *V. cholerae* is controlled by the high expression OmpW gene. Two nanoparticles, gold nanoparticle (AuNPs) and magnetic nanoparticles (MNPs), are used in the preparation of the biosensor. MNPs binds to one end of the gene DNA probe 1 and AuNP binds to the other end of DNA-FAM probe 2. On the release of FAM-probe 2, there is fluorescence emission that is detected by fluorescent spectroscopy. This biosensor can detect as low as 2.34 ng/mL and is highly sensitive [23].

6.7.1.2 Escherichia coli

SPR-based label-free detection of analytes is done to detect *E. coli* O157: H7 at 10^7 CFU/mL [24]; Oh 2003, detected 10^2 CFU/mL and Su reported 10^5–10^8 CFU/mL. A fiber optic sensor was also developed to detect *E. coli* O157:H7 3–30 CFU/mL in spiked ground beef samples [25]. A biconical tapered optical waveguide could detect *E. coli* as low as 70 cells/mL in the buffer even in the presence of other bacteria [26]. A U-bent optical fiber increases the penetration depth of the evanescent wave and thus is highly sensitive [27].

Coherent Anti-stroke Raman Spectroscopy (CARS) is a very precise and sensitive instrument as it is based upon scattered photons. Recently CARS imaging approach was used to detect single *E.coli* cells in milk and urine samples. More recently, droplet-based fluorescence immunoassay was done to detect LOD of *E. coli* 10^2 CFU/mL. Further it could be used to detect *E. coli* in 12 min using smartphones with proper filters. With anti-*E. coli* analytical reagent, SPR Spectroscopy could detect *E. coli* in pure culture and from spinach leaves with LOD 10^{3-4} CFU/mL [28]. *E. coli* (10^3 CFU/mL) flowing through an optical fiber could be detected by using immobilized T4 Bacteriophage as a bioreceptor [29]. Surface-immobilized BP 14 bacteriophage for methillin resistant *S. aureus* and T4 bacteriophage for *E. coli* could detect bacteria upto 10^3 CFU /mL in either case. Parallel bacterial sensing of more than one foodborne pathogens can be done by using multichannel SPR sensing and affinity antibodies. Wang [30] designed grating coupled long range SPR biosensors with antibody-labeled magnetic nanoparticles for identifying *E. coli* O157:H7 LOD of 50 CFU/mL.

6.7.1.3 *Salmonella*

Salmonella is a rod-shaped Gram negative bacteria that belongs to Enterobacteriaceae family. It contains two main types, *Salmonella enterica* and *Salmonella bongori*, with more than 2500 serotypes, all capable of causing disease in humans (WHO, 2018). According to WHO, *Salmonella, S. aureus, Campylobacter and Clostridium perfringens* are the four major bacterial pathogens that cause illness. The infection is caused due to consumption of contaminated food and is marked by fever, nausea, diarrhea, stomach cramps, headache and vomiting. Optical biosensors that are used for detection of *Salmonella* are of four different variants: colorimetric, SERS, SPR and fluorescent-based devices. AuNPs possess a unique optical property that shows a color shift from red (dispersion) to blue/purple (aggregation). AuNPs hybridize to a specific DNA probe designed to bind to *invA* gene of *S. enterica*. Hybridization results in aggregation and there is a color change from red to blue indicating the presence of *S. enterica* [31]. A whole cell of *Salmonella typhimurium* was detected by a simple label-free colorimetric method using an antibody conjugated to AuNPs [32]. Zhu et al. [33] used catalase and AuNPs in an immunoassay for the detection of *S. typhimurium*. *S. typhimurium* was labeled with catalase and sandwiched between two aptamers. Catalase utilized H_2O_2 in solution, slowing down the aggregation of AuNPs whereas in the absence of target, reduction of AuNPs took place at a faster rate resulting in well-dispersed gold nanoparticles. It could detect LOD of 10 CFU/mL. Urease is said to lose its catalytic activity upon adhering to AgNPs. A colorimetric sensor is designed to study competitive binding between *S. typhimurium* and urease for antibody-modified AgNPs that can detect 100 cells/mL [34]. Many gold and silver nanoparticles are used in label free SERS biosensors for the detection of Salmonella. In novel SERS-labeled biosensors, tag or probes are present that carry metal nanostructures to increase the signal, biorecognition element and Raman dyes to modify output signals. A biotinylated single-stranded oligonucleotide probe targeting the *invA* gene of Salmonella was immobilized on a streptavidin-coated dextran sensor surface. The genosensor could detect *Salmonella* LOD 10^2 CFU/mL within 4.5 h [35]. Melanie (2017) innovated an SPR imaging biosensor for multiplex detection of 16S r DNA of *S. typhimurium*, *P. aeruginosa* and *Legionella pneumophila* using AuNPs with LOD as low as 10 pg/mL. Compared to other biosensors, SPR biosensors are used for real time monitoring of Salmonella in food samples. Though there are plenty of innovations in optical biosensors for detecting target bacteria but their use is restricted to laboratory and commercialization is still a big challenge.

6.7.1.4 Campylobacter

There are three foremost species of *Campylobacter*, *Campylobacter coli*, *Campylobacter lari* and *Campylobacter jejuni* that are the causative agent in foodborne infection resulting in gastroenteritis. The most virulent amongst them is *C. jejuni*, that is responsible for 95% of sporadic illness and foodborne diseases. It is a Gram negative bacterium belonging to *Campylobacteraceae*, with 0.2–0.8 μm by 0.5–5 μm in size and spiral, rod or curved shaped. They are obligate aerophiles, theromotolerant and require blood in culture media for growth. It is difficult to culture them in laboratory conditions as they acquire a viable but non-culturable state. It is also difficult to characterize them through biosensors as it changes it shape and becomes coccoid. The main source of *Campylobacter* are uncooked meat and to some extent milk, fruits and vegetables. The majority of optical sensors for detection of bacteria are genosensors that use DNA probes. Organic Light emission electrode was used by Manzano et al. [36] to detect *Campylobacter* spp. A specific DNA probe modified at 5' position was used as capture element with Alexa Fluor 430 (Thermo Fisher Scientific, Waltham, MA, United States) as a detection probe. The biosensor was specific to a 16s rRNA sequence of *Campylobacter*, showing high sensitivity and no cross reactivity. Upon target detection, OLED produced a specific emission spectrum that excited the Alexa Fluor 430 molecule. In another method, DNA probes specific for *CadF* of *C. coli* and *C. jejuni* were fixed to gold nano rods. They measured the variation in SPR before and after hybridization and could detect at LOD of 100 directly in stool samples [37]. Two different methods, SPR and DOT (diffraction optical technology), were done for the quantification of the target hippuricase (*HipO*) gene. Thiolated DNA probes were used for SPR and biotinylated probes for DOT. In SPR, the detection level was as low as 2.5×10^{-9} mol/L and the sensor chip could be reutilized [38]. Wei et al. [39] used an antibody specific for *C. jejuni* whole cells immobilized on a gold surface of SPR chip. They tested in wash samples of broiler and could detect as low as 10^3 CFU/mL. Masdor [40] used a polyclonal antibody in optical biosensors for the detection of *C. jejuni* in food samples. Gold nanoparticles functionalized with aptamers were used for the detection of *C. jejuni* and *C. coli*. Dispersed gold nanoparticles were red in color and after binding to target bacteria, aggregated and the color changed to purple. The sample was tested on chicken samples LODs of 5.6×10^5 CFU/ mL for *C. coli* and 7.2×10^5 CFU/mL for *C. jejuni*.

6.7.1.5 *Bacillus*

The members of the *B. cereus* group (*Bacillus anthracis*, *B. cereus*, *Bacillus pseudomycoides*, *Bacillus mycoides*, *Bacillus thuringiensis*, *Bacillus medusa and Bacillus weihenstephanensis*) are pathogenic to humans. *B. anthracis* is known to cause the deadly anthrax disease in human and animals. Consumption of food contaminated with *B. cereus* may lead to gastrointestinal diseases characterized by non-bloody diarrhea and abdominal pain within 4–16 h after eating. These symptoms are due to the release of diarrheal toxin hemolysin BL, cytotoxin Cytk and nonhemolytic enterotoxin, Nhe [41]. Microbiological methods like pour plate or plate count agar, used for the detection of spores requires two days for detection. The spore-specific Dipicolinic acid, is present as Ca2 + chelate complex and is a specific biomarker of aerobic spore-forming bacteria. Several optical methods such as photoluminescence, fluorescence, infrared spectroscopies and Raman/SERS methods are used for dipicolinic acid detection. In one method, terbium chloride was used to color a suspension of bacterial spores that reacts with calcium dipicolinate. Formed terbium dipicolinate can be detected by photoluminescence [42]. The LOD was 4.4×10^5 CFU/mL. Using Raman spectroscopy/SERS, a single spore in a confined electromagnetic field generated a highly specific spectral signature within seconds. Raman spectroscopy combined with SERS active gold nanoparticles was used to detect and discriminate among five different types of *Bacillus* spores [43]. A SERS portable device was innovated by Cowcher [44] that could detect 1100 spores/200 μL of sample.

6.7.1.6 *Staphylococcus*

Though optical biosensors have been reported to have high sensivity and selectivity, it is still a challenge to accurately identify bacteria up to the strain level. Some researchers have reported using optical biosensors to identify methicillin-sensitive *S. aureus* (MSSA) from methicillin-resistant *S. aureus* (MRSA) in less than 3 h. A biosensor was developed based on the *S. aureus* proteolytic activity. The sensor was functionalized with pathogen proteases that acted on a specific peptide substrate placed at the center of two magnetic beads. The dissociation of the magnetic nanobeads from the peptide moieties results in change in color. Tip Enhanced Raman Spectroscopy (TERS) was used for the first time for bacterial biosensing by Neugebauer et al. [45] to detect a single bacterium of *Staphylococcus epidermidis*. He simultaneously obtained AFM images and TERS spectra of a single bacterium with 10^4-10^5 enhancement factor. Recent development in light scattering sensors uses BARDOT to create a signature library. Nearly 90% efficiency is seen in the screening of various microorganisms like *Salmonella*, *Listeria*, *Bacillus*, *Staphylococcus* and *Vibrio*. Lu [46] used a microfluidic chip functionalized with SERS technology to identify MRSA. The results were similar to that obtained from traditional PCR, serological tests, colony screening and other methods. In another method, single cell detection of *S. aureus* was done using a magnetically assisted SERS biosensor [47]. Some clinically important microorganisms like *Enterococcus faecium, E. coli, S. aureus, S. epidermidis, and Candida albicans* were detected using Raman spectroscopy.

6.8 Virus pathogen

A virus can grow and multiply only in a host cell like plants, animals and bacteria. Once a virus spreads in the general population it can create a pandemic like the ongoing COVID-19 pandemic. One of the major reasons for the rapid spread of the disease is the lack of effective POC testing methods. The conventional virus detection methods include: viral culture, RT-PCR, and immunoassays like ELISA. All of these methods are labor-intensive, expensive, time consuming and with complicated operation. Many nanostructured-based biosensors, with label-free and real-time detection have been coupled to optical sensors (e.g., SERS, fluorescence and colorimetric sensors) and are used for virus detection. Optical nanobiosensors play a vital role in the early diagnosis of viruses and have many benefits like low cost of the portable laser, high sensitivity and a very low limit of detection.

6.8.1 Influenza

Influenza is the causative agent of many worldwide epidemics, starting with; the Spanish flu (H1N1) in 1918, then Asian flu (H2N2) in 1957 followed by Hong Kong influenza (H3N2) in 1968 and swine influenza (H1N1) in 2009. This virus affects school children, elderly and immunocompromised patients. This virus belongs to the *orthomyxovirus* family with three prominent subtypes, influenza A, B, and C. It has segmented RNA and two glycoproteins cover the viral surface, haemagglutinin (HA) and neuraminidase (NA), in a ratio one to four, respectively. Human populations do not have immunity against HA and NA antigens on the virion surface. Most of the new viruses are mutants created due to antigenic drift. Most common optical sensor methods used for the detection are UV/Vis Spectroscopy, SERS and Fluoresence spectroscopy. In one method, Liu [48] used a macrocyclic dye substrate (squaraine-derived core blocked on one side with sialic acid) reacted with viral NA and the blocks were removed. In the presence of NA, squarine showed a red shift in fluorescence emission and absorption. Results could be checked with the naked eye. With slight optimization, this principle can be used to design portable sensors for influenza detection. In another technique, SERS was used in magnetic immunosensors for avian influenza detection. Chemisorption of 4-mercaptobenzoic acid (4-MBA) molecules on the gold nanoparticles was used for the immobilization of influenza A IgG antibodies. H3N2 was identified with LOD at 10^2 $TCID_{50}$/mL. In another study, SERS- based assay was used for the detection influenza antibodies. Au nanoparticles were exchanged with SERS active nanotags. A complex (Au-nanotags-influenza-antibodies) was trapped on an immobilized antibodies strip with probes. A positive test will give two red lines (control line and SERS line) while negative will give only one control line. Influenza haemagglutinin (HA) expressed by Avian Influenza Virus binds to α-2,3 sialic acid present in the intestine while HA proteins of human infecting virus binds to α-2,6 sialic acid glycan, mainly expressed in the epithelial cells of the upper respiratory tract. Based on the difference in sialic acid linkages biosensors were developed that could differentiate between AIV and Human Influenza virus. HA binding to sialic acid attached to gold nanoparticles is used for the detection of influenza virus without amplification and pretreatment. Most of the portable commercial devices used to detect influenza virus were based on lateral flow tests. Samples to be tested were added on sample pads that were conducted through chromatography. In the specific test influenza virus marker is recognized by antibodies carried by gold nanoparticles fixed to adsorbent pad. Influenza virus attached to gold nanoparticles reaches the test lines pre-immobilized with antibodies specific to influenza A and B virus. When the virus has its complementary antibody, then a prominent red line is seen. Usually, antibodies specific to HA protein are used as an immobilized antibody in the test line. The non-bound immune gold nanoparticles reach the control line which harbors the secondary antibody showing the second red line. By this method, we can distinguish between influenza A and B virus but not its subtypes. Anti M1 antibodies conjugated to gold nanoparticles were used in a Plasmon-assisted fluroimmunoassay for the detection of influenza virus. A fluoresence signal was produced on the addition of cadmium telluride quantum rods. In another method, fabricated biotinylated aptamers, were coated with streptavidin and immobilized on the biosensor gold surface. The aptamers captured AIV H5N1 virus in the sample and caused a change in the refractive index. The refractive index value was directly proportional to the concentration of AIV in the sample.

6.8.2 Human immunodeficiency virus

HIV is a member of a subset of retroviruses known as lentiviruses. HIV infects a person's immune system especially the $CD4^+$ T cells and completely destroys it. Thus the person becomes susceptible to tuberculosis, fungal infections, bacterial diseases and some types of cancers. The final stage of the disease is Acquired Immunodeficiency Syndrome (AIDS). According to WHO, around 37.7 million people were affected with HIV at end of 2020, of which 25.4 million were from Africa [49]. Nearly 680,000 died from HIV-related causes. There are two types of HIV viruses, and HIVI is the most common variant that causes the disease. Glycoprotein 41 (Gp41) is a transmembrane protein that binds virus and the infected cells. The extent of AIDS progression could be determined by monitoring the level of Gp41. Lu et al. [50] developed a biosensor by coating the surface of quartz crystal with a synthetic peptide analogous to 579−613 residues of Gp41. This film had a strong affinity towards the Gp41 protein and LOD was 2 ng/mL. In another optical sensing platform, the virus could be detected from biological samples like serum and even a phosphate buffered saline. When the intact virus was adsorbed to the surface, a shift in resonance peak was observed and it could be detected with 10 pm wavelength resolution [51].

6.8.3 Hepatitis

Hepatitis B is one of the most dreadful diseases affecting humans, causing 800,000 deaths every year. Nearly 15%−40% of infected patients develop liver cirrhosis, liver failure or liver cancer [52]. In one method, DNA hybridization was done to detect Hepatits B virus using non-faradic impendance spectroscopy. DNA probes were biotinylated on streptavidin-based magnetic nanoparticle and then nanoparticles were immobilized on a gold electrode using a magnet. By this method, up to 50 pm of HBV DNA could be detected in 20 μL sample [53]. Hepatitis B surface antigen and antibody reaction was studied using a SPR biosensor. The reaction was more sensitive and had nearly seven folds higher limit of detection as compared to ELISA. In this method, no cross-reactivity with other antibodies was seen [54]. In another report, Hepatitis B surface antigen was detected by SPR. They used Hepatitis B positive antibody human serum samples for the test and non-immunized serum samples as controls. The biosensor had a low detection value of 208 mIU/mL. Zengin et al. [55] reported another method for detection of Hepatitis B virus DNA sequence using sandwich assay and SERS. They first immobilized the DNA strand on a gold nanoparticle surface, and by using sandwich assay, target DNA was detected using surface enhance Raman scattering signals. In this method, the lowest level of hepatitis virus detection was up to fm level at different temperatures and without labeling [55].

6.8.4 Ebola

Ebola virus disease (EVD) was earlier known as Ebola hemorrhagic fever is a rare but fatal human ailment. Average EVD fatality rate is 50% but sometimes it has been reported to go as high as 90%. The virus family *Filoviridae* has three genera Ebolavirus, Marburgvirus and Cuevirus. It is transmitted from wild animals and spreads in human population through human to human transmission. An electrochemical biosensor was designed for the detection of Ebola virus DNA by using an enzyme complex. The biotinylated streptavidin phosphate conjugate was used to enhance the sensitivity and specificity that could detect the pathogen as low as 4.7 nm. Ilkhani et al. [56] designed the above biosensor and used electrochemical impendance spectroscopy for detection. An optofluidic biosensor was used for the detection of the whole virus in biological media by Yanik et al. [57]. In these biosensors, immobilized antibodies against Ebola glycoprotein were developed. In another method, a microfluidic chip was designed for concentration and single nucleotide detection of Ebola virus. This method had high specificity and could detect sample as low as 0.2 pfu/mL [58].

6.8.5 Dengue

Dengue is a mosquito-transmitted viral infection prevalent in tropical and subtropical Asian and African countries. The virus causing the infection is known as dengue virus (DENV). It has four serotypes, DENV-1, -2, -3, -4, and each serotype has three structural proteins and seven nonstructural proteins. There is no specific treatment for dengue, hence early diagnosis of dengue is essential for prevention of dengue shock syndrome and dengue hemorrhagic fever. According to WHO, 3.9 billion people are at the risk of getting the disease with 96 million showing the severity of the disease. In one of the methods, gold/iron nanoparticles were fabricated and IgM was immobilized to trap DENV protein. This SPR method for DENV detection has high sensitivity can detect in the concentration range of 0.0001−10 nM [59]. The Faradaic electrochemical impedance method is used to characterize the impedance variation of a nanoporous alumina biosensor while detecting dengue serotype 2 (Denv2) viral particles on binding to its serotype 2-specific immunoglobulin G antibody trapped in the thin alumina layer [60]. In another method, SPR was used for the detection of anti-dengue virus samples in the human serum. Four dengue serotypes were used as ligands on the biosensor. Only 1 μL of the patient serum was sufficient to show variation in the SPR angle; based on different angle variation the four different serotypes could be identified [61].

6.8.6 COVID-19 pandemic

Currently, many viruses have the potential to cause pandemics due to different factors like expeditious dissemination, rapid transmission rate of mutated variants, inability to develop sensitive diagnostic techniques and the absence of proper vaccines and safe drugs. The current pandemic caused by SARS-CoV-2 belongs to positive strand RNA viruses. Like any other viral outbreak, early diagnosis of the disease is primarily essential in controlling the spread of the disease. However, the current situation became uncontrollable because of the specific mode of transmission and the unavailability of the drugs to check the spread of the disease. Another peculiarity is that 30% of the confirmed cases are asymptomatic, thus it is difficult to check the transmission of the disease. Presently, RT-PCR is the most suitable technique used for detecting SARS-CoV-2 infections [62]. However, this method is time consuming, requires skilled scientific personnel, is labor intensive and unavailable in rural areas. In the case of SARS-COV-2 virus, the spike glycoprotein binds to receptor angiotennsin converting enzyme 2 (ACE2) present on the surface of human cell and enters the cell. The RT-PCR technique used for detection has several pitfalls like RNA extraction, position of nasopharyngeal swab collection, enzyme inhibitors, transport of samples to the test centers. Recent development in the field of photonics-based appliances, imaging devices, highly reproducible spectral reading, measurement and viral detection has become easier. A group of researchers from Samsung Strategy and Innovation Center are working on developing SERS-based technology for rapid detection of SARS-COV-2 from human saliva [63]. Botanisol Analytics at the University of Arizona are working on a Raman Spectroscopy based COVID-19 detector for rapid and cost effective detection as compared to conventional RT-PCR. In another method, a protein was created comprised of a gold-binding protein (GBP) and SARS-recognizing element SARS coronaviral surface antigen (SCVme) that was used as in a SPR biosensor. SPR sensograms showed highly specific binding of the anti-SARS SCVme to the GBP-SCVme coated biosensors. A packing fraction of 10 g/mL of the above fusion protein has a detection limit of 200 ng/mL anti-SCVme in less than 10 min [64]. In another recent study, real-time SPR assays were done to detect the interactions between SARS-RBD (Receptor binding entity of SARS-CoV) and SARS-CoV-2-CTD (Receptor binding entity of SARS-CoV-2) with its hACE2 receptor. The binding affinity was four times higher in SARS-CoV-2-CTD as compared to SARS-RBD with the hACE2 receptor. The C-terminal spike protein of the antigen was different in the two variants [65]. Similarly, Shang [66] also found that SARS-CoV-2 has higher binding affinity for hACE2 receptors compared to SARS-CoV spike binding protein. The viral genome after sequencing showed three conserved regions; enveloped protein gene (E gene), nucleocapsid phosphoprotein gene (N gene) and NRA-dependent RNA polymerase gene (RdRP gene). In this work, simultaneously two target sequences were tapped by four antisense oligonucleotide sequences. In the presence of the target RNA sequence of SARS-CoV-2, the thiolated capped AuNP start to aggregate and that further increases on addition of RNaseH, finally leading to visual precipitation of the plasmonic nanoparticles and detectable color change. This aggregation resulted in 40 nm red shift in UV-VIS absorption spectral studies [67].

6.9 Recent development in the optical biosensors

Nanotechnology based optical biosensors are gradually gaining a lot of importance as they can directly detect molecular targets in real time. Optical nanobiosensors are used for the detection of pathogens which are based on QD, upconversion nanoparticles (UCNPs), noble metal nanoparticles, and chemiluminescence assays. Recent advances in plasmonics and nanofabrication has further helped in the design of new biosensors. The design of optical biosensors depends upon the properties of plasmonic nanostructures but also on its further application. That is whether it has to be used as high-end laboratory instruments or robust POC instruments. Two developments in the field of plasmonic biosensors, localized surface plasmons (LSPs) and propagating surface plasmon (PSPs), completely revolutionized the plasmonic biosensors. SPR biosensors have been used for rapid label-free detection of biological macromolecules in numerous fields like medical diagnostics, food safety, and environmental samples. Recent developments in the field of fabrication like electron beam and ion beam lithography and characterization of plasmonic structures with scanning electron and atomic force microscopy have greatly influenced the designing of a new generation of optical biosensors. Thus optical bionanosensors based on plasmonic nanostructures are the focus of extensive research. Plasmonic nanostructures used in many optical biosensors can be excited with attenuated total reflection, illumination, diffractive coupling or coupling with an optical waveguide. A typical biosensor when compared with its counterparts (other biosensors available in the market) is preferred only if it has the ability to detect low concentrations of target analyte; expressed as LOD. The LOD of a biosensor depends upon several components, like the optical platform, microfluidic system, functional coating and plasmonic nanostructure. These optical biosensors can detect a wide range of target analytes starting from small chemicals, drugs, and biomolecules like toxins, proteins, and nucleic acids to bacterial and viral pathogens. It has been reported that target analytes as low as 10^{-5} ng/mL has been detected in buffers while LOD in blood serum is $1-10$ ng/mL.

6.9.1 Nanostructured optical biosensors

Quantum dot-based biosensors are made up of semiconductor nanocrystals, used for bioimaging such as cell imaging, tumor diagnosis, gene therapy and biosensing (biomarker detection). These have been used for multiplex virus detection as well as in test strips for the identification of HIV target DNA. Due to their photostability, luminous efficiency, and energy transfer efficiency they are currently used in many biosensor devices.

Upconversion nanoparticle-based biosensors: in comparision to conventional dye-based sensing methods for analyte detection, these biosensors have the advantage of minimal background fluorescence, high photostability, low toxicity and minimal harm to the oligonucleotide probe. Further, the availability of a portable and cheap infrared laser source as the triggering source facilitates onsite detection and reduces damage to the analyte, as excitation is due to shorter wavelength. This has been used for the detection of Ebola virus gene sequence using heterogeneous gene assay.

2D materials based fluorescence nanobiosensors are two dimensional materials due to their large surface area, good metallic conductivity, biocompatibility, and specific absorption peaks makes them best option for biological applications. In this method, dye-labeled single-stranded DNA probes with specific complementarity are used to bind to target DNA or RNA forming a fluorescence quencher pair. The nanosheets having fluorescence-quenching ability are used for fabrication of biosensors. These can adsorb ssDNA probe on their surface and quench its fluorescence. If the ssDNA probe finds its complementary target DNA or RNA it detaches from the surface of the nanosheets resulting in a fluorescence signal.

Noble nanomaterial-based biosensors Due to their intrinsic chemical physical and optical properties, noble metals like gold, silver, palladium are used for coating of sensors. The surface stability of the nanoparticles has been used for various biological molecules. Recently COVID-19 diagnosis was done using plasmonic photothermal effect with LSPR. In another method, specific LSPR induced immunoassay was done to detect influenza virus. LOD of this LSPR induced biosensor for influenza virus was 0.03 pg/mL in water and 0.4 pg/mL for human serum. Smartphone-based POC biosensors were developed for the detection avian influenza virus by Xia and his co-authors 2019. After capture of the virus by antibodies, the Au/Ag NPs can detect the virus by colorimetric reaction that could be detected by the naked eye. The LOD is 2.7×10^4 embryonic infectious dose/mL.

6.10 Conclusion

The last seven decades has witnessed tremendous development in the field of biosensors. Though many methods have been reported for faster and real-time detection of disease pathogens, all of them are still at the laboratory level. In order to be used as diagnostic tool, further research is required in terms of sensitivity, simultaneous analysis of multiple biomarkers, reusability, and integration various functions by the same chip. Though optical biosensors are used in routine clinical biochemistry laboratories for analysis of blood parameters like lactate, urea, creatinine and glucose, the sensitivity of immunosensors is still in its infancy. POC molecular testing requires highly specific transducer technology, simple, easy to operate, interchangeable bioreceptors, cost effective, reduced sample and reagent volume and most importantly, complete validation of the device with other assay methods. Thus the major challenge is to make a device that can be afforded by all groups of people without compromising on its quality and features. Further, the capability of the biosensor to detect pathogens from complex food matrices, drains, floor drains, water samples and environmental samples is a big challenge. Thus the amalgamation of optical sensors with nanotechnology can definitely bring a change in sensing technology for the detection of pathogens. Optical nanobiosensors are still in the infancy stage of development and detection of pathogens. The size of the nanoparticles has to be carefully manipulated as all the properties are size dependent. While designing commercial biosensors that can be used at the POC, many factors have to be standardized. It should have minimal background fluorescence, photostability, low toxicity, oligonucleotide hybridization with a probe. With the development and ongoing research in this field, very soon, low cost, reliable, sensitive, robust biosensors will be available in the market.

References

[1] Y. Tsao, T. Yao-Hung, L. Wan-Ting, S. Ching-Ju, S. Ching-Fen, C. Chao-Min, Trends Mol. Med. 26 (12) (2020). Available from: https://doi.org/10.1016/j.molmed.2020.09.004.

[2] O.I. Guliy, B.D. Zaitsev, O.S. Larionova, I.A. Borodina, Virus detection methods and biosensor technologies, Biophysics 64 (2019) 890–897.
[3] M.S. Cheng, J.S. Ho, C.H. Tan, J.P.S. Wong, L.C. Ng, C.-S. Toh, Development of an electrochemical membrane-based nanobiosensor for ultrasensitive detection of dengue virus, Anal. Chim. Acta 725 (2012) 74–80.
[4] F. Malvano, R. Pilloton, D. Albanese, A novel impedimetric biosensor based on the antimicrobial activity of the peptide nisin for the detection of Salmonella spp, Food Chem. 325 (2020).
[5] R.E. Ionescu, Biosensor platforms for rapid detection of *E. coli* bacteria, in: A. Samie (Ed.), Recent Advances on Physiology, Pathogenesis and Biotechnological Applications, IntechOpen, London, 2017 [CrossRef].
[6] P. Yagupsky, F.S. Nolte, Quantitative aspects of septicemia, Clin. Microbiol. Rev. 3 (1990) 269–279.
[7] S.O. Kelley, What are clinically relevant levels of cellular and biomolecular analytes? ACS Sensors 2 (2017) 193–197.
[8] K.D. Hue, T.V. Tuan, H.T. Thi, C.T. Bich, H.H. Anh, B.A. Wills, et al., Validation of an internally controlled one-step real-time multiplex RT-PCR assay for the detection and quantitation of dengue virus RNA in plasma, J. Virol. Methods 177 (2011) 168–173.
[9] V.M. Corman, A. Rasche, C. Baronti, S. Aldabbagh, D. Cadar, C.B.E.M. Reusken, et al., Clinical Comparison, Standardization, and Optimization of Zika Virus Molecular Detection, Bull. WHO 94 (2016) 880.
[10] B. Nakstad, et al., Early detection of neonatal group B streptococcus sepsis and the possible diagnostic utility of IL-6, IL-8, and CD11b in a human umbilical cord blood in vitro model, Infect. Drug Resistant 9 (2016) 171–179.
[11] F. Bian, et al., Variation in the levels of IL-6 in pediatric patients with severe bacterial infectious diseases and correlation analysis between the levels of IL-6 and procalcitonin, Exp. Ther. Med. 13 (2017) 3484–3488.
[12] A. Aliu-Bejta, et al., Presepsin values as markers of severity of sepsis, Intern. J. Infect. Dis. 95 (2020) 1–7.
[13] M. Magliulo, P. Simoni, M. Guardigli, E. Michelini, M. Luciani, R. Lelli, et al., A rapid multiplexed chemiluminescent immunoassay for the detection of *Escherichia coli* O157: H7, *Yersinia enterocolitica, Salmonella typhimurium*, and *Listeria monocytogenes* pathogen bacteria, J. Agric. Food Chem. 55 (2007) 4933–4939.
[14] M.E. Ilahi, S.,C. Mukhopadhyay, Detection methodologies for pathogen and toxins: a review, Sensors 17 (2017) 1885. Available from: https://doi.org/10.3390/s17081885.
[15] D. Dey, T. Goswami, Optical biosensors: a revolution towards quantum nanoscale electronics device fabrication, J. Biomed. Biotechnol. 348218 (2011).
[16] T.E. Rohr, T. Cotton, N. Fan, P.J. Tarcha, Immunoassay employing surface-enhanced Raman spectroscopy, Anal. Bochem. 182 (1989) 388–398. Available from: https://doi.org/10.1016/0003-2697(89)90613-1.
[17] B. Liedberg, C. Nylander, I. Lundsrom, Surface-plasmon resonance for gas detection and biosensing, Sens. Actuators 4 (1983) 229–304.
[18] V. Canalejas-Tejero, S. Herranz, A. Bellingham, M.C. Moreno-Bondi, C.A. Barrios, Passivated aluminum nanohole arrays for label-free biosensing applications, ACS Appl. Mater. Interfaces 6 (2) (2014) 1005–1010.
[19] Maddali, H., Miles, C.E., Kohn, J., O'Carroll, D.M. Optical biosensors for virus detection: prospects for SARS-CoV-2/COVID-19. Chembiochem. 22(7), 2021, 1176–1189. Available from: https://doi.org/10.1002/cbic.202000744. Epub 2020 Dec 9. PMID: 33119960, PMCID: PMC8048644.

[20] M.J. Lochhead, K. Todorof, M. Delaney, J.T. Ives, C. Greef, K. Moll, et al., Rapid multiplexed immunoassay for simultaneous serodiagnosis of HIV-1 and coinfections, J. Clin. Microbiol. 49 (2011) 3584–3590.
[21] S. Rau, U. Hilbig, G. Gauglitz, Label-free optical biosensor for detection and quantification of the non-steroidal anti-inflammatory drug diclofenac in milk without any sample pretreatment, Anal. Bioanal. Chem. 406 (2014) 3377–3386.
[22] T. Kumeria, M.D. Kurkuri, K.R. Diener, L. Parkinson, D. Losic, Label-free reflectometric interference microchip biosensor based on nanoporous alumina for detection of circulating tumour cells, Biosens. Bioelectron. 35 (2012) 167–173.
[23] A. Narmani, M. Kamali, B. Amini, H. Kooshki, A. Amini, L. Hasani, Highly sensitive and accurate detection of *Vibrio cholera* O1 OmpW gene by fluorescence DNA biosensor based on gold and magnetic nanoparticles, Process. Biochem. 65 (2018) 46–54.
[24] P.M. Fratamico, Comparison of culture, polymerase chain reaction (PCR), *Taqman salmonella*, and *Transia card* salmonella assays for detection of salmonella spp. in naturally-contaminated ground chicken, ground turkey, and ground beef, Mol. Cell Probes 17 (2003) 215–221.
[25] D.R. DeMarco, D.V. Lim, Detection of *Escherichia coli* O157:H7 in 10-and 25-gram ground beef samples with an evanescent-wave biosensor with silica and polystyrene waveguides, J. Food Prot. 65 (2002) 596–602.
[26] K. Rijal, A. Leung, P.M. Shankar, R. Mutharasan, Detection of pathogen *Escherichia coli* O157:H7 at 70 cells/mL using antibody-immobilized biconical tapered fiber sensors, Biosens. Bioelectron. 21 (2005) 871–880.
[27] R. Bharadwaj, V. Sai, K. Thakare, A. Dhawangale, T. Kundu, S. Titus, et al., Evanescent wave absorbance based fiber optic biosensor for label-free detection of *E. coli* at 280 nm wavelength, Biosens. Bioelectron. 26 (2011) 3367–3370.
[28] H. Baccar, M.B. Mejri, I. Hafaiedh, T. Ktari, M. Aouni, A. Abdelghani, Surface plasmon resonance immunosensor for bacteria detection, Talanta 82 (2) (2010) 810–814.
[29] S.M. Tripathi, W.J. Bock, P. Mikulic, R. Chinnappan, A. Ng, M. Tolba, et al., Long period grating based biosensor for the detection of *Escherichia coli* bacteria, Biosens. Bioelectron. 35 (1) (2012) 308–312.
[30] Y. Wang, W. Knoll, J. Dostalek, Bacterial pathogen surface plasmon resonance biosensor advanced by long range surface plasmons and magnetic nanoparticle assays, Anal. Chem. 84 (19) (2012) 8345–8350.
[31] A.K. Singh, R. Drolia, X. Bai, A.K. Bhunia, Streptomycin induced stress response in *Salmonella enterica* serovar Typhimurium shows distinct colony scatter signature, PLoS One 10 (2015) e0135035.
[32] O.R. Miranda, X. Li, L. Garcia-Gonzalez, Z.-J. Zhu, B. Yan, U.H.F. Bunz, et al., Colorimetric bacteria sensing using a supramolecular enzyme–nanoparticle biosensor, J. Am. Chem. Soc. 133 (25) (2011) 9650–9653.
[33] C. Zhu, G. Yang, H. Li, D. Du, Y. Lin, Electrochemical sensors and biosensors based on nanomaterials and nanostructures, Anal. Chem. 87 (2014) 230–249.
[34] Singh, A.K., Bettasso, A.M., Bae, E., Rajwa, B., Dundar, M.M., Forster, M.D., et al. Laser optical sensor, a label-free on-plate *Salmonella enterica* colony detection tool. mBio 5(1), 2014, e01019–01013.
[35] H. Zhang, X. Ma, Y. Liu, N. Duan, S. Wu, Z. Wang, et al., Gold nanoparticles enhanced SERS aptasensor for the simultaneous detection of *Salmonella typhimurium* and *Staphylococcus aureus*, Biosens. Bioelectron. 74 (2015) 872–877.
[36] M. Manzano, F. Cecchini, M. Fontanot, L. Iacumin, G. Comi, P. Melpignano, OLED-based DNA biochip for *Campylobacter* spp. detection in poultry meat samples, Biosens. Bioelectron. 66 (2015) 271–276.

[37] S. Shams, B. Bakhshi, T. Tohidi, M. Behmanesh, M. Behmanesh, A sensitive gold-nanorods-based nano-biosensor for specific detection of *Campylobacter jejuni* and *Campylobacter coli*, J. Nanobiotechnol 17 (2019) 43 [CrossRef].

[38] T.J. Gnanaprakasa, O.A. Oyarzabal, E.V. Olsen, V.A. Pedrosa, A.L. Simonian, Tethered DNA scaffolds on optical sensor platforms for detection of hipO gene from *Campylobacter jejuni*, Sens. Actuators B 156 (2011) 304–311.

[39] D. Wei, O. Oyarzabal, T.S. Huang, S. Balasubramanian, S. Sista, A.L. Simonian, Development of a surface plasmon resonance biosensor for the identification of *Campylobacter jejuni*, J. Microbiol. Methods 69 (2007) 78–85.

[40] N.A. Masdor, Z. Altintas, I.E. Tothill, Surface plasmon resonance immunosensor for the detection of *Campylobacter jejuni*, Chemosensors 5 (2017) 16.

[41] J. Vidic, P. Vizzini, M. Manzano, D. Kavanaugh, N. Ramarao, M. Zivkovic, et al., Point-of-need DNA testing for detection of foodborne pathogenic bacteria, Sensors 19 (2019) 1100.

[42] D.L. Rosen, Luminescence dynamics of terbium monodipicolinate in ethyl ammonium nitrate, Appl. Spectrosc. 60 (12) (2006) 1453–1460. Available from: https://doi.org/10.1366/000370206779321409.

[43] W. He, W.A. Henne, Q. Wei, et al., Two-photon luminescence imaging of *Bacillus* spores using peptide-functionalized gold nanorods, Nano Res. 1 (2008) 450–456. Available from: https://doi.org/10.1007/s12274-008-8047.

[44] D.P. Cowcher, Y. Xu, R. Goodacre, Portable, quantitative detection of Bacillus bacterial spores using surface-enhanced Raman scattering, Anal. Chem. 85 (6) (2013) 3297–3302. Available from: https://doi.org/10.1021/ac303657k. Epub 2013 Mar 1. PMID: 23409961.

[45] U. Neugebauer, U. Schmid, et al., Towards a detailed understanding of bacterial metabolism—spectroscopic characterization of *Staphylococcus epidermidis*, ChemPhysChem 8 (1) (2006) 124–137. Available from: https://doi.org/10.1002/cphc.200600507.

[46] X. Lu, D.R. Samuelson, Y. Xu, H. Zhang, S. Wang, B.A. Rasco, et al., Detecting and tracking nosocomial methicillin-resistant *Staphylococcus aureus* using a microfluidic SERS biosensor, Anal. Chem. 85 (2013) 2320–2327.

[47] J. Wang, X. Wu, C. Wang, N. Shao, P. Dong, R. Xiao, et al., Magnetically assisted surfaceenhanced Raman spectroscopy for the detection of *Staphylococcus aureus* based on aptamer recognition, ACS Appl. Mater. Interfaces 7 (2015) 20919–20929.

[48] W. Liu, C.F.A. Gómez-Durán, B.D. Smith, Fluorescent neuraminidase assay based on supramolecular dye capture after enzymatic cleavage, J. Am. Chem. Soc. 139 (18) (2017) 6390–6395. Available from: https://doi.org/10.1021/jacs.7b01628.

[49] HIV/AIDS. 2021. Available online: https://www.who.int/en/news-room/fact-sheets/detail/hiv-aids.

[50] C.H. Lu, Y. Zhang, S.F. Tang, Z.B. Fang, H.H. Yang, X. Chen, et al., Sensing HIV related protein using epitope imprinted hydrophilic polymer coated quartz crystal microbalance, Biosens. Bioelectron. 31 (2012) 439–444.

[51] H. Shafiee, E.A. Lidstone, M. Jahangir, F. Inci, E. Hanhauser, T.J. Henrich, et al., Nanostructured optical photonic crystal biosensor for HIV viral load measurement, Sci. Rep. 4 (2014) 4116–4123.

[52] D. Lavanchy, M. Kane, Global epidemiology of hepatitis B virus infection, Hepatitis B Virus in Human Diseases, Humana Press, Cham, Switzerland, 2016, pp. 187–203.

[53] W.M. Hassen, C. Chaix, A. Abdelghani, F. Bessueille, D. Leonard, N. Jaffrezic-Renault, An impedimetric DNA sensor based on functionalized magnetic nanoparticles for HIV and HBV detection, Sens. Actuators B—Chem. 134 (2008) 755–760.

[54] Y.J. Tam, N.A. Zeenathul, M.A. Rezaei, N.H. Mustafa, M.L.M. Azmi, A.R. Bahaman, et al., Wide dynamic range of surface-plasmon-resonance-based assay for

hepatitis B surface antigen antibody optimal detection in comparison with ELISA, Biotechnol. Appl. Biochem. 64 (2017) 735–744.
[55] A. Zengin, U. Tamer, T. Caykara, SERS detection of hepatitis B virus DNA in a temperature-responsive sandwich hybridization assay, J. Raman Spectrosc. 48 (2017) 668–672.
[56] H. Ilkhani, S. Farhad, A novel electrochemical DNA biosensor for Ebola virus detection, Anal. Biochem. 557 (2018) 151–155.
[57] A.A. Yanik, M. Huang, O. Kamohara, A. Artar, T.W. Geisbert, J.H. Connor, et al., An optofluidic nanoplasmonic biosensor for direct detection of live viruses from biological media, Nano Lett. 10 (2010) 4962–4969 [CrossRef] [PubMed].
[58] H. Cai, J.W. Parks, T.A. Wall, M.A. Stott, A. Stambaugh, K. Alfson, et al., Optofluidic analysis system for amplification-free, direct detection of Ebola infection, Sci. Rep. 5 (2015) 14494–14502.
[59] N.A.S. Omar, Y.W. Fen, J. Abdullah, et al., Development of an optical sensor based on surface plasmon resonance phenomenon for diagnosis of dengue virus E-protein, Sens. Bio-Sens. Res. 20 (2018) 16–21. Available from: https://doi.org/10.1016/j.sbsr.2018.06.001.
[60] B.T.T. Nguyen, et al., Electrochemical impedance spectroscopy characterization of nanoporous alumina dengue virus biosensor, Bioelectrochemistry 88 (2012) 15–21. Available from: https://doi.org/10.1016/j.bioelechem.2012.04.006.
[61] P. Jahanshahi, E. Zalnezhad, S.D. Sekaran, F.R.M. Adikan, Rapid immunoglobulin M-based dengue diagnostic test using surface plasmon resonance biosensor, Sci. Rep. 4 (2014) 3851–3858.
[62] T. Ji, Z. Liu, G. Wang, X. Guo, S. Akbar khan, C. Lai, et al., Detection of COVID-19: a review of the current literature and future perspectives, Biosens. Bioelectron. (2020) 112455.
[63] Devpost 2020. <https://devpost.com/software/samsung-research-for-sars-cov-2-detection-with-spectroscopy>.
[64] T.J. Park, M.S. Hyun, H.J. Lee, S.Y. Lee, S. Ko, A self-assembled fusion protein-based surface plasmon resonance biosensor for rapid diagnosis of severe acute respiratory syndrome, Talanta 79 (2) (2009) 295–301.
[65] Q. Wang, Y. Zhang, L. Wu, S. Niu, C. Song, Z. Zhang, et al., Structural and functional basis of SARS-CoV-2 entry by using human ACE2, Cell 181 (2020) 894–904.
[66] J. Shang, G. Ye, K. Shi, Y. Wan, C. Luo, H. Aihara, et al., Structural basis of receptor recognition by SARS-CoV-2, Nature (2020) 1–4.
[67] P. Moitra, M. Alafeef, K. Dighe, M. Frieman, D. Pan, Selective naked-eye detection of SARS-CoV-2 mediated by N gene targeted antisense oligonucleotide capped plasmonic nanoparticles, ACS Nano 14 (6) (2020) 7617–7627.

Further reading

S.K. Metkar, K. Girigoswami, Diagnostic biosensors in medicine—a review, Biocatal. Agric. Biotechnol. 17 (2019) 271–283.

CHAPTER 7

Biosensor as quick analytic tool in pandemic!

Mirna Garhnayak[1], Abhijita Mahapatra[1], Loknath Garhnayak[2], Shakti Rath[3] and Aswini Kumar Kar[4]

[1]Department of Prosthodontics, Institute of Dental Sciences, Siksha O Anusandhan (Deemed to be University), Bhubaneswar, Odisha, India
[2]Department of Prosthodontics, SCB Dental College & Hospital, Cuttack, Odisha, India
[3]Central Research Laboratory, Institute of Dental Sciences, Siksha O Anusandhan (Deemed to be University), Bhubaneswar, Odisha, India
[4]Department of Prosthodontics, Kalinga Institute of Dental Sciences, Bhubaneswar, Odisha, India

7.1 Introduction

Pandemics are when widespread infectious diseases affect a substantial number of people worldwide on multiple continents. History suggests that several pandemics have reduced human populations, from time to time signaling the cul mination of whole civilizations. In the year 1918, Spanish flu had infested more than one-third of the global population, killing nearly 50 million people. Since then, the world has been affected by influenza pandemics several times like: "Severe Acute Respiratory Syndrome (SARS) in 2002, H5N1 Influenza (avian FLU) in 2004, H1N1 Influenza (swine FLU) in 2009, the Middle East Respiratory Syndrome (MERS) in 2012, H7N9 Influenza in 2013, the Enterovirus D68 in 2014" [1]. Apart from influenza, there have been other pandemics like the bubonic plague (the Black Death) in the 14th century, HIV/AIDS, and the Ebola virus outbreak in 2014 [2]. The ongoing, coronavirus disease of 2019 (COVID-19) commenced by severe acute respiratory syndrome coronavirus 2 (SARS-CoV-2) has created chaos worldwide, since its origin in Hubei, China at the end of 2019. Because of the rapid upsurge in the number of cases around the world, the World Health Organization (WHO) declared it a pandemic in March 2020.

Normally, viruses proliferate in host cells and establish the genetic materials for replication. The breakdown influences the host immune reaction and under the adverse situation, can mutate very fast. Therefore to develop proficient therapies and vaccines, it is imperative to recognize the virus protein structure, genome sequence, pathogenicity, transmission, infectivity, and the host immunologic response to it.

Biosensors for Emerging and Re-emerging Infectious Diseases
DOI: https://doi.org/10.1016/B978-0-323-88464-8.00012-9

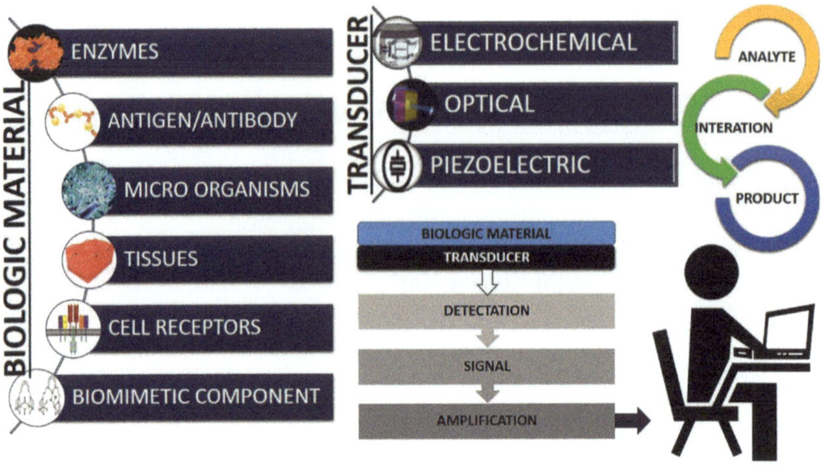

Figure 7.1 A general illustration of a biosensor. *From B.V. Ribeiro, A.R.T. Cordeiro, G.R. O. e Freitas, et al., Biosensors for the detection of respiratory viruses: a review, Talanta Open 2 (2020) 100007.*

Social isolation has been recommended by the WHO as a preventive measure, to mitigate virus transmission worldwide during the COVID-19 pandemic [3].

Different methodologies like a polymerase chain reaction (PCR) [4], reverse transcription-PCR (RT-PCR) [5], real-time quantitative PCR (RT-qPCR), enzyme-linked immunosorbent assay (ELISA) [6], and biosensors [7–9] have been used for the diagnosis of pathogenic agents in point-of-care management of the pandemic [10]. Among these, RT-qPCR [11] is the most effective, whereas methods involving antibodies (IgG/IgM) are used as auxiliary tools [12]. However, they are costly and time-consuming detection tools requiring expertise management. The biosensor can be a suitable alternative, as it is not only inexpensive but can detect the disease quickly and authentically even in a patient with asymptomatic condition [13]. Moreover, it can be used in a remote location by any person. The role of biosensors for quick detection of pathogens in pandemics particularly, COVID-19 is emphasized in this chapter (Figs. 7.1 and 7.2).

7.2 Biosensors

A biosensor is a compact analytical device with biological or biologically-derived sensing elements integrated within a physicochemical transducer [14].

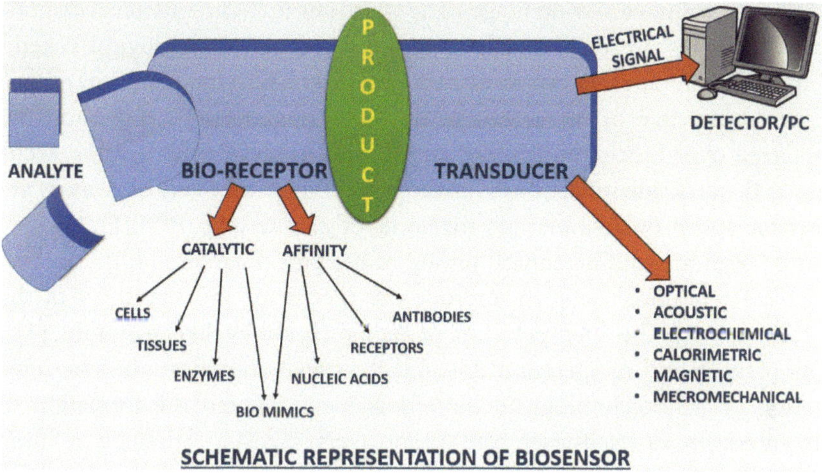

Figure 7.2 Schematic representation of a biosensor. *From M. Mascini, S. Tombelli, Biosensors for biomarkers in medical diagnostics, Biomarkers 13(78) (2008) 637657.*

These devices convert biological responses into computable signals and comprise sensing bioreceptors, a transducer, and a digital detector. When the target analytes in the solution like viruses [15,16], bacteria [17,18] nucleic acids [18], proteins [19], cancer biomarkers [20] and toxins [21] interconnect with the bioreceptor [22] attached to the transducer, then biochemical responses are produced. Bioreceptors are biotic material like microorganisms, antibodies, enzymes, cell receptors, tissues, or a biomimetic component, that are put out of action over the transducer (Fig. 7.1). Subsequently, the biochemical response is converted into a quantifiable signal which is computed with a digital detector [23,24]. The commonly available transducing systems are optical, electrochemical, and piezoelectric types. Electrochemical biosensors are sensitive to the variations in the charge dispersal on the transducer surface, depending on amperometric [25–27], potentiometric [28,29] or impedimetric [30–32] transduction principles. Optical biosensors offer multiplexed sensors within a single device. They measure the optical changes and features of the transducer surface during the bioreceptor and analyte interaction [33–35].

Therefore, they are multipurpose tools for analytical purposes. Piezoelectric biosensors estimate the alterations in the frequency of resonance in the course of bioreceptor and the mass of the crystal interaction. They use transducers that vibrate in the presence of an external alternating electrical field.

Biosensors have a wide range of applications including medical diagnosis, where they are used to sense the target analytes in biological systems helping in the rapid detection of pathogens [36,37]. Thus, they are pivotal in the deterrence of the spread of pandemic disease [38]. Biosensors are preferred over biochemical assays and immunoassays because of the accurate and quick identification of pathogens [39]. Moreover, biosensors are portable and provide a method for point-of-care testing (POCT). So they can be used for routine clinical testing of biological samples [40].

Researches are trying to improve the specificity and sensitivity of the biosensor with the incorporation of nanotechnology like nanofilm [41], nanoparticle [42] or quantum dot [43] for signal amplification. The integration of nanotechnology in biosensors has facilitated the formation of bio-interfaces of "self-assembled monolayers (SAMs)," thereby refining the resistance to nonspecific adsorption and biocompatibility [44—47]. Thus the effectiveness of POCT of infectious disease has improved. In POCT, a diagnostic test is performed along the patient bedside for quick diagnosis and effective management of rapidly progressing infections. Apart from being convenient, they are inexpensive instruments [48,49]. Rapid and accurate diagnosis plays a vital role to keep the spread of pandemic diseases in check. In the current scenario, early detection of COVID-19 disease has become vital to control the morbidity and mortality rate. Being non-invasive, convenient, and cost-effective, saliva can be used as a POCT platform for fast detection of COVID-19 disease [50]. Based on the analytes or reactions monitored by the biosensors, they have been further categorized into immune sensors (antigen-antibody interaction), enzymatic biosensors (target analyte-enzyme interaction), whole-cell biosensors, and DNA biosensors (hybridization).

7.2.1 Immunosensors

Immunosensors are biotic sensors dependent on precise antigen-antibody interactions. The immune system of a human body gets activated and antibodies are produced by lymphocyte B upon contact with the pathogens which act as antigens [51]. Immunosensors are used for around-the-clock monitoring of analytes using point-of-care tools. They are portable, low-cost, highly sensitive, and fully automated devices with fast response, and precision [52]. Their importance has been stressed in the detection of pathogenic bacteria [53,54] viruses [55,56] biomarkers [57] hormones [58,59] and toxins [60,61].

7.2.2 Enzymatic biosensors

These biosensors can precisely sense the alteration in catalytic properties during the interaction of biorecognition particles and enzymes. The transducer in the enzymatic biosensor, upon interaction with the analyte, generates an electrochemical [62], piezoelectric [63], or optical [64] signal. These signals are proportional to the analyte concentrations in the sample [65]. Because they are precise, these devices are recommended for point-of-care analysis on an extensive array of analytes [66].

7.2.3 Genosensors

The genetic information of any pathogens, virus or living organisms is distinct and determined by their deoxyribonucleic acid (DNA) sequencing. Genosensors detect human pathogens and various diseases by identifying these genetic sequences. In these biosensors, strands of the DNA or ribonucleic acid (RNA) (probe) corresponding to pathogens are incorporated on the surface of a physical transducer. Upon interaction with the corresponding (target) sequence, duplexes are developed through hybridization that is analyzed by the hybridization indicator [67]. Genosensors used for recognizing various diseases and pathogens have evolved with time [68].

7.2.4 Whole-cell biosensors

Whole-cell biosensors can use either surface antigens on the cell envelope (proteins, lipopolysaccharides, glycoproteins, and peptidoglycan) or the whole cell as an analyte for biorecognition. Biosensors for complete microorganisms are challenging and require the organism to be altered in the laboratory so that it can be fixed over the transducer for increased sensitivity [69,70]. These biosensors are becoming increasingly popular in biomedical diagnostics for their high sensitivity and selectivity [71,72].

7.3 Biosensors and medical diagnosis

Since their invention many years ago, biosensors have been used in variety of fields like: diagnosis [73] doping analysis [74], and laboratory medicine [75], etc. Due to their profitability, biosensors have further been explored for accurate sensing and fast clinical diagnosis. Emphasis has been put on new recognition elements, their arrangements, and the utilization of nanotechnology to enhance the analytical performance of biosensor systems. These improvements have made biosensors appropriate for quick, multi-analyte detection in point-of-care diagnostics [76]. Optical [77,78], magnetic [79,80] piezoelectric [81,82], micromechanical [83,84] electrochemical [85,86] and thermal [87–90] type biosensors are commonly used for medical diagnosis.

7.4 Applications of biosensors in pandemics

7.4.1 Human immunodeficiency virus

Human immunodeficiency virus (HIV) is a deadly lentivirus that results in acquired immunodeficiency syndrome (AIDS). Since its discovery in the year 1983, the number of cases is increasing around the world [91,92]. Literature shows around 36.9 million individuals were affected by HIV and more than 940,000s deaths were due to HIV-linked disease globally by the end of 2020. This explains the potential threat of this virus to mankind and hence requires prompt and accurate diagnosis.

7.4.1.1 Structure of human immunodeficiency virus

HIV-1 belongs to the deadly lentivirus group (Fig. 7.3). It is composed of two single-stranded RNA encircled with a conical capsid. The capsid is made up of the viral protein p24. A typical RNA element comprises 9749 nucleotides, a 5′ cap (Gppp), a 3' poly(A) tail, and numerous open reading frames (ORFs).

7.4.1.2 Laboratory diagnosis of human immunodeficiency virus

Routinely, ELISA test is used for the diagnosis of HIV. It requires the detection of HIV antibodies (such as IgM/IgG Abs) in the saliva or blood [93]. Though sensitive, a window period of several months is required before the test can be commenced [94,95]. A suitable alternative in the recent years has been the ultrasensitive biosensing mechanism. It determines the disease promptly by sensing either the HIV biomarkers, CD4 + cells, or the HIV genes [96]. Moreover, these tools are simple and cost-effective. Though many biosensors have been suggested for detecting HIV pathogens, simultaneous detection of different analytes using multiplexed biosensors was a daunting task [97].

Nanotechnology-based biosensors have overcome these difficulties. The new generation of biosensors apart from being sensitive and selective, give a very fast response [81]. They count on transcription-mediated amplification (TMA), PCR, branched-chain DNA methods, nucleic acid sequence-based amplification (NASBA), and point-of-care (POC) testing [98].

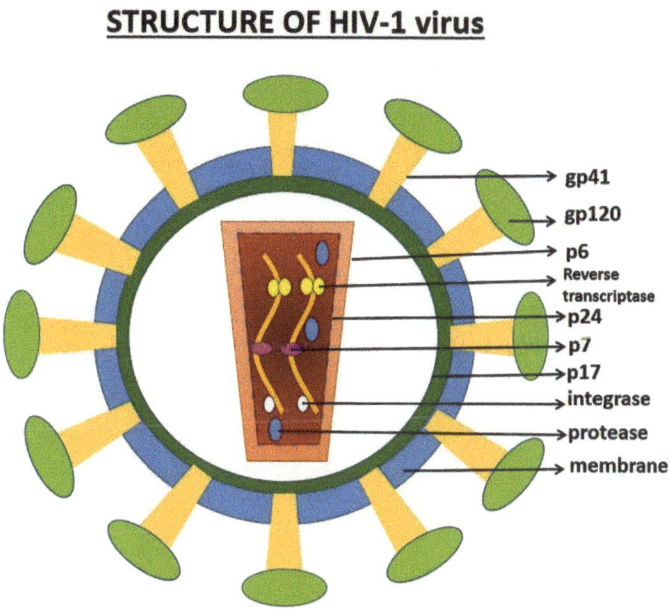

Figure 7.3 Structure of HIV. *From F.R. Cleghorn, M.S. Reitz, M. Popovic, R.C. Gallo, Human immunodeficiency viruses, in: G.L. Mandell, J.E. Bennett, R. Dolin (Eds.), Principles and Practice of Infectious Diseases (6th ed.), Churchill Livingstone, Philadelphia, 2005, pp. 2119–2133.*

7.5 Human immunodeficiency virus: biosensing approaches

Biosensing technologies for the detection of HIV are constantly evolving. Usually, electrical or optical sensors are used for the detection of binding actions that occur on the sensing surface like "antigen-antibody, target-aptamer, nucleic acid hybridization, and enzyme-cofactor coupling." However, researchers are focused on the development of POC diagnostics or resource-limited sets based on nanoscale technologies to screen HIV infections with precision.

7.6 Viral markers in human immunodeficiency virus infection and acquired immunodeficiency syndrome

The virus, as well as immune biomarkers present in the host body, are looked for to know both the stage of HIV disease and the effectiveness of antiviral therapy to treat it. Biomarkers are used in automated tests because they are measurable. Literature shows the available biomarkers for HIV are as follows: "HIV p24 antigen (HIV p24 core protein, after acid glycine dissociation), anti-p24 antibody titers, HIV p17 peptide, HIV-1 Tat protein, CD4 + cells, HIV gene, HIV-1 virus-like particles (VLPs), HIV-related enzymes and quantitative DNA and RNA PCR performed on cells and plasma, and HIV isolate phenotype" [99–102].

7.7 Nanomaterial-based biosensor and human immunodeficiency virus

These biosensors are divided into direct and indirect categories based on the type of transducers used. In the direct variety, the intact virus is the analyte whereas in the indirect variety, the antibodies developed in response to the virus are the analytes.

7.7.1 Direct detection

In this method, VLPs are used on the platforms of biosensors for virus detection [103]. Biosensors based on nanomaterials are very sensitive to detect HIV-based VLP quickly and in real-time. Lee et al. in their study have detected the signal of direct electron transfer from the HIV-1 virus by electrochemical method. They have electrodeposited the AuNPs on the electrode, made up of indium tin oxide (ITO) coated glass for enhanced electron-transfer kinetics thereby, producing a greater background charging current [104].

In yet another study, Lee and co-workers have developed an immunosensor based on the LSPR method for sensing HIV-1 [105]. For this, they uniformly coated the ITO glass substrate with nanopattern of round Au-dots (10–20 nm) by simple electrochemical deposition.

7.7.2 Indirect detection

Most of the biosensors for HIV detection count on indirect detection of virus-related proteins or genes. The biosensor for indirect biosensing of HIV is categorized into Electrochemical biosensors, Fluorescent biosensors, Electromagnetic piezoelectric biosensors, SERS biosensors, SPR biosensors, and Visual biosensors.

7.8 Review of literatures

7.8.1 Point-of-care detection of human immunodeficiency virus

Simple, swift, and precise detection of the pathogen for diagnosis and treatment of the disease is the need of the hour. However, the laboratory setting of the developing countries is not able to cope with the load.

The development of simple, accurate, and inexpensive one-to-one care tools, such as rapid POC can be the answer to this. This will help in the diagnosis, bedside or at remote care centers without the need for trained personnel [106]. Efforts are ongoing to develop POC maneuvers based on a mobile phone [107–110]. Mobile phones being low-cost, portable devices capable of exchanging data through telecommunication networks worldwide, they can be incorporated into the detection platform to provide a precise diagnosis. Li et al. had developed an inexpensive, simple, and compact immuno-testing platform with an assimilated device using a mobile phone and micro pit array chip (μPAC) [111] (Fig. 7.4).

Research is ongoing for the diagnosis of infection using HIV-related proteins assay or DNA hybridization technique. Though biosensors for the diagnosis of HIV have improved significantly, the POC technologies for the diagnosis will be of great help particularly in remote areas.

7.9 Biosensor for hepatitis

Hepatitis B virus (HBV)-related disease is one of the most serious health issues worldwide; it mostly ends with chronic hepatitis, cirrhosis, and

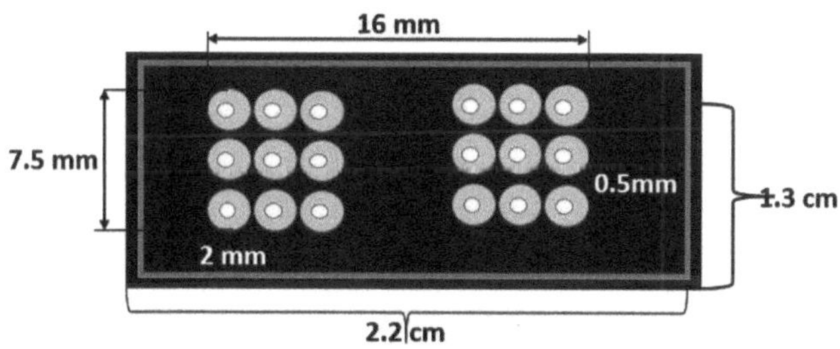

Figure 7.4 Photograph of the μPAC device mediated by a mobile phone. *μPAC*, Micro pit array chip; *POC*, point-of-care. *From L. Farzin, M. Shamsipur, L. Samandari, S. Sheibani, HIV biosensors for early diagnosis of infection: the intertwine of nanotechnology with sensing strategies, Talanta 206 (2020) 120201. doi: 10.1016/j.talanta.2019.120201.*

primary liver cancer with nearly 800,000 deaths every year [112]. Reports suggest the prevalence of the HBV in humans is even more than the combined populations affected with human HIV and Hepatitis C Virus (HCV) [113]. This warrants urgent attention for early diagnosis and treatment. Nano-based materials and technologies are commonly used in the detection of the HBV DNA [114].

Currently, PCR or immunoassay methods are used for the diagnosis of HBV infections. The immunoassay method targets the Hepatitis B surface (HBs) antibody [115,116]. Quantitative PCR measures the quantity of amplified product to calculate the virus DNA. On the contrary, real-time PCR relies on the assessment of the threshold cycle (CT) once a magnified PCR product is first found.

Matsunaga et al. have engineered handy whole-cell biosensors that target the HBV surface antigen. They have modified the conventional synNotch technology proposed by Morsut et al., 2016 which senses extracellular membrane-bound ligands and offers high input-output programmability. Matsunaga et al. found that the synNotch receptor is capable of recognizing and responding to encoded membrane antigens of the virus. The infectious whole viral Dane particles derived from HepG2.2.15.7 cells stating a complete HBV genome and recombinant HBs protein in liposomes (rHBs) were used as the ligand analytes for the purpose. While the synNotch technology provides high customization of input-output response, the platform is unable to provide signal amplification [117].

Shariati et al. had fabricated a very sensitive DNA biosensor for detecting hepatitis B virus that counts on tin-doped WO_3/In_2O_3 nanowires as the heterojunction photoelectrode. The physical vapor deposition method was used to prepare the tin-doped WO_3/In_2O_3 nanowires of length 50 nm. The nanowire surface was fixed with a single-strand DNA probe covalently hybridized with the complementary DNA. The reduction-oxidation responses in the electrochemical impedance spectrum are used to detect the hybridization even at very low concentrations. These nanowire biosensors are very stable as they show 96% of their early reactions even after 6 weeks of use [118].

Tian et al. had suggested a new platform made up of hybrid nanosheets for multiplexed recognition of HBV DNA, hepatitis A virus (HAV) DNA, and observed the intracellular K^+ level and the distribution of potassium ions (K^+) in living cells. Two-dimensional covalent organic framework nanosheets (Au NPs/COF NSs) were infiltrated with small Au nanoparticles to prepare these hybrid nanosheets [119].

7.10 Ebola

Ebola virus (EBOV) created substantial fear globally. After an incubation period (3–21 days), patients diseased with the EBOV generally develop influenza-like side effects that end in hemorrhage, shock-like syndrome, and multiple organ failure [120], thus getting infected with EBOV is a horrible outcome. This virus easily transmits via body fluids and was the cause of the epidemic in West Africa (2013–2016), leading to over 11,300 deaths. Hence, first POC detection, as well as treatment of the Ebola virus, is essential to avert such devastation in the future. This method is particularly important in countries lacking infrastructure [121]. Though EBOV is detected by the RT-PCR, it is not apt for field testing. Generalov et al. have developed the nanowire-based on silicon-on-insulator films (NW SOI biosensors) for the detection of Ebola virus VP40 protein. This film can sense the immune complexes formed with specific monoclonal antibodies (MCAs) instantaneously (~200–300 s per test sample) [122]. Additionally, Cai et al. had proposed another amplification-free method for sensing and quantifying the Ebola virus [123]. They have used a microfluidic chip for sample preparation. This silicone chip can detect the fluorescence of a single nucleic acid with high specificity, within minutes; but it has a low limit of detection value (0.2 pfu/mL). Ilkhani et al. had invented an electrochemical biosensor based on enzyme-amplified detection for sensing the Ebola virus DNA [124]. Once the conjugate of the biotinylated hybrid with streptavidin–alkaline phosphatase is formed, electrochemical impedance spectroscopy is used to measure this. It has a limited sensing ability (4.7 nM) (Fig. 7.5).

7.11 Zika

ZIKA virus (ZIKV) has become a global health menace with the involvement of more than fifty countries around the world since its outbreak in 2015 [125]. This disease is spread through the Aedes genus mosquitoes of tropical regions. The infected patients mimic the symptoms of dengue virus (DENV) and complain about a headache, fever, joint pain, muscle pain, rash, and conjunctivitis [126]. Particularly, pregnant patients infected with this virus often deliver babies having acute neurological syndromes or microcephaly. Apart from this, it has been associated with Guillain-Barré syndrome and ophthalmological disorders. Hence, fast and precise

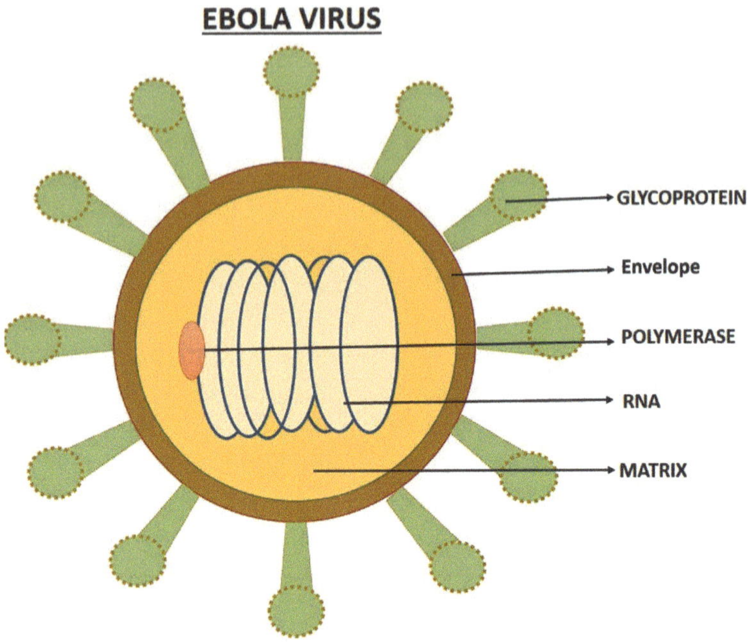

Figure 7.5 Ebola virus structure. *From https://www.labpedia.net/ebola-virus-haemorrhagic-fever/.*

field diagnosis is essential [127]. Genosensors are the equipment for accurate field diagnosis of ZIKV. It differentiates ZIKV from the chikungunya (CHIKV), DENV, and other homologous arboviruses. Few electrochemical genosensors for ZIKV are as follows: a square-wave voltammetry method used for detection of ZIKV DNA using a pencil carbon graphite electrode that is fortified with a 3-amino-4-hydroxybenzoic acid derivative; but this has a low limit of detection (LOD) of 25.4 pmol/L [128]. An impedimetric genosensor having disposable polyethene terephthalate-based electrodes protected with a nanometric gold layer is also used for ZIKV detection and it has an LOD of 25 nmol/L [129]. Afsahi et al. had developed a cost-effective, compact graphene-based biosensor for sensing the Zika virus using a monoclonal antibody [130]. Similarly, an electrochemical biosensor was invented by Kaushik et al. for Zika virus protein identification [131]. This biosensor uses electrochemical impedance spectroscopy for Zika virus protein detection, in the range of 10 pM to 1 nM. It has a detection limit below 10 pM with high sensitivity (Fig. 7.6).

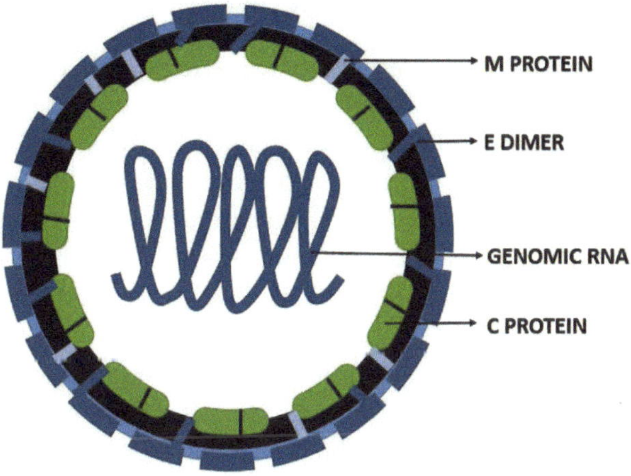

Figure 7.6 Zika virus. *From R. Mittal, D. Nguyen, L.H. Debs, A.P. Patel, G. Liu, V.M. Jhaveri, S.-I.S. Kay, J. Mittal, E.S. Bandstra, R.T. Younis, P. Chapagain, D.T. Jayaweera, X.Z. Liu, Zika Virus: An Emerging Global Health Threat, 2017.*

7.12 Norovirus

Foodborne viruses are liable for acute gastroenteritis outburst globally. They often result in necrotizing enterocolitis, irritable bowel syndrome, lethal dehydration, and even worsening of Crohn's disease [132]. This outbreak can be precluded with the use of a nanotechnology-based biosensors for the detection of foodborne viruses at the POC [133].

Namely, the electrochemical biosensing platform was assimilated with an aptamer for sensing norovirus accurately [134]. Chung et al. had invented another automatic microfluidic chip for sensing the norovirus in the oyster. All the procedures, like RNA extraction, cell concentration, amplification, and detection were carried out in the chip [135]. Neethirajan et al. had incorporated the graphene-gold nanoparticles in a microfluidic chip for sensing norovirus electrochemically [136].

Ashiba et al. had considered the biosensor with surface plasmon resonance to detect particles similar to the norovirus [137]. Further, Weerathunge et al. had proposed a biosensor with a colorimetric platform for fast and accurate detection of the infective murine norovirus [138]. This biosensor combines the murine norovirus aptamer with the enzyme-mimic catalytic activity of gold nanoparticles and emits blue color in the presence of norovirus.

7.13 Influenza

Influenza viruses are responsible for most acute respiratory infections (ARI). Because of the frequencies of pandemics caused by this virus, it has been named 'the mother of all pandemics'. Influenza virus belongs to the Orthomyxovirus family. It has three subtypes: A, B, C with type A being mostly accountable for 20th–21st-century pandemics. Literature shows it has been accountable for a pandemic four times: "The Spanish flu (H1N1) in 1918, Asian flu (H2N2) in 1957, Hong Kong influenza (H3N2) in 1968 and swine influenza (H1N1) in 2009." Hence there is a high possibility of recurrence of this pandemic in the future [139]. Early detection is crucial for the accurate management of the disease when vaccines are not well-devised and the conventional methods lack the in-field detection of the infections [140]. Few such techniques are enumerated below. Sayhi et al. had invented a biosensor for isolation of the influenza-A virus H9N2 sub-type from the allantoic fluid. Here they have immobilized the anti-matrix protein 2 antibodies on the iron magnetic nanoparticles [141].

This biosensor can directly detect the influenza virus A/H9N2 type in a lesser amount of 16 Hemagglutinin Units (HAU) titer. Tam et al., similarly, had formulated another biosensor that can detect influenza virus by immobilizing the DNA with carbon multi-walled nanotubes [142]. The interaction between the DNA probe attached to the biosensor and the target DNA was identified with help of Fourier transform infrared and Raman spectrometry analyses. Pang et al. had devised a biosensor for the H5N1 influenza virus detection in human serum. This fluorescent sensor can sense the recombinant hemagglutinin protein of the virus with a limitation of 2 and 3.5 ng/mL [143]. Again, Bai et al. had designed a compact biosensor with surface plasmon resonance for the detection of influenza virus H5N1 in birds. It counts on the aptamer [144].

Recently, Gouma et al. have devised the Novel Isoprene Sensor for influenza virus detection. It is a compact instrument with a 3-sensor array microsystem. The combination of nitric oxide released from the lungs and volatile organic compounds (VOCs) is used as biomarkers for the detection in this non-invasive screening method [145]. Rapid Influenza Diagnostic Tests (RIDTs) have been the golden standard for rapid detection of the influenza virus up until now. This immunoassay identifies the nucleoprotein antigens of the influenza virus in respiratory samples. Here, interactions of the monoclonal antibodies and viral nucleoproteins are depicted by the color changes [146]. Out of several biosensors, the electrochemical one delivers the fast and exact diagnosis of influenza. It has a colossal number of electrode materials with target detection molecules that have endless adaptation options.

7.14 COVID-19

The current coronavirus disease 2019 (COVID-19) pandemic is a transmissible disease allied with grave respiratory discomfort [147]. It is caused by a beta coronavirus (SARS-CoV-2), having a single positive strand RNA genome that is depicted as SARS-CoV-2 [148]. The SARS-CoV-2 genome encodes four basic proteins like nucleocapsid (N), matrix (M), envelope (E), and spike (S) (Fig. 7.7).

SARS-CoV-2 is extremely contagious with the rate of transmission far quicker than those of SARS and MERS. Even the asymptomatic cases of COVID-19 are capable of spreading the infection [149]. COVID-19 cases are routinely diagnosed by RT-PCR, CT scan, and lateral flow immunoassays technique. Though the RT-PCR method is sensitive, it is time-consuming and at times gives false-negative results [150,151].

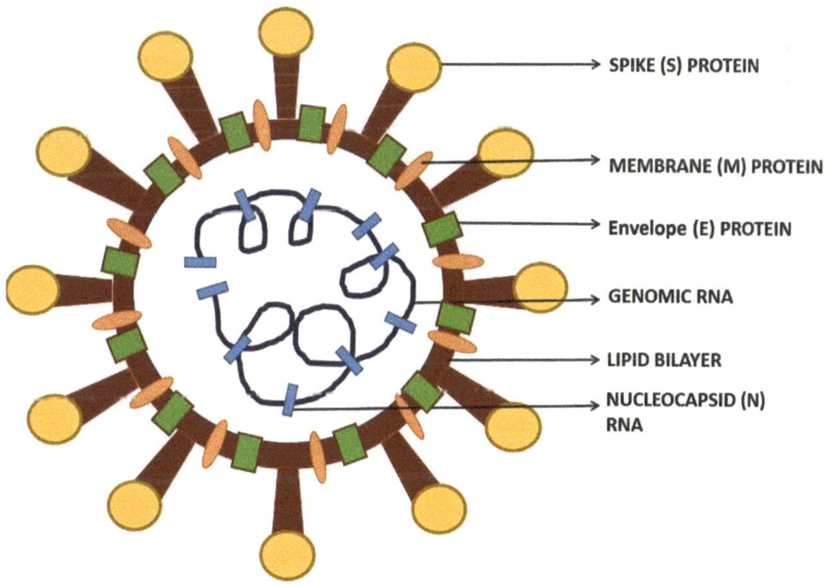

Figure 7.7 Structure of Covid-19 virus. *From S.K. Das, The pathophysiology, diagnosis and treatment of corona virus disease 2019 (COVID-19), Ind. J. Clin. Biochem. 35 (2020) 385–396.*

Efficient POCT assays are essential for early and accurate detection of SARS-CoV-2-infected patients to break the chain of transmission [152]. This requires the development of newer cost-effective and reliable diagnostic techniques with fast response.

A field-effect transistor (FET)-based biosensor is one such device with high sensitivity capable of POCT, and on-site detection [153]. Seo et al. had devised a graphene-based FET biosensor (COVID-19 FET sensor) to detect the SARS-CoV-2 virus. This SARS-CoV-2 spike antibody-laced biosensor is capable of sensing SARS-CoV-2 antigen, with a limit of detection (LOD) of 1 fg/mL [154].

Few other prospective biosensors for point-of-care sensing of SARS-CoV-2 virus in that manner are: "Lateral flow assay (LFA), Electrochemical biosensors and Chip-based nucleic acid detection." [155].

7.14.1 Lateral flow assay

Lateral flow assay for COVID-19 is based on the colloidal gold anti-SARS-CoV-2 antibody. Though having very low sensitivity, it can still be used for the detection of SARS-CoV-2 antigens and RNA in remote areas. Huang et al., 2016 had upgraded the sensitivity of LFA, using nanomaterial immunolabels such as: "quantum dots, up-conversion nanoparticles, and magnetic nanoparticles." [156]. Similarly, C. Hu et al. had developed an LFA based on ultra-bright fluorescence nanomaterials as immunolabels. The longer fluorescence lifetimes of this sensor enhanced the sensitivity and decreased the noise [157]. Kim et al. and Liu et al. in their research have designed an LFA based on lanthanide-laced nanoparticles (UCNP). In this way, the nitrocellulose membranes, stool, and blood show no interference with fluorescence absorbance [158,159]. As a result, sensitivity and accuracy of the biosensor are improved. Magnetic nanoparticle-based LFA is found to be of superior sensitivity in COVID-19 detection. Phillips et al. and Reboud et al. think that in-field detection of SARS-CoV-2 RNA can be done rapidly with high sensitivity using a combination of nucleic acids or microfluidics isothermal exponential amplification [160,161].

7.14.2 Electrochemical biosensors

An electrochemical biosensor is a viable alternative for point-of-care detection of SARS-CoV-2. It senses the analyte using DNA, RNA, antigen-antibody reaction, aptamers-based binding, or peptide nucleic acid (PNAs) hybridization [162]. Dai and Liu et al. have found this smart sensing system to be simple, cost-effective, and highly precise [163].

Label-free EISs developed by Faria et al. and Zhang et al., in particular, are fast and ultrasensitive. It is capable of identifying the virus antigens, antibodies, and RNA [164,165]. Chen et al. have found impedimetric sensors to be highly effective against the antigen and antibodies of the influenza virus [166]. Recently, Seo et al. used the highly sensitive G-FET for the SARS-CoV-2 virus particles detection in human NS specimens [167].

7.14.3 Chip-based nucleic acid detection

The development of nucleic acid detection platforms requires a series of steps such as: "lysis of cell or virus, nucleic acid extraction and amplification." [168]. Microfluidic technology-based "lab on a chip" has simplified this procedure. It does the sample preparation to the detection of nucleic acid automatically in inexpensive portable devices. This is a cost-effective disposable device with a short testing time that can detect the virus in smaller samples in limited settings [168].

7.15 Future perspectives

Biosensors are capable of detecting the SARS-CoV-2 RNA quickly and precisely, but human antibodies against this virus take 1–2 weeks to appear [169]. However, POC screening of virus particle biomarkers and SARS-CoV-2 NP are essential for the diagnosis of potential COVID-19 patients at an early stage. Hence, further research is required in the field of POCT immunosensors like near infrared (NIR)-emitting nanoparticles, magnetic nanoparticle-based LFAs, and electrochemical biosensors, for refining their anti-NP, anti-SP antibody, virus particle detection assays, and SARS-CoV-2 NP antigen [169–171].

7.16 Conclusion and future outlook

Fast, supersensitive, and precise analytic techniques are required for the timely diagnosis of pathogens in a pandemic. It will not only help in the acceleration of the treatment but also in the isolation of infected individuals, thereby breaking the chain of a pandemic. Biosensors in association with clinical observations can be a game-changer in the diagnosis of pathogens at an early stage. This will particularly be helpful in rural areas, where trained medical staff and advanced medical facilities are a far cry.

References

[1] (a) Centres for Disease Control and Prevention, Initial page, CDC. <https://www.cdc.gov/>, 2020 (accessed March 28, 2020);
(b) World Health Organization (WHO), Disease outbreaks by year, WHO. <https://www.who.int/csr/don/archive/year/en/>, 2020 (accessed April 1, 2020).

[2] Y.K.; Chan, M.U. Gack, Viral evasion of intracellular DNA and RNA sensing, Nat. Rev. Microbiol. 14 (2016) 360–373.

[3] World Health Organization (WHO) WHO Director-General's Opening Remarks at the Media Briefing on COVID-19 - 16 March 2020, WHO, 2020.

[4] A.C. Gourinat, O. O'Connor, E. Calvez, C. Goarant, M. Dupont-Rouzeyrol, Detection of Zika virus in urine, Emerg. Infect. Dis. 21 (2015) 84–86.

[5] S. Payungporn, S. Chutinimitkul, A. Chaisingh, S. Damrongwantanapokin, C. Buranathai, A. Amonsin, et al., Single-step multiplex real-time RT-PCR for H5N1 influenza A virus detection, J. Virol. Methods 131 (2006) 143–147.

[6] R. Viswanathan, S. George, M.V. Murhekar, A.M. Abraham, M.P. Singh, S.M. Jadhav, et al., Comparison of two commercial ELISA kits for detection of rubella specific IgM in suspected congenital rubella syndrome cases and rubella IgG antibodies in a serosurvey of pregnant women: Comparison of two commercial ELISA kits for detection of rubella specific IgM and IgG antibodies, Diagn. Microbiol. Infect. Dis. (2019). in press.

[7] V.D. Krishna, K. Wu, A.M. Perez, J.P. Wang, Giant magnetoresistance-based biosensor for detection of influenza A virus, Front. Microbiol. 7 (2016) 400–408.

[8] K. Navakul, C. Warakulwit, P.T. Yenchitsomanus, A. Panya, P.A. Lieberzeit, C. Sangma, A novel method for dengue virus detection and antibody screening using a graphene-polymer based electrochemical biosensor, NBM 13 (2017) 549–557.

[9] M. Khater, A. de la Escosura-Muñiz, A. Merkoçi, Biosensors for plant pathogen detection, Biosens. Bioelectron. 93 (2017) 72–86.

[10] A. Afzal, A. Mujahid, R. Schirhagl, S. Bajwa, U. Latif, S. Feroz, Gravimetric viral diagnostics: QCM based biosensors for early detection of viruses, Chemosensors 5 (2017) 7.

[11] V.M. Corman, O. Landt, M. Kaiser, R. Molenkamp, A. Meijer, D.K.W. Chu, et al., Detection of 2019 novel coronavirus (2019-nCoV) by real-time RT-PCR, Euro Surveill. 25 (2020) 2000045. Available from: https://doi.org/10.2807/1560-7917.

[12] J. Xiang, M. Yan, H. Li, T. Liu, C. Lin, S. Huang, et al., Evaluation of enzyme-linked immunoassay and colloidal gold- immunochromatographic assay kit for detection of novel coronavirus (SARS-Cov-2) causing an outbreak of pneumonia (COVID-19), MedRxiv (2020) 1–13, doi: 10.1101/2020.02.27.20028787.

[13] C.I. Justino, A.C. Freitas, R. Pereira, A.C. Duarte, T.A.R. Santos, Recent developments in recognition elements for chemical sensors and biosensors, Trends Anal. Chem. 68 (2015) 2–17.

[14] J.D. Newman, L.J. Tigwell, A.P.F. Turner, P.J. Warner, Biosensors – A Clearer View, Cranfield University Publication, 2004. 2004.

[15] U. Anik, Y. Tepeli, M.F. Diouani, Fabrication of electrochemical model influenza a virus biosensor based on the measurements of neuraminidase enzyme activity, Anal. Chem. 88 (2016) 6151–6153.

[16] M. Bäcker, C. Koch, S. Eiben, F. Geiger, F. Eber, H. Gliemann, et al., Tobacco mosaic virus as enzyme nanocarrier for electrochemical biosensors, Sens. Actuators B: Chem 238 (2017) 716–722.

[17] N. Cheeveewattanagul, E. Morales-Narváez, A.R.H. Hassan, J.F. Bergua, W. Surareungchai, M. Somasundrum, et al., Straightforward immunosensing platform based on graphene oxide-decorated nano paper: a highly sensitive and fast biosensing approach, Adv. Funct. Mater. 27 (2017) 1702741–1702749.

[18] K. Bartold, A. Pietrzyk-Le, K. Golebiewska, W. Lisowski, S. Cauteruccio, E. Licandro, et al., Oligonucleotide determination via peptide nucleic acid macromolecular imprinting in an electropolymerized CG-rich artificial oligomer analogue, ACS Appl. Mater. Int. 10 (2018) 27562–27569.

[19] H.B. Wang, H.D. Zhang, Y. Chen, Y.M. Liu, A fluorescent biosensor for protein detection based on poly (thymine)-templated copper nanoparticles and terminal protection of small molecule-linked DNA, Biosens. Bioelectron 74 (2015) 581–586.

[20] S. Atay, K. Pişkin, F. Yılmaz, C. Çakır, H. Yavuz, A. Denizli, Quartz crystal microbalance based biosensors for detecting highly metastatic breast cancer cells via their transferrin receptors, Anal. Methods 8 (2016) 153–162.

[21] I. Cunha, R. Biltes, M.G.F. Sales, V. Vasconcelos, Aptamer-based biosensors to detect aquatic phycotoxins and cyanotoxins, Sensors 18 (2018) 2367.

[22] J.A. Goode, J.V.H. Rushworth, P.A. Millner, Biosensor regeneration: a review of common techniques and outcomes, Langmuir 31 (2015) 6267–6276.

[23] A. Goode, J.V.H. Rushworth, P.A. Millner, Biosensor regeneration: a review of common techniques and outcomes, Langmuir 31 (2015) 6267–6276. Available from: https://doi.org/10.1021/la503533g.

[24] F. Yang, Y. Ma, S.G. Stanciu, A. Wu, Transduction process-based classification of biosensors, in: A. Wu, W.S. Khan (Eds.), Nanobiosensors From Design to Applications, Wiley -VCH, 2020. Available from: 10.1002/9783527345137.ch2.

[25] N. German, A. Ramanavicius, A. Ramanaviciene, Amperometric glucose biosensor based on electrochemically deposited gold nanoparticles covered by polypyrrole, Electroanalysis 29 (2017) 1267–1277. Available from: https://doi.org/10.1002/elan.201600680.

[26] N. Labban, L. Hughes, M. Wayu, J. Pollock, M. Leopold, MON 182 adaptable amperometric biosensor platforms for the diagnosis of endocrine disorders, J. Endocr. Soc. 3 (2019). Available from: https://doi.org/10.1210/js.2019-MON-182.

[27] A.B. Lima, L.F. Ferreira, S.L. Barbosa, E. de, S. Gil, R.A.B. da Silva, et al., Selective determination of verapamil in pharmaceutics and urine using a boron-doped diamond electrode coupled to flow injection analysis with multiple-pulse amperometric detection, Electroanalysis 30 (2018) 1880–1885. Available from: https://doi.org/10.1002/elan.201800206.

[28] K. Saeedfar, L. Yook Heng, M. Rezayi, Fabricating long shelf life potentiometric urea biosensors using modified MWCNTs on screen printed electrodes, Sens. Lett. 15 (2017) 97–103. Available from: https://doi.org/10.1166/sl.2017.3780.

[29] J. Ding, W. Qin, Recent advances in potentiometric biosensors, TrAC. Trends Anal. Chem. 124 (2020) 115803. Available from: https://doi.org/10.1016/j.trac.2019.115803.

[30] E.B. Aydın, Highly sensitive impedimetric immunosensor for determination of interleukin 6 as a cancer biomarker by using conjugated polymer containing epoxy side groups modified disposable ITO electrode, Talanta 215 (2020) 120909. Available from: https://doi.org/10.1016/j.talanta.2020.120909.

[31] A. Chakraborty, D.N. Tibarewala, A. Barui, Impedance-based biosensors, Bioelectron. Med. Devices, Woodhead Publishing, 2019, pp. 97–122. Available from: 10.1016/B978-0-08-102420-1.00005-4.

[32] T.A.R. Cordeiro, M.V.C. Gonçalves, D.L. Franco, A.B. Reis, H.R. Martins, L.F. Ferreira, Label-free electrochemical impedance immunosensor based on modified screen-printed gold electrodes for the diagnosis of canine visceral leishmaniasis, Talanta 195 (2019) 327–332. Available from: https://doi.org/10.1016/j.talanta.2018.11.087.

[33] A. Soni, R.K. Surana, S.K. Jha, Smartphone-based optical biosensor for the detection of urea in saliva, Sens. Actuat. B: Chem. 269 (2018) 346–353. Available from: https://doi.org/10.1016/j.snb.2018.04.108.

[34] V. Garzón, D.G. Pinacho, R.-H. Bustos, G. Garzón, S. Bustamante, Optical biosensors for therapeutic drug monitoring, Biosensors 9 (2019) 132. Available from: https://doi.org/10.3390/bios9040132.
[35] M. Soler, L.M. Lechuga, Boosting cancer immunotherapies with optical biosensor nanotechnologies, Eur. Med. J. Oncol. 4 (4) (2019) 124–132.
[36] A.I. Jeffree, S. Karman, S. Ibrahim, M.S.A. Karim, S. Rozali, Biosensors approach for lung cancer diagnosis—a review, in: A.P.P.A. Majeed, J.A. Mat-Jizat, M.H.A. Has-san, Z. Taha, H.L. Choi, J. Kim (Eds.), RITA 2018, Springer, Singapore, 2019, pp. 425–435. Available from: 10.1007/978-981-13-8323-6.
[37] M. Pohanka, Piezoelectric biosensor for the determination of tumor necrosis factor-alpha, Talanta 178 (2018) 970–973. Available from: https://doi.org/10.1016/j.talanta.2017.10.031.
[38] F. Ricci, G. Adornetto, G. Palleschi, A review of experimental aspects of electrochemical immunosensors, Electrochim. Acta. 84 (2012) 74–83. Available from: https://doi.org/10.1016/j.electacta.2012.06.033.
[39] F. Saylan, E. Yilmaz, A. Özgür, H. Derazshamshir, A. Yavuz, Denizli, Molecular imprinting of macromolecules for sensor applications, Sensors 17 (2017) 898. Available from: https://doi.org/10.3390/s17040898.
[40] C.I.L. Justino, A.C. Duarte, T.A.P. Rocha-Santos, Immunosensors in clinical laboratory diagnostics, in: G.S. Makowski (Ed.), Advances in Clinical Chemistry, Elsevier, 2016, pp. 65–108. Available from: 10.1016/bs.acc.2015.10.004.
[41] K.H. Cho, D.H. Shin, J. Oh, J.H. An, J.S. Lee, J. Jang, Multidimensional conductive nanofilm-based flexible aptasensor for ultrasensitive and selective HBsAg detection, ACS Appl. Mater. Interfaces 10 (2018) 28412–28419.
[42] L. La Spada, L. Vegni, Electromagnetic nanoparticles for sensing and medical diagnostic applications, Materials 11 (2018) 603.
[43] Y. Pang, J. Jian, T. Tu, Z. Yang, J. Ling, Y. Li, et al., Wearable humidity sensor based on porous graphene network for respiration monitoring, Biosens. Bioelectron. 116 (2018) 123–129.
[44] C.-J. Huang, Advanced surface modification technologies for biosensors, in: K. Mit- subayashi, O. Niwa, Y. Ueno (Eds.), Chem. Gas, Biosens. Internet Things Relat. Appl, Elsevier, 2019, pp. 65–86. Available from: 10.1016/B978-0-12-815409-0.00005-X.
[45] W.-C. Lan, T.-S. Huang, Y.-C. Cho, Y.-T. Huang, C.J. Walinski, P.-C. Chiang, et al., The potential of a nanostructured titanium oxide layer with self-assembled monolayers for biomedical applications: surface properties and biomechanical behaviours, Appl. Sci. 10 (2020) 590. Available from: https://doi.org/10.3390/app10020590.
[46] S. Acosta, L. Quintanilla, M. Alonso, C. Aparicio, J.C. Rodríguez-Cabello, Recombinant AMP/polypeptide self-assembled monolayers with synergistic antimicrobial properties for bacterial strains of medical relevance, ACS Biomater. Sci. Eng. 5 (2019) 4708–4716. Available from: https://doi.org/10.1021/acsbiomaterials.9b00247.
[47] H. Lv, L. Ye, Q. Liu, S. Li, T. Li, N. Huang, et al., S-S-PEG-COOH self-assembled monolayer on gold surface enabled a combined assay for serological EBV antibody isotypes, Proteom. Clin. Appl. 13 (2018). Available from: https://doi.org/10.1002/prca.201800067.
[48] K. Baryeh, S. Takalkar, M. Lund, G. Liu, Introduction to medical biosensors for point of care applications, in: R.J. Narayan (Ed.), Med. Biosens. Point Care Appl, Woodhead Publishing, 2017, pp. 3–25. Available from: http://doi.org/10.1016/B978-0-08-100072-4.00001-0.
[49] L.C. Brazaca, L. Ribovski, B.C. Janegitz, V. Zucolotto, Nanostructured materials and nanoparticles for point of care (POC) medical biosensors, in: R.J. Narayan (Ed.), Med. Biosens. Point Care Appl, Woodhead Publishing, 2017, pp. 229–254. Available from: 10.1016/B978-0-08-100072-4.00010-1.

[50] (a) R. Sabino-Silva, A.C.G. Jardim, W.L. Siqueira, Coronavirus COVID-19 impacts to dentistry and potential salivary diagnosis, Clin. Oral Investig. 24 (2020) 1619–1621. Available from: https://doi.org/10.1007/s00784-020-03248-x.
(b) T. Nguyen, D.D. Bang, A. Wolff, Novel Coronavirus disease (COVID-19): paving the road for rapid detection and point-of-care diagnostics, Micromachines (Basel) 11 (2020) (2019) 306. Available from: https://doi.org/10.3390/mi11030306.
[51] F. Li, Y. Li, J. Feng, Y. Dong, P. Wang, L. Chen, et al., Ultrasensitive amperometric immunosensor for PSA detection based on Cu2O@CeO2-Au nanocomposites as integrated triple signal amplification strategy, Biosens. Bioelectron. 87 (2017) 630–637. Available from: https://doi.org/10.1016/j.bios.2016.09.018.
[52] E.B. Bahadır, M.K. Sezgintürk, Applications of electrochemical immunosensors for early clinical diagnostics, Talanta 132 (2015) 162–174. Available from: https://doi.org/10.1016/j.talanta.2014.08.063.
[53] S. Jampasa, P. Laengee, K. Patarakul, N. Ngamrojanavanich, O. Chailapakul, N. Rodthongkum, Electrochemical immunosensor based on gold-labelled monoclonal anti-LipL32 for leptospirosis diagnosis, Biosens. Bioelectron. 142 (2019) 111539. Available from: https://doi.org/10.1016/j.bios.2019.111539.
[54] D. Juronen, A. Kuusk, K. Kivirand, A. Rinken, T. Rinken, Immunosensing system for rapid multiplex detection of mastitis-causing pathogens in milk, Talanta 178 (2018) 949–954. Available from: https://doi.org/10.1016/j.talanta.2017.10.043.
[55] H. Zhang, B.L. Miller, Immunosensor-based label-free and multiplex detection of influenza viruses: state of the art, Biosens. Bioelectron. 141 (2019) 111476. Available from: https://doi.org/10.1016/j.bios.2019.111476.
[56] C. da, C. Santos, P.C.M. Santos, K.L.S. Rocha, R.L. Thomasini, D.B. de Oliveira, et al., A new tool for dengue virus diagnosis: optimization and detection of anti-NS1 antibodies in serum samples by impedimetric transducers, Microchem. J. 154 (2020) 104544. Available from: https://doi.org/10.1016/j.microc.2019.104544.
[57] F. Li, Y. Li, J. Feng, Y. Dong, P. Wang, L. Chen, et al., Ultrasensitive amperometric immunosensor for PSA detection based on Cu2O@CeO2-Au nanocomposites as integrated triple signal amplification strategy, Biosens. Bioelectron. 87 (2017) 630–637. Available from: https://doi.org/10.1016/j.bios.2016.09.018.
[58] Li, A.M. Chow, H.V.S. Ganesh, M. Ratnam, I.R. Brown, K. Kerman, Diazonium-modified screen-printed electrodes for immunosensing growth hormone in blood samples, Biosensors 9 (2019) 88. Available from: https://doi.org/10.3390/bios9030088.
[59] J. Kelch, A. Delaney, F. Kelleher, P. Baker, E. Iwuoha, E. Dempsey, Impedi-metric and electrochemical evaluation of a new redox-active steroid derivative for hormone immunosensing, Biosens. Bioelectron. 150 (2020) 111876. Available from: https://doi.org/10.1016/j.bios.2019.111876.
[60] H.-P. Cheng, H.-S. Chuang, Rapid and sensitive nano-immunosensors for botulinum, ACS Sens. 4 (2019) 1754–1760. Available from: https://doi.org/10.1021/acssensors.9b00644.
[61] O.C. Ozoemena, T. Maphumulo, J.L. Shai, K.I. Ozoemena, Electrospun carbon nanofibers as an electrochemical immunosensing platform for vibrio cholerae toxin: ageing effect of the redox probe, ACS Omega 5 (2020) 5762–5771. Available from: https://doi.org/10.1021/acsomega.9b03820.
[62] A. Liu, Q. Lang, B. Liang, J. Shi, Sensitive detection of maltose and glucose based on dual enzyme-displayed bacteria electrochemical biosensor, Biosens. Bioelectron. 87 (2017) 25–30. Available from: https://doi.org/10.1016/j.bios.2016.07.050.
[63] T. Zhao, Y. Fu, H. He, C. Dong, L. Zhang, H. Zeng, et al., Self-powered gustation electronic skin for mimicking taste buds based on piezoelectric–enzymatic reaction coupling process, Nanotechnology 29 (2018). Available from: https://doi.org/10.1088/1361-6528/aaa2b9.

[64] M.A. Tabrizi, J. Ferré-Borrull, L.F. Marsa, An optical biosensor for the determination of cathepsin B as a cancer-associated enzyme using nanoporous anodic alumina modified with human serum albumin-thionine, Microchim. Acta. 187 (2020) 230. Available from: https://doi.org/10.1007/s00604-020-4188-9.
[65] P. Bollella, L. Gorton, Enzyme based amperometric biosensors, Curr. Opin. Electrochem. 10 (2018) 157–173. Available from: https://doi.org/10.1016/j.coelec.2018.06.003.
[66] L. Wang, M. Tianjiao, D. Zhao, H. Jia, S. An, X. Yang, et al., An enzyme-free electrochemical biosensor based on well monodisperse Au nanorods for ultra-sensitive detection of telomerase activity, Biosens. Bioelectron. 148 (2020) 111834. Available from: https://doi.org/10.1016/j.bios.2019.111834.
[67] P.A. Rasheed, N. Sandhyarani, Electrochemical DNA sensors based on the use of gold nanoparticles: a review on recent developments, Microchim. Acta 184 (2017) 981–1000. Available from: https://doi.org/10.1007/s00604-017-2143-1.
[68] A.A. Saeed, J.L.A. Sánchez, C.K. O'Sullivan, M.N. Abbas, DNA biosensors based on gold nanoparticles-modified graphene oxide for the detection of breast cancer biomarkers for early diagnosis, Bioelectrochemistry 118 (2017) 91–99. Available from: https://doi.org/10.1016/j.bioelechem.2017.07.002.
[69] C. Roggo, J.R. der Meer, Miniaturized and integrated whole-cell living bacterial sensors in field applicable autonomous devices, Curr. Opin. Biotechnol. 45 (2017) 24–33. Available from: https://doi.org/10.1016/j.copbio.2016.11.023.
[70] C. Brutesco, S. Prévéral, C. Escoffier, E.C.T. Descamps, E. Prudent, J. Cayron, et al., Bacterial host and reporter gene optimization for genetically encoded whole cell biosensors, Environ. Sci. Pollut. Res. Int. 24 (2017) 52–65. Available from: https://doi.org/10.1007/s11356-016-6952-2.
[71] P. Riangrungroj, C.S. Bever, B.D. Hammock, K.M. Polizzi, A label-free optical whole-cell *Escherichia coli* biosensor for the detection of pyrethroid insecticide exposure, Sci. Rep. 9 (2019) 12466. Available from: https://doi.org/10.1038/s41598-019-48907-6.
[72] N. Kylilis, P. Riangrungroj, H.-E. Lai, V. Salema, L.Á. Fernández, G.-B.V. Stan, et al., Whole-cell biosensor with tunable limit of detection enables low-cost agglutination assays for medical diagnostic applications, ACS Sens. 4 (2019) 370–378. Available from: https://doi.org/10.1021/acssensors.8b01163.
[73] J.E. Chang, D.S. Lee, S.W. Ban, J. Oh, M.Y. Jung, S.H. Kim, et al., Analysis of volatile organic compounds in exhaled breath for lung cancer diagnosis using a sensor system, Sens. Actuators B: Chem 255 (2018) 800–807.
[74] A. Saini, N. Kaur, N. Singh, A highly fluorescent sensor based on hybrid nanoparticles for selective determination of furosemide in an aqueous medium, Sens. Actuators B: Chem. 228 (2016) 221–230.
[75] D.H. Choi, A. Thaxton, I.; Cheol Jeong, K.; Kim, P.R.; Sosnay, G.R.; Cutting, et al., Sweat test for cystic fibrosis: wearable sweat sensor vs. standard laboratory test, J. Cyst. Fibros. 17 (2018) e35–e38.
[76] C.I.L. Justino, T.A.P. Rocha-Santos, S. Cardoso, A.C. Duarte, Strategies for enhancing the analytical performance of nanomaterial-based sensors, Trends Anal. Chem. 47 (2013) 27–36.
[77] M. Loyez, J.C. Larrieu, S. Chevineau, M. Remmelink, C. Leduc, B. Bondue, et al., In situ cancer diagnosis through online plasmonics, Biosens. Bioelectron. 131 (2019) 104–112.
[78] H.T. Theint, J.E. Walsh, S.T. Wong, K. Von, M. Shitan, Development of an optical biosensor for the detection of *Trypanosoma evansi* and *Plasmodium berghei*, Spectrochim. Acta A 218 (2019) 348–358.
[79] D. Cheng, M. Yu, F. Fu, W. Han, G. Li, J. Xie, et al., Dual recognition strategy for specific and sensitive detection of bacteria using aptamer-coated magnetic beads and antibiotic-capped gold nanoclusters, Anal. Chem. 88 (2016) 820–825.

[80] X. Liu, C. Zhang, K. Liu, H. Wang, C. Lu, H. Li, et al., Multiple SNPs detections based on lateral flow assay for phenylketonuria diagnostic, Anal. Chem. 90 (2018) 3430–3436.
[81] M. Jenik, R. Schirhagl, C. Schirk, O. Hayden, P. Lieberzeit, D. Blaas, et al., Sensing picornaviruses using molecular imprinting techniques on a quartz crystal microbalance, Anal. Chem. 81 (2009) 5320–5326.
[82] R. Schirhagl, P.A. Lieberzeit, F.L. Dickert, Chemosensors for viruses based on artificial immunoglobulin copies, Adv. Mater. 22 (2010) 2078–2081.
[83] Z. Zhang, M. Sohgawa, K. Yamashita, M. Noda, A micromechanical cantilever-based liposome biosensor for characterization of protein-membrane interaction, Electroanalysis 28 (2016) 620–625.
[84] M. Tardivo, V.; Tooli, G.; Fracasso, D.; Borin, S.; Dal Zilio, A.; Colusso, et al., Parallel optical read-out of micromechanical pillars applied to prostate specific membrane antigen detection, Biosens. Bioelectron. 72 (2015) 393–399.
[85] R. Abolhasan, A. Mehdizadeh, M.R. Rashidi, L. Aghebati-Maleki, M. Yousefi, Application of hairpin DNA-based biosensors with various signal amplification strategies in clinical diagnosis, Biosens. Bioelectron. 129 (2019) 164–174.
[86] N. Dükar, S. Tunç, K. Öztürk, S. Demirci, M. Dumangöz, M. Sönmez Çelebi, et al., Highly sensitive and selective dopamine sensing in biological fluids with one-pot prepared graphene/poly(*o*-phenylenediamine) modified electrodes, Mater. Chem. Phys. 228 (2019) 357–362.
[87] S. Casadio, J.W. Lowdon, K. Betlem, J.T. Ueta, C.W. Foster, T.J. Cleij, et al., Development of a novel flexible polymer-based biosensor platform for the thermal detection of noradrenaline in aqueous solutions, Chem. Eng. J. 315 (2017) 459–468.
[88] Z. Wang, L. Jinlong, Z. An, M. Kimura, T. Ono, Enzyme immobilization in completely packaged freestanding SU-8 microfluidic channel by electro click chemistry for compact thermal biosensor, Process. Biochem. 79 (2019) 57–64.
[89] M.S. Khan, S.K. Misra, K. Dighe, Z. Wang, A.S. Schwartz-Duval, D. Sar, et al., Electrically-receptive and thermally-responsive paper-based sensor chip for rapid detection of bacterial cells, Biosens. Bioelectron. 110 (2018) 132–140.
[90] B. van Grinsven, K. Eersels, O. Akkermans, S. Ellermann, A. Kordek, M. Peeters, et al., Label-free detection of *Escherichia coli* based on thermal transport through surface imprinted polymers, ACS Sens. 1 (2016) 1140–1147.
[91] C.R. Suri, Immunosensors for pesticide monitoring, Advances in Biosensors, Perpectives in Biosensors, Vol. 5, Elsevier Science, Amsterdam, The Netherlands, 2003, pp. 161–176.
[92] F.R. Cleghorn, M.S. Reitz, M. Popovic, R.C. Gallo, Human immunodeficiency viruses, in: G.L. Mandell, J.E. Bennett, R. Dolin (Eds.), Principles and Practice of Infectious Diseases, 6th (ed.), Churchill Livingstone, Philadelphia, 2005, pp. 2119–2133.
[93] B. Sciacca, A. François, P. Hoffmann, T.M. Monro, Multiplexing of radiative-surface plasmon resonance for the detection of gastric cancer biomarkers in a single optical fiber, Sens. Actuators B: Chem 183 (2013) 454–458.
[94] M.A.M. Rodrigo, Z. Heger, N. Cerneil, A.M.J. Jinemez, O. Zitka, V. Adam, et al., HIV biosensors—the potential of the electrochemical way, Int. J. Electrochem. Sci. 9 (2014) 3449–3457.
[95] L.J. Urio, M.A. Mohamed, J. Mghamba, A. Abade, S. Aboud, Evaluation of HIV antigen/antibody combination ELISAs for diagnosis of HIV infection in Dar Es Salaam, Tanzania, Pan, Afr. Med. J. 20 (2015) 196.
[96] R. Bayer, C. Edington, HIV testing, human rights, and global AIDS policy: exceptionalism and its discontents, J. Health Polit. Policy Law 34 (2009) 301–323.

[97] M.A. Lifson, M.O. Ozen, F. Inci, S.Q. Wang, H. Inan, M. Baday, et al., Advances in biosensing strategies for HIV detection, diagnosis, and therapeutic monitoring, Adv. Drug. Deliv. Rev. 103 (2016) 90–104.
[98] L. Farzin, M. Shamsipur, L. Samandari, S. Sheibani, Advances in the design of nanomaterial-based electrochemical affinity and enzymatic biosensors for metabolic biomarkers: a review, Microchim. Acta 185 (2018) 276.
[99] J. Ma, J. Wang, M. Wang, G. Zhang, W. Peng, Y. Li, et al., Preparation of cuprous oxide mesoporous spheres with different pore sizes for non-enzymatic glucose detection, Nanomaterials 8 (2018) 73.
[100] J. Zhao, L. Chang, L. Wang, Nucleic acid testing and molecular characterization of HIV infections, Eur. J. Clin. Microbiol. Infect. Dis. 38 (2019) 829–842.
[101] A.L. Cunningham, D.E. Dwyer, D.N. Dowton, Viral markers in HIV infection and AIDS, J. Acquir. Immune. Defic. Syndr. 6 (1993) 32–35.
[102] M. Minunni, S. Tombelli, A. Gullotto, E. Luzi, M. Mascini, Development of biosensors with aptamers as bio-recognition element: the case of HIV-1 Tat protein, Biosens. Bioelectron. 20 (2004) 1149–1156.
[103] J. Schupbach, Viral RNA and p24 antigen as markers of HIV disease and antiretroviral treatment success, Int. Arch. Allergy Immunol. 132 (2003) 196–209.
[104] L.X. Doan, M. Li, C. Chen, Q. Yao, Virus-like particles as HIV-1 vaccines, Rev. Med. Virol. 15 (2005) 75–88.
[105] J.H. Lee, B.K. Oh, J.W. Choi, Highly sensitive localized surface plasmon resonance immunosensor for label-free detection of HIV-1, Nanomedicine: NBM 9 (2013) 1018–1026.
[106] Z. Shirbaghaee, A. Bolhassani, Different applications of virus-like particles in biology and medicine: vaccination and delivery systems, Biopolymers 105 (2016) 113–132.
[107] J.H. Lee, B.K. Oh, J.W. Choi, Electrochemical sensor based on direct electron transfer of HIV-1 Virus at Au nanoparticle modified ITO electrode, Biosens. Bioelectron. 49 (2013) 531–535.
[108] L. Belec, J.P. Bonn, Challenges in implementing HIV laboratory monitoring in resource-constrained settings: how to do more with less, Future Microbiol. 6 (2011) 1251–1260.
[109] L.J. Wang, R. Sun, T. Vasile, Y.C. Chang, L. Li, High-throughput optical sensing immunoassays on smartphone, Anal. Chem. 88 (2016) 8302–8308.
[110] K.L. Lee, M.L. You, C.H. Tsai, E.H. Lin, S.Y. Hsieh, M.H. Ho, et al., Nanoplasmonic biochips for rapid label-free detection of imidacloprid pesticides with a smartphone, Biosens. Bioelectron. 75 (2016) 88–95.
[111] F. Li, H. Li, Z. Wang, J. Wu, W. Wang, L. Zhou, et al., Mobile phone mediated point-of-care testing of HIV p24 antigen through plastic micro-pit array chips, Sens. Actuators B: Chem 271 (2018) 189–194.
[112] (a) C. Seeger, W.S. Mason, Molecular biology of hepatitis B virus infection, Virology 479 (2015) 672–686.
(b) D. Lavanchy, M. Kane, Global epidemiology of hepatitis B virus infection, Diseases, Humana Press, Cham, Switzerland, 2016, pp. 187–203.
[113] C. Gomes, R.J. Wong, R.G. Gish, Global perspective on hepatitis B virus infections in the era of effective vaccines, Clin. Liver Dis. 23 (2019) 383–399.
[114] M. Shariati, The field effect transistor DNA biosensor based on ITO nanowires in label-free hepatitis B virus detecting compatible with CMOS technology, Biosens. Bioelectron. 105 (2018) 58–64.
[115] J.W. Chung, S.D. Kim, R. Bernhardt, J.C. Pyun, Application of SPR biosensor for medical diagnostics of human hepatitis B virus (hHBV), Sens. Actuators B Chem. 111-112 (2005) 416–422.

[116] D. Paraskevis, C. Haida, N. Tassopoulos, M. Raptopoulou, D. Tsantoulas, H. Papachristou, et al., Development and assessment of a novel real-time PCR assay for quantitation of HBV DNA, J. Virol. Methods 103 (2002) 201–212.
[117] S. Matsunaga, S.S. Jeremiah, K. Miyakawa, D. Kurotaki, S. Shizukuishi, K. Watashi, H. Nishitsuji, H. Kimura, T. Tamura, N. Yamamoto, K. Shimotohno, T. Wakita, A. Ryo, Engineering cellular biosensors with customizable antiviral responses targeting hepatitis B virus, iScience 23 (3) (2020) 100867.
[118] M. Shariati, M. Sadeghi, Ultrasensitive DNA biosensor for hepatitis B virus detection based on tin-doped WO_3/In_2O_3 heterojunction nanowire photoelectrode under laser amplification, Anal. Bioanal. Chem. 412 (22) (2020) 5367–5377.
[119] Y. Tian, Q. Lu, X. Guo, S. Wang, Y. Gao, L. Wang, Au nanoparticles deposited on ultrathin two-dimensional covalent organic framework nanosheets for in vitro and intracellular sensing, Nanoscale. 12 (14) (2020) 7776–7781.
[120] X. Qiu, G. Wong, J. Audet, A. Bello, L. Fernando, J.B. Alimonti, et al., Reversion of advanced Ebola virus disease in nonhuman primates with ZMapp, Nature 514 (2014) 47–62.
[121] J.T. Baca, V. Severns, D. Lovato, D.W. Branch, R.S. Larson, Rapid detection of Ebola virus with a reagent-free, point-of-care biosensor, Sensors 15 (2015) 8605–8614.
[122] V.M. Generalova, O.V. Naumovab, B.I. Fominb, S.A. P'yankova, I.V. Khlistuna, A.S. Safatova, et al., Detection of Ebola Virus VP40 Protein using a Nanowire SOI Biosensor, Optoelectronics, Instrum. Data Process. 55 (6) (2019) 618–622.
[123] H. Cai, J.W. Parks, T.A. Wall, M.A. Stott, A. Stambaugh, K. Alfson, et al., Optofluidic analysis system for amplification-free, direct detection of Ebola infection, Sci. Rep. 5 (2015).
[124] H. Ilkhani, S. Farhad, A novel electrochemical DNA biosensor for Ebola virus detection, Anal. Biochem. 557 (2018) 151–155.
[125] Gulland, Zika virus is a global public health emergency, declares WHO, BMJ 657 (2016) i657.
[126] Camacho, M. Paternina-Gomez, P.J. Blanco, J.E. Osorio, M.T. Aliota, Detection of autochthonous zika virus transmission in Sincelejo, Colombia, Emerg. Infect. Dis. 22 (2016) 927–929.
[127] Musso, C. Roche, T.X. Nhan, E. Robin, A. Teissier, V.M. Cao Lormeau, Detection of Zika virus in saliva, J. Clin. Virol. 68 (2015) 53–55.
[128] R. da Fonseca Alves, D.L. Franco, M.T. Cordeiro, E.M. deOliveira, R.A. FiremanDutra, M. Del, PilarTaboadaSotomayor, Novel electrochemical genosensor for Zika virus based on a poly-(3-amino-4-hydroxybenzoicacid)-modified pencil arbongraphite electrode, Sens. Actuators B: Chem 296 (2019) 126681.
[129] A.M. Faria, V. Zucolotto, Label-free electrochemical DNA biosensor for zika virus identification, Biosens. Bioelectron. 131 (2019) 149–155.
[130] S. Afsahi, M.B. Lerner, J.M. Goldstein, J. Lee, X. Tang, D.A. Bagarozzi Jr, et al., Novel graphene-based biosensor for early detection of Zika virus infection, Biosens. Bioelectron. 100 (2018) 85–88.
[131] A. Kaushik, A. Yndart, S. Kumar, R.D. Jayant, A. Vashist, A.N. Brown, et al., A sensitive electrochemical immunosensor for label-free detection of Zika-virus protein, Sci. Rep. 8 (2018) 9700–9705.
[132] H. Ashiba, Y. Sugiyama, X. Wang, H. Shirato, K. Higo-Moriguchi, K. Taniguchi, et al., Detection of norovirus virus-like particles using a surface plasmon resonanceassisted fluoroimmunosensor optimized for quantum dot fluorescent labels, Biosens. Bioelectron. 93 (2017) 260–266.
[133] M. Bierho, W. Arvelo, A. Estevez, J. Bryan, J.P. McCracken, M.R. López, et al., Incidence and clinical profile of norovirus disease in Guatemala, 2008–2013, Clin. Infect. Dis. 67 (2018) 430–436.

[134] R. Beier, C. Pahlke, P. Quenzel, A. Henseleit, E. Boschke, et al., Selection of a DNA aptamer against norovirus capsid protein VP1, FEMS Microbiol. Lett. 351 (2014) 162–169.
[135] S.H. Chung, C. Baek, V.T. Cong, J. Min, The microfluidic chip module for the detection of murine norovirus in oysters using charge switchable micro-bead beating, Biosens. Bioelectron. 67 (2015) 625–633.
[136] R. Chand, S. Neethirajan, Microfluidic platform integrated with graphene-gold nano-composite aptasensor for one-step detection of norovirus, Biosens. Bioelectron. 98 (2017) 47–53.
[137] H. Ashiba, Y. Sugiyama, X. Wang, H. Shirato, K. Higo-Moriguchi, K. Taniguchi, et al., Detection of norovirus virus-like particles using a surface plasmon resonance-assisted fluoroimmunosensor optimized for quantum dot fluorescent labels, Biosens. Bioelectron. 93 (2017) 260–266.
[138] P. Weerathunge, R. Ramanathan, V. Torok, K. Hodgson, Y. Xu, R. Goodacre, et al., Ultrasensitive colorimetric detection of murine norovirus using NanoZyme aptasensor, Anal. Chem. 91 (2019) 3270–3276.
[139] C. Paules, K. Subbarao, Influenza, Lancet 390 (2017) 697–708.
[140] A. Moulick, L. Richtera, V. Milosavljevic, N. Cernei, Y. Haddad, O. Zitka, et al., Advanced nanotechnologies in avian influenza: current status and future trends—a review, Anal. Chim. Acta 983 (2017) 42–53.
[141] M. Sayhi, O. Ouerghi, K. Belgacem, M. Arbi, Y. Tepeli, A. Ghram, et al., Electrochemical detection of influenza virus H9N2 based on both immunomagnetic extraction and gold catalysis using an immobilization-free screen printed carbon microelectrode, Biosens. Bioelectron. 107 (2018) 170–177.
[142] P.D. Tam, N.; Van Hieu, N.D.; Chien, A.T.; Le, M.A. Tuan, DNA sensor development based on multi-wall carbon nanotubes for label-free influenza virus (type A) detection, J. Immunol. Methods 350 (2009) 118–124.
[143] Y. Pang, Z. Rong, J. Wang, R. Xiao, S. Wang, A fluorescent aptasensor for H5N1 influenza virus detection based-on the core–shell nanoparticles metal-enhanced fluorescence (MEF), Biosens. Bioelectron. 66 (2015) 527–532.
[144] H. Bai, R. Wang, B. Hargis, H. Lu, Y. Li, A SPR aptasensor for detection of avian influenza virus H5N1, Sensors 12 (2012) 12506–12518.
[145] P.-I. Gouma, L. Wang, S.R. Simon, M. Stanacevic, Novel isoprene sensor for a flu virus breath monitor, Sensors (2017) 17.
[146] WHO, Use of Influenza Rapid Diagnostic Tests, WHO, Rome, Italy, 2010.
[147] WHO. Coronavirus disease 2019 (COVID-19): situation report – 52. 2020.
[148] Coronaviridae Study Group of the International Committee on Taxonomy of Viruses, The species severe acute respiratory syndrome-related coronavirus: classifying 2019-nCoV and naming it SARS-CoV-2, Nat. Microbiol. 5 (2020) 536–544.
[149] (a) Y. Bai, L. Yao, T. Wei, F. Tian, D.Y. Jin, L. Chen, et al., Presumed asymptomatic carrier transmission of COVID-19, JAMA 323 (2020) 1406–1407.
(b) L. Zou, F. Ruan, M. Huang, L. Liang, H. Huang, Z. Hong, J. Yu, M. Kang, Y. Song, J. Xia, Q. Guo, T. Song, J. He, H.L. Yen, M. Peiris, J. Wu, SARS-CoV-2 Viral Load in Upper Respiratory Specimens of Infected Patients, N. Engl. J. Med. 382 (2020) 1177–1179.
[150] WHO, Coronavirus disease (COVID-19) technical guidance: laboratory testing for 2019-nCoV in humans. 2019.
[151] P. Zhou, X.L. Yang, X.G. Wang, B. Hu, L. Zhang, W. Zhang, et al., A pneumonia outbreak associated with a new coronavirus of probable bat origin, Nature 579 (7798) (2020) 270–273.

[152] (a) Z. Du, X. Xu, Y. Wu, L. Wang, J. B, L. C, et al., Serial interval of COVID-19 from publicly reported confirmed cases, Emerg. Infect. Dis. 26 (6) (2020) 1341–1343.
(b) M.J. Loeffelholz, Y.W. Tang, Laboratory diagnosis of emerging human coronavirus infections - the state of the art, Emerg. Microb. Infect. (2020) 1–26.
[153] J. Liu, X. Chen, Q. Wang, M. Xiao, D. Zhong, W. Sun, et al., Ultrasensitive monolayer MoS2 field-effect transistor based DNA sensors for screening of down syndrome, Nano Lett. 19 (2019) 1437–1444.
[154] G. Seo, G. Lee, M.J. Kim, S.H. Baek, M. Choi, K.B. Ku, et al., Rapid detection of COVID-19 causative virus (SARS-CoV-2) in human nasopharyngeal swab specimens using field-effect transistor-based biosensor, ACS Nano 14 (4) (2020) 5135–5142.
[155] T. Ji, Z. Liu, G.Q. Wang, X. Guo, S. Akbar khan, C. Lai, et al., Detection of COVID-19: a review of the current literature and future perspectives, Biosens. Bioelectron. 166 (2020) 112455.
[156] X. Huang, Z.P. Aguilar, H. Xu, W. Lai, Y. Xiong, Membrane-based lateral flow immunochromatographic strip with nanoparticles as reporters for detection: a review, Biosens. Bioelectron. 75 (2016) 166–180.
[157] C. Hu, S. Kalsi, I. Zeimpekis, K. Sun, P. Ashburn, C. Turner, et al., Ultra-fast electronic detection of antimicrobial resistance genes using isothermal amplification and Thin Film Transistor sensors, Biosens. Bioelectron. 96 (2017) 281–287.
[158] J. Kim, J.H. Kwon, J. Jang, H. Lee, S. Kim, Y.K. Hahn, et al., Rapid and background-free detection of avian influenza virus in opaque sample using NIR-to-NIR upconversion nanoparticle-based lateral flow immunoassay platform, Biosens. Bioelectron. 112 (2018) 209–215.
[159] Q. Liu, S. Cheng, R. Chen, J. Ke, Y. Liu, Y. Li, et al., Near-infrared lanthanide-doped nanoparticles for a low interference lateral flow immunoassay test, ACS Appl. Mater. Interfaces 12 (4) (2020) 4358–4365.
[160] E.A. Phillips, T.J. Moehling, S. Bhadra, A.D. Ellington, J.C. Linnes, Strand displacement probes combined with isothermal nucleic acid amplification for instrument-free detection from complex samples, Anal. Chem. 90 (11) (2018) 6580–6586.
[161] J. Reboud, G. Xu, A. Garrett, M. Adriko, Z. Yang, E.M. Tukahebwa, et al., Paper-based microfluidics for DNA diagnostics of malaria in low resource underserved rural communities, Proc. Natl. Acad. Sci. U.S.A 116 (11) (2019) 4834–4842.
[162] M. Labib, E.H. Sargent, S.O. Kelley, Electrochemical methods for the analysis of clinically relevant biomolecules, Chem. Rev. 116 (16) (2016) 9001–9090.
[163] Y. Dai, C.C. Liu, Recent advances on electrochemical biosensing strategies toward universal point-of-care systems, Angew. Chem. 58 (36) (2019) 12355–12368.
[164] H.A.M. Faria, V. Zucolotto, Label-free electrochemical DNA biosensor for zika virus identification, Biosens. Bioelectron. 131 (2019) 149–155.
[165] H. Zhang, B.L. Miller, Immunosensor-based label-free and multiplex detection of influenza viruses: state of the art, Biosens. Bioelectron. 141 (2019) 111476.
[166] X. Chen, T. Dong, X. Wei, Z. Yang, N.M. Matos Pires, J. Ren, et al., Electrochemical methods for detection of biomarkers of Chronic Obstructive Pulmonary Disease in serum and saliva, Biosens. Bioelectron. 142 (2019) 111453.
[167] G. Seo, G. Lee, M.J. Kim, S.H. Baek, M. Choi, K.B. Ku, et al., Rapid detection of COVID-19 causative virus (SARS-CoV-2) in human nasopharyngeal swab specimens using field-effect transistor-based biosensor, ACS Nano 14 (4) (2020) 5135–5142.

[168] M. Mauk, J. Song, H.H. Bau, R. Gross, F.D. Bushman, R.G. Collman, et al., Miniaturized devices for point of care molecular detection of HIV, Lab. Chip 17 (3) (2017) 382–394.
[169] L. Lv, X. Xie, Q. Gong, R. Feng, X. Guo, B. Su, et al., 2020. Transcriptional difference between SARS-COV-2 and other human coronaviruses revealed by subgenomic RNA profiling. bioRxiv.
[170] L. Pan, M. Mu, P. Yang, Y. Sun, R. Wang, J. Yan, et al., Clinical characteristics of COVID-19 patients with digestive symptoms in Hubei, China: a descriptive, cross-sectional, multicenter study, Am. J. Gastroenterol. 115 (5) (2020) 766–773.
[171] Y. Pan, L. Long, D. Zhang, T. Yan, S. Cui, P. Yang, et al., Potential false-negative nucleic acid testing results for severe acute respiratory syndrome coronavirus 2 from thermal inactivation of samples with low viral loads, Clin. Chem. 66 (6) (2020) 794–801.

Further reading

M.T. Chorsi, E.J. Curry, H.T. Chorsi, R. Das, J. Baroody, P.K. Purohit, et al., Piezoelectric biomaterials for sensors and actuators, Adv. Mater. 31 (2019). Available from: https://doi.org/10.1002/adma.201802084.
Ö. Erdem, Y. Saylan, N. Cihangir, A. Denizli, Molecularly imprinted nanoparticles based plasmonic sensors for real-time *Enterococcus faecalis* detection, Biosens. Bioeltron. 126 (2019) 608–614.

CHAPTER 8

Trends in biosensors for the detection of *Staphylococcus aureus* as a key cause of foodborne illnesses worldwide

Amina Antonacci[1], Fabiana Arduini[2] and Viviana Scognamiglio[1]
[1]Department of Chemical Sciences and Materials Technologies, Institute of Crystallography, National Research Council, Rome, Italy
[2]Department of Chemical Science and Technologies, University of Rome Tor Vergata, Rome, Italy

8.1 Introduction

A widespread variety of pathogenic bacteria are present in the environment, food, water, and soil and may transmit foodborne infections to humans with millions of cases annually. According to the data published by European food safety authority (EFSA) and European Center for Disease Prevention and Control (ECDC), bacterial toxins were the second cause of foodborne outbreak in 2017, and in Italy they show an increasing trend [1]. Among pathogenic bacteria, *Staphylococcus aureus* is considered as one of the most remarkable risk for public health.

S. aureus is a facultative, anaerobic, non-spore-forming, Gram-positive bacterium, widely spread in the environment, capable of surviving in hot and dry conditions and in fresh and saline environments.

As a potential pathogen, *S. aureus* causes foodborne infections ranging from mild skin infections to post-operative wound infections. It is also associated with dermatitis, gastrointestinal tract infections, bacteremia and infective endocarditis. Some strains of *S. aureus* can develop resistance to beta-lactam antibiotics such as penicillin, which are widely used to treat infections, these strains are thus known as methicillin-resistant *S. aureus* (MRSA).

Therefore, the analysis of the presence of *S. aureus* in food and environmental matrices is a crucial issue for minimizing the risks associated with public health and food safety. The EU system for monitoring and collecting information on zoonoses is based on the Zoonoses Directive 2003/99/EC1, which obliges European Union (EU) Member States (MS)

Biosensors for Emerging and Re-emerging Infectious Diseases
DOI: https://doi.org/10.1016/B978-0-323-88464-8.00013-0

to collect relevant and, when applicable, comparable data on zoonoses, zoonotic agents, antimicrobial resistance and foodborne outbreaks [2].

Conventional analysis for the detection of *S. aureus* involves culture methods, which require a laboratory setup as well as lengthy procedures (3–5 days) to produce results. Polymerase chain reaction (PCR), enzyme linked immunosensor assay (ELISA), and nucleic acid-based molecular biology methods are capable of reducing the response time from days to hours, but require trained personnel, extensive sample preparation, and instrumentations. Considering these limitations, biosensor technology can furnish cost-effective and reliable diagnostic tools capable of sensitive analysis for the in-field screening of *S. aureus* also in limited resource settings as in developing countries. This chapter reports the last trends on the biosensors for *S. aureus* detection available in literature, described according to the biological recognition elements, the detection mechanism, and to the available vanguard technologies which can be exploited for the enhancement of the analytical capability of such diagnostics, including rational design, nanotechnology, and microfluidics.

8.2 Biological recognition elements

8.2.1 Immunosensors

Antibodies are considered the most exploited bioprobes for different applications (e.g., food safety, veterinary medicine, clinical diagnosis, and environmental monitoring) [3]. Antibodies can be produced by biological approaches, such as the, immunization of rabbits, mice and goats against foreign substances (antigens). In the context of biosensors, the main advantage of antibody (monoclonal and polyclonal)-based devices is the high sensitivity and selectivity of detection. In the case of monoclonal antibodies, the advantage is due to selectivity and the unique interaction between the antibody and antigen. Moreover, these kind of biorecognition elements boast ad hoc immobilization (orientation on the surface), strategic labelling and desiderable size, with significant effects on the sensor biolayer [3]. In this context, many notable examples are available in literature. For example, Wu and colleagues [4] proposed an immunosensor with a mussel-inspired material (ε-poly-L-lysine-3,4-dihydroxy benzaldehyde, EPD) based on its Schiff base structure. This device has been tested on real milk samples for the detection of *S. aureus*, and the results were directly compared with results from the conventional plate count method. The analytical data revealed a good detection limit (28.55 CFU/mL) and linear range with on-demand antibacterial activity.

Instead, Yang et al. [5] exploited ZnS-Ag_2S/polydopamine (PDA) as a novel photoelectric material and Cu_2O as the peroxidase mimic tag to design a sensitive photoelectrochemical immunosensor for *S. aureus* (Fig. 8.1). In detail, the metal sulfides have been chosen as photoelectric materials and deposited on indium tin oxide (ITO) via electrodeposition and silver ion exchange. Moreover, the authors included the PDA use, as a mussel-inspired polymer coating material, since it claims quinone groups particularly suitable for covalent immobilization of amine-terminated biomolecules, such as the antibody for *S. aureus*. The analytical data reported a linear range between 10 and 10^7 CFU/mL and a low detection limit of 2 CFU/mL.

Another sensitive antibody-based electrochemical immunosensor has been described by Wang and colleagues [6]. The authors designed a hierarchical mesoporous silica (HMS) inspired by a biotemplate method based on butterfly wings and amino-functionalized it to cross-link the antibody with glutaraldehyde. The bioconjugates were immobilized on a glassy carbon electrode (GCE), and *S. aureus* was detected by changes in the peak currents after the antigen-antibody complex formation. Differential pulse voltammetry was

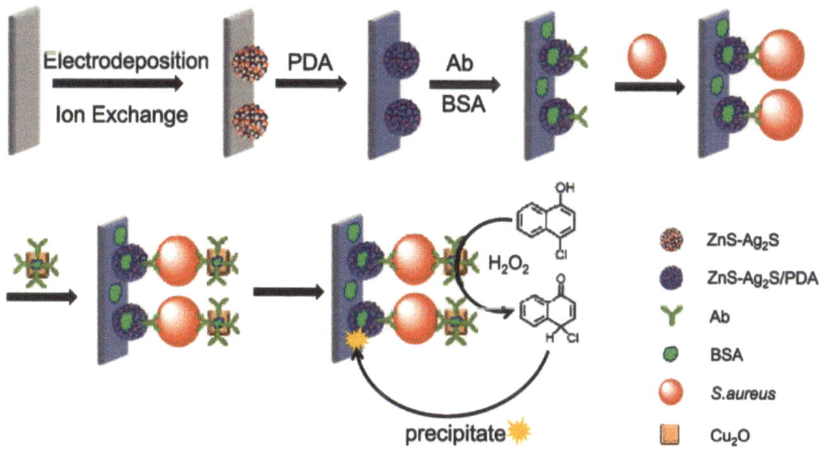

Figure 8.1 Illustration of modification steps of the photoelectrochemical immunosensor. *Reprinted with permission from H. Yang, H. Chen, L. Cao, H. Wang, W. Deng, Y. Tan, Q. Xie, An immunosensor for sensitive photoelectrochemical detection of* Staphylococcus aureus *using ZnS-Ag_2S/polydopamine as photoelectric material and Cu_2O as peroxidase mimic tag, Talanta, 212 (2020) 120797, Elsevier.*

carried out (bacterial concentrations ranging from 10 to 2×10^3 CFU/mL), and a good linear relationship was obtained ($R^2 = 0.9759$), as well as a performant detection limit (11 CFU/mL) within 20 min.

The study by Escamilla-Gómez and colleagues [7] contributed to the development of electrochemical immunosensors for the detection of *S. aureus*. In particular, the blending of traditional immunochemical methods with modern electrochemical systems allowed authors to analyze milk samples with a detection limit of 2.3×10^3 cells/mL, according to the regulations for *S. aureus* monitoring in foodstuffs. The biosensor involved the co-immobilization of rabbit immunoglobulin G and tyrosinase on a mercaptopropionic acid self-assembled monolayer modified gold electrode.

Within the immunosensor scenario, many studies have focused on the oriented immobilization of antibodies on the surface of the sensor chip. In fact, this is a critical aspect of the sensitivity and specificity of a sensing tool. Frequently, the antibody immobilization is carried out by chemical crosslinking, without orientation control. Iijima and colleagues [8] proposed an innovative method based on adsorbing immunoglobulin proteins onto 32-nm bio-nanocapsules (BNCs) by IgG Fc-binding domains derived from *S. aureus* protein A (ZZ-BNC). The reported results revealed that ZZ-BNC are a promising scaffold for displaying antibodies on a solid phase, demonstrated by the increased sensitivity (50-fold).

Moreover, several studies are available in the literature attesting to the commitment to improve analytical performances of a disposable electrochemical immunosensor including the critical aspect of stability [9–11].

8.2.2 Aptasensors

Synthetic biology has imbued different sectors including the biosensor field. As reported in our review published on Biosensors and Bioelectronics, 2015:

Recent advances in both disciplines, in fact, allow to re-design the configuration of the sensing elements—either by modifying toggle switches and gene networks, or by producing synthetic entities mimicking key properties of natural molecules. Indeed, the role of synthetic biology in sustaining biosensor technology has well reported by several examples from the literature and reflecting on the features that make it a useful tool for designing and constructing engineered biological systems for sensing application [12].

For the detection *of S. aureus*, aptamers have attracted a huge amount of attention as synthetic biocomponents for the development of smart biosensors, which do not require animals for the production of the recognition element [13]. Xu et al. [14] developed an aptamer-based biosensor using a selected aptamer as a probe capable of specifically binding to *S. aureus*. The binding between the aptamer and *S. aureus* allows for the release of the probe aptamer, which starts the DNA amplification reactions producing a G-quadruplex/hemin complex. The formed complex has horseradish peroxidase (HRP)-mimic activity, catalyzing the chemiluminescence reaction of luminol, mediated by H_2O_2 (Fig. 8.2). This multireaction system showed a linear range comprised between $5-10^4$ CFU/mL with the detection limit of 5 CFU/mL.

Hao et al. [15] developed a chemiluminescence resonance energy transfer aptasensor for the detection of *S. aureus* with Co^{2+}-enhanced *N*-(aminobutyl)-*N*-(ethylisoluminol) functional flowerlike gold nanoparticles as the donor and WS_2 nanosheet as the acceptor. In the presence of *S. aureus*, rolling circle amplification starts forming a duplex which is less adsorbed onto the WS_2 nanosheet, thus attenuating the quenching.

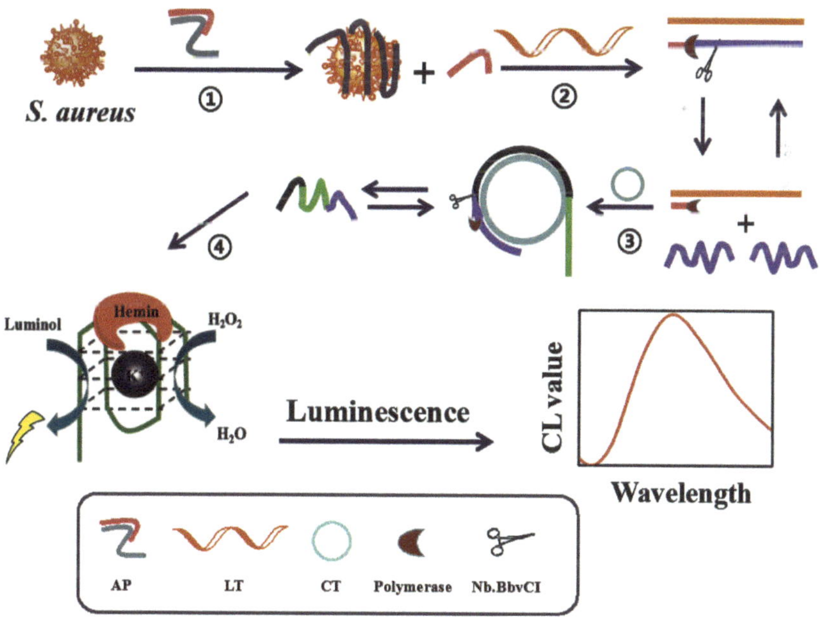

Figure 8.2 The chemiluminescent aptasensor for *Staphylococcus aureus* detection that was conceived in this study. *Reprinted with permission from J. Xu, J. Guo, S.W. Maina, Y. Yang, Y. Hu, X. Li, Xin, Z. An aptasensor for* Staphylococcus aureus *based on nicking enzyme amplification reaction and rolling circle amplification, Anal. Biochem., 549 (2018) 136–142, Elsevier.*

Under optimized conditions, the linear range was $50-1.5 \times 10^5$ CFU/mL with a limit of detection of 15 CFU/mL for *S. aureus*, and tested applicability in meat samples.

An electrochemiluminescence aptasensor was designed by Han et al. [16] based on a sensitive "on-off" approach using gold nanoparticles as nanomaterial and hemin as the regenerable enhancers of $S_2O_8^{2-}/O_2$ and MoS_2-PtNPs as quencher. The GCE was used as a platform by the modification with reduced graphene oxide, gold nanoparticles, and a thiol-terminated aptamer. In the absence of *S. aureus*, a high $S_2O_8^{2-}/O_2$ electrochemiluminescence intensity was obtained thanks to the cooperative response of reduced graphene oxide, gold nanoparticles, and hemin ("signal-on"). In the presence of *S. aureus*, the combination with MoS_2-Pt-Vancomycin allows for a decreased response due to the quenching effect of MoS_2-Pt-Van on the $S_2O_8^{2-}/O_2$ ECL system. Using this "on-off" system, the device showed a linearity between 1.5×10^2 and 1.5×10^8 CFU/mL, with a detection limit of 28 CFU/mL. The suitability of this aptasensor was evaluated by measuring *S. aureus* in spiked urine samples.

Zelada-Guillén et al. [17] developed an aptasensor using potentiometric transduction technique. They used a network of single-walled carbon nanotubes as ion-to-electron potentiometric transducers and anti-*S. aureus* aptamer as the recognition element. The analytical features of this type of aptasensor were investigated by immobilizing aptamers on carbon nanotubes through non-covalent adsorption or a covalent bond (i.e. between amine from aptamers and carboxylic groups from oxidated single-wall carbon nanotubes). This device was able to detect *S. aureus* with a sensitivity of 0.36 mV/decade and detection limit of 8×10^2 CFU/mL in the case of covalent immobilization, and sensitivity of 1.52 mV/decade and detection limit of 10^7 CFU/mL in the case of non-covalent adsorption immobilization. The suitability of the aptasensor was tested by evaluating real samples from freshly excised pig skin contaminated with *S. aureus*.

Other optical based-aptasensors were developed by using fluorescence or surface-enhanced Raman scattering as technique. For instance, Cai et al. [18] developed a fluorescent aptasensor by using a strand displacement approach, in which in the presence of *S. aureus*, a partially complementary strand of *S. aureus* aptamer is competitively released from cDNA/aptamer duplex immobilized on magnetic beads. The large number of single-stranded DNA production induces the opening of a hairpin probe with subsequent self-assembly to form a hexagonal structure with a fluorescence signal. This multireaction aptasensor exhibits a linear range from 7 to 7×10^7 CFU/mL, with a detection limit of 1.7 CFU/mL for *S. aureus*. The suitability of this aptasensor was demonstrated by successfully measuring *S. aureus* in milk samples.

Lu et al. [19] developed an aptasensor consisting of two DNA sequences: the aptamer SA43 used as a specific recognition component and a molecular beacon labeled with 6-carboxyfluorescein (FAM) fluorophores and a black hole quencher 1 (BHQ1) to evaluate the binding. The mechanism of the enzymatic cleavage aptasensor relies on the enzymatic protection effect of aptamer-pathogen cell binding complex. The molecular beacon hybridizes with the aptamer SA43, and no additional optimization of the molecular beacon sequence was needed as long as MB could be ensured to hybridize stably with the aptamer. In the presence of *S. aureus*, the strain can efficiently bind with the aptamer, and prevent it from cleavage by the exonuclease, thus sequentially, aptamer SA43 stably hybridized with the molecular beacon, and turns on the fluorescence. On the contrary, in the absence of *S. aureus*, aptamer SA43 will be free in the solution and cut by the exonuclease, resulting in a quenched fluorescence state (Fig. 8.3). The proposed aptasensor showed a linear range between 10^2 and 10^7 CFU/mL with a detection limit of 64 CFU/mL.

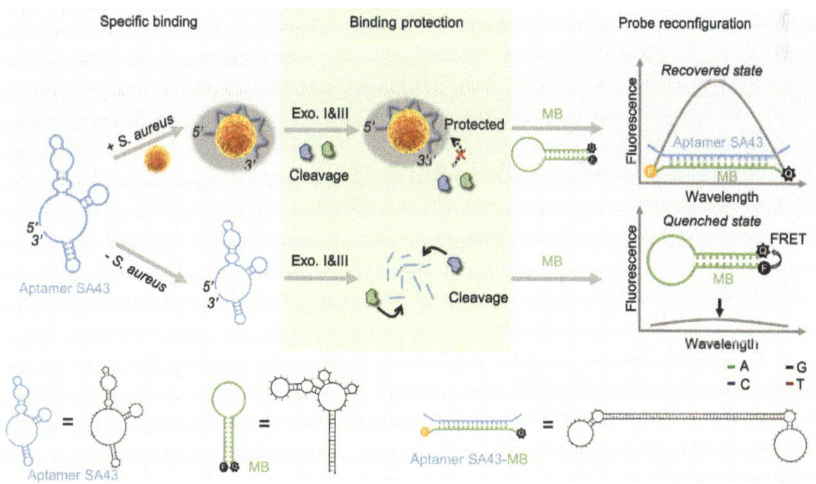

Figure 8.3 Working principle of the enzymatic cleavage aptasensor-based biosensing platform for *Staphylococcus aureus* detection. *Reprinted with permission from Y. Lu, Z. Yuan, J. Bai, Q. Lin, R. Deng, A. Luo, Q. He, Directly profiling intact* Staphylococcus aureus *in water and foods via enzymatic cleavage aptasensor, Anal. Chim. Acta, 1132 (2020) 28–35, Elsevier.*

This developed aptasensor was successfully applied for the detection of *S. aureus* in complex samples namely tap water, milk, and porker.

Zhu et al. [20] designed a surface-enhanced Raman scattering for the detection of *S. aureus* using a polydimethylsiloxane coated with gold nanoparticles to immobilize the aptamer through a gold-sulfur bond. For the detection of *S. aureus*, a signal molecular probe constituted of gold-silver core–shell nanoflowers modified with mercaptobenzoic acid was used beyond the immobilized aptamer, delivering a sandwich structure: "capture substrate-target-signal molecular probe." The developed aptasensor was able to detect *S. aureus* concentration in the range of $4.3 \times 10–4.3 \times 10^7$ CFU/mL with a detection limit of 13 CFU/mL.

8.2.3 Molecular imprinting polymers

Molecular imprinting polymers (MIPs) emerged in the scientific community in the 1930s [ref], thanks to the research of Polyakov, but only in the last 20 years have they achieved a huge success [21]. Several studies shown their high selectivity, stability and durability also in harsh conditions (e.g., thermal, mechanical, and highly acidic and basic conditions) for various applications (e.g., detection of mycotoxins, illegal food additives, allergenic proteins, bacterial exotoxins, as well as pathogenic bacteria and viruses). MIP production is possible through the polymerization of monomers, crosslinkers and other required components in the presence of the template target molecule. After the removal of the template, the free cavities obtained, which are capable of preserving the size, shape and chemical functionality acquired in the presence of the analyte template, start the specific recognition of the analyte target.

Regarding the development of alternative approaches for bacterial detection, MIPs-based biosensors represent promising candidates to obtain sensitive, selective and fast responses. In this context, two methods for surface imprinting have been described, whole cell and cell membrane molecular imprinting, according to the type of bacterium to be recognized [22]. In particular, for the detection of *S. aureus*, MIP technology has proven to be extremely valid, as described by Andersson and colleagues [23]. They designed a novel MIP- and surface plasmon resonance (SPR)-based biosensor for the α-hemolysin detection scheme, to determine its potential usability as a tool for early clinical testing. The authors prepared the MIPs on gold-coated sensor chips and subsequently combined the molecular imprinting with SPR-technology. The device identified α-hemolysin in septic blood samples ($n = 9$) with high affinity

(K_D = 2.75×10^{-7} M), a limit of detection of 0.022 μM and with a statistically significant difference ($P < .0001$).

Furthermore, Bezdekova's team [24] described another easy imprinting approach, exploiting dopamine to form the MIP layer, being this catecholamine is able to undergo oxidative polymerization and generates a polydopamine layer on a wide variety of supporting materials. Therefore, the MIP layer was created on the surfaces of magnetic particles (MPs) to enhance the device sensitivity, as well as to facilitate the final collection of target molecules by an external magnetic field (Fig. 8.4). This sensing tool

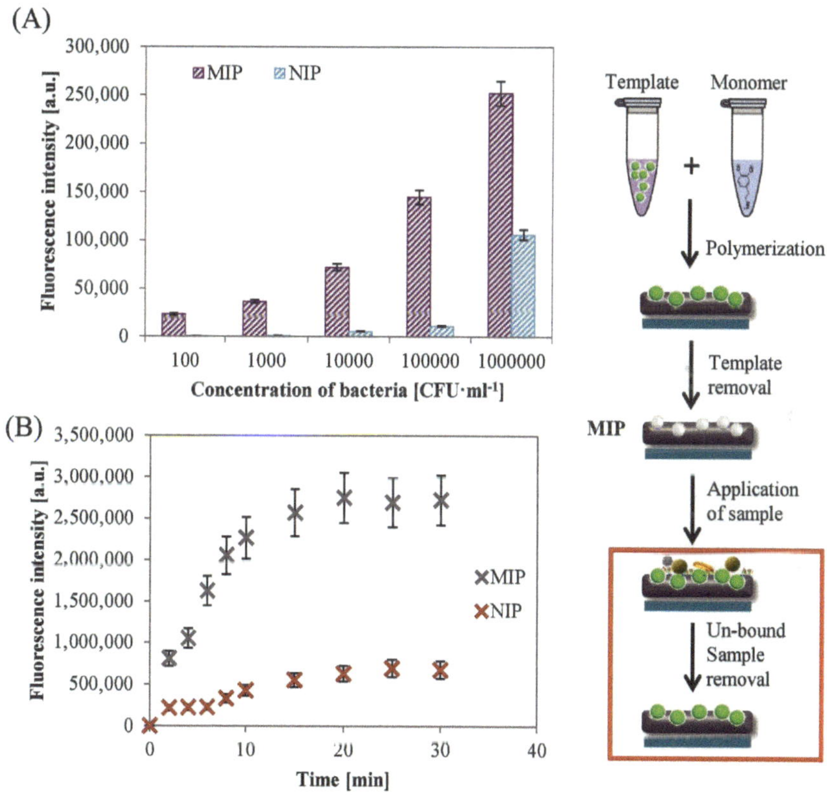

Figure 8.4 Adsorption kinetics (A) and isotherms (B) of MIPs and *NIPs*, non imprinted polymers towards bacteria. The points represent mean values of three measurements. On the left side of figure for a better idea the schema that demonstrated in which step the process is located is shown. From schema it is seen that process is in step in which the analyte is bound to selective cavities on the surface of MIP. *MIPs*, Molecular imprinting polymers. *Reprinted with permission from J. Bezdekova, K. Zemankova, J. Hutarova, S. Kociova, K. Smerkova, V. Adam, M. Vaculovicova, Magnetic molecularly imprinted polymers used for selective isolation and detection of* Staphylococcus aureus, *Food Chem. 321 (2020) 126673, Elsevier.*

has been tested for the detection of *S. aureus* in raw milk from cows with mastitis with a limit of detection (LOD) of 1×10^3 CFU/mL, thereby in full compliance with the regulations set out by the European Union Legislation for microbial control of food.

Recently, several authors underlined the criticalities related to imprinting large structures (i.e. whole bacterial cells), defining them as still very challenging due to the difficulty in the complete removal of micrometer-sized templates, as well as the slow diffusion of large-sized template bacteria to and from the imprinted sites [25]. In this context, Wang and colleagues [26] proposed a novel bacteria-imprinted conductive poly(3-thiopheneacetic acid) (BICP) film-based impedimetric sensor for *S. aureus* detection. The sensing device has been developed by paying great attention to environmental sustainability, excluding the use of toxic organic solvents and crosslinkers. Moreover, the authors faced the issue related to imprinting, suggesting a BICP film with an innovative structure without *S. aureus*-shaped cavities, exploiting the electropolymerization approach to provide a good microenvironment for bacterial embedment, hence the imprinted sites were more accessible on the surface of poly(3-thiopheneacetic acid) matrices. A rapid output has been recorded (within 10 min) and a very good low LOD (2 CFU/mL) has been achieved.

Another prominent study on this topic has been proposed by Guo and colleagues [27] and it is based on bacteria imprinting on polydimethylsiloxane film, easily produced by corresponding specific sites on the polydimethylsiloxane surface via stamp imprinting using *S. aureus*. Moreover, the proposed device was also equipped with a fluorescence resonance energy transfer platform for turn-on fluorescence sensing, obtained by the electrostatic interaction between citrate-functional copper clusters and dopamine-stabilized gold nanoparticles. The authors described a complete detection process within 135 min, a wide linear calibration response (from 10 to 1×10^7 CFU/mL) with a low detection limit (11.12 CFU/mL). Moreover, optimal recoveries from spiked samples have been achieved (from 97.7% to 101.90%) with relative standard derivations lower than 10%.

8.2.4 Bacteriophages

In the last ten years, another interesting type of bioreceptor, the bacteriophages, have appeared on the panorama of pathogen detection, as well as in *S. aureus* identification. These phages are viruses with pivotal roles of microbial balance on earth. Moreover, they boast several advantages in the detection and identification of bacterial pathogens. In detail, bacteriophages (1) show specificity of the interaction with target host cell, (2) have a characteristic lytic ability, (3) can discern between live and dead cells, and (4) claim a distinct capacity to multiply during the infection process. In addition, these viruses have important proprieties in the context of bioreceptor generation, being robust, easy to produce, and cost-effective. Tawil and colleagues [28] described the exploitation of surface plasmon resonance (SPR) for the detection of MRSA using a new and highly specific phage, BP14. This approach allowed for the label-free, real-time, specific, rapid and cost effective detection of bacteria, with a detection limit of 10^3 CFU/mL within 20 min. The authors chose the SPR method to avoid the limitations of direct impedimetric detection due to the mandatory optimization of the medium for electrical measurements, as well as the inability of certain microorganisms to produce sufficient electrolytes for detection. A home-built biosensor has been designed concerning the immobilization of recognition elements on sensor surfaces, a physisorption of bacteriophages has been carried out. Finally, a covalent immobilization of phages via L-cysteine, glutaraldehyde and self-assembled monolayers was made.

Similar research was also proposed by Balasubramanian and colleagues [29], who described a lytic phage as a specific and selective probe for detection of *S. aureus* through the use of an SPR-based SPREETA sensor as a detection platform. The virus 12,600 was selected from the commercial mixture of phages by incubation with the *S. aureus* ATCC 12,600 strain. Subsequently, the bacteriophages were immobilized on the gold surface of the SPREETA sensor with direct physical adsorption, a flow rate of 100 μL/min, and a detection limit of 10^4 CFU/mL [30].

Another representative example in the use of phages for *S. aureus* biosensing is the research of Bhardwaj and colleagues, which described for the first time, an innovative interface between bacteriophages and a water-dispersible and environmentally stable metal–organic framework (MOF) through the use of glutaraldehyde as the cross-linker. The sensing devices showed good sensitivity (detection limit of 31 CFU/mL), regeneratibility, specificity, extended range of linear detection, and stability for long-term storage.

Recently, Rees and Barr [31] reported a novel approach for the detection of MRSA and the determination of antibiotic susceptibility by phage amplification combined with matrix-assisted laser desorption/ionization mass spectrometry (MALDI-MS). In particular, the authors described an easy procedure involving sample incubation with bacteriophage K (ATCC-19685-B1) in the presence and absence of cefoxitin, followed by a rapid trypsin digestion, and finally by MALDI-Ms analysis. The working method is based on the capability of MALDI-Ms to detect the tryptic peptides derived from amplified phage proteins, furthermore validated by time-of-flight (TOF)/TOF analysis of each peptide combined with database searching. The detection limit for bacteriophage K was determined by a serial dilution to be 5×10^8 to 7.8×10^6 pfu/mL.

8.2.5 Antimicrobial resistance-related proteins

The appearance of MRSA is a serious health problem worldwide. For this reason, many efforts are currently directed towards the distinction between the resistance patterns from those of susceptible strains. This investigation area is sharpened by research focused on the exploitation of the antimicrobial resistance-related protein, as a specific target for pathogenic bacteria detection. Tawil and colleagues [32] reported a mechanism based on the penicillin-binding protein 2a (PBP2a) detection by a surface plasmon resonance biosensor. In particular, the PBP2a protein presents low affinity toward β-lactams, and it is the cause of resistance to this specific class of antibiotics. Hence, the authors proposed a rapid and accurate detection of MRSA within 20 min, discriminating between the susceptible to methicillin and the borderline strains, with a detection limit of 10 CFU/mL.

Moreover, Wang and colleagues [33] developed a multiple cross displacement amplification approach (m-MCDA) (at 63°C for 40 min) for the simultaneous detection of *S. aureus* strains and identification of MRSA based on *nuc* (*S. aureus*-specific gene) and *mecA* (encoding penicillin-binding protein-2′) gene. The authors used disposable lateral flow biosensors with simultaneous visual detection of the two targets in a single test. This simple, rapid, sensitive, and reliable technique showed trusted analytical sensitivity for the target genes (100 fg DNA/reaction in the pure cultures and 10 CFUs per tube with spiked blood samples) in an easy-to-use and disposable way, as well as it suggested a different approach to identify a MRSA infection and consequently identify an appropriate antibiotic therapy.

Instead, Nemr and colleagues [34] exploited a membrane-bound protein marker, that confers β-lactam antibiotic resistance to detect MRSA, directly from patient nasal swabs. The described procedure did not involve prior culture, but had minimal processing steps, using a microfluidic device and antibody-functionalized magnetic nanoparticles. In fact, the bacteria were specifically captured by a strain-specific antibody functionalized with alkaline phosphatase and was electrochemical sensed, with a limit of detection of 845 CFU/mL within 4.5 h.

8.3 Detection mechanism

8.3.1 Electrochemical detection

The electrochemical-based biosensors represent an important method for earlier detection of bacteria due to their high sensitivity, inexpensive detection process as well as the versatility to propose analytical platforms. The assurance of these devices is based on the linear relationship between the current intensity and the analyte concentration. For all of these features, the electrochemical detection impressed the clinical and food industries, testified by interesting configurations proposed in the scientific literature. In particular, several applications reported electrochemical detection exploitation for the detection of *S. aureus*. For example, de Avila and colleagues [35] designed a disposable amperometric magneto-immunosensor for the identification and quantification of *S. aureus* based on protein A (up-to 1 CFU/mL) detection in milk, without sample pretreatments, within 2 h. The methodology is based on functionalized magnetic beads with anti-ProtA antibody and gold screen-printed electrodes with an amperometric response at -0.15 V.

Another study related to the amperometric immunosensor based on protein A antibody of *S. aureus* has been proposed by Majumdar and colleagues [36]. In this investigation, a platinum electrode surface was exploited, following the modification with polyethylenimine polymer via cross-linkage by glutaraldehyde and finally tested on milk, cheese, and meat samples. The analytical data reported a detection limit down to 10 CFU/mL and these results are supported by Scanning Electron Microscopy analyses.

However, some authors highlighted several critical issues related to the use of electrochemical biosensors, namely the possibility of stumbling in false current values generation, and therefore in a potential interference due to active electrolytes. In these cases, a very useful workaround is to use selective membranes, in order to discern the electrolytes on the basis of their weight or charge [37].

In the wide plethora of electrochemical devices, there are also biosensors based on electrochemical impedance spectroscopy, having demonstrated intriguing responses to the detection of foodborne pathogens. In this context, Tan and colleagues [38] provided a PDMS microfluidic immunosensor for rapid detection of *Escherichia coli* O157:H7 and *S. aureus*. In particular, antibodies for the bacteria have been covalently immobilized on the nanoporous alumina membranes exploiting the self-assembled approach. The electrochemical impedance spectrum has been recorded in the range of 1 Hz to 100 kHz and the maximum impedance amplitude has been noted at 100 Hz. A fast detection limit (within 2 h) of 10^2 CFU/mL has been reported and cross-bacteria experiments validated the specificity of this system.

Moreover, Braiek and colleagues [39] designed an electrochemical immunosensor for the detection of *S. aureus* based on antibody recognition. The authors exploited the spontaneous self-assembled monolayer immobilization of antibodies on the functionalized gold electrode and characterized the functionalization using both cyclic voltammetry and electrochemical impedance spectroscopy. The performance of the device has been verified in terms of sensitivity (127 Ω/decade), detection limit (10 CFU/mL), correlation coefficient (0.994) and reproducibility (average standard deviation of 8%). Finally, negative tests injecting several concentrations of non-specific bacteria targets (e.g., *E. coli* and *S. epidermidis*) have been carried out and a good specificity has been verified (Fig. 8.5). A similar result was published by Bekir and colleagues [40] with the development of an electrochemical impedance immunosensor for rapid detection of *S. aureus*. Also, these authors immobilized polyclonal anti-*S. aureus* onto a modified gold electrode by self-assembled molecular monolayer method, and the analytical data reported a sensitivity of 15 kω/decade and a low detection limit of 10^1-10^7 CFU/mL.

A different configuration of potentiometric biosensor for real-time *S. aureus* detection has been reported in the study by Zelada-Guillén and colleagues [17], in which a single-walled carbon nanotube represents the ion-to-electron potentiometric transducers and anti-*S. aureus* aptamers play the role of recognition element. In this research, the authors investigated two different approaches for the functionalization of aptamers (non-covalent adsorption of drop-casted and covalent bond formation between amine-modified aptamers and carboxylic groups), underlining significant differences in the minimum-detectable bacteria concentration and sensitivity values. In particular, the study concluded with evidence of multiple advantages of the covalent method in terms of simplicity, analysis time, sensitivity (0.36 mV/decade) and for a good detection limit (8×10^2 CFU/mL). Subsequent studies contributed positively to the

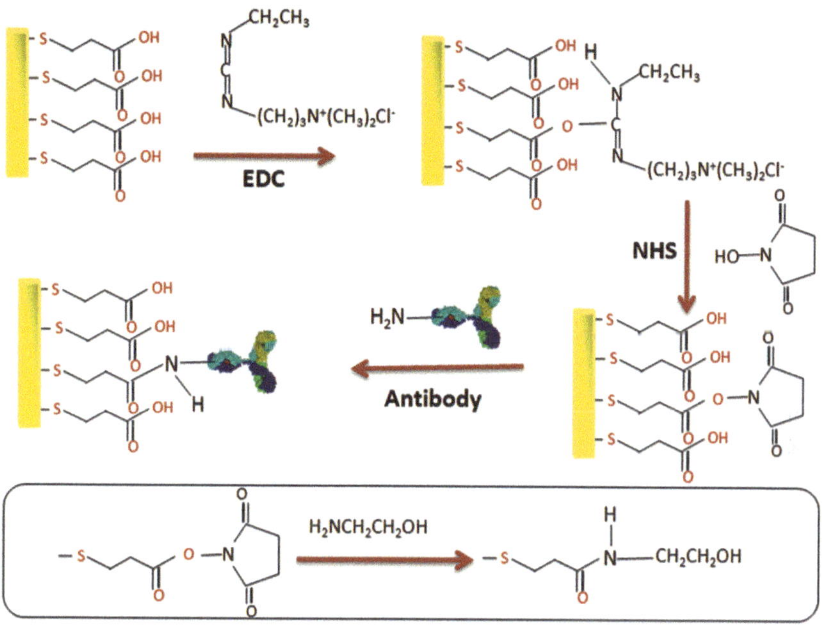

Figure 8.5 Different stages in the development of the biofilm. *Reprinted with permission from M. Braiek, K.B. Rokbani, A. Chrouda, B. Mrabet, A. Bakhrouf, A. Maaref, N. Jaffrezic-Renault, An electrochemical immunosensor for detection of* Staphylococcus aureus *bacteria based on immobilization of antibodies on self-assembled monolayers-functionalized gold electrode, Biosensors, 2(4) (2012) 417–426. MDPI.*

development of these potentiometric aptasensors for the detection of *S. aureus* by the introduction of graphene-modified electrodes. This particular improvement allowed for the achievement of detection values lower in comparison with the previous one (1 CFU/mL), as well as very short response times (60 min) and easily regenerated and reused (at least 10 times) [41,42].

8.3.2 Optical detection

The detection of pathogens through the technology of optical biosensors is a concrete reality already available in the international market. In particular, optical devices, such as colorimetric sensors, are rapid, easy-to-use, cost-effective, sensitive, and portable. Additionally, optical tools enable the integration with microfluidic systems, as well as their versatility allows the designing of a smartphone-based system with a light detector, for a simple user interface and for rapid sensing. For all these reasons, it is very likely that the importance of optical biosensors will increase for pathogen diagnostics and point-of-care (POC) monitoring in clinical and environmental fields. These important issues have led to the development of several ingenious optical sensors, also for *S. aureus* and it is highlighted in scientific literature.

Recently, Pohanka and colleagues [43] reported the design of a biosensor based on quartz crystal microbalance (QCM) for a simple and reliable determination of *S. aureus*. The working mechanism is based on the specific activity of the antibody anti-*S. aureus* immobilized on 10 MHz QCM sensors, with an interdigitated layer stabilized by iron nanoparticles. The limit of detection on samples without any pre-treatment was 145 μg/mL (5.18×10^8 CFU/mL), while test of storage stability reached 90 days in laboratory conditions with a sensitivity of 90%.

A representative example of flat substrate-based formats with paper and lateral flow immunoassays has been proposed. Li and colleagues [44] developed a paper-based POC testing immuno-disk sensor for the specific detection of *Pseudomonas aeruginosa* and *S. aureus*. In particular, three different antibodies, which are specific to the whole-cell of target bacteria, were conjugated with gold nanoparticles and the analyte was identified within 5 min by the color change in the testing region onto the paper disk. The detection range reported was of 500−5000 CFU/mL for each of the infectious agents tested.

The rational of these colorimetric methods based on a lateral flow immunoassay combined with solution based on nanoparticles is to provide fast processing time and increased sensitivity [45].

In this context, Sung and colleagues [46] tested the reliability of a colorimetric biosensor using antibody/gold nanoparticle/magnetic nanoparticle nanocomposites for the detection of *S. aureus* in phosphate buffer saline and milk, and the capture efficiency of antibody/AuNP/MNPs increased significantly compared to antibody-coupled magnetic nanoparticles (MNPs) without AuNP (Fig. 8.6). The detection limit observed was 1.5×10^3 and 1.5×10^5 CFU/mL in PBS and milk, respectively with analysis time of 40 min.

Another intriguing example of a calorimetric approach from the selective and sensitive detection of *S. aureus* has been described by Liu and colleagues [47]. The authors identified and purified an *S. aureus*-specific pVIII fusion protein by phage display, and subsequently assembled it on AuNPs to design an innovative CS-AuNPs@fusion-pVIII probe. The analytical tests revealed an improved detection limit of 19 CFU/mL with faster analyses (within 30 min).

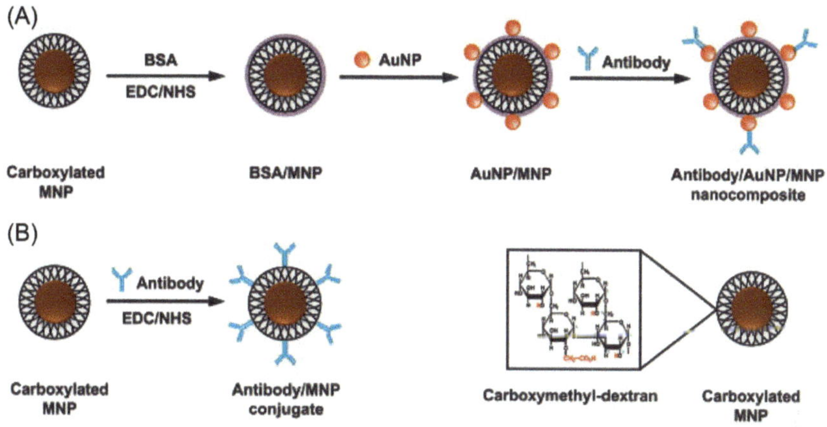

Figure 8.6 Schematic illustration of: (A) Bovine serum albumin, BSA/MNPs, antibody/AuNP/MNP nanocomposites and (B) antibody/MNP conjugates synthesis procedure. *MNPs,* Magnetic nanoparticles. *Reprinted with permission from Y.J. Sung, H.J. Suk, H.Y. Sung, T. Li, H. Poo, M.G. Kim, Novel antibody/gold nanoparticle/magnetic nanoparticle nanocomposites for immunomagnetic separation and rapid colorimetric detection of* Staphylococcus aureus *in milk. Biosens. Bioelectron., 43 (2013) 432–439, Elsevier.*

Zuo and colleagues [48] reported a sophisticated polydimethylsiloxane (PDMS)/paper/glass hybrid microfluidic system integrated with aptamer-functionalized graphene oxide (GO) nano-biosensors for a simple, one-step, multiplexed pathogen detection. In the beginning, the authors exploited this sensing device for *Lactobacillus acidophilus* as a bacterium model to develop the microfluidic platform, and subsequently, successfully extended this approach for the detection of *S. aureus* with a detection limit of 8×10^2 CFU/mL. This particular sensing device boasts an advantageous microfluidic system with the following: (1) it is simple, (2) the presence of paper allows for facilitated immobilization, and avoidance of complicated surface treatment, (3) the outputs are fast, and (4) the potential of this detection can be extended to a wide variety of pathogens, including plant, animal, foodborne, biodefense and other infectious diseases.

He and colleagues [49] described another strategy for *S. aureus* detection exploiting the aptamer recognition and fluorescent silica nanoparticles (FSiNPs). In particular, the aptamer was covalently immobilized onto chloropropyl-functionalized FSiNPs by a click chemistry approach to generate aptamer-nanoparticle bioconjugates, and then a dual-color flow cytometric detection was obtained with FSiNPs and SYBR Green I dye. The detection limit of 1.5×10^2 and 7.6×10^2 CFU/mL was observed for *S. aureus* in buffer and milk samples, respectively.

In this context, Shangguan and colleagues [50] reported another optical sensor based on an aptamer with high affinity to *S. aureus*. In this case, the authors combined positive dielectrophoresis (pDEP) driven technology (for accumulating the Apt*S.aureus*/FNP labeled *S. aureus*) and aptamer-fluorescent silica nanoparticle (Apts/FNPs) label. The procedure involved a sample incubation in Apts/FNPs, then an injection into the microfluidic channel for the pDEP on-line enrichment and, by applying an AC voltage in a pDEP frequency region, finally the fluorescence detection. The device showed within 2 h, a detection limit of 9.3×10^1 and 2.7×10^2 CFU/mL of *S. aureus* in deionized water and spiked water samples, respectively.

8.4 Nanomaterial-based biosensors

As mentioned above, the use of biosensors for the detection of pathogenic microorganisms is a particularly fervent sector. Despite the recent great advances in research, the marketing of these sensing devices has lagged behind, especially compared to the great success of glucose meters. However, the advent of nanomaterials allowed for the design of novel biosensors for foodborne pathogen detection, as well as in *S. aureus* identification, significantly contributing to their evolution. In particular, these new biosensors boast exceptional properties in terms of sensitivity, specificity, speed, and accuracy, thanks to innovative nanometer-sized materials. In detail, many of the nanomaterials exploited in bio-sensing technology are: noble metal nanoparticles, MNPs, quantum dots (QDs), carbon-based nanomaterials (further subdivided according to their size and shape) and nanoarrays.

Concerning the electrical conductive proprieties, a representative example can be the study of Sun and colleagues [51], who exploited a carbon material, multiwalled carbon nanotubes (MWCNTs) and gold nanoparticles. This strategic choice is useful for the design of a novel electrochemical DNA biosensor for *S. aureus* in tap water, by a single-stranded DNA probe anchored by gold-thiol affinity. The positive responses due to the hybridization between the target and ssDNA sequences have been measured by using the reduction peak current of methylene blue as an electrochemical indicator. In this study, the authors highlighted satisfactory results obtained thanks to the high surface area, favorable electronic properties, and strong electrocatalytic activity of the exploited nanomaterials, which reached the detection limit of 3.3×10^{-an} M (3σ) (within the concentration range 1.0×10^{-15}–1.0×10^{-1} M) and 10 CFU/mL of *S. aureus*.

Regarding the important contribution of nanomaterials related to their high efficiency as fluorescence quenchers and labels, several studies in the literature are already available. For example, Shi and colleagues [52] describes the important impact of gold nanoparticle (AuNP) as the energy donor-acceptor pairs and graphene quantum dot (GQD) as fluorescence labels (with high brightness, long fluorescence lifetime, and good photostability), in the design of FRET biosensor for *S. aureus* detection. In this case, the large surface area to volume ratio, easy surface functionalization, and strong surface plasma absorption in NIR-to-IR region have been crucial features for the reaching of good selectivity and sensitivity (1 nM for *S. aureus* gene sequence detection).

A noteworthy example is the study by Gopal and colleagues [53], since it was the first research highlighting the intriguing potentiality of nanoparticles, and in particular of TiO_2, working as affinity probes for MALDI-Ms studies of bacteria from real samples. The authors proved the positive role of anatase TiO_2, at optimum concentrations of 1 mg/mL, to selectively enrich the bacterial peaks in real samples; discerning in the samples from the nasal passage among many proteins. The limit of detection was 4.6×10^2 CFU/mL.

Another useful application of nanomaterials in the design of biosensors is their significant changes in color due to the size of the nanoscale. Shahbazi and colleagues [54] described this helpful property of AuNP during the development of a colorimetric method for the determination of *S. aureus*. In detail, the authors exploited the principle that, if the genomic DNA is present in the solution, and the probes are complemented by them, the addition of salt can determine the accumulation of AuNPs because the color changes. On the contrary, when the genomic DNA is absent in the sample, the probes with the AuNPs are in the solution, no color change occurs with the addition of salt. This method was accomplished within 10–15 min and the limit of detection of the target DNA was 8.73 ng/μL.

8.5 Microfluidics devices

In the last decade, microfluidics has been considered one of the cutting-edge technologies most promising to deliver smart (bio)sensors. As reported in our recent review,

"*Microfluidics allows to project platforms consisting of a network of channels and wells embedded into glassy or polymer chips, enabling sample handling, mixing, dilution, electrophoresis and chromatographic separation, staining and detection. In addition, microfluidics can be considered an attractive approach in conjunction with electrochemistry, being both compliant for easily miniaturization with minimal loss of performance at reduced scale*" [55].

In this context, Song et al. [56] developed a microfluidic device based on the use of immune spheres functionalized with *S. aureus* antibody to bind the bacteria (Fig. 8.7). Once optimized, the experimental conditions, namely flow rate and detection time, *S. aureus* was quantitatively monitored reaching a detection limit of 1.5×10^1 CFU/μL.

Tan et al. [38] used alumina nanoporous membrane to immobilize antibody for *S. aureus* as well as *E. coli O157:H7*. In detail, antibodies were immobilized on the nanoporous alumina membranes covalently via self-assembled 3-glycidoxypropyl)trimethoxysilane silane and the impedance spectrum was recorded in the frequency range of 1 Hz to 100 kHz for monitoring of these bacteria. This microfluidic immunosensor allowed for the detection of the target analytes using 2 h of analysis time, obtaining a sensitivity of 10^2 CFU/mL.

Another interesting microfluidic device was designed by Soares et al. [57]. They developed an immunosensor using interdigitated electrodes as a platform to immobilize the selected antibody with layer-by-layer films

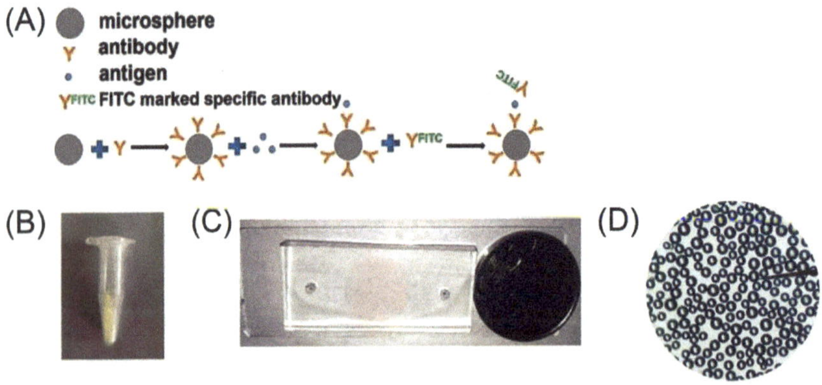

Figure 8.7 (A) The binding mode of antibody and microspheres. (B) The immune microspheres prepared were completed. (C) The diameter of sample hole and outlet of microfluidic chip is 3 mm, and the diameter of detection hole is 15 mm. The prepared chip was compared with the size of one dollar coin, (D) The microspheres under a microscope. *Reprinted with permission from B. Song, J. Wang, Z. Yan, Z. Liu, X. Pan, Y. Zhang, X. Zhang, Microfluidics for the rapid detection of* Staphylococcus aureus *using antibody-coated microspheres, Bioengineered 11(1) (2020) 1137–1145. Taylor & Francis Group.*

of chitosan and chondroitin sulfate. The conceived biosensor was able to detect *S. aureus* with a limit of detection of 2.83 CFU/mL by using 10-bilayer layer-by-layer films.

8.6 Paper-based biosensors

In the era of sustainability, paper-based biosensors have garnered much attention in the sensing field thanks to their cost-effectiveness, easiness to use, and miniaturization. Indeed, as reported by Arduini et al. [58].

"The features of paper such as foldability and porosity have opened new unprecedented electrochemical (bio)sensor configurations allowing for reagent-free measurements, origami-like set-up, and the absence of sample treatment. In addition, after the measure, the device can be burned reducing the waste management with relevant decrease of analysis costs in case of biological fluids or warfare agents."

The outstanding features of paper-based devices including optical and electrochemical transduction were also exploited in the case of *S. aureus* detection. Bhardwaj et al. [59] developed a paper-based electrochemical immunosensor using antibody-single walled carbon nanotube bio-conjugates as the biocomponent. In detail, the functionalized biocomponent was immobilized on the working electrode, and the presence of *S. aureus* was detected by analyzing the change in peak current following antigen-antibody complex formation by using the differential pulse voltammetry technique (Fig. 8.8). The electrochemical aptasensor demonstrated the capability to detect *S. aureus* in a linear range comprised between 10 and 10^7 CFU/mL. The applicability in real samples was evaluated by testing spiked milk samples, demonstrating high sensitivity and rapid measurement time (30 min).

A paper-based aptasensor has been proposed by Suaifan and colleagues [60]. The sensing mechanism was based on the proteolytic activity of *S. aureus* proteases on a specific peptide substrate, sandwiched between magnetic nanobeads and gold surface on top of a paper support. Due to the presence of a magnet in the back of the sensor, the split reaction between magnetic nanoparticles/peptides and sensor has been accelerated. The result of the analysis, visible to the naked eye, has been however elaborated through ImageJ software for a quantitative determination. The detection limits reported were 7, 40 and 100 CFU/mL for *S. aureus* in pure broth culture, and inoculated in food products (ground beef, turkey sausage, lettuce and milk) and environmental samples.

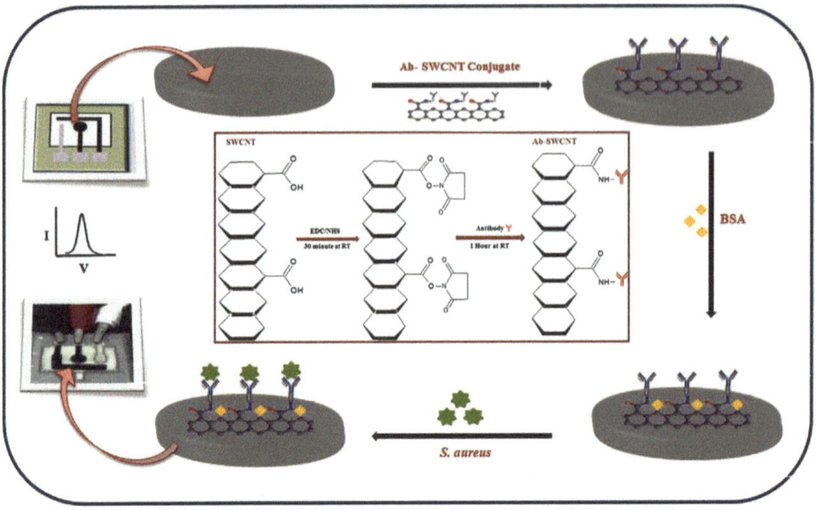

Figure 8.8 Schematic of the Ab-SWCNTs (single walled carbon nanotubes) bio-conjugate: attachment of antibodies to SWCNTs using EDC/NHS coupling reagent, electrochemical biosensor fabrication, and assay procedure. *Reprinted with permission from J. Bhardwaj, S. Devarakonda, S. Kumar, J. Jang, Development of a paper-based electrochemical immunosensor using an antibody-single walled carbon nanotubes bio-conjugate modified electrode for label-free detection of foodborne pathogens, Sens. Actuators B: Chem. 253 (2017) 115–123, Elsevier.*

An additional paper-based immunosensor using magnetic beads was reported by Pang et al. [61] using urease as the label of immunocomplex, it was able to catalyze the decomposition of urea. In the presence of *S. aureus*, the increase in pH was detected using phenolphthalein test paper which undergoes a color change from white to pink within 20 min, observing a detection limit of 476 CFU/mL in standard solution. In the case of real samples such as contaminated Chinese cabbage and beef samples, 1000 CFU/mL was the level accurately detected.

A colorimetric microfluidic device was developed by Hosseini [62], in which the bioconjugate with peroxidase like activity was prepared by using DNA as a template. The developed device showed a linear dynamic range between 10^8 and 10^2 CFU/mL with a detection limit of 80 CFU/mL, and the advantages of very fast measurement (i.e. 5 min combined), cost-effectiveness and ease of use (Fig. 8.9).

Another interesting colorimetric paper-based device using the lateral flow approach was reported by Li et al. [44]. In this case, we reported multiplex immunosensing by measuring *S. aureus* and *P. aeruginosa* using antibody-conjugated gold nanoparticles as signal agents with the detection

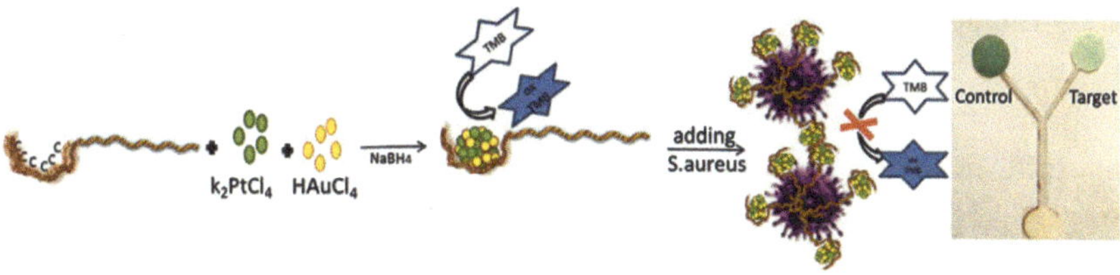

Figure 8.9 Schematic representation mechanism of the formation of Au/Pt NCs, peroxidase mimetic and paper based detection of *Staphylococcus aureus. Reprinted with permission from M. Hosseini, Fast and selective whole cell detection of* Staphylococcus aureus *bacteria in food samples by paper based colorimetric nanobiosensor using peroxidase-like catalytic activity of DNA Au/Pt bimetallic nanoclusters, Microchem. J. 159 (2020) 105475.*

within 500–5000 CFU/mL. The advantage of this device is the use of an immuno-disk sensor which does not require any processing of the sample by measuring directly the whole cell bacteria.

8.7 Conclusion and future perspectives

Biosensor technology, despite its declinations, for several years has been awakening a strong interest in the detection of microorganisms dangerous to humans as the causes of foodborne illnesses. In particular, some representative examples in the detection of *S. aureus*, one of the most common causes of hospital-acquired infections and food-associated disease, have been herein collected with the intent to highlight the successes and drawbacks of the issue.

Undoubtedly, an important strength of biosensors is their perfect fusion with other avant-garde disciplines, ability to enhance the potential of sensing (*e.g.*, nanomaterials, molecular biology, microfluidics and the use of new substrates). In particular, a critical reading of the scientific literature shows that electrochemical biosensors have several advantages in this context because they are not affected by sample turbidity, exhibit excellent sensitivity, have a large linear detection range with a low detection limit, and are also in a widespread variety of solvents, electrolytes, and temperatures. Additionally, it should be noted that the use of optical biosensors is extremely simple and intuitive.

Nevertheless, some challenges still need to be solved in pathogen detection, such as the possibility to find a sufficient amount of samples from patients in order to test the practicality and usefulness of these sensing tools. Moreover, the detection of antimicrobial-resistant pathogens is becoming an important research topic, and for this reason there is a strong urgency to discover new biomarkers for the development of more diverse receptors specific to resistance-relevant molecules. Interesting insights are emerging from the use of new approaches based on the systematic evolution of ligands by exponential enrichment (SELEX) and click chemistry, from which new promising receptors are likely to come.

Acknowledgment

This work is supported by *NanoSWS* project EraNetMed-RQ3–2016 and *AdSWiM* Interreg Project Italy-Croatia 2019/2020.

References

[1] EFSA and ECDC, The European Union summary report on trends and sources of zoonoses, zoonotic agents and food-borne outbreaks in 2017, EFSA J. 16 (2018). Available from: https://efsa.onlinelibrary.wiley.com/doi/full/10.2903/j.efsa.2018.5500.

[2] EU summary report on trends and sources of zoonoses, zoonotic agents and food-borne outbreaks in 2017, (2018). <https://efsa.onlinelibrary.wiley.com/doi/full/10.2903/j.efsa.2018.5500>.

[3] P.J. Conroy, S. Hearty, P. Leonard, R.J. O'Kennedy, Antibody production, design and use for biosensor-based applications, Seminars in Cell & Developmental Biology, Vol. 20, Academic Press, 2009, pp. 10–26. No. 1.

[4] W. Wu, Y. Yang, L. Wang, T. Xu, R. Wang, Electrochemical immunosensor based on mussel inspired coating for simultaneous detection and elimination of *Staphylococcus aureus* in drinks, RSC Adv. 11 (30) (2021) 18252–18258.

[5] H. Yang, H. Chen, L. Cao, H. Wang, W. Deng, Y. Tan, et al., An immunosensor for sensitive photoelectrochemical detection of *Staphylococcus aureus* using ZnS–Ag2S/polydopamine as photoelectric material and Cu2O as peroxidase mimic tag, Talanta 212 (2020) 120797.

[6] H. Wang, Y. Xiu, Y. Chen, L. Sun, L. Yang, H. Chen, et al., Electrochemical immunosensor based on an antibody-hierarchical mesoporous SiO_2 for the detection of *Staphylococcus aureus*, RSC Adv. 9 (28) (2019) 16278–16287.

[7] V. Escamilla-Gómez, S. Campuzano, M. Pedrero, J.M. Pingarrón, Immunosensor for the determination of *Staphylococcus aureus* using a tyrosinase–mercaptopropionic acid modified electrode as an amperometric transducer, Anal. Bioanal. Chem. 391 (3) (2008) 837–845.

[8] M. Iijima, H. Kadoya, S. Hatahira, S. Hiramatsu, G. Jung, A. Martin, et al., Nanocapsules incorporating IgG Fc-binding domain derived from *Staphylococcus aureus* protein A for displaying IgGs on immunosensor chips, Biomaterials 32 (6) (2011) 1455–1464.

[9] S. Centi, S. Laschi, M. Mascini, Improvement of analytical performances of a disposable electrochemical immunosensor by using magnetic beads, Talanta 73 (2) (2007) 394–399.

[10] F. Conzuelo, M. Gamella, S. Campuzano, A.J. Reviejo, J.M. Pingarrón, Disposable amperometric magneto-immunosensor for direct detection of tetracyclines antibiotics residues in milk, Anal. Chim. Acta 737 (2012) 29–36.

[11] F. Ricci, G. Adornetto, G. Palleschi, A review of experimental aspects of electrochemical immunosensors, Electrochim. Acta 84 (2012) 74–83.
[12] V. Scognamiglio, A. Antonacci, M.D. Lambreva, S.C. Litescu, G. Rea, Synthetic biology and biomimetic chemistry as converging technologies fostering a new generation of smart biosensors, Biosens. Bioelectron. 74 (2015) 1076–1086.
[13] M. Shahdordizadeh, S.M. Taghdisi, N. Ansari, F.A. Langroodi, K. Abnous, M. Ramezani, Aptamer based biosensors for detection of *Staphylococcus aureus*, Sens. Actuators B Chem. 241 (2017) 619–635.
[14] J. Xu, J. Guo, S.W. Maina, Y. Yang, Y. Hu, X. Li, et al., An aptasensor for *Staphylococcus aureus* based on nicking enzyme amplification reaction and rolling circle amplification, Anal. Biochem. 549 (2018) 136–142.
[15] L. Hao, H. Gu, N. Duan, S. Wu, X. Ma, Y. Xia, et al., An enhanced chemiluminescence resonance energy transfer aptasensor based on rolling circle amplification and WS2 nanosheet for *Staphylococcus aureus* detection, Anal. Chim. Acta 959 (2017) 83–90.
[16] D. Han, Y. Yan, J. Wang, M. Zhao, X. Duan, L. Kong, et al., An enzyme-free electrochemiluminesce aptasensor for the rapid detection of *Staphylococcus aureus* by the quenching effect of MoS2-PtNPs-vancomycin to S2O82 − /O2 system, Sens. Actuators B Chem. 288 (2019) 586–593.
[17] G.A. Zelada-Guillén, J.L. Sebastián-Avila, P. Blondeau, J. Riu, F.X. Rius, Label-free detection of *Staphylococcus aureus* in skin using real-time potentiometric biosensors based on carbon nanotubes and aptamers, Biosens. Bioelectron. 31 (1) (2012) 226–232.
[18] R. Cai, F. Yin, H. Chen, Y. Tian, N. Zhou, A fluorescent aptasensor for *Staphylococcus aureus* based on strand displacement amplification and self-assembled DNA hexagonal structure, Microchim. Acta 187 (5) (2020) 1–9.
[19] Y. Lu, Z. Yuan, J. Bai, Q. Lin, R. Deng, A. Luo, et al., Directly profiling intact *Staphylococcus aureus* in water and foods via enzymatic cleavage aptasensor, Anal. Chim. Acta 1132 (2020) 28–35.
[20] A. Zhu, S. Ali, Y. Xu, Q. Ouyang, Q. Chen, A SERS aptasensor based on AuNPs functionalized PDMS film for selective and sensitive detection of *Staphylococcus aureus*, Biosens. Bioelectron. 172 (2021) 112806.
[21] S.A. Zaidi, Molecular imprinting polymers and their composites: a promising material for diverse applications, Biomater. Sci. 5 (3) (2017) 388–402.
[22] J. Zhang, Y. Wang, X. Lu, Molecular imprinting technology for sensing foodborne pathogenic bacteria, Anal. Bioanal. Chem. (2021) 1–18.
[23] T. Andersson, A. Bläckberg, R. Lood, G.E. Bergdahl, Development of a molecular imprinting-based surface plasmon resonance biosensor for rapid and sensitive detection of *Staphylococcus aureus* alpha hemolysin from human serum, Front. Cell. Infect. Microbiol. 10 (2020).
[24] J. Bezdekova, K. Zemankova, J. Hutarova, S. Kociova, K. Smerkova, V. Adam, et al., Magnetic molecularly imprinted polymers used for selective isolation and detection of *Staphylococcus aureus*, Food Chem. 321 (2020) 126673.
[25] Y. Saylan, F. Yilmaz, E. Özgür, A. Derazshamshir, H. Yavuz, A. Denizli, Molecular imprinting of macromolecules for sensor applications, Sensors 17 (4) (2017) 898.
[26] R. Wang, L. Wang, J. Yan, D. Luan, J. Wu, X. Bian, Rapid, sensitive and label-free detection of pathogenic bacteria using a bacteria-imprinted conducting polymer film-based electrochemical sensor, Talanta 226 (2021) 122135.
[27] Y. Guo, J. Li, X. Song, K. Xu, J. Wang, C. Zhao, Label-free detection of *Staphylococcus aureus* based on bacteria-imprinted polymer and turn-on fluorescence probes, ACS Appl. Bio Mater. (2020).

[28] N. Tawil, E. Sacher, R. Mandeville, M. Meunier, Surface plasmon resonance detection of *E. coli* and methicillin-resistant *S. aureus* using bacteriophages, Biosens. Bioelectron. 37 (1) (2012) 24–29.
[29] N. Bhardwaj, S.K. Bhardwaj, J. Mehta, K.H. Kim, A. Deep, MOF–bacteriophage biosensor for highly sensitive and specific detection of *Staphylococcus aureus*, ACS Appl. Mater. Interfaces 9 (39) (2017) 33589–33598.
[30] S. Balasubramanian, I.B. Sorokulova, V.J. Vodyanoy, A.L. Simonian, Lytic phage as a specific and selective probe for detection of *Staphylococcus aureus*—a surface plasmon resonance spectroscopic study, Biosens. Bioelectron. 22 (6) (2007) 948–955.
[31] J.C. Rees, J.R. Barr, Detection of methicillin-resistant *Staphylococcus aureus* using phage amplification combined with matrix-assisted laser desorption/ionization mass spectrometry, Anal. Bioanal. Chem. 409 (5) (2017) 1379–1386.
[32] N. Tawil, F. Mouawad, S. Lévesque, E. Sacher, R. Mandeville, M. Meunier, The differential detection of methicillin-resistant, methicillin-susceptible and borderline oxacillin-resistant *Staphylococcus aureus* by surface plasmon resonance, Biosens. Bioelectron. 49 (2013) 334–340.
[33] Y. Wang, W. Yan, S. Fu, S. Hu, Y. Wang, J. Xu, et al., Multiple cross displacement amplification coupled with nanoparticles-based lateral flow biosensor for detection of *Staphylococcus aureus* and identification of methicillin-resistant *S. aureus*, Front. Microbiol. 9 (2018) 907.
[34] C.R. Nemr, S.J. Smith, W. Liu, A.H. Mepham, R.M. Mohamadi, M. Labib, et al., Nanoparticle-mediated capture and electrochemical detection of methicillin-resistant *Staphylococcus aureus*, Anal. Chem. 91 (4) (2019) 2847–2853.
[35] B.E.F. de Ávila, M. Pedrero, S. Campuzano, V. Escamilla-Gómez, J.M. Pingarrón, Sensitive and rapid amperometric magnetoimmunosensor for the determination of *Staphylococcus aureus*, Anal. Bioanal. Chem. 403 (4) (2012) 917–925.
[36] T. Majumdar, R. Chakraborty, U. Raychaudhuri, Development of PEI-GA modified antibody based sensor for the detection of *S. aureus* in food samples, Food Biosci. 4 (2013) 38–45.
[37] P. Arora, A. Sindhu, N. Dilbaghi, A. Chaudhury, Biosensors as innovative tools for the detection of food borne pathogens, Biosens. Bioelectron. 28 (1) (2011) 1–12.
[38] F. Tan, P.H. Leung, Z.B. Liu, Y. Zhang, L. Xiao, W. Ye, et al., A PDMS microfluidic impedance immunosensor for *E. coli* O157: H7 and *Staphylococcus aureus* detection via antibody-immobilized nanoporous membrane, Sens. Actuators B Chem. 159 (1) (2011) 328–335.
[39] M. Braiek, K.B. Rokbani, A. Chrouda, B. Mrabet, A. Bakhrouf, A. Maaref, et al., An electrochemical immunosensor for detection of *Staphylococcus aureus* bacteria based on immobilization of antibodies on self-assembled monolayers-functionalized gold electrode, Biosensors 2 (4) (2012) 417–426.
[40] K. Bekir, H. Barhoumi, M. Braiek, A. Chrouda, N. Zine, N. Abid, et al., Electrochemical impedance immunosensor for rapid detection of stressed pathogenic *Staphylococcus aureus* bacteria, Environ. Sci. Pollut. Res. 22 (20) (2015) 15796–15803.
[41] R. Hernández, C. Vallés, A.M. Benito, W.K. Maser, F.X. Rius, J. Riu, Graphene-based potentiometric biosensor for the immediate detection of living bacteria, Biosens. Bioelectron. 54 (2014) 553–557.
[42] Y. Lian, F. He, H. Wang, F. Tong, A new aptamer/graphene interdigitated gold electrode piezoelectric sensor for rapid and specific detection of *Staphylococcus aureus*, Biosens. Bioelectron. 65 (2015) 314–319.
[43] M. Pohanka, QCM immunosensor for the determination of *Staphylococcus aureus* antigen, Chem. Pap. 74 (2) (2020) 451–458.

[44] C.Z. Li, K. Vandenberg, S. Prabhulkar, X. Zhu, L. Schneper, K. Methee, et al., Paper based point-of-care testing disc for multiplex whole cell bacteria analysis, Biosens. Bioelectron. 26 (11) (2011) 4342–4348.
[45] S.M. Yoo, S.Y. Lee, Optical biosensors for the detection of pathogenic microorganisms, Trends Biotechnol. 34 (1) (2016) 7–25.
[46] Y.J. Sung, H.J. Suk, H.Y. Sung, T. Li, H. Poo, M.G. Kim, Novel antibody/gold nanoparticle/magnetic nanoparticle nanocomposites for immunomagnetic separation and rapid colorimetric detection of *Staphylococcus aureus* in milk, Biosens. Bioelectron. 43 (2013) 432–439.
[47] P. Liu, L. Han, F. Wang, V.A. Petrenko, A. Liu, Gold nanoprobe functionalized with specific fusion protein selection from phage display and its application in rapid, selective and sensitive colorimetric biosensing of *Staphylococcus aureus*, Biosens. Bioelectron. 82 (2016) 195–203.
[48] P. Zuo, X. Li, D.C. Dominguez, B.C. Ye, A PDMS/paper/glass hybrid microfluidic biochip integrated with aptamer-functionalized graphene oxide nano-biosensors for one-step multiplexed pathogen detection, Lab Chip 13 (19) (2013) 3921–3928.
[49] X. He, Y. Li, D. He, K. Wang, J. Shangguan, H. Shi, Aptamer-fluorescent silica nanoparticles bioconjugates based dual-color flow cytometry for specific detection of *Staphylococcus aureus*, J. Biomed. Nanotechnol. 10 (7) (2014) 1359–1368.
[50] J. Shangguan, Y. Li, D. He, X. He, K. Wang, Z. Zou, et al., A combination of positive dielectrophoresis driven on-line enrichment and aptamer-fluorescent silica nanoparticle label for rapid and sensitive detection of *Staphylococcus aureus*, Analyst 140 (13) (2015) 4489–4497.
[51] Y. Sun, X. He, J. Ji, M. Jia, Z. Wang, X. Sun, A highly selective and sensitive electrochemical CS–MWCNTs/Au-NPs composite DNA biosensor for *Staphylococcus aureus* gene sequence detection, Talanta 141 (2015) 300–306.
[52] J. Shi, C. Chan, Y. Pang, W. Ye, F. Tian, J. Lyu, et al., A fluorescence resonance energy transfer (FRET) biosensor based on graphene quantum dots (GQDs) and gold nanoparticles (AuNPs) for the detection of mecA gene sequence of *Staphylococcus aureus*, Biosens. Bioelectron. 67 (2015) 595–600.
[53] J. Gopal, J.L. Narayana, H.F. Wu, TiO_2 nanoparticle assisted mass spectrometry as biosensor of *Staphylococcus aureus*, key pathogen in nosocomial infections from air, skin surface and human nasal passage, Biosens. Bioelectron. 27 (1) (2011) 201–206.
[54] R. Shahbazi, M. Salouti, B. Amini, A. Jalilvand, E. Naderlou, A. Amini, et al., Highly selective and sensitive detection of *Staphylococcus aureus* with gold nanoparticle-based core–shell nano biosensor, Mol. Cell. Probes 41 (2018) 8–13.
[55] F. Arduini, S. Cinti, V. Scognamiglio, D. Moscone, G. Palleschi, How cutting-edge technologies impact the design of electrochemical (bio) sensors for environmental analysis. A review, Anal. Chim. Acta 959 (2017) 15–42.
[56] B. Song, J. Wang, Z. Yan, Z. Liu, X. Pan, Y. Zhang, et al., Microfluidics for the rapid detection of *Staphylococcus aureus* using antibody-coated microspheres, Bioengineered 11 (1) (2020) 1137–1145.
[57] A.C. Soares, J.C. Soares, V.C. Rodrigues, O.N. Oliveira, L.H.C. Mattoso, Controlled molecular architectures in microfluidic immunosensors for detecting *Staphylococcus aureus*, Analyst 145 (18) (2020) 6014–6023.
[58] F. Arduini, D. Moscone, Multifarious aspects of electrochemical paper-based (bio) sensors, Comprehensive Analytical Chemistry, Vol. 89, Elsevier, 2020, pp. 139–161.
[59] J. Bhardwaj, S. Devarakonda, S. Kumar, J. Jang, Development of a paper-based electrochemical immunosensor using an antibody-single walled carbon nanotubes bioconjugate modified electrode for label-free detection of foodborne pathogens, Sens. Actuators B Chem. 253 (2017) 115–123.

[60] G.A. Suaifan, S. Alhogail, M. Zourob, Rapid and low-cost biosensor for the detection of *Staphylococcus aureus*, Biosens. Bioelectron. 90 (2017) 230–237.
[61] B. Pang, Y. Zheng, J. Wang, Y. Liu, X. Song, J. Li, et al., Colorimetric detection of *Staphylococcus aureus* using gold nanorods labeled with yolk immunoglobulin and urease, magnetic beads, and a phenolphthalein impregnated test paper, Microchim. Acta 186 (9) (2019) 1–9.
[62] M. Hosseini, Fast and selective whole cell detection of *Staphylococcus aureus* bacteria in food samples by paper based colorimetric nanobiosensor using peroxidase-like catalytic activity of DNA-Au/Pt bimetallic nanoclusters, Microchem. J. 159 (2020) 105475.

CHAPTER 9

Rapid detection and diagnosis of emerging and re-emerging infectious diseases using biosensing techniques

Priyanka Samal[1] and Shahani Begum[2]
[1]Department of Clinical Hematology-Hemato-oncology and Stem Cell Transplant, IMS & SUM Hospital, Bhubaneswar, Odisha, India
[2]Plant Biotechnology Laboratory, Post Graduate Department of Botany, Utkal University, Vani-Vihar, Bhubaneswar, Odisha, India

9.1 Introduction

The supersensitive diagnosis of causative pathogens responsible for infectious diseases is required for the rapid management and prevention of their spread. Several emerging diseases such as human immunodeficiency virus (HIV), tuberculosis, influenza, emerging nonpolio enterovirus disease, ebola, lassa fever, etc., cause long-term medical emergencies for the public. They are even deemed major threats and extremely contagious by the World Health Organization [1]. Several viral and nonviral contagious diseases have driven public health emergencies in the past decade. In order to monitor the dynamics of an epidemic or pandemic situation, it is necessary to focus on the infectious agents' plausible mechanism of transmission with constancy. Many developing countries rely on the polymerase chain reaction (PCR) and enzyme-linked immunosorbent assay (ELISA) techniques for the amplification of genetic material of infectious agents [2]. Though the abovementioned processes are sensitive and reliable but they also include the use of unsafe components (Etbr and UV light) and the analytical steps are time consuming.

In today's medical management program, the phenomenon of drug resistance remains an alarming situation worldwide that needs to be addressed with appropriate disease management strategies. Recently, point-of-care (POC) diagnosis methods are expanding for clinical and biomedical applications. Designing biosensors is one such approach that have been increasing for the fast detection of PCR or DNA amplicons. Self-contained inexpensive

Biosensors for Emerging and Re-emerging Infectious Diseases
DOI: https://doi.org/10.1016/B978-0-323-88464-8.00017-8

biosensors are influential analytical tools and the major applications of the sensor system are monitoring environmental pollution, detection of disease-causing agents, studying physiological changes in the biological system and the recognition of potential pathogens in agricultural and food industries. In the past few years, biosensing strategies are extended to capture the pathogenic agents, which are primarily responsible for the contagious diseases. Further, rapid development of conjugated models such as aptamer-based, nanomaterials-based and quantum dots biosensors are increasing in production for their reliable, fast and tunable detection of infectious agents. A series of biosensor variants have been developed that pose high quenching ability which determines the sensitivity of the assay. The basic principle of the biosensor is signal transduction and the major elements of the process involve a biological recognition element, transducing device and the combination of amplifier and display machine (Fig. 9.1 and Table 9.1).

9.2 Emerging and re-emerging diseases epidemics

9.2.1 Historical perspective

History has witnessed the extraordinary potential, uniqueness and unpredictability of infectious agents, which have an strong impact on human health. Re-emerging diseases continue to reappear in new mutant forms

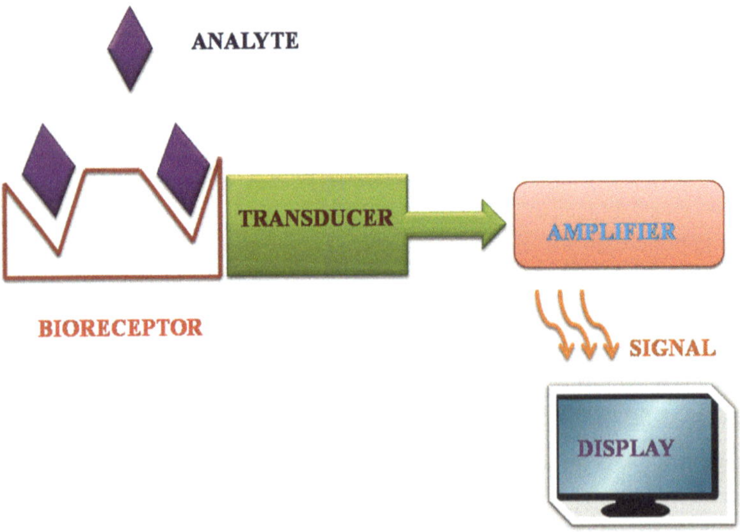

Figure 9.1 Schematic diagram of biosensor components.

Table 9.1 List of major infectious diseases and biosensor formats.

Agents	Disease	Biosensor format	References
Human immunodeficiency viruses	AIDS	DNA sensor	[3]
HCV	HCV	DNA-based sensor	[4]
Bovine viral diarrhea virus	Bovine viral diarrhea	Sandwich immunoassay	[5]
Picornavirus	Foot and mouth disease	immunoassay	[6]
Kaposi sarcoma-associated herpes virus	Castleman's disease	Enzyme-labeled immunosensor	[7]
Influenza virus (H1N1)	Influenza	DNA	[8]
Ebola virus	Ebola hemorrhagic fever	DNA sensor	[9]
Legionella pneumophila	Legionella pneumophilia	DNA sensor	[10]
Mycobacterium tuberculosis	Tuberculosis	Immunosensor/ DNA aptasensor	[11]
Plasmodium falciparum	Malaria	Immunosensor	[12]
Corona virus	Severe acute respiratory disease	Immunoassay/ peptide binding	[13]
Avian influenza virus (H9N2)	Avian influenza	DNA sensor	[14]
	dengue	Immunoassay	[15]
Zika virus	Zika virus disease	DNA sensor/ Immunosensor	[16]
Chikungunya virus	Chikungunya	DNA sensor	[17]
Corona virus	COVID disease	Immunosensor	[18]

AIDS, Acquired immunodeficiency syndrome; HCV, hepatitis C virus.

even after apparent eradication under atypical circumstances. In the late 1900s, the successful identification of one of the first ever emerging diseases, Hantavirus pulmonary syndrome caused by phlebovirus, was carried out by the newer molecular approach at the time: PCR [19]. The critical breakthrough event was the development of antiviral vaccines and drugs in the 1960s. Smallpox, rinderpest, poliomyelitis, dracunculiasis, etc., were eradicated within five decades since their development [20–23]. It has been postulated by Wolfe et al. [24] that about 11,000 years ago, pathogens of domesticated animals had the ability to spread to humans and multiply within them. With the evolution of humans, historians have realized that major human infectious diseases originated from the old world microbes and mostly from the disease-causing microbes of warm-blooded animals, especially primates [24]. It is interesting to note that the major EID's have vertebrate origin, for instance, parentage of the existing falciparum malaria is from a bird parasite, which infects both humans and chimpanzees [25].

Not only animals or other biotic factors, but experimental evidence suggests that global climate change also affects the expansion and geographic distribution pattern of infectious diseases. Warm and humid temperatures permits higher metabolism, diversity and growth rate of pathogens. The microorganism can adapt to hot or cold temperatures depending upon their characteristics features and habitat. High temperature and humidity generally elevate the rate of transmission of malaria and on the other hand, there is also evidence that the dry and cold climate conditions lead to increased transmission of influenza virus and the virus is somewhat more stable at lower relative humidity [26,27].

In the last fifty years, the biosensor concept has evolved without a doubt. For the first time in 1952, Professor Clarkin developed a sensing material called an oxygen probe which was further improved to quantify different analytes by transforming electrodes with biological enzymes. Later, for the quantification of glucose in the sample, glucose oxidase enzyme was fixed with a probe/Clark oxygen electrode and changes in enzyme activity was directly corresponded to the concentration of glucose. Simultaneously, amperometric-based and ferrocene-conjugated glucose oxidase was developed in 1984 to quantify glucose in blood samples. First described, optical-fiber pH sensor for the detection of oxygen, carbon dioxide and other blood gases were based on the principle of wavelength surface plasmon resonance (SPR) shifts. Since then, construction and commercial application of novel biosensors has been broadly adopted for the diagnosis of various infections [28].

9.2.2 Infectious diseases and cancer

Several hypotheses confirm a direct causal link between microbial infections and the transformation of a normal gene into an oncogene. Microbes are believed to affect human metabolism and have been held responsible for 10%–20% of cancer cases [29]. Potential pathogens such as Epstein-Barr Virus (EBV) and *Helicobacter pylori* have a credible role in the development of Lymphomas. Bacterial multiplication, inflammation reaction and toxin production are mainly responsible for the creation of oncogenic conditions [30]. Moreover, when cytotoxic necrotizing factor is present as a microbial toxin, it enhances cellular proliferation and division [31]. Viruses mainly escape cell cycle checkpoints by modulating the role of tumor suppressing genes like TP53 and retinoblastoma gene (RB1). In addition, EBV promotes neoplastic transformation by downregulating apoptotic protein, caspase-3. Receptors of natural killer (NK) cells play a prominent role in viral infection and exhibit toxic activity against tumor cells. Defective NK cells increase the incidence of virus-induced cancer in patients. *Human papilloma virus* (HPV) and EBV affect the activating receptors of the NK cells and transfect their gene into a host genome. The defect in NK cells ($CD56^+CD3^-$) along with GATA2 (responsible for the maintenance of hematopoietic stem cells), and MCM4 (chromosome maintenance protein) mutations result in hematopoietic malignancies [32].

In addition, herpes simplex virus (HSV) has the ability to infect any body parts and reports have suggested that the virus is associated with the development of uterine cervical cancer [33,34]. Further, it causes Kaposi sarcoma and bladder cancer in immunocompromised patients. Another life-threatening virus, cytomegalovirus (CMV) deregulates many cellular process like the formation of new blood vessels (angiogenesis), cell cycle progression (gene products of the virus activates the cyclins and kinases) and programmed cell death pathways. The examination of patients with adenocarcinomas and colorectal polyps revealed the presence of CMV proteins in the affected cells, which indicates the ability of the virus to transform the normal cells to malignant ones [35]. Heng et al. [36] demonstrated the presence of HPV DNA sequences using the *in situ* PCR in the nucleus of the different breast cancer cell lines. In addition, the normal cells surrounding the cancerous cells were also detected with the presence of a low titer of viral proteins. The oncogenic mechanism of HPV-related breast cancer is very comparable to the HPV-associated cervical cancer [36]. In summary, some groups of viruses are capable of altering the cell metabolism and affecting the genetic makeup thus leading to development of cancer.

9.3 Review of some biosensors developed to detect emerging infectious diseases

9.3.1 Electrochemical sensing techniques

The recent pandemic caused by novel coronavirus strain SARS-CoV-2 constitutes a risk to human health. Real-time reverse transcriptase PCR has been used as a diagnostic tool to detect the presence of viral nucleic acid. It is imperative to mention that this method of COVID-19 detection has some drawbacks and gives false negative results. Highly sensitive and user-friendly supersandwich-electrochemical (EC) biosensors with no requirement for nucleic acid amplification are reported to detect the infection by using portable smartphone. The detectable ratio of viral infection was >80% when the RNA extract from the COVID patients was targeted using calixarene-fuctionalized graphene oxide [37].

A direct and alternating current (AC)-based EC biosensor is one of the valuable and favorable diagnostic tools for the detection of bacterial infections. For detection applications, conventional methods need sophisticated equipment, trained operators and multistep processes. Generally, cell-secreted metabolism derivatives which respond to electrical stimuli in nature or the interaction ability of the microbes with the specific recognition component of the electrode enables the biosensor to detect the presence of bacteria. Though the direct current biosensors over the years are technically modified which includes signal and sensitivity enhancement and label-free detection of microbes, the recognition element mostly gets damaged by the used voltage. This contributes to false positives and low specificity. However, in an AC biosensor, such drawbacks are avoided as small voltage is enforced thus cell-probe reciprocal activity is not disturbed. Further, to improve the performance of the AC biosensor, recent development strategies support the inclusion of metal nanotubes, nanoparticles or nanofibres that help to amplify the impedance signal and increase the surface conduction [38].

Recently, different conductive metals (carbon, silver, graphene, etc.) are coupled with EC DNA biosensors to be used as signaling agents [39–41] but some compounds like carbides (carbon compound) offer higher electrical conduction and better EC sensing ability. Early clinical investigation of a deadly disease caused by *Mycobacterium tuberoculosis*, was reported to be carried out successfully using an EC biosensor comprised of 16S rDNA (the most conserved sequence across species) as the desired biomarker, a probe and a signal amplifier (titanium carbide). The proposed technique has a limit of detection (LOD) of about 20 CFU/mL with 2 h of detection time [42].

Infections in wounds can lead to ulceration, late healing process and chronic inflammation reactions. Infected wounds recruit neutrophils that release heme-containing enzyme myeloperoxidase (MPO). Hypochlorous oxidant produced by MPO is the strongest antibacterial agent, which plays a crucial role in controlling the spread of infections associated with acute or chronic wounds. Thus enzyme MPO is considered a potential marker for the early diagnosis and predicting status of wound infection. An EC hydrogen peroxide sensor system was designed for the estimation of MPO chlorination action in fluid collected from a good healing wound and an infected wound. In the study, the MPO activity was significantly higher in the infected wound fluid and with the aid of this sensor, the wounds were categorized as infected, unaffected or critical [43].

Moreover, a Nanocoaxial electrode EC biosensor was constructed for the detection of cholera toxin, a waterborne contagious disease. This novel nanomaterial has the ability to recognize multiple markers and enhance the electrode's surface area. The detection profile of the sensor exhibited the LOD of 2 ng/mL that was comparable to the conventional ELISA readout (1 ng/mL).

9.3.2 Piezoelectric sensing techniques

Hepatitis B is a potentially chronic infectious disease caused by a DNA virus that damages the liver. The WHO estimated that in 2015, the viral disease caused the death of 887,000 people and around 257 million people in 2016 were infected with the virus. In his study, Zhou et al. [44] reported the development of piezoelectric (PZE) DNA sensor for the detection of hepatitis B virus. The sensor effectively recorded the viral DNA even at a low concentration of 0.02−0.14 μg/mL. Gold deposited PZE crystal was used for the immobilization of nucleic acid and the crystal could be reused around five times for the detection purpose without affecting the sensitivity of the sensor [44].

One of the highly contagious diseases cause by airborne bacteria *Francisella tularensis* is tularemia. Several methods of bacteria identification such as culture isolation followed by immunofluorescence assay are tedious and require skilled personnel. PZE sensors offer label-free detection of microbes and target analyte in a prompt manner. PZE immunosensor gold electrodes were used to immobilize *F. tularensis* LVS (live vaccine strain) antigen which was collected from tularemic infected mouse serum. The incubation time was 1−3 days and antibodies against bacteria were detected within 10 min of analysis. The result of the immunosensor was compared with a standard dot blot method. Surprisingly, the standard method showed positive results only in the late phase of infection [45].

For the rapid detection of *Salmonella* infection, a biosensor system containing PZE crystal coated with an antibody was developed. This biosensor was the merger of highly-sensitive gold coated quartz crystal and specific anti-salmonella antibody. The analysis time was very short compared to other conventional methods and the lowest frequency change was 20 Hz [46].

A new generation of safe and reliable biosensors are a prerequisite for the detection of bloodborne pathogens for emergency medical situations. Transfusion of contaminated blood is one of the main causes for the transmission of HIV. Therefore, use of a biosensor with POC approach is an important aspect to help victims in need of blood and especially when stocked resources are depleted. Highly portable PZE biosensor coupled with HIV-specific antibodies was demonstrated to detect the infection in the blood. Additionally, the sensitivity of the sensor chips was unaffected even after third measurements but the performance diminished after repeated elution [47].

9.3.3 Magnetic sensing techniques

Microchip-based sensors have magnetic beads on their surface which helps in detecting antibodies or protein antigens. These versatile and inexpensive sensors were used for the detection of antibodies (IgG) against dengue virus in twelve clinical samples from infected individuals. A strong correlation was observed (0.966) between the magnetic sensor chip assay and the standard conventional method (ELISA). The sensitivity of this biological assay was improved by combining the basic principle of ELISA (antigen-antibody reactions) and the restrained magnetic force of beads present on the sensor arrays. This setup enhanced the detection range and improved the sensitivity of the immunological assay [48].

Single-stranded RNA virus, porcine reproductive and respiratory syndrome virus, causes infectious diseases in pigs. Indirect ELISA immunoassay detects the antibodies against the highly immunogenic protein "open reading frame 7" of virus. However, the limitations of the abovementioned technique have forced researchers to establish some novel approaches of detection. A prerequisite to prevent the spread of deadly virus is the timely detection of infection status. Magnetic immunoassay technique was developed for the rapid diagnosis of the infection in the host serum samples. The on-site assay analytical measurement didn't get disturbed by the obstruction caused by nonparticular binding of the heterophilic immunoglobulin [49].

Foodborne infections are equally dangerous for the outbreaks of contagious illness. Severe abdominal pain, dehydration, diarrhea and loss of consciousness are symptoms of an infection. In recent years, a newer approach of biomolecular detection has been developed which relies on nuclear magnetic resonance (NMR) technology for the screening of foodborne pathogens. The fast detecting protocol was adopted which includes the use of contrasting agents for the enhancement of the resolution of the NMR output. Milk samples containing *Salmonella* species were also analyzed with a detection sensitivity of 2.4×10^4 cfu/mL [50].

9.3.4 Optical sensing techniques

For the detection of bacteria (*E. coli*), an optical biosensor hybrid with porous silica oxide nanostructure (transducer) was used to identify the infection. The active constituent of the biosensor was the polyacrylamide (hydrogel) which was combined with the antibody-IgG. Reflective Interferometric Fourier Transform Spectroscopy (RIFTS) studies confirmed the immobilization of the antibody on the hydrogel without affecting IgG sensitivity and specificity present on the surface of the sensor. The immunosensor had the detection limit of 10^3 cells/mL within few minutes [51].

Various optical sensor platforms are being tested for the detection of dengue virus by calibrating the SPR. The monoclonal antibody was immobilized on the thin surface of gold/iron nanocellulose composite and directed against the E-protein of virus. The change in the SPR angle is directly proportional to the concentration of the antigen. Atomic force microscopy was used to demonstrate the binding of antibody to antigen [52].

Human adenovirus is the double stranded DNA virus which mainly affects the GI and respiratory tracts. To manage the spread of the infectious agent, a new silicon optical biosensor has been developed to detect viral genetic material. A microfluidic-based genetic material isolation method was adopted for viral DNA extraction. The isolation process was completed within 30 min and the optical biosensor offered rapid detection of 10 copies of human adenovirus virus in the supplied sample. This diagnostic assay was 100 times more sensitive than quantitative PCR [53].

In addition, Zika virus NS1 protein (highly conserved nonstructural protein) was targeted with another protein that has the affinity for the antigen. Emulsion agglutination assay validated the protein-ligand interactions as the antigen binding protein was coupled to emulsion droplets. The agglutination of droplets occurred when the multivalent antigen bound to an antigen binding protein and sensor showed the limit of detection of 100 nM [54].

9.3.5 Aptamer based biosensor

Transmission of pathogens also occurs through contaminated food. A strong analytical strategy, fluorescence resonance energy transfer (FRET)-based aptamer biosensor was tested to confirm the presence of *Vibrio parahaemolyticus* and *Salmonella typhimurium* in food samples. Two different aptamers were used in conjugation with light-emitting quantum dots to identify the separate sample pathogens. The concentration of the infectious agents could easily be deduced from the fluorescence level of the quantum dots and in the reported study, the result of the aptamer based assay was quite comparable to the plate count method. The novel and sensitive biosensor platform had the detection limit of 25 cfu/mL for *Vibrio* species and of 35 cfu/mL for *Salmonella* species [55].

Gold-based magnetic nanoparticles linked with an aptamer/antibiotic were constructed for the detection of specific pathogenic bacteria. In addition, surface-enhanced Raman scattering (SERS) biosensor recognizes the presence of the infectious bacterial strain. The assay could achieve the limit of detection of 3 cells/mL within 1 h, free of any nonspecific binding activity. By following the simple operational method of SERS, a wide range of microbes can be identified by using various aptamers [56]. The abovementioned quantification assay was also attempted for the analysis of real infected samples.

Apart from human life, different food industries like the fishery industry gets hampered by the infection caused by microbial entity. For instance, viral nervous necrosis caused by betanodavirus affects wide range of fishes and due to lack of potent remedial antiviral treatment, the pathogen causes heavy economic loss. As usual, the conventional methods, immunizer-based assessment, ELISA and PCR have limitations like the requirement of a sophisticated laboratory, expensive chemicals and instruments. For the detection of the virus in aquaculture, an effective aptamer-based lateral flow biosensor has been proposed which is a single-stranded nucleotide that binds to the coat protein of the virus. In the experiment, two variants of aptamers were developed which enhanced the sensitivity of the assay and the outcome of study was within 5 minutes [57].

9.4 Coronavirus and biosensors

SARS-CoV-2 is a novel strain of coronavirus whose outbreak has severely hampered the lives of people around the world. This virus is reported to be highly contagious and generally spreads though physical contact, virus-containing droplets and contaminated surfaces. For the early detection of this deadly agent, several assays and promising biosensors are being developed and investigated. Some crucial parameters such as the identification of a biological analyte, the development of an effective transducer and signal amplification system should be taken into consideration before designing biosensor platforms. An optical sensor comprising the techniques of surface plasmon fluorescence and immunoassay has been developed to detect the presence of a protein called N protein which is an important element of the capsid structure of SARS-CoV-2. It is notable that the sensitivity level was enhanced significantly in contrast to the classic conventional method (ELISA). The biosensor setup was also able to detect very low concentrations (1 pg/mL approx) of protein in buffer solution [58].

The spike proteins present on the virus surface are mostly targeted because they have crucial role in establishment of viral infections and spike-targeting drug could be an ideal candidate for preventing spread of diseases. Protein S1 is one such spike protein which is detected by the S1 antibody developed in the clinical laboratory. As soon as the attachment occurs, a marked change is noticed in the biosensor bioelectric readings within the 5 min with the detection limit of approximately 1 fg/mL. In addition, the results of the assay were configured and operated by a smartphone permitting an ultra-fast test of a large number of infected individuals [59].

Another sensor called field-effect transistor (FET) biosensor was constructed for the early diagnosis of COVID-19 to monitor the increased rate of infection. Antigen-specific antibody was used to coat the graphene sheets of the transistor. An extremely sensitive biosensor was reported to detect the presence of spike protein in buffer solution (1 fg/mL) and in clinical culture (100 fg/mL) medium, respectively. It is interesting to note that no treatment of the sample swab specimen was needed prior to analysis [60].

A remarkable experimental setup consisting of nanoparticle-based biosensors showed 100% specificity and sensitivity while diagnosing COVID-19 infected patients. The amalgamation of the biosensing assay and reverse transcription loop-mediated isothermal amplification (RT-LAMP) resulted in accurate and fast detection. It took only an hour to interpret the results after initializing the diagnostic steps. With the aid of primers, the rapid amplification of the gene of interest (nucleoprotein) was carried out and no cross-reaction was noticed during the assay [61]. Timely diagnosis of the presence of coronavirus 2 requires equipment that are costly and highly sophisticated such as RT-PCR. However, with advancements of technology, exceptionally sensitive EC biosensors have been developed. The assay involved techniques of rolling circle amplification (RCA) with modified probes that were easily detected by a coupled biosensor. It was demonstrated that the sensor detected the nucleocapsid proteins (N and S) in 2 h and the resultant similarity index was 100% with that of RT-PCR [62].

9.5 Future challenges

The major concern of the ongoing fields of biosensor research is to develop a point of care diagnosis model for the early identification of infections. The specificity of the biosensor device can be improved by modulating the ratio of signal strength and noise by multiplying biosensor elements, without increasing analysis time. Further, the construction of the modern biosensor prototype is needed with reduced background sound turbulence resulting from transistor system. To fully realize the sensitivity of the detection ability of the sensors, the magnetic field strength could be enhanced by incorporating high quality silicon sensors platforms [63]. Recent advances also focus on detection of the particular microorganism from the complex population in unprocessed samples such as food, blood or other body fluids.

9.6 Conclusion

Further key improvements are required for the development of a reliable testing kit for the detection of latent infections in samples. Biosensors are a promising rapid detection technique with minimal preconditioning of samples. Better specificity and potent anti-blocking ability are the main parameters for the effective quantification of biological agents. Further work should include the designing of a highly responsive probe and compact detectors making the assay relatively uncomplicated and quicker. Food matrices, body fluid diagnosis and environmental monitoring are the main fields where the biosensor has maximum utility. Besides this, the specificity of the antibody against an antigen can also be studied. The aim during an outbreak is to develop the vaccine after properly realizing the epizootiological characteristics of the pathogen and therefore an imperative POC approach with inclusion of conjugated biosensors and modern display technique such as smartphones, should be encouraged to combat the contagious diseases.

References

[1] M.S. Mehand, F. Al-Shorbaji, P. Millett, B. Murgue, The WHO R&D Blueprint: 2018 review of emerging infectious diseases requiring urgent research and development efforts, Antivir. Res. 159 (2018) 63–67. Nov 1.

[2] H. Li, Y.W. Tang, In vitro nucleic acid amplification: an introduction, Advanced Techniques in Diagnostic Microbiology, Springer, Boston, MA, 2006, pp. 158–165.

[3] A.I. Tamayo, L.S. Rizo, M.B. de Armas, A.A. Ferreira, D. Manzani, H. Yamanaka, et al., Biotin self-assembled monolayer for impedimetric genosensor for direct detection of HIV-1, Microchemical J. 153 (2020) 104462. Mar 1.

[4] J. Chen, Z. Liu, Y. Zheng, Z. Lin, Z. Sun, A. Liu, et al., B/C genotyping of hepatitis B virus based on dual-probe electrochemical biosensor, J. Electroanalyt. Chem. 785 (2017) 75–79. Jan 15.

[5] Z. Muhammad-Tahir, E.C. Alocilja, D.L. Grooms, Rapid detection of bovine viral diarrhea virus as surrogate of bioterrorism agents, IEEE Sens. J. 5 (4) (2005) 757–762.

[6] P. Gomes, E. Giralt, D. Andreu, Surface plasmon resonance screening of synthetic peptides mimicking the immunodominant region of C-S8c1 foot-and-mouth disease virus, Vaccine 18 (3–4) (1999) 362–370.

[7] J.L. Gong, F.C. Gong, G.M. Zeng, G.L. Shen, R.Q. Yu, An amperometric immunosensor for the Newcastle disease antibody assay, Anal. Lett. 36 (2) (2003) 287–302.

[8] H. Mohan, P.S. Gill, A. Kumar, Hemagglutinin gene based biosensor for early detection of swine flu (H1N1) infection in human, Int. J. Biol. Macromol. 130 (2019) 720–726.

[9] H. Ilkhani, S. Farhad, A novel electrochemical DNA biosensor for Ebola virus detection, Anal. Biochem. 557 (2018) 151–155.

[10] A. Mobed, M. Hasanzadeh, P. Babaie, M. Agazadeh, A. Mokhtarzadeh, M.A. Rezaee, DNA-based bioassay of legionella pneumonia pathogen using gold nanostructure: a new platform for diagnosis of legionellosis, Int. J. Biol. Macromol. 128 (2019) 692–699.

[11] H. Thakur, N. Kaur, P. Sabherwal, D. Sareen, N. Prabhakar, Aptamer based voltammetric biosensor for the detection of *Mycobacterium tuberculosis* antigen MPT64, Microchim. Acta 184 (7) (2017) 1915–1922.

[12] G.V. Soraya, C.D. Abeyrathne, C. Buffet, D.H. Huynh, S.M. Uddin, J. Chan, et al., Ultrasensitive and label-free biosensor for the detection of *Plasmodium falciparum* histidine-rich protein II in saliva, Sci. Rep. 9 (1) (2019) 1.

[13] S. Mahari, A. Roberts, D. Shahdeo, S. Gandhi, eCovSens-ultrasensitive novel in-house built printed circuit board based electrochemical device for rapid detection of nCovid-19 antigen, a spike protein domain 1 of SARS-CoV-2. BioRxiv. 2020 Jan 1.

[14] K. Kurzątkowska, A. Sirko, W. Zagórski-Ostoja, W. Dehaen, H. Radecka, J. Radecki, Electrochemical label-free and reagentless genosensor based on an ion barrier switch-off system for DNA sequence-specific detection of the avian influenza virus, Anal. Chem. 87 (19) (2015) 9702–9709.

[15] T.Z. Wu, C.C. Su, L.K. Chen, H.H. Yang, D.F. Tai, K.C. Peng, Piezoelectric immunochip for the detection of dengue fever in viremia phase, Biosens. Bioelectron. 21 (5) (2005) 689–695.

[16] R. da Fonseca Alves, D.L. Franco, M.T. Cordeiro, E.M. de Oliveira Junior, R.A. Dutra, M.D. Sotomayor, Novel electrochemical genosensor for Zika virus based on a poly-(3-amino-4-hydroxybenzoic acid)-modified pencil carbon graphite electrode, Sens. Actuators B: Chem. 296 (2019) 126681.

[17] C. Singhal, M. Khanuja, N. Chaudhary, C.S. Pundir, J. Narang, Detection of chikungunya virus DNA using two-dimensional MoS 2 nanosheets based disposable biosensor, Sci. Rep. 8 (1) (2018) 1–11.

[18] G. Seo, G. Lee, M.J. Kim, S.H. Baek, M. Choi, K.B. Ku, et al., Rapid detection of COVID-19 causative virus (SARS-CoV-2) in human nasopharyngeal swab specimens using field-effect transistor-based biosensor, ACS Nano 14 (4) (2020) 5135–5142.
[19] B. Hjelle, S. Jenison, N. Torrez-Martinez, T. Yamada, K. Nolte, R. Zumwalt, et al., A novel hantavirus associated with an outbreak of fatal respiratory disease in the southwestern United States: evolutionary relationships to known hantaviruses, J. Virol. 68 (2) (1994) 592–596.
[20] F. Fenner, Smallpox: emergence, global spread, and eradication, History Philosophy Life Sci. (1993) 397–420.
[21] D.M. Morens, E.C. Holmes, A.S. Davis, J.K. Taubenberger, Global rinderpest eradication: lessons learned and why humans should celebrate too, J. Infect. Dis. 204 (4) (2011) 502–505.
[22] J.F. Modlin, The bumpy road to polio eradication, N. Engl. J. Med. 362 (25) (2010) 2346–2349.
[23] D.R. Hopkins, E. Ruiz-Tiben, P. Downs, P.C. Withers Jr, S. Roy, Dracunculiasis eradication: neglected no longer, Am. J. Tropical Med. Hyg. 79 (4) (2008) 474–479.
[24] N.D. Wolfe, C.P. Dunavan, J. Diamond, Origins of major human infectious diseases, Nature 447 (7142) (2007) 279–283.
[25] A.P. Waters, D.G. Higgins, T.F. McCutchan, *Plasmodium falciparum* appears to have arisen as a result of lateral transfer between avian and human hosts, Proc. Natl Acad. Sci. USA 88 (8) (1991) 3140–3144.
[26] A.C. Lowen, S. Mubareka, J. Steel, P. Palese, Influenza virus transmission is dependent on relative humidity and temperature, PLoS Pathog. 3 (10) (2007) e151.
[27] S.I. Hay, R.W. Snow, The Malaria Atlas Project: developing global maps of malaria risk, PLoS Med. 3 (12) (2006) e473.
[28] J.D. Newman, A.P. Turner, Historical Perspective of Biosensor and Biochip Development. Handbook of Biosensors and Biochips. 2008 Mar 15.
[29] C. De Martel, J. Ferlay, S. Franceschi, J. Vignat, F. Bray, D. Forman, et al., Global burden of cancers attributable to infections in 2008: a review and synthetic analysis, Lancet Oncol. 13 (6) (2012) 607–615.
[30] R.M. Peek Jr, J.E. Crabtree, Helicobacter infection and gastric neoplasia, J. Pathol. A J. Pathol. Soc. Gt. Br. Irel. 208 (2) (2006) 233–248.
[31] J.P. Nougayrède, F. Taieb, J. De Rycke, E. Oswald, Cyclomodulins: bacterial effectors that modulate the eukaryotic cell cycle, Trends Microbiol. 13 (3) (2005) 103–110.
[32] W.Y. Moon, S.J. Powis, Does natural killer cell deficiency (NKD) increase the risk of cancer? NKD may increase the risk of some virus induced cancer, Front. Immunol. 10 (2019) 1703.
[33] S.K. Kjaer, et al., Human papillomavirus, herpes simplex virus and other potential risk factors for cervical cancer in a high-risk area (Greenland) and a low-risk area (Denmark)—a second look, Br. J. Cancer 67 (4) (1993) 830–837.
[34] R. Schmauz, P. Okong, E.M. De Villiers, R. Dennin, L. Brade, S.K. Lwanga, et al., Multiple infections in cases of cervical cancer from a high-incidence area in tropical Africa, Int. J. Cancer 43 (5) (1989) 805–809.
[35] L. Harkins, A.L. Volk, M. Samanta, I. Mikolaenko, W.J. Britt, K.I. Bland, et al., Specific localisation of human cytomegalovirus nucleic acids and proteins in human colorectal cancer, Lancet 360 (9345) (2002) 1557–1563.
[36] B. Heng, W.K. Glenn, Y. Ye, B. Tran, W. Delprado, L. Lutze-Mann, et al., Human papilloma virus is associated with breast cancer, Br. J. Cancer 101 (8) (2009) 1345–1350.

[37] H. Zhao, F. Liu, W. Xie, T.C. Zhou, J. OuYang, L. Jin, et al., Ultrasensitive supersandwich-type electrochemical sensor for SARS-CoV-2 from the infected COVID-19 patients using a smartphone, Sens. Actuators B Chem. 327 (2021) 128899.

[38] T.B. Tran, S.J. Son, J. Min, Nanomaterials in label-free impedimetric biosensor: current process and future perspectives, BioChip J. 10 (4) (2016) 318–330.

[39] S.J. Park, T.A. Taton, C.A. Mirkin, Array-based electrical detection of DNA with nanoparticle probes, Science 295 (5559) (2002) 1503–1506.

[40] Y. Fan, X. Chen, A.D. Trigg, C.H. Tung, J. Kong, Z. Gao, Detection of microRNAs using target-guided formation of conducting polymer nanowires in nanogaps, J. Am. Chem. Soc. 129 (17) (2007) 5437–5443.

[41] A. Bonanni, M. Pumera, Graphene platform for hairpin-DNA-based impedimetric genosensing, ACS Nano 5 (3) (2011) 2356–2361.

[42] J. Zhang, Y. Li, S. Duan, F. He, Highly electrically conductive two-dimensional Ti3C2 Mxenes-based 16S rDNA electrochemical sensor for detecting *Mycobacterium tuberculosis*, Analytica Chim. Acta 1123 (2020) 9–17.

[43] M. Hajnsek, D. Schiffer, D. Harrich, D. Koller, V. Verient, J. vd Palen, et al., An electrochemical sensor for fast detection of wound infection based on myeloperoxidase activity, Sens. Actuators B Chem. 209 (2015) 265–274.

[44] X. Zhou, L. Liu, M. Hu, L. Wang, J. Hu, Detection of hepatitis B virus by piezoelectric biosensor, J. Pharm. Biomed. Anal. 27 (1-2) (2002) 341–345.

[45] M. Pohanka, O. Pavliš, P. Skládal, Diagnosis of tularemia using piezoelectric biosensor technology, Talanta 71 (2) (2007) 981–985.

[46] J. Ye, S.V. LETCHER, A.G. RAND, Piezoelectric biosensor for detection of *Salmonella typhimurium*, J. Food Sci. 62 (5) (1997) 1067–1086.

[47] M. Bisoffi, V. Severns, D.W. Branch, T.L. Edwards, R.S. Larson, Rapid detection of human immunodeficiency virus types 1 and 2 by use of an improved piezoelectric biosensor, J. Clin. Microbiol. 51 (6) (2013) 1685–1691.

[48] T. Aytur, J. Foley, M. Anwar, B. Boser, E. Harris, P.R. Beatty, A novel magnetic bead bioassay platform using a microchip-based sensor for infectious disease diagnosis, J. Immunol. Methods 314 (1-2) (2006) 21–29.

[49] Y. Luo, E.C. Alocilja, Portable nuclear magnetic resonance biosensor and assay for a highly sensitive and rapid detection of foodborne bacteria in complex matrices, J. Biol. Eng. 11 (1) (2017) 1–8.

[50] B. Wu, T. Yang, D. Zou, L. Jin, X. Liang, T. Li, et al., Nuclear magnetic resonance biosensor based on streptavidin–biotin system and poly-L-lysine macromolecular targeted gadolinium probe for rapid detection of Salmonella in milk, Int. Dairy J. 102 (2020) 104594.

[51] N. Massad-Ivanir, G. Shtenberg, T. Zeidman, E. Segal, Construction and characterization of porous SiO2/hydrogel hybrids as optical biosensors for rapid detection of bacteria, Adv. Funct. Mater. 20 (14) (2010) 2269–2277.

[52] N.A. Omar, Y.W. Fen, J. Abdullah, C.E. Chik, M.A. Mahdi, Development of an optical sensor based on surface plasmon resonance phenomenon for diagnosis of dengue virus E-protein, Sens. Bio-sensing Res. 20 (2018) 16–21.

[53] C.E. Jin, T.Y. Lee, B. Koo, H. Sung, S.H. Kim, Y. Shin, Rapid virus diagnostic system using bio-optical sensor and microfluidic sample processing, Sens. Actuators B Chem. 255 (2018) 2399–2406.

[54] Q. Zhang, L. Zeininger, K.J. Sung, E.A. Miller, K. Yoshinaga, H.D. Sikes, et al., Emulsion Agglutination assay for the detection of protein–protein interactions: an optical sensor for Zika virus, ACS Sens. 4 (1) (2019) 180–184.

[55] N. Duan, S. Wu, S. Dai, T. Miao, J. Chen, Z. Wang, Simultaneous detection of pathogenic bacteria using an aptamer based biosensor and dual fluorescence

resonance energy transfer from quantum dots to carbon nanoparticles, Microchim. Acta 182 (5) (2015) 917–923.

[56] Y. Pang, N. Wan, L. Shi, C. Wang, Z. Sun, R. Xiao, et al., Dual-recognition surface-enhanced Raman scattering (SERS) biosensor for pathogenic bacteria detection by using vancomycin-SERS tags and aptamer-Fe3O4@ Au, Anal. Chim. Acta 1077 (2019) 288–296.

[57] J. Liu, Q. Qin, X. Zhang, C. Li, Y. Yu, X. Huang, et al., Development of a novel lateral flow biosensor combined with aptamer-based Isolation: application for rapid detection of grouper nervous necrosis virus, Front. Microbiol. 11 (2020) 886.

[58] J.C. Huang, Y.F. Chang, K.H. Chen, L.C. Su, C.W. Lee, C.C. Chen, et al., Detection of severe acute respiratory syndrome (SARS) coronavirus nucleocapsid protein in human serum using a localized surface plasmon coupled fluorescence fiber-optic biosensor, Biosens. Bioelectron. 25 (2) (2009) 320–325.

[59] S. Mavrikou, G. Moschopoulou, V. Tsekouras, S. Kintzios, Development of a portable, ultra-rapid and ultra-sensitive cell-based biosensor for the direct detection of the SARS-CoV-2 S1 spike protein antigen, Sensors 20 (11) (2020) 3121.

[60] G. Seo, G. Lee, M.J. Kim, S.H. Baek, M. Choi, K.B. Ku, et al., Rapid detection of COVID-19 causative virus (SARS-CoV-2) in human nasopharyngeal swab specimens using field-effect transistor-based biosensor, ACS Nano 14 (4) (2020) 5135–5142.

[61] X. Zhu, X. Wang, L. Han, T. Chen, L. Wang, H. Li, et al. Reverse Transcription Loop-Mediated Isothermal Amplification Combined with Nanoparticles-based Biosensor for Diagnosis of COVID-19. MedRxiv. 2020 Jan 1.

[62] T. Chaibun, J. Puenpa, T. Ngamdee, N. Boonapatcharoen, P. Athamanolap, A.P. O'Mullane, et al., Rapid electrochemical detection of coronavirus SARS-CoV-2, Nat. Commun. 12 (1) (2021) 1–10.

[63] P.A. Besse, G. Boero, M. Demierre, V. Pott, R. Popovic, Detection of a single magnetic microbead using a miniaturized silicon Hall sensor, Appl. Phys. Lett. 80 (22) (2002) 4199–4201.

CHAPTER 10

Development of optical biosensor for diagnosis of microbial pathogens

Abhijita Mohapatra, Mirna Garhnayak, Pradyumna Sahoo and Sangram Panda
Department of Prosthodontics, Institute of Dental Sciences, Siksha O Anusandhan (Deemed to be University), Bhubaneswar, Odisha, India

10.1 Introduction

A pathogen is an infectious organism that can cause disease. These organisms may be bacteria, virus, fungus or other microorganism, which are the key step in the sequence of infection [1]. The pathogenicity of the disease depends upon the immune response of the host.

These pathogens can be divided into human, animal, and environmental pathogens [2,3]. The persistence capability and spread of the disease in the host are the key difference among these pathogens.

Because of the newer development of different types of vaccines and antibiotics for treating the disease, multidrug resistance (MDR) among the pathogens is continuously developing [4].

Broad spectrum antibiotics used to treat the infectious disease without identifying the causative organism, leads to damage of normal microbiota and the spreading of drug resistant bacteria. This issue has become a serious threat, causing "annually at least 23,000 deaths in the United States and about 25,000 deaths in the European Union" [5].

The gold standard microbial culture-based assays method requires a prolonged time for the investigation and is labor intensive. This difficulty is intensified by the MDR bacteria which reduces the efficacy of culture-based systems [6].

Thus biosensors integrated with nanobiotechnology as a fruitful pathogen-diagnostic systems with enhanced multiplex capacity, sensitivity, selectivity, speed, and cost-effectiveness can overcome the existing issues [7].

Biosensors for Emerging and Re-emerging Infectious Diseases
DOI: https://doi.org/10.1016/B978-0-323-88464-8.00016-6

According to the International Union of Pure and Applied Chemistry (IUPAC) [8,9] a biosensor is defined as "an independently integrated receptor transducer device, which is capable of providing selective quantitative or semi-quantitative analytical information using a biological recognition element" [10].

Biosensors can be classified depending upon methods of signal transduction, they are optical, electrochemical, thermometric, piezoelectric or magnetic.

10.2 Optical biosensor: a trend toward detection of pathogens

The optical biosensor is the most developing trend in pathogen detection for the point-of-care (POC) testing. There are various reasons for development of POC systems including: reduced turnaround time, improved morbidity and mortality, reduced clinical visits, eliminates specimen transport, contributes to rapid diagnosis, and is low cost [11].

Optical biosensors are the most widely tested class of biosensors. This biosensor is based on integration of the optical field with a biorecognition component. The optical biosensing unit is commonly divided into two general means. They are

"label-free"

"label-based"

The advantages of these sensing units are high accuracy, sensitivity, small mass and cost-effectiveness. Several advanced concepts and multidisciplinary approaches like microelectronics, microelectromechanical systems (MEMSs), micro/nanotechnologies, molecular biology, biotechnology and chemistry are applied in the execution of new optical biosensors.

10.3 Basic principles of optical biosensors

An optical biosensor is a compact systematic device which contains a biorecognition sensing element integrated with an optical transducer system. A signal is produced in the optical biosensor equivalent to the concentration of calibrated substance (analyte) [12].

We can use various biological analyte like enzymes, antibodies, antigens, receptors, nucleic acids, whole cells and tissues as biorecognition elements (Fig. 10.1).

10.4 Different types of optical biosensors for diagnosis of pathogens

An optical biosensor measures the change in light absorption as reactants are converted to product.

Colorimetric biosensors Optical biosensors allow for the certain and prompt observation of the presence of pathogenic microorganisms in the sample through color change seen by the naked eye without an analytical instrument.

Colorimetric biosensors are based on two system formats (1) flat substrate-based and, (2) solution-based. As an example, lateral flow assay (LFA)-based biosensors are globally available.

10.5 Principles of lateral flow assay

LFA consists of a (1) sample pad, (2) conjugate pad, (3) nitrocellulose strip which contains test and control lines and, (4) wicking pad. Each component overlaps by at least 1–2 mm for capillary flow of sample (Fig. 10.2) [13].

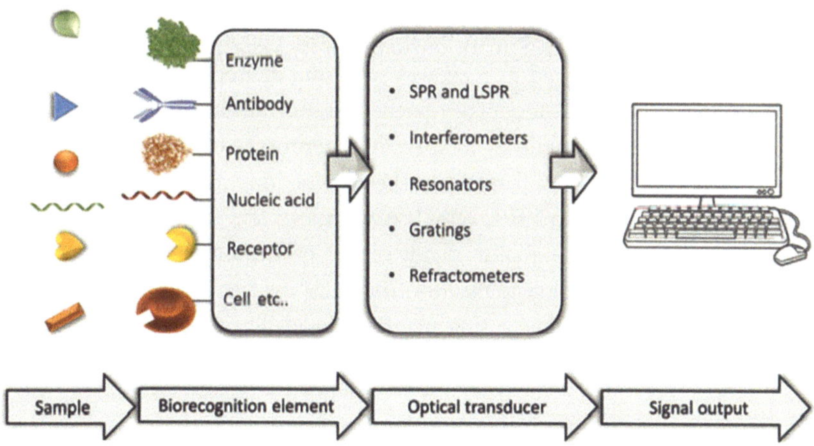

Figure 10.1 Line diagram of optical biosensor. *From: P. Damborský, J. Svitel, J. Katrlík, Optical biosensors. Essays Biochem. 60 (2016) 91–100.*

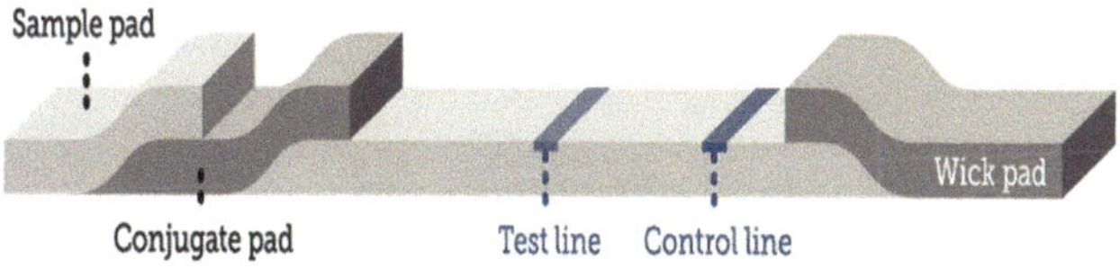

Figure 10.2 (Lateral flow assay) Introduction to lateral flow rapid test diagnostics https://nanocomposix.com/.

A sample such as blood, serum, plasma, urine, saliva, or solubilized solids, is added directly to the sample pad where neutralization and filtration of unwanted particle of the samples takes place. The sample can then flow to the conjugate pad which consists of fluorescent nanoparticles on the surface. When the liquid reaches the conjugate pad, these dried nanoparticles (antibody) are released and mix with the sample. The antibody recognizes the target analyte and binds to the surface of analyte. The conjugation of analyte-bound nanoparticles take place then flow through a nitrocellulose membrane which crosses the test and control line. The test line consists of immobilized proteins which bind the nanoparticle producing a signal correlating to the presence of the analyte in the sample. The fluid continues to flow across the strip until it reaches the control line. The control line (labeled C in the image above) contains ligands that will bind the nanoparticle conjugate with or without the analyte present in solution to check that the assay is working properly. After the control line, the fluid flows into the wicking pad to absorb all of the sample liquid to ensure that there is consistent flow across the test and control lines.

In the lateral flow technique the color change takes place due to the accumulation of gold nanoparticles (Au NPs) Fig. 10.3. The injected sample spreads toward the Au NPs attached to receptors capturing desired pathogens [14].

Then, the Au NP–receptor–pathogen complex streams toward a membrane region and attaches to specific receptors. On binding of pathogens by receptors, a line forms on the device due to the clumping of pathogen– receptor–Au NPs, which is visible to the naked eye [15].

Low sensitivity is the major disadvantage of the LFA based biosensor. However, there are signal amplification methods used to overwhelm the disadvantages [16]. For example, using magnetic beads in preconcentrating cells.

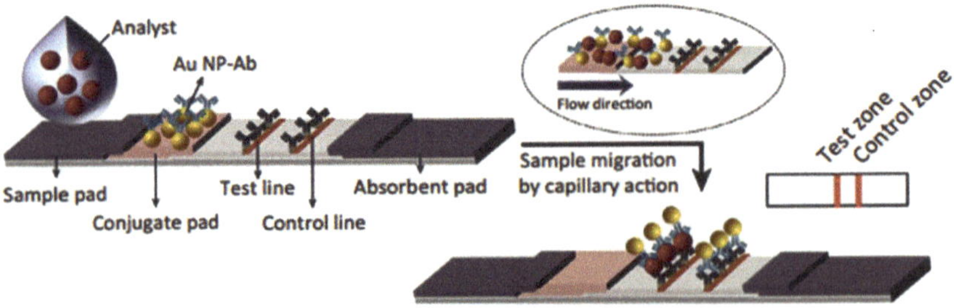

Figure 10.3 LFA-based biosensor. *LFA*, Lateral flow assay. *From: Cz. Li, et al., Paper-based point-of-care testing disk for multiplex whole cell bacterial analysis, Biosens. Bioelectron. 26 (2011) 4342–434.*

Another major disadvantage is the limited multiplex detection of target analytes, which can be overcome by substrate patterning through photolithography, inkjet printing plasma etching, wax printing, etc. [17].

In Solution-based biosensors, colloidal solution of Au NPs is used where the aggregation of receptor and pathogens leads to visible color changes which can be identified with the naked eye. This solution-based biosensor is based on nanotechnology.

A new type of solution-based colorimetric method has two ends, one end is an RNA-containing aptazyme which binds magnetic beads and the other end is where urease attached to a pH-sensitive paper. RNA-containing aptazyme facilitated capturing bacterial cells, and fragmented into two. The part of the aptazyme which contains the magnetic beads is separated from samples and the urease-containing part is transferred to a urea-containing solution [18].

The urease enzyme hydrolyzes urea Fig. 10.4 in solution resulting in increased pH induced color changes due to presence of pH-sensitive dye like phenol red [19]. Although this method is simple and inexpensive, its sensitivity is relatively low.

10.6 Plasmonic biosensors for detection of pathogens

Surface Plasmon Resonance (SPR)
Surface Enhanced Raman spectroscopy (SERS)

10.7 Surface plasmon resonance

SPR is a spectroscopic biosensors which are popularly used in the detection of pathogens because of its simple instrumentation, high sensitivity, label-free and real-time detection (Figs. 10.4 and 10.5).

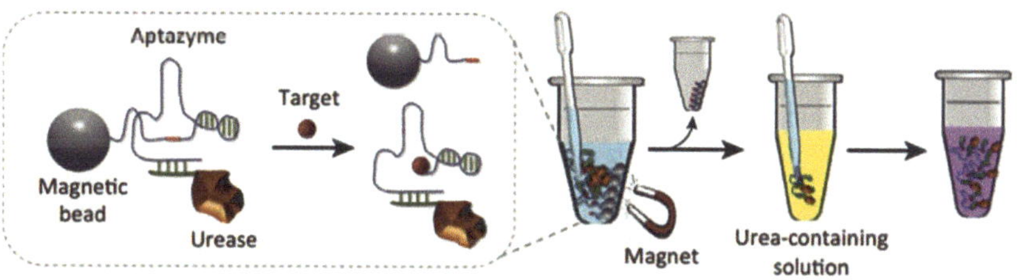

Figure 10.4 Solution-based colorimetric biosensor. *From: K. Tram, et al., Translating bacterial detection by DNAzymes into a litmus test. Angew. Chem. Rev. Int. (Ed.) Engl. 53 (2014) 12799–127802.*

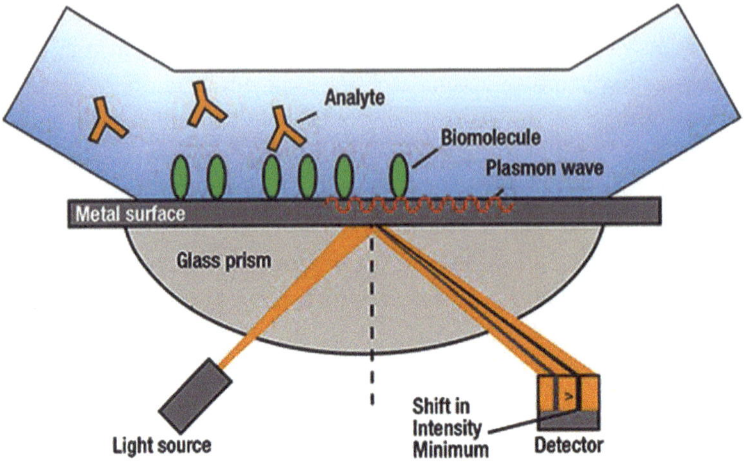

Figure 10.5 Principles of surface plasmon resonance. *From: Katie Thoren, PhD // Date: MAY.1.2019 //. Clinical Laboratory News.*

10.7.1 Principles

The surface sensitive spectroscopic method measures changes in the refractive index of the medium directly in contact with the sensor surface that is gold, silver, etc. The medium in contact with the surface is commonly an aqueous sample containing an analyte protein.

SPR is an electromagnetic wave that can be excited at certain metal interfaces mostly gold and silver. The surface plasmons are the electromagnetic waves that propagate parallel to metal or dielectric surface. From that interface, the plasmons are created; when light energy from polarized incident photons is coupled in the oscillation mode of free electron density which is present in metal surfaces. From these gold surfaces, the plasmons are generated at the boundary of the metal and an external medium, usually air. These are very sensitive to any changes of this boundary like adsorption of biomolecules to the metal. In SPR, a light beam impinges between metal and media at a defined angle known as a resonance angle. The resonance angle depends upon the immediate vicinity of gold surface. When materials bind to the gold surface, the refractive index increases and the SPR curve shifts towards the higher angle. These changes in the angle of reflection of light caused by the binding of the probe to the immobilized protein are measured for the characterization biomolecular interaction in real time.

SPR is an optical phenomenon which is based on "refractive index changes" on metal surfaces and can display in real time and in a label-free manner. The restrained biomolecules like (Fig. 10.5), proteins, antibodies, DNA, or RNA, are constantly flowing across the metal surface. A single wavelength polarized light is directed through a glass prism to the bottom of a metal surface and is reflected back to the detector. The electrons on the metal surface are known as surface plasmons. These electrons are very sensitive to the environment in the metal surface.

10.7.2 Principles of SPR: [20–22]

When an analyte like RNA, DNA, proteins, etc. binds to the restrained biomolecules, a signal change occurs which allows for the interaction to be measured in label-free manner. The metal surface washes out in order to remove the analyte-bound molecules, as a result to make the binding site available following the reaction. The data generated in this technique can be recorded in a sensogram (Fig. 10.6) [20–22].

Advantages of surface plasma resonance:

1. Label-free
2. No need for a tedious or expensive labeling protocol
3. Avoid a potential labeling artifact.
4. Direct
5. Measures binding of the actual analyte
6. Measures binding kinetic and affinity
7. Real Time
8. Allows user to watch the experiment as it happens
9. Not an "end point assay."

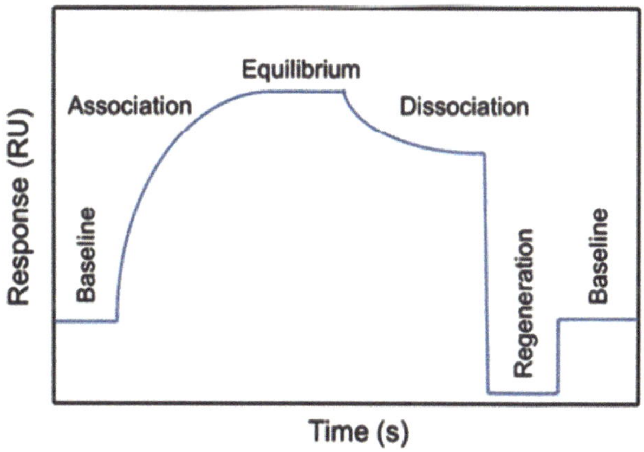

Figure 10.6 ***Sensogram in SPR.*** *SPR*, Surface plasmon resonance. *From: P. Damborský, J. Svitel, J. Katrlík, Optical biosensors. Essays Biochem. 60 (2016) 91–100.*

10. SPR is particularly well suited for measuring the antibody titer of a patient with autoimmune disease.

It also measures clinically-relevant proteins, like C-reactive protein, prostate-specific antigen, brain natriuretic peptide, cytokines, myoglobin, etc. It also has superior sensitivity, multiplexing capability, label free detection, and quantification capability.

10.8 Limitation

Some limitations are, it is an incomplete detection of the whole cell and a complex system due to less penetration depth because of the similarity in refractive index of metal and bacterial cytoplasm [23]. In addition, it is limited in choice of metal like gold, silver, the collected sample must be homogenous, and sample preparation and probe attachment to metal surface can be difficult.

The limitataions of the SPR can be overcome by: magnetic separation method using Aucoated magnetic NPs [24–27].

In localized SPR (LSPR) on the Au NP, copper can be used instead of gold. A shell material on a silica NP core structure can be orderly arranged on an Au surface for the ultrasensitive detection of pathogenic DNAs. The sensitivity of LSPR sensors can be maximized by increasing their reflective index sensitivity (Fig. 10.7).

10.9 Optical biosensors integrated with microfluids

Biosensors with lab or chip technology using microfluidic systems have tremendous benefits in the biosensor technology field. Integration of

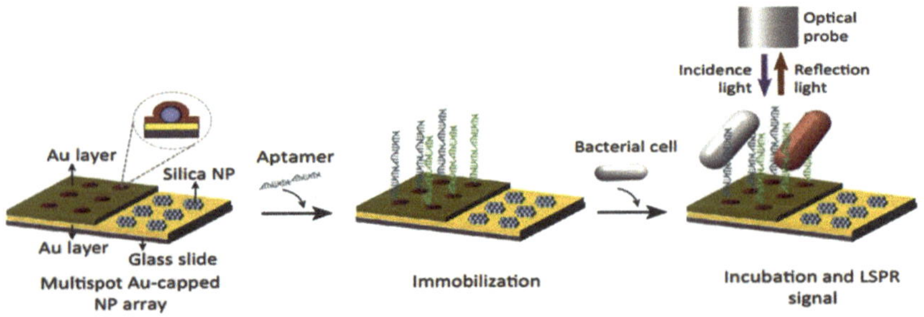

Figure 10.7 LSPR-based biosensor. *LSPR*, Localized surface plasmon resonance. *From: S.M. Yoo, et al., Aptamer-Functionalized Localized Surface Plasmon Resonance Sensor for the Multiplexed Detection of Different Bacterial Species, 2015.*

biosensors with fluidic systems offers a combined and diminished substitute to out-dated laboratory method. These integrated biosensors have several advantages like reduction in sample size, reagent, energy consumption and waste production. It also decreases the cost and increases the specificity and detection sensitivity limit compared to regular detection methods. A single microfluidic biosensor can perform the full analysis including continuous sampling, sample separation and mixing. The integration of a microfluidic system is a powerful analytical devices that will be considered an advanced home testing approach which will benefit developing countries.

Several new microfluidic separation methods using a propagating surface acoustic wave, inertial focusing using spiral micro channels, deterministic lateral displacement and dielectrophoresis-based separation have been developed [28,29]. However, careful isolation of pathogens from a complex sample for flawless detection of pathogens using an integrated biosensor is a sensitive technique.

10.10 Smartphone-based portable surface plasmon resonance biosensor

A portable low-cost biosensor is comprised of a custom scaffold, a disposable polymeric chip for SPR sensing, and a smartphone suited for future medical diagnostic application. Smartphones provides a light source and serves as a detector, while fiber optic SPR is used as the sensing element [30].

As all optical components are connected by optical fibers, this SPR sensor can be easily installed and removed from a smartphone. Biosensors integrated with localized SPR (L)SPR that use a white light source can be used for rapid diagnostic assays [31].

10.11 Future research

Future research should develop in the field of wireless communication biosensors with more powerful pathogen-diagnostic systems which enable remote monitoring and provide real-time feedback information in various practical settings [32]. The capacity for wireless remote sensing of the device would allow for real-time and continuous monitoring of bacteria from various samples from teeth to muscle, etc.

Furthermore, all the results recorded and analyzed by biosensors could communicate to the doctors or pharmacists via personal electronic devices with approval of user. In the near future, such connected biosensor devices will have increasingly important roles in all patient care environments.

It is expected that optical biosensors for pathogen diagnosis will become increasingly important, impacting the fields of clinical research, forensics, biodefense, food safety, animal healthcare, pathology, and drug discovery. Several methods have been developed, and the advantages of each can be combined to develop more advanced diagnostic devices.

10.12 Conclusion

Optical biosensors are gaining momentum in the healthcare, diagnostic, biomedical and biopharmaceutical sectors. Optical biosensors are very versatile platforms with new analytical tools and reduced size which facilitates large-scale high-throughput sensitive screening of a very wide range of samples in POC. The main challenge for further research and innovation in the field of optical biosensors are the construction of sensitive, simple and cheap analytical devices with a wide variety of possible applications in portable biosensor systems for in situ screening and monitoring in developing and developed countries.

References

[1] F.Y. Ramírez-Castillo, A. Loera-Muro, M. Jacques, P. Garneau, F.J. Avelar-González, J. Harel, et al., Waterborne pathogens: detection methods and challenges, Pathogens (Basel, Switz.) 4 (2015) 307–334. Available from: https://doi.org/10.3390/pathogens4020307.

[2] J.W.-F. Law, N.-S. Ab Mutalib, K.-G. Chan, L.-H. Lee, Rapid methods for the detection of foodborne bacterial pathogens: principles, applications, advantages and limitations, Front. Microbiology 5 (2014) 1–19. Available from: https://doi.org/10.3389/fmicb.2014.00770.

[3] Cangelosi GA, Freitag NE, Buckley MR. From Outside to Inside: Environmental Microorganisms as Human Pathogens. Washington, DC: American Academy of Microbiology. Book NLM ID: 101249346. 2004:1-16.

[4] T. Maxson, D.A. Mitchell, Targeted treatment for bacterial infections: Prospects for pathogen-specific antibiotics coupled with rapid diagnostics, Tetrahedron 72 (2016) 3609–3624. Available from: https://doi.org/10.1016/j.tet.2015.09.069.

[5] Fournier, P.E. et al. (2013) Modern clinical microbiology: new challenges and solutions. Nat. Rev. Microbiol.

[6] S.M. Yoo, et al., Microarray-based identification of bacterial and fungal pathogens in bloodstream infections, Mol. Cell. Probes 24 (2010) 44–52.

[7] P. Tissari, et al., Accurate and rapid identification of bacterial species from positive blood cultures with a DNA-based microarray platform: an observational study, Lancet 375 (2010) 224–230.

[8] D.R. The, K. Toth, R.A. Durst, G.S. Wilson, Technical report Electrochemical biosensors: recommended definitions and classification, Biosens. Bioelectron. 16 (2001) 121–131.
[9] IUPAC Compendium of Chemical Terminology, International Union of Pure and Applied Chemistry: Research Triangle Park, NC, USA 2nd Edition (1997, 1992).
[10] S.P. Mohanty, (IEEE POTENTIALS Elias Kougianos, MARCH/APRIL 2006) Biosensors, A Tutorial Review) (2006).
[11] Murugan Veerpandian Application of Nanobiosensors and Biochips in Health Care: A Review Pharmainfo.net, 2008.
[12] P. Damborský, J. Svitel, J. Katrlík, Optical biosensors, Essays Biochem. 60 (2016) 91–100.
[13] Introduction to Lateral Flow Rapid Test Diagnostics <https://nanocomposix.com/>. Accessed on July 2021.
[14] Cz Li, et al., paper based point care testing disc for multiplex whole cell bacterial analysis, Biosens. Bioelectron. 26 (2011) 4342. 434.
[15] C. Pohalman, I. Dieser, L. Sprinzl, A lateral flow assay for identification of Escherichia coli by ribosomal RNA hybridisation, Analyst 139 (5) (2014) 1067–1074.
[16] S.M. Hossain, et al., multiplex paper strip for quantitative bacterial detection, Anal. Bioanal. Chem. 403 (2012) 1567–1576.
[17] H. Ko, et al., active digital microfluidic paper chip inkjet printing patterned electrode, Adv. Mater. 26 (2014) 2335–2340.
[18] Shafee, et al., paper and flexible materials for biosensing platform for multiple biotarget, Sci. Rep. 5 (2015) 8719.
[19] K. Tram, et al., Translating bacterial detection by DNAzymes into a litmus test. Angew. Chem. Rev. Int. (Ed.) Engl. 53 (2014) 12799–127802.
[20] F. Real-Fernández, R. Cimaz, G. Rossi, et al., Surface plasmon resonance-based methodology for anti-adalimumab antibody identification and kinetic characterization, Anal. Bioanal. Chem. 407 (2015) 7477. 85.
[21] S. Mariani, M. Minunni, Surface plasmon resonance applications in clinical analysis, Anal. Bioanal. Chem. 406 (2014) 2303.
[22] J. Metzger, P. Landenberg, M. von Kehrel, et al., Biosensor analysis of β2-glycoprotein I–reactive autoantibodies: Evidence for isotype-specific binding and differentiation of pathogenic from infection-induced antibodies, Clin. Chem. 53 (2007) 1137–1143.
[23] Tokel, et al., Advanced in plasmonic technology for point of care application, Chem. Rev. 114 (2014) 5728–5752.
[24] Y. Wang, et al., Bacterila pathogens surface plasmon resonance biosensor advanced by range surface plasmon and magnetic nanaoparticle assay, Anal. Chem. 84 (2012) 8345–8350.
[25] D.M. Kim, et al., Plasmonic properties of multisport copper capped nanoparticle array chips and its application to detection of bacterial pathogens detection of multiplex DNA, Anal. Chem. 83 (2011) 6215–6222.
[26] F. Liu, et al., Effects of nanoparticle size and cell type on high sensitivity cell detection using a localized surface plasmon resonance biosensor, Biosens. Bioelectron. 55 (2014) 141–148.
[27] N. Nath, A. Chilkoti, Label-free biosensing by surface plasmon resonance of nanoparticles on glass: optimization of nanoparticle size, Anal. Chem. 76 (2004) 5370–5378.
[28] Z. Liu, et al., Rapid isolation of cancer cells using microfluidic deterministic lateral displacement structure, Biomicrofluidics 7 (2013) 1180.

[29] W. Lee, et al., 3D-printed microfluidic device for the detection of pathogenic bacteria using size-based separation in helical channel with trapezoid cross-section, Sci. Rep. 5 (2015) 7717.
[30] Y. Liu, et al., Surface plasmon resonance biosensor based on smart phone platforms, Sci. Rep. 5 (2015) 1286438.
[31] c, S.S. Acimovi, et al., LSPR chip for parallel, rapid, and sensitive detection of cancer markers in serum, Nano Lett. 14 (2014) 2636–2641.
[32] M.S. Mannoor, et al., Graphene-based wireless bacteria detection on tooth enamel, Nat. Commun. 3 (2012) 763.

CHAPTER 11

Magnetic and colorimetric point-of-care biosensors for infectious disease diagnosis

Madhappan Santhamoorthy[1], Dinakaran Thirumalai[2] and Kokila Thirupathi[3]

[1]Department of Biomaterials and Polymer Engineering, Yeongnam University, Gyeongsan, Republic of Korea
[2]Department of Cogno-Mechatronics Engineering, College of Nanoscience and Nanotechnology, Pusan National University, Busan, Republic of Korea
[3]Department of Physics, Sri Moogambigai College of Arts and Science, Dharmapuri, Tamil Nadu, India

11.1 Introduction

Infectious diseases distress the routine lives of millions of people all over the world, and are responsible for numerous deaths, mostly in the developing world. Although most of these major infectious diseases are treatable, the early detection of individual diseases requiring treatment remains a challenging concern. Immense research efforts have been undertaken, focused on replacing standard clinical diagnostic methods, such as invasive detection techniques (biopsy or endoscopy) or expensive diagnostic and monitoring methods, with affordable and sensitive tests based on novel biomarkers. Diagnosis of an infectious disease comprises the detection of the infecting agent, either directly or indirectly. The following are common methods including, microbial culture, microscopy, biochemical tests, immunoassays, and molecular diagnostics that are currently used for the diagnosis of infectious diseases, and these methods are either used independently or in combination with each other. Infectious disease is a clinically evident disease resulting from the presence of a pathogenic agent which can either be a pathogenic virus, bacteria, fungi, or a parasite. These kinds of diseases are also termed communicable diseases owing to their ability to be transmitted from one person or specious to another like some unknown diseases i.e. severe acute respiratory syndrome and avian influenza. These diseases cause more than half of the deaths occurring worldwide. Instant, accurate, and early detection of trace amounts of pathogenic bacteria and viruses is important and critical for clinical

Biosensors for Emerging and Re-emerging Infectious Diseases
DOI: https://doi.org/10.1016/B978-0-323-88464-8.00007-5

diagnosis and prevention. The main drawback is that most of these techniques are laborious and time-consuming as they involve the pathogen amplification step. For instance, the existing technique for bacteria detection requires microorganisms to grow into visible colonies, which takes 1−5 days.

Alternatively, the developing nanodiagnostic technology has strong potential in introducing reliable, rapid, cost-effective, easy-to-use diagnostic methods that could be beneficial in the near future [1]. Therefore, in order to find alternative methodologies, recent research has focused on using biosensors in biomedical applications based on the advantages of these methods including short-time results, excellent sensitivity, and suitable selectivity [2]. In area, biosensors are considered to be highly accurate, sensitive, and specific measurement systems that can determine very low analyte concentrations in biological samples. A biosensor is an analytical device made up of biological elements such as microorganisms, organelles, cell receptors, enzymes, antibodies, nucleic acids, etc. that produces signals (electrical, optical, or thermal) based on the interaction with tested element and a transducer that converts these signals into a measurable electrical parameter [3].

Nanoparticles (NPs)-based nanodiagnostics is an alternative that involves the use of nanotechnology in clinical diagnosis to meet the demands for increased sensitivity and early detection in less time. The advantage of NP formulations compared to conventional systems is that they can increase the efficacy of treatment and reduce side effects, due to specific targeting action [4]. The large surface area of nanomaterials enables the attachment of a large number of target-specific molecules of interest for ultra-sensitive detection. With such capability, diagnosis at the molecular and single-cell level is possible. Because of high sensitivity, nanotechnology enables the detection of a few microorganisms or target molecular analytes specific to pathogens. Conventional methods are limited to achieving this ultra-sensitivity. In addition, the unique properties of nanomaterials could allow for rapid (as short as a few minutes) and real-time detection of pathogens. Moreover, relatively small sample volumes are required for the tests to implement nanotechnology for the diagnosis of infectious diseases.

11.2 Magnetic biosensors

11.2.1 Magnetic nanoparticle-based biosensors

A growing field is in the area of magnetic-based biosensors utilizing magnetic NPs (MNPs) for the detection of infectious disease agents. This technique commonly uses binding affinities between MNP conjugated antibodies and the target pathogen [5]. There are two main uses of MNPs for the detection of infectious diseases.

1. The magnetic relaxation properties: in this method, the MNPs conjugated with antibodies are placed in a solution with the target pathogens and they bind to one another, causing aggregates. This aggregation alters the spin-spin relaxation time (T2) of the surrounding water protons which can be quantified by a magnetic resonance spectroscopy machine.
2. The antibody is fixed to a surface and the sample is washed over it with the pathogen antigen, resulting in binding. The antibody-conjugated MNPs are then washed over the same surface and then bound to the antigen. Thus, the second MNP binding changes the magnetic field and produces an electrical signal that can be observed [6]. This type of assay is very simple and takes less than an hour to detect and can be used effectively for point-of-care (POC) settings because the ability of MNPs to benefit from solution-phase assays is significantly faster than diffusion-dependent surface structure-based assays.

MNPs have been used in a range of studies to fabricate potential POC diagnostic platforms for the detection of *Escherichia coli*, *Salmonella typhimurium*, Staphylococcus *aureus*, and influenza [7]. Biosensors capable of detecting and differentiating between influenza glycoproteins can be built using superparamagnetic iron oxide NPs (SPIONs). The MNPs are coated with poly-poly in this approach (acrylic acid). The desired polymer coating provides stability and the target molecules are further functionalized through ionic or covalent conjugation with NPs. These target molecules in the nanomolar concentration may interact with specific pathogens such as influenza glycoproteins. In addition, the identification of viral pollutants was carried out within 30 min and is simple and relatively compact, requiring only a magnetic relaxometer for sample analysis on the bench-top. Also, the magnetic resonance spectrometry method is stable and cost-effective to produce and is a worthy candidate for POC diagnostics.

MNPs-coupled detectors for biosensing can be used for electrochemical signal amplification with the advantage that they are amenable to use in solution phase sandwich assays such as diagnostic magnetic resonance [8].

Compared with diffusion-dependent surface structure-based assays, the primary benefit of solution-phase assays is substantially faster assay times. Both the capture and detection functionalities are in solution and conjugated to the MNPs in magnetic-based sensors. The MNPs as the antibodies, bind the analyte when the target analyte is present. In dephasing nuclear spin-spin relaxation time, the MNPs are therefore more effective, resulting in a quantifiable signal. Magnetic-based DNA platforms are capable of profiling 13 bacterial species from clinical samples including urine, biliary fluid, ascorbic acid, pleural fluid, and blood from bacterial 16S rRNAs [9].

Three signal amplification steps, including transcription-PCR amplification of 16S rRNA, polymeric bead capture and enrichment of target DNA, and magnetic amplification with magnetically-labeled beads connected to target DNA, allow the developed system to detect single bacteria.

Furthermore, MNPs have been established to have promising applications in medical diagnostics and immunoassay-based diagnostics and therapy [10]. Due to their specific magnetic properties, high surface-to-volume ratio, thermal, chemical, and mechanical stability, simple surface modification of a variety of reagents, and biocompatibility, MNPs are used as labels in most magnetic-based immunoassay platforms, such as magnetic resonance sensors or contrast enhancers. In addition, there were other platforms that used MNPs as auxiliary methods for virus and pathogen detection. With H5N2, viral antibodies targeting the hemagglutinin protein, the MNPs are modified and capped with functionalized ethylene glycol methoxy end group to inhibit non-specific binding. These types of MNPs demonstrate successful isolation of H5N2 from lysate without any elution measures for a direct matrix-assisted laser deposition time-of-flight mass spectrometry reading. Within the diagnostic time of one hour, a detection limit can be reached in the range of 104.5−105.5 TCID50. Similarly, a homogeneous and volumetric DNA detection ferromagnetic resonance-based biosensor was developed to detect and quantify target DNA based on isothermal amplification followed by hybridization of modified MNPs with a detection antibody [11]. Magnetic immunoassay technique is based on surface-enhanced Raman scattering (SERS) spectroscopy to detect influenza virus H3N2. In this system, the magnetic Fe_3O_4/Au NPs are modified with SERS tags, targeting the influenza virus for selective capturing of the target virus [12]. The magnetic nanocomposite was fabricated by combining with green-emitting CdTe quantum dots and two kinds of labels for the effective sensing of virus antibody and antigen, separately. When applying different kinds of fluorescent nanocomposites to the antigens, multiplexed detection of equine influenza virus and equine infectious anemia virus can be detected with a sensitivity of about 1.3 and 1.2 ng/mL, respectively.

For a specific capture and detection of the virus through a shift in the magnetic resonance light scattering signal, a virus-printed magnetic-polymer complex has been developed. Through this method, the target virus with a detection limit of 1.3 pM could be identified in human serum by the device developed. Similarly, the MNPs are employed as a platform for multiplexed detection of hepatitis B virus, hepatitis C virus, and human immunodeficiency virus [13,14].

11.2.2 Magnetoresistive biosensors

Biosensor-based technology is gaining enormous research interest in the production of POC devices and the ongoing development of technologies for microfabrication. In traditional sensing techniques, they offer numerous key features, such as the ability to speed up binding interactions for the rapid detection of target molecules without displaying any interference or noise during signal capture. Magnetic field sensors were historically used in biological diagnostics, specifically superconducting quantum interference instruments, but their use was constrained by large sizes, low sensitivity, and high-power consumption. The use of magnetoresistive materials has solved these limitations. They provide key benefits such as small size, low power, low cost, high sensitivity, and biocompatibility. Therefore, many researchers have extensively focused on magnetic biosensors as an alternative method for overcoming the shortcomings of optical biosensors. Magnetoresistivity can be defined by the change in material resistivity due to the magnetic field applied. Currently, magnetoresistive technology has been shown to be a promising medium for the realization of highly sensitive biosensors in combination with superparamagnetic particles. The superparamagnetic particles are used in this method to generate biosensors by inserting magnetic particles on the surface at the places where the analyte was attached to the sensing sheet. In addition, the embedded magnetic particles are magnetized using an external magnetic field that allows their presence to be sensed/detected by magnetic sensors (e.g., magnetoresistive sensor).

In addition, a heterogeneous protease assay using giant magnetoresistive sensors has been developed by researchers. As biosensors, these types of magnetoresistive sensors transduce shifts in the local magnetic field into electrical signals in the local magnetic field.

These are highly versatile, cost-effective, and reproducible types of giant magnetoresistive sensors. A number of samples with minimal sample preparation and therefore convenient for use in POC applications are intensive to the sample matrix and effective to use magnetic sensors and these types of sensors continuously quantify local magnetic field changes allowing them to be used in real-time monitoring [14,15].

To detect H1N1 and H3N2v from spiked nasal swab samples, the giant magnetoresistive sensor system was developed. This form of wash-free immunoassay approach helps non-technicians with limited training requirements to handle sensing [16]. Schematic representation of the magnetic detection for protease activity. A biotinylated peptide is immobilized on the GMR SV sensors and placed in a magnetic field. The addition of streptavidin-coated MNPs causes an increase in magnetoresistance (Mr) as they are orientated close to the sensor surface via the streptavidin-biotin interaction. When a biofluid sample containing a protease is added, cleavage of the peptide causes a time-dependent change in the Mr as the MNPs are enzymatically released from the sensor surface. Similarly, for the detection of the Hepatitis B virus, giant magnetoresistive sensors were designed, using a GMR bioship integrated with microfluidic channels with a detection sensitivity of 200 IU/mL for HBV DNA molecules (Fig. 11.1). GMR-based sensors have also been developed for the detection of bacteria specifically, *E. coli* O157:H7 antigen and Mycobacterium tuberculosis-specific antigen—ESAT-6 [17]. The key take-away point is that several experimental

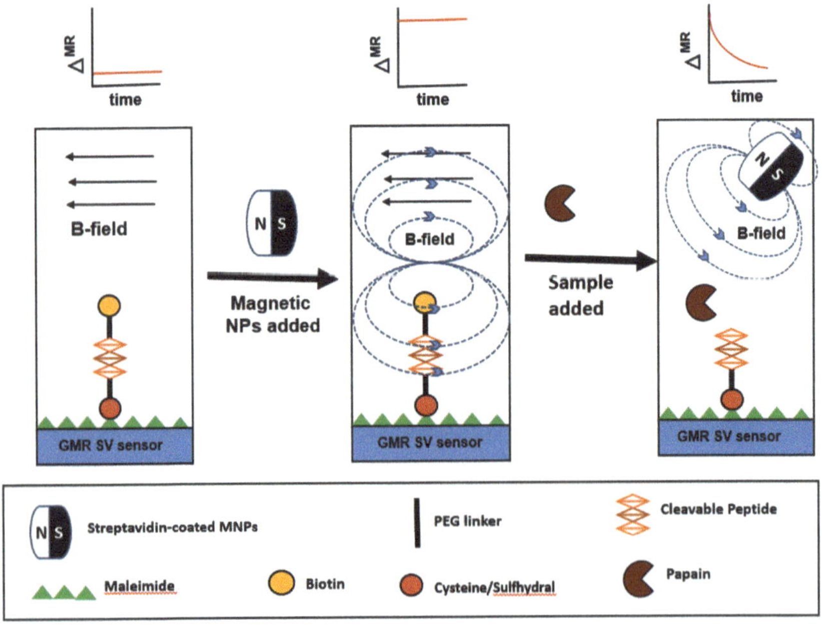

Figure 11.1 Schematic representation of the magnetic detection for protease activity.

demonstrations of the magnetic assays for GMRs based on virus detection and the recorded LOD suggest that one of the promising candidates for on-site, quick, and sensitive detection of COVID-19 is the GMR-based bioassay.

11.3 Colorimetric biosensors

11.3.1 Loop-mediated isothermal amplification

A type of novel gene amplification process, loop-mediated isothermal amplification (LAMP) is an auto cycling and strand displacement DNA synthesis method involving the use of the large BstDNA polymerase fragment and a set of four specially designed primers. Because of its benefits, such as an expensive thermocycler is not needed for LAMP because all reactions can be performed at a constant temperature ranging from 60°C to 65°C; (2) the specificity of the amplification is extremely high because the LAMP reaction needs a collection of four oligopeptide primers that recognize the target DNA in six distinct regions; (3) With a shorter detection time, the detection limit of LAMP is expected to be equal to or greater than that of PCR; (4) the LAMP response can be accelerated by using two specially-built loop primers [18]. One of the leading causes of community-acquired pneumonia (CAP) of all ages, especially in school-age children, is Mycoplasma pneumonia (MP). MP was responsible for 40% of childhood CAP cases, and as many as 18% of hospitalized patients. Only via clinical presentations, was it hard to confirm MP infection for clinicians, as it could also be seen with other common pathogens. Laboratory research is therefore very critical for implementing the right MP infection drug strategy. Three available techniques, including culture-based methods such as serological assay and nucleic acid amplification technique, can be used to detect MP at this level. The culture-based approach for MP detection, however, is an insensitive and long process and hence is not recommended in clinical settings for traditional diagnosis. Alternatively, LAMP is a basic isothermal amplification test that has been successfully applied to MP recognition with high sensitivity and specificity [19]. This technique is particularly specific to foodborne pathogens, which typically need amplification to achieve the required sensitivity prior to detection. The result can be visualized by this technique by the formation of magnesium pyrophosphate as a white precipitate. A LAMP technique has been developed to detect *sea, seb*, and *sed* genes, and is reported that is more sensitive or specific than PCR by an order of magnitude.

11.3.2 Nanoparticle-based colorimetric biosensors

Recently, immunoassay detection techniques including fluorescence, adsorption, and electrochemical sensing approaches are commonly used. In comparison, due to its recent advances in biosensor growth, emerging disease diagnosis has attracted much interest. Biosensors are miniaturized instruments used in the detection of antigen, antibody, and enzyme and nucleic acid biological analytes. In addition, in the assessment of many drugs, chip-scale instruments are useful. Therefore, it is an urgent demand to develop a simple, low-cost, highly sensitive, and portable biosensor that is useful to detect and quantify biological analytes. The production of assays that do not rely on microscopes, a reliable supply of energy, or costly and bulky equipment is one of the key objectives of POC diagnostic research. A variety of tests have been developed by considering these points, which can be defined by the naked eye by or a minimal detector technologies and often depend on colorimetric modalities. Nitrocellulose, nylon, paper, or other composite membrane is commonly used in an antigen-capture assay for this test. These membranes are capable, and therefore self-sufficient, of capillary action. Labeled antibodies are added to the membrane and the sample of antigen is allowed to spread along the membrane. It results in a color shift in the colorimetric molecule being used when the antigens bind to the labeled antibodies. It is possible to imagine this transition with the naked eye or with limited assistance. These types of diagnostic techniques are very attractive because, with minimal preparation, they often use very little to no instrumentation, are cost-effective, simple, and can be easily interpreted. On the downside, this technique is solely quantitative and cannot indicate the severity of the infection [20].

A colorimetric analysis is a quantitative analysis focused on variations in color caused by bacteria that are visible to the naked eye. A number of plasmonic NPs have been developed and used for biological analyte colorimetric sensing. Colorimetric sensor platforms based on MNPs have been developed for multiple analytes, including bacteria. Recent studies have confirmed that MNPs modified with monoclonal antibodies have been specifically used to detect sensitive *S. typhimurium* and *Listeria monocytogenes* on the basis of color changes due to magnetic NPs being aggregated under the magnetic fields applied [21,22].

Similarly, superparamagnetic NPs may also be added by an enzyme-substrate reaction for color-change virus detection. Superparamagnetic NPs

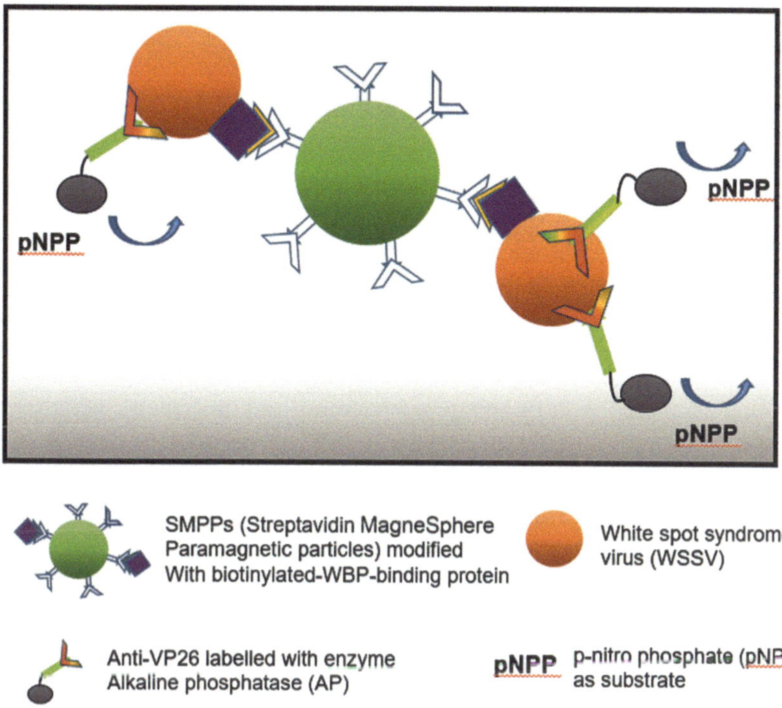

Figure 11.2 A paramagnetic particle with enzyme-substrate colorimetric for WSSV. *WSSV*, White spot syndrome virus.

modified with streptavidin are used, for example and the device detects protein-binding antiglutathione Stransferase-VP26 and WSSV together (Fig. 11.2). In addition, the biotinylated WBP can be used for WSSV-WPB recognition grafted on MNPs. The MNPs can be isolated inside the test tube in this process, in order to wash away unbound molecules. As the para nitrophenyl phosphate substrate is added, alkaline phosphate hydrolyzes the colorless para nitrophenyl phosphate on the MNPs into yellow colored para nitrophenol and inorganic phosphate [23].

11.4 Conclusions

As the need for fast and reliable POC diagnostic tools continues to become more apparent, using various methods and modalities, different platforms are being developed. While many show impressive abilities to detect pathogens or biomarkers in a controlled laboratory condition, it is important to take into account the features needed for practical implementation across the entire laboratory development process. It is possible to use the magnetic activity of such metal NPs in opaque and non-processed solutions. Moreover, the ability of MNPs to be manipulated by a magnetic field enables various diagnostic tools to be developed. The current lack of portability and machinery costs, however, restricts the usage of many of these platforms as diagnostic tools for POC. Before they achieve their full potential, these obstacles would have to be overcome. Colorimetric biosensors are widely used to detect biological analytes identified by analyzing their change in color, but the LAMP technique has been more thoroughly studied. There are other new methods, however, including the use of NPs, phages, enzymes, and polymers, which are capable of successfully detecting analytes and can be used in selected circumstances. Colorimetric assays based on NPs are most appropriate for manufacturing in lab-on-chip devices and can be coupled with LAMP or a portable POC method for highly sensitive pathogen detection on site. It is also critical that these color-changing detection techniques are easily applied for successful use outside the laboratory bench for commercial purposes in practical testing platforms or testing kits. A focus on such qualities as timelines, affordability, sensitivity, and portability will increase the rate at which these novel technologies leave the lab and enter the real world.

Acknowledgment

Madhappan Santhamoorthy expresses his thanks to the department of organic and biopolymer and biomaterials at Yeungnam University for its kind financial support. Also, Madhappan Santhamoorthy, Kokila Thirupathi, and Dinakaran Thirumalai are thankful for the collaboration and scientific contribution.

References

[1] D. Mabey, R.W. Peeling, A. Ustianowski, M.D. Perkins, Diagnostics for the developing world, Nat. Rev. Microbiol. 2 (2004) 231–240.
[2] S.S. Iqbal, M.W. Mayo, J.G. Bruno, B.V. Bronk, C.A. Batt, J.P. Chambers, A review of molecular recognition technologies for detection of biological threat agents, Biosens. Bioelectron. 15 (2000) 549–578.
[3] P. Pashazadeh, A. Mokhtarzadeh, M. Hasanzadeh, M. Hejazi, M. Hashemi, M. de la Guardia, Nano-materials for use in sensing of salmonella infections: recent advances, Biosens. Bioelectron. 87 (2017) 1050–1064.

[4] A.P. Jallouk, R.U. Palekar, H. Pan, P.H. Schlesinger, S.A. Wickline, Modifications of natural peptides for nanoparticle and drug design, Adv. Protein Chem. Struct. Biol. 98 (2015) 57–91.
[5] K.H. Lee, H. Zeng, Aptamer-based ELISA assay for highly specific and sensitive detection of zika NS1 protein, Anal. Chem. 89 (2017) 12743–12748.
[6] G. Lin, D. Makarov, O.G. Schmidt, Magnetic sensing platform technologies for biomedical applications, Lab. Chip 17 (2017) 1884–1912.
[7] H. Kearns, R. Goodacre, L. Jamieson, D. Graham, K. Faulds, SERS detection of multiple anti-microbial resistant pathogens using nanosensors, Anal. Chem. 89 (2017) 12666.
[8] S.D. Soelberg, R.C. Stevens, A.P. Limaye, C.E. Furlong, Surface plasmon resonance detection using antibody-linked magnetic nanoparticles for analyte capture, purification, concentration, and signal amplification, Anal. Chem. 81 (2009) 2357–2363.
[9] H.J. Chung, C.M. Castro, H. Im, A magneto-DNA nanoparticle system for rapid detection and phenotyping of bacteria, Nat. Nanotechnol. 8 (2013) 369–375.
[10] E. Duguet, S. Vasseur, S. Mornet, J.M. Devoisselle, Magnetic nanoparticles and their applications in medicine, Nanomedicine 1 (2006) 157–168.
[11] B. Tian, X. Liao, P. Svedlindh, M. Strömberg, E. Wetterskog, Ferromagnetic resonance biosensor for homogeneous and volumetric detection of DNA, ACS Sens. 3 (2018) 1093–1101.
[12] Y. Sun, L. Xu, F. Zhang, Z. Song, Y. Hu, Y. Ji, et al., A promising magnetic SERS immunosensor for sensitive detection of avian influenza virus, Biosens. Bioelectron. 89 (2017) 906–912.
[13] L. Luo, J. Yang, K. Liang, C. Chen, X. Chen, C. Cai, Fast and sensitive detection of Japanese encephalitis virus based on a magnetic molecular imprinted polymer–resonance light scattering sensor, Talanta 202 (2019) 21–26.
[14] Z. Ali, J. Wang, Y. Tang, B. Liu, N. He, Z. Li, Simultaneous detection of multiple viruses based on chemiluminescence and magnetic separation, Biomater. Sci. 5 (2017) 57–66.
[15] S. Adem, S. Jain, M. Sveiven, X. Zhou, A.J. O'Donoghue, D.A. Hall, Giant magnetoresistive biosensors for real-time quantitative detection of protease activity, Sci. Rep. 10 (2020) 7941. Available from: https://doi.org/10.1038/s41598-020-62910-2.
[16] D. Su, K. Wu, V. Krishna, T. Klein, J. Liu, Y. Feng, et al., Detection of influenza a virus in swine nasal swab samples with a wash-free magnetic bioassay and a handheld giant magnetoresistance sensing system, Front. Microbiol. 10 (2019) 1077. Available from: https://doi.org/10.3389/fmicb.2019.01077.
[17] S. Gupta, V. Kakkar. DARPin based GMR Biosensor for the detection of ESAT-6 Tuberculosis Protein Tuberculosis. Tuberculosis (Edinb) 118, 2019, 101852. Available from: https://doi.org/10.1016/j.tube.2019.07.003.
[18] M. Goto, H. Hayashidani, K. Takatori, Y.H. Kudo, Rapid detection of enterotoxigenic *Staphylococcus aureus* harbouring genes for four classical enterotoxins, SEA, SEB, SEC and SED, by loop-mediated isothermal amplification assay, Lett. Appl. Microbiol. 45 (2007) 100–107.
[19] F. Zhao, L. Liu, X. Tao, L. He, F. Meng, J. Zhang, Culture-independent detection and genotyping of Mycoplasma pneumoniae in clinical specimens from Beijing, China, PLoS One 10 (10) (2015) e0141702.
[20] K.M. Koczula, A. Gallotta, Lateral flow assays, Essays Biochem. 60 (2016) 111–120.
[21] S. Sharma, J. Zapatero-Rodriguez, P. Estrela, R. O'Kennedy, Point-of-care diagnostics in low resource settings: present status and future role of microfluidics, Biosensors 5 (2015) 577–601.

[22] W.-B. Shim, J.-E. Song, H. Mun, D.-H. Chung, M.-G. Kim, Rapid colorimetric detection of *Salmonella typhimurium* using a selective filtration technique combined with antibody-magnetic nanoparticle nanocomposites, Anal. Bioanal. Chem. 406 (2014) 859–866.
[23] S. Loyprasert-Thananimit, A. Saleedang, P. Deachamag, T. Waiyapoka, M. Neulplub, W. Chotigeat, Development of an immuno-based colorimetric assay for white spot syndrome virus, Biotechnol. Appl. Biochem. 61 (2014) 357–362.

CHAPTER 12

Nucleic acid based biosensor as a cutting edge tool for point of care diagnosis

Sejal Shah[1], Anjisha Maharshi[2], Medha Pandya[3], M. Dhanalakshmi[4] and Kajari Das[5]

[1]Department of Computer Science and Bioscience, Faculty of Engineering and Technology, Marwadi University, Rajkot, Gujarat, India
[2]Department of Microbiology, Faculty of Science, Marwadi University, Rajkot, Gujarat, India
[3]Department of LifeSciences, Moharaja Krishnakumarsinhji Bhavnagar University, Bhavnagar, Gujarat, India
[4]Research and Development Centre, Bharathiar University, Coimbatore, Tamil Nadu, India
[5]Department of Biotechnology, College of Basic Science and Humanities, Odisha University of Agriculture and Technology, Bhubaneswar, Odisha, India

12.1 Introduction

Biomolecules exist in nature with a vast size range and unique structures and functions, binding of these molecules among each other being one important function. These properties of binding to any molecule have been leveraged for designing biosensors with innumerable applications. Among all biomolecules, DNA has proved to have the best molecular recognition properties as it can self-assemble readily for the invention of various nanoscale architectures for biosensing. In this chapter, we discuss the possibility of using nucleic acids (NAs) (DNA/RNA) with a combination of other materials for designing biosensors. Various types of analytical approaches have been used thus far, such as western blots, ELISAs and other immunochemical methods that dominate molecular pathology. Still, there is a gap for the development of new innovations using a single-step approach as all these analytical methods are expensive, time consuming, and require multiple steps [1]. Biosensor-based devices are able to reduce the escalating expenditures of healthcare systems with in vitro diagnostics, and as a point-of-care (POC), decentralized, rapid, and low-cost alternative. Various diagnostic modalities are incorporated to actualize POC diagnostic technologies, and this article emphasizes the prospective effectiveness in resource-limited settings. Broadly, the main approaches are optical-, electrochemical-, magnetic-, and colorimetric-based methods pursued in diagnostic technologies

Biosensors for Emerging and Re-emerging Infectious Diseases
DOI: https://doi.org/10.1016/B978-0-323-88464-8.00014-2

for infectious diseases. These modalities with their individual features and pros and cons when considering application in POC settings, are discussed and the anticipated future looks promising in this field of technological development. Optical biosensors are highly sensitive, robust, reliable, and have the potential to be integrated into one chip without difficulty in the pre-treatment phase [2]. Biosensor applications could be more acceptable if a multiplexed DNA sequence assisted electrode device simultaneously analyzes multiple biomolecules without cross-interference [3].

Recently, due to the miniaturization and multianalyte detection capability, electrochemical biosensors are accepted for use with untreated body fluids; such devices add value to smart measurement strategies and could provide instant corrections and adjustment of signal drift and time-resolution for continuous drug metabolite or important biomolecule monitoring [4]. Along with discussing the detailed mechanism of NA biosensors also known as "genosensors," the chapter has incorporated the updated trends in the analysis of biologically appropriate biomarkers (ions, small molecules, proteins, cells and NA) from major research outcomes of NA sensors starting from lab-on-chip/ DNA array to CRISPR technique through electrochemical as well as optical sensing, in single measurement POC devices and continuous/immediate monitoring wearable devices. Such an overview will provide readers with a comprehensive dataset on NA biosensor devices and their applications which is the prime focus of the chapter.

12.2 Biosensor

A biosensor device is built with three basic elements, (1) a biorecognition element that binds with a target and generates a physico-chemical signal, (2) a transducing element and, (3) a detection element that captures the transduced signal to quantify. These recognition elements can be natural (antibodies, enzymes, membranes, NA, hormone) artificial (synthetic polymers or aptamers of biological materials), or chemical compounds with binding properties for some specific targets to be analyzed. The energy produced from the interaction between the receptor and target in a specific microenvironment is transduced to a signal. The transduced signals are results of changes in the microenvironment like gas exchanges and variations in proton concentration, reflected light absorption, light and heat emission, mass changes, etc. Essentially, it can be used to analyze at both ends: disease-causing elements biomarker (metabolites, NAs or protein) or the disease-curing drug concentrations [5,6]. Thus types of biosensors are determined by either their transduction mode (mechanical, optical, electrochemical, etc.) or the property of the biorecognition element (affinity, catalytic, etc.).

12.3 POC diagnostics

POC is a very essential for the procurement of a patient's clinical data with quick acquisition of reports, following which the patients are cured at the earliest stage of process. First reported in 1962, with the device that could measure blood sugar levels fast [7], then in 1977, a rapid pregnancy test kit was developed and used successfully. These have established the trend for personalized diagnostics.

The biomolecules were being conventionally analyzed earlier by using various methods, like polymerase chain reaction (PCR) [8], enzyme-linked immunosorbent assay (ELISA) [9] spectroscopies [10], electrochemistry [11], and onwards. Recent developments in POC testing devices have designed them to measure biological agents for prompt diagnosis even at remote places (near a patient) with inadequate healthcare facilities, instead of only in hospitals and laboratories across the globe. This has immensely changed the approach in the area of diagnostics.

Newly developed materials are being used for making such biosensors. Carbon-based nanomaterials, molecularly-imprinted polymers, quantum dots, dendrimers, etc. are immensely employed in POC devices. Overall, the idea of POC diagnostic techniques can be categorized as follows.

12.3.1 Paper-based

Paper-based is another promising method in which the surface of cellulose was modified to use it as analytical sensors and sensor arrays. It fabricates noninvasive recognition of biomarkers for diverse diseases. The excellent properties of cellulose/nano-cellulose fiber-based papers make the cellulose polymer amongst the highest captivating and inventive resources with abundant applications such as synthesized from trees, plants and non-pathogenic bacteria. Bacterial cellulose nano-paper are used for optical biosensing applications due to their enormous unique specifications such as thermal conductivity, optical transparency, high porosity, extraordinary mechanical strength, biodegradability and very good flexibility in nature. Nanoparticles or bio nanomaterials, carbon nanotubes or graphene-based nanostructures facilitate non-covalent attachment of DNA/RNA for genosensing, designed with enhanced signal amplification [12]. Nowadays, paper-based chip or devices are widely used for various diagnostic tools based on different mi RNA as a material, and saliva also used for microfluidic paper-based analytical devices (mPADs) [13]. Paper-based microfluidic devices emerged to be convenient for their capillary fluid-flow (power-free), due to less time consumption, economic, high reagent storage capacity and high surface-area-to-volume ratio [14,15]. Recently, a new technique is being introduced to fabricate microminiaturized devices containing channels and chambers for 1 mm or less scale or a dimension which is able to function similar to capillary fluid-flow named as microfluidics [16,17]. Considering this, several applications of microfluidics can be envisioned in many fields like analytical chemistry, biotechnology and chemical synthesis. A representation of the compiled microfluidic POC device is shown in Fig. 12.1, applied for the rapid and accurate detection of malaria. A paper strip is the sole platform on which DNA is extracted and processed before detection [18].

12.3.2 Chip-based

Incorporation of MEMS (Microelectromechanical systems) in a POC device allows reductions of mechanical and electrical apparatuses to swap laboratory apparatus and devices for the investigation of several biological

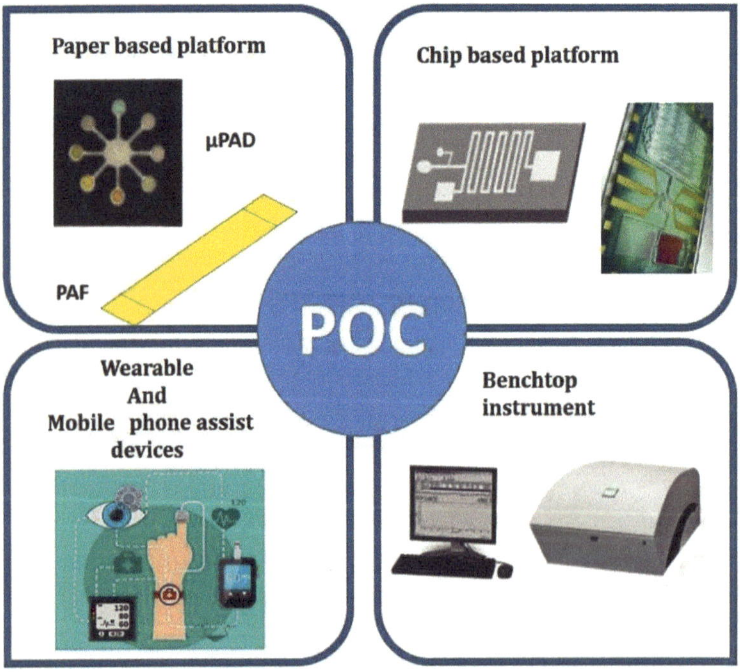

Figure 12.1 The outline of the nucleic acid-based POC testing device. *POC*, Point-of-care.

analytes. Micro-fabrication aims to minuteness of devices commanding a minimum reaction volume with increased proficiency. The advantages of lab-on-a-chip (LOC) format is that it embraces low power consumption, small reagent, size, quality analysis, and mainly, able to incorporate innumerable devices and features. Though, plenty of the POC devices have been designed with the principles that were familiarized decades ago [19]. Inexpensive, portable, and environmentally friendly, microfluidic devices with microchip electrophoresis and temperature gradient focusing (TGF) methodologies were developed [20,21].

12.3.3 Mobile phone-assisted

The widespread use of mobile healthcare technologies has already been established. Among these technologies, the highly assuring one is a cellphone-based device with display of colorimetric, electrochemical, fluorescent, lateral flow, chemiluminescent, and label-free assays, with detection of biomolecules, microorganisms, cells and nanoparticles possible. Realizing that the number of cellphone users is increasing day after day and has already passed 7.4 billion, this device has definitely a huge potential for the future. Amongst them, 70% belong to developing countries, having the critical need for POC devices. Numerous cellphone-based equipment and keen applications have been commercialized for the supervising and organization of basic health parameters, such as physical activity, weight, pulse rate, blood glucose, body analysis, blood pressure, and electrocardiogram. Additional approaches include incorporating WiFi and Bluetooth with cell phones in systematized healthcare structures [22–24].

12.3.4 Wearable POC

Conventional biosensors were developed for in vitro diagnostics, providing a fast turnaround time, portability, small amount of sample requirements, low costs and low power as added advantages. In recent times, epidermal electronics gained ample importance embracing wearable devices (Fig. 12.1). These biosensors consent for constant real-time capture of biological data and simultaneous access to the internet for advance recording and transformation into clinical data. The most advanced applications being noninvasive or minimally invasive chemical analysis of samples comprising saliva, tears, sweat, or interstitial fluid (ISF) instead of blood. ISF analytes can be obtained noninvasively (via iontophoresis or sonophoresis) or with least invasion (using microdialyser or microneedles). For neonates or small children this facet can be of particular value. Microneedles have been merged with sensors for the detection of antibiotics and glucose.

12.3.5 Portable benchtop instruments

For certain on-location experiments, advancement of the general laboratory bench top appliances into a portable size is helpful, specifically for when pandemics happen and performing the tests at the site must be more feasible. A PCR detection device developed by Cepheid company is a game-changer in this scenario. It is able to do RT-PCR in around 45 min by using the Xpert Xpress SARS-CoV-2 test in contrast to the traditional detection method used worldwide to detect the SAR-CoV-2 virus. The portable Xpert Xpress system can easily accomplish the task in any location with less sample preparation steps.

12.3.6 POC platforms in optical and electrochemical modalities

There are plenty of reviews published in recent years (considering the strategies for POC testing [25–27]. Among the above-mentioned modalities, very few are considered important, thus the reviewed techniques are highlighted. POC platforms applying optical modalities have potential prospects, due to the availability of a variety of natural/synthetic fluorescent molecules. The advanced POC diagnostic technologies are focused to allow untrained and effortless usage of these devices by any individual.

Electrochemical POC diagnostic systems have the greatest utilization for reduced size and portability, but its improvement alone is time-consuming and sometimes may not be cost-effective. The glucometer is the best example that has successfully marketed electrochemical diagnostic devices, and similar platforms are still evolving and reaching the frontlines of POC detection. Mathematical algorithms are being used to estimate the electrochemical signals from fasting blood glucose levels [28] (Table 12.1).

12.4 Nucleic acid biosensors

12.4.1 Types and application (applicable for electrochemical, optical and multiplex)

The detection of biomarkers of diseases in their early stages is of vital importance. ssDNA, dsDNA, Peptide DNA, mRNA, Hairpin loop

Table 12.1 Overview of different point-of-care (POC) platform devices.

Analytes	Sample	Detection	LOD (Limit of detection)	References
Paper-based POC device				
DNA	Serum	Electrochemistry	0.002 fM	[29]
DNA	Serum	Chemiluminescence	3.35×10^{17} M	[30]
DNA-TAMRA	Serum	Distance-based measurment	10 mM	[31]
EBER	Blood	Fluorescence	66.4 nM	[32]
Chip-based POC device				
miRNAs	Serum	Fluorescence	35–39 pmol/L	[33]
cfmDNA	Serum	Fliorescence	0.5 ng/mL	[34]
DNAs of *Salmonella typhimurium* and *Pseudomonas aeruginosa*	Defatted milk	Electrochemistry	15 and 5 CFU/mL	[35]
miRNA	Urine sample	Distance-based measurement	11 fM	[36]
Wearable POC devices				
Zika virus RNA	Serum (indirect)	RPA	$10-10^6$ copies/mL	[37]
HIV-1 DNA	Serum (indirect)	RPA	10^2-10^5 copies/mL	[38]

cfmDNA, Cell-free mitochondrial DNA; EBER, Epstein–Barr encode RNA.

DNA, aptamer, CRISPR RNA etc. are receptors commonly used in electrochemical as well as optical biosensors (Fig. 12.4). The most recent biosensor-based therapeutic drug monitoring (TDM) practices are some wearable biosensors.

It is always preferable to make a very simple readout system for POC testing. Aptamer-based biosensors are considered the simplest, for this purpose. Aptamers can be artificially synthesized, as they are made of DNA or RNA single strands and being flexible, they can be shaped into various complex structures as required by the target molecules. Complementary interaction between the target and the aptamer occurs by a number of bonds like electrostatic, hydrogen, shape complementarity, stacking interactions, etc. [39]. When a target binds the redox-tagged receptor molecule, the electrochemical electrode system can detect the alteration of charge results due to the change in conformation of oligonucleotide (redox-labeled). The efficiency of the E-AB sensors can be escalated by improvising the fabrication of the electrode surface. A thiolated electrode surface is commonly used for gold electrodes where NAs are layered using the thiol group as spacer. Ferrocene and methylene blue are employed as the reversible redox-active tag frequently in the majority of E-AB sensors. The electric charge is produced from the reduction of two electrons and one proton in methylene blue, and reversible oxidation of one electron in ferrocene to form ferrocenium in optimum pH. To prevent nucleophilic attack on ferrocenium, an appropriate buffer is selected that does not contain strong nucleophiles. Some less frequently used oligonucleotide-bound redox markers are iron oxide particles and anthraquinones. Modifications of the electrode surface are chosen to increase the signaling efficacy. RNA aptamers are also used though in a proper buffer and RNase inhibitor-treated serum to confirm stability. The scope of aptamer in biosensing is very exciting as it can target a huge range of small molecules along with nucleic acid (NA), such as toxins, hormones and pesticides selected in proper combinations.

12.4.2 Research and development (applicable for electrochemical, optical and multiplex)

12.4.2.1 Electrochemical biosensors

Generation of an electrical signal resulting from the interaction of the target analyte with its complementary biological recognizing molecule, is the basis of an electrochemical biosensor. The magnitude of the electrical signal thus is proportional to the quantity of the analyte under study [40]. The signal is then transduced to an electrical charge by a signal processing unit where the signal is amplified. Amplified signal is then converted to an easily readable form. Electrochemical biosensors are becoming popular and are being focused to be modernized with time so that they can be adapted for various commercial uses. A substantial number of such biosensors are available in the market and successfully used in environmental and agricultural applications apart from clinical laboratory and industrial analysis. Designing and use of these biosensors based on NAs are summarized in Table 12.2.

12.4.2.1.1 Electrochemical nucleic acid biosensors

There are plenty of research papers explaining the working mechanism of devices for detecting the target DNA sequences based on successful approval by physicians and patients. Most of these follow the principle of the present-day popular glucometer [4].

Most of biosensors developed up until now are electrochemical devices [96]. An electrochemical (EC) biosensor consists of a biological agent as a recognition element associated with the electrode transducers where a chemical reaction with the target analyte takes place, generating an electric current. Targeted to create better biosensors with improved overall performances, [42] Li et al. have reported various nanotechnology-based DNA biosensors with controllable sizes and dimensions. These include three-dimensional (3D) framework NAs (FNAs), tetrahedral DNA frameworks (TDNs), electrochemical miRNA biosensors (EMRS), etc. FNAs can provide multiple probing facilities on the electronic interface. The simplest forms of working mechanisms with all of the above type receptors, along with ssDNA, dsDNA is explained in Fig. 12.2. In comparison to traditional electrochemical biosensors consisting of single stranded DNA as a receptor, DNA nanostructures are proven to have more potential showing an interference-free system with higher protein resistance.

There are plenty of research papers explaining the working mechanism of devices for detecting target DNA sequences based on successful approval by physicians and patients. Most of these follow the principle of the present-day popular glucometer [4].

Designing a nucleic acid (NA) biosensor:

1. Selection of the electrode surface: Silver, gold, graphene oxide, carbon nanotubes, glassy carbon, cellulose, etc. solely or with various combinations employed according to the microenvironment required for sensitivity, stability, signal generation time, target concentration, etc.

Table 12.2 Overview of different electrochemical nucleic acid based biosensors.

Bio-receptors: specific DNA-based

Biosensor design	Target	Linear range	LOD (Limit of detection)	References
MWCNT-NH2-IL-rGO	HIV gene sequences	8.5 nM to 10.7 μ M	1.3 nM	[41]
Magnetic rGO-CuNCs	HCV subtype gene sequences	0.5–10 nM	405.0 pM	[42]
APTMS-ZnO/c- GO	*Escherichia coli* O157: H7 DNA	10^{-16}–10^{-6} M	0.1 fM	[43]
ssDNA/PAN/GN	HIV gene sequences	5.0×10^{-16}–1.0×10^{-10} M	1.0×10^{-16} M	[44]
CTS-GR/AuNPs/CILE	*Staphylococcus aureus* nuc gene sequences	1.0×10^{-12}–1.0×10^{-6} M	3.3×10^{-13} M	[45]
ssDNA/Au/RGO	*Mycobacterium tuberculosis* DNA	0.1 fM to 10^{-6} M	0.1 fM	[46]
Microfluidic integrated rGO	H5N1 avian influenza virus gene	10 pM to 100 nM	5 pM	[47]
ssDNA/Graphene-Nafion/Fc	HIV gene sequences	1.0×10^{-13}–1.0×10^{-10} M	2.3×10^{-14} M	[48]
ssDNA/Pencil graphite electrode (PGE)-S	*Bacillus cereus*	10^{0}–10^{7}/mL	9.4 pM	[49]
Graphene-AuNCs/GCE/C-rich base captureProbe	HIV gene sequences	0.1 fM to 100 nM	30 aM	[50]
G/Au NR/PT/[Ru(phen)3]$^{2+}$	HPV gene sequences	1.0×10^{-8}–1.0×10^{-13} M	4.03×10^{-14} M	[51]
ssDNA/Gold disk electrode	*Salmonella typhimurium*	10–400 pM and 1–400 pM	0.15 pM	[52]
ssDNA-GNAs-ITO	*S. typhimurium* DNA	4 aM to 24 fM	4 aM	[53]

CdSNPs /MWCNT– GCE	*Staphylococcus*, *E. coli* O157:H7 DNA	–	1.97×10^{-14} M	[54]
Thiol – Au interaction	*Bacillus subtilis* DNA	0.1–20 fM	0.08 fM	[55]
pDNA/CysFl/Au bioelectrode	*E. coli* O157:H7 DNA	10^{-6}–10^{-15} M	1×10^{-15} M	[56]
Carbon paste electrode (CPE)-MWCNTs	*Aeromonas hydrophila* DNA	–	0.8 μg/cm	[57]
pDNA/RGO/AuNP-GCE	*S. aureus*	10–10^{6} CFU /mL	10 CFU/ mL	[58]
ssDNA/GO-CHI/Indium tin oxide (ITO)	*S. typhimurium* DNA	10 fM to 50 nM	10 fM	[59]
GE-Au-thiol cDNA/Exonuclease III	*Enterobacteriaceae* bacteria DNA	0.01 pM – 1nM	8.7 fM	[60]
Au/GR/Carbon ionic liquid electrode (CILE)	*Listeria* monocytogenes DNA	10^{-12}–10^{-6} M	2.9 × 10 – 13 M	[61]
ssDNA/CTS-V(2)O(5)-MWCNTs/CILE	*Yersinia enterocolitica* DNA	1 uM – 10 PM	1.76 pM	[61]
GCE/Fe_3O_4/PSS/PDDA/Au	27-mer sequence DNA	–	100 aM	[62]
Bio-receptors: Aptamers				
ssDNA/rGO-CNT/GCE	*S. typhimurium*	10–10^{8} CFU/mL	10 CFU/mL	[63]
GE/Aptamer/methylene blue, MB	Rat brain ATP	0.1 nM to 1 mM		[64]
PANI-modified ITO SPE	*P. aeruginosa*	1.2–1.2 × 10^{7} CFU/mL	12 CFU/mL	[65]
rGO-AuNPs/GCEHexacyanoferrate; Magnesium carbon quantum dot	LPS from *E. coli* 055: B5	–	30 fg/mL	[66]
Competitive/UiO-67/GrapheneApt-AuNP-HRP/cDNA	*E. coli*	2 × 10^{1}–2 × 10^{8} CFU/mL	5 CFU/mL	[67]
Porous reduced graphene oxide (prGO/MoS2)	L1-major capsid protein of HPV	0.2–2 ng/mL	0.1 ng/mL	[68]
B. cereus spores-binding aptamer (BAS-6R)/Fc	*B. cereus* spore simulant	10^{4-5} × 10^{6} CFU/mL	3000 CFU/mL	[69]

(*Continued*)

Table 12.2 (Continued)

Biosensor design	Target	Linear range	LOD (Limit of detection)	References
Bio-receptors: specific DNA-based				
ssDNA/MWCNT/ITO aptasensor	*S. typhimurium*	–	10 CFU/mL	[70]
Graphene-AuNPs with the viral capsid-specific aptamer	Norovirus	100 pM to 3.5 nM	100 pM	[71]
rGO-azophloxine (AP) nanocomposite aptasensor	*S. typhimurium*	$10-10^8$ CFU/mL	10 CFU/mL	[72]
Graphene/Pyrene phosphoramidite linker-tagged DNA aptamers	*E. coli* O78:K80:H11	10^2-10^6 CFU/mL	10^2 CFU/mL	[73]
Bridged Rebar Graphene/Cell-SELEX; DNA aptamer	*E. coli* O78:K80:H11	$10-10^6$ CFU/mL	10 CFU/mL	[74]
AP-Pr/ExoIII/MB (Au-S)	*Salmonella*	$72-7.2 \times 10^6$ CFU/mL	28 CFU/mL	[75]
GCE/pyrrole-co-3-carboxyl-pyrrole	*Salmonella* DNA	10–400 and 1–400 pM	2.1 and 0.15 pM	[76]
GE/aptamer-biotinylation of ssDNA/streptavidin–alkaline phosphatase,	*Salmonella*	$2 \times 10^2-2 \times 10^6$ CFU/mL	200 CFU/mL	[77]
MWCNT-NH_2-IL-rGO/GCE	*Salmonella*	$75-7.5 \times 10^5$ CFU/mL	25 CFU/mL	[78]
Pt/Irelectrodes-Aptamer	*Listeria* monocytogenes	–	100 CFU/mL	[79]
AntiVEGF165 aptamer/SPE/OMC-Au nano	Vascular endothelial growth factor (VEGF 165) /tumor marker	10.0–300.0 pg/mL	1.0 pg/mL	[80]
Bio-receptors: Peptide nucleic acids				
PNA probe-modified (NH_2-rGO)/TEMPO-NCC	*M. tuberculosis* DNA	$10^{-13}-10^{-8}$ M	3.14×10^{-14} M	[81]

Graphene-polyaniline/anthraquinone-labeled pyrrolidinyl peptide nucleic acid (acpcPNA) probe (AQ-PNA)	HPV type 16 DNA	10–200 nM	2.3 nM	[82]
GO-CdS QDs/SPCE	*M. tuberculosis* DNA	10^{-11}–10^{-7} M	8.948 × 10^{-13} M	[83]
Peptide clavanin A (ClavA)/CNTs-Au	*Escherichia coli, K. pneumonia*	10^{2}–10^{6} CFU/mL	100 CFU/mL	[84]
Peptide clavanin A (ClavA)/CNTs-Au	*E. faecalis, B. subtilis*	10^{3}–10^{6} CFU/mL	1000 CFU/ mL	[84]
Bio-receptors: 3D DNA nanostructure(Tetrahedral DNA/Framework DNA)				
SPE/TSPs/BRCA1/PolyA mediated AuNPs-DNA-r/SA-HRP	Human tumor-suppressor gene (BRCA1)	1 fM to 1 nM	0.1 fM	[85]
DNA biosensor based on a DNA tetrahedron,		1 fM to 10 pM	0.2 fM	[86]
AFB1/AFB1 aptamer /TDNs/3DOM MoS2-AuNPs		0.1 fg/mL to 0.1 μg/mL	1.51 fM	[87]
DNA nanostructure-based gold electrode	miRNA in esophageal squamous cell carcinoma (ESCC)		10 aM	[88]
LFTS, Nanomaterial-enhanced paper-based biosensors	microRNA		10 aM	[89]
Gold electrode (AuNPs) mediated by poly (diallyldimethylammonium chloride) (PDDA).	microRNA		2 aM	[90]

(*Continued*)

Table 12.2 (Continued)

Bio-receptors: specific DNA-based				
Biosensor design	**Target**	**Linear range**	**LOD (Limit of detection)**	**References**
Gold electrode (GE)2D stem-loop DNA -3D DNA origami	Lung cancer specific microRNA	100 pM to 1 μM	10 pM	[91]
AuNP/DNA nanostructure based scaffold/ capture antibodies	Prostate specific antigen		1 pg/mL	[92]
GE/guanine nanowire tetrahedral DNA	microRNA	500 f. to 10 nM	176 fM	[93]
Multibranched HCR	Cancer cell		4 cells	[94]
EACS/GE/probe-pendant tetrahedral DNA nanostructure/ anticocaine aptamer	Cocaine		33 nM	[95]

GCE, Glassy carbon electrode; *GE*, gold electrode; *HCR*, hybridization chain reaction; *LFTS*, lateral-flow test-strip; *rGO*, reduced graphene oxide; *SPCE*, screen-printed carbon electrode.

2. Non-covalent attachment of NA to the electrode surface: This is a critical part of most biosensors.
3. Redox marker and electron-transfer system: An appropriate electron transporting system from the receptor-target complex to the electrode surface is carefully selected. Frequently used redox reporters could be ferrocene or derivatives due to their specific characteristics such as low oxidation potential, stability in any redox state, and comparatively higher rate in electron transfer. Apart from ferrocene, other commonly used redox reporters are, methylene blue (MB), quantum dot, biotin, etc. Anthraquinone and ruthenium are also specifically used for this purpose. Biotin generates a signal by conformational change of the probe while other reporters generate signal from electron flow.
4. The signal detection system: Amperometric and potentiometric techniques are commonly used for EC biosensors. In potentiometric devices, a potential difference is generated by ionic discharge from the target analyte which is detected by the specific electrode. Whereas, amperometric biosensors recognize the current generated from an electroactive species on oxidation or reduction. An amperometric biosensor is preferable, as it performs better with its high sensitivity and wide range of linearity [97,98].

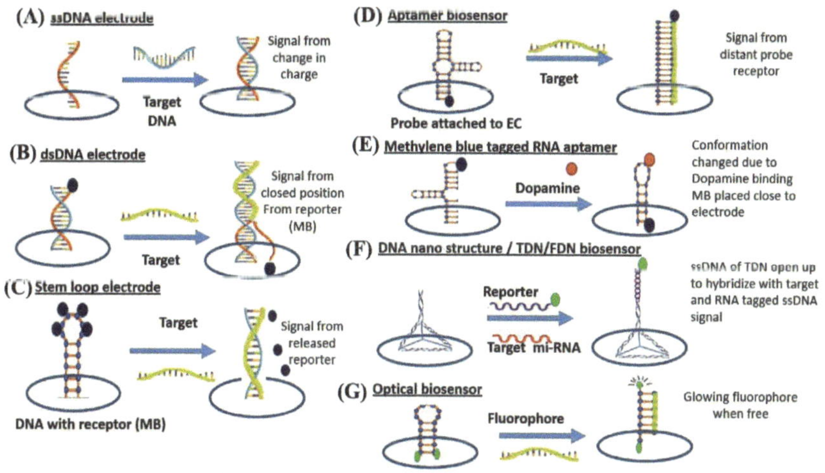

Figure 12.2 Working mechanisms of (A) ssDNA electrode, (B) dsDNA electrode, (C) Stem loop electrode (D) Aptamer biosensor, (E) Methylene blue tagged RNA aptamer, (F) DNA nano structure/TDN/FDN biosensor, (G) Optical biosensor.

12.4.2.1.2 Designing electrochemical DNA biosensors for specific molecular detection

Cancer-associated genes have great use as biomarkers and are applied in different biosensors [99]. Poly-amidoamine (PAMAM) dendrimers were first noticed in 1985. They have large surface areas tagged with large numbers of cancer gene complementary NA which can be used to probe target analytes such as enzymes, aptamers, and nucleotides. Ferrocene the redox marker is associated with dendrimers to enhance sensitivity.

TDM in conventional methods is a time consuming process where blood plasma is analyzed for the study of drug levels using immunoassays or HPLC-Ms. Researchers [100] determined vancomycin levels in unprocessed whole blood collected from the finger prick using vancomycin-binding aptamer sensor which is proved to be rapid, convenient (single-step, calibration-free) and accurate (±20%), providing hour-to-hour information on drug metabolism.

L-tryptophan is an important indicator for many diseases detected in urine in the micromolar range. In one study, an E-AB sensor with an aptamer of DNA was able to recognize tryptophan when it was complexed with a rhodium-based amino-acid receptor [101].

A specific mutation in an oxidoreductase gene responsible for anthracycline drug resistance to breast cancer was identified by E-DNA [102].

Electrochemical multiplexed DNA sensors for interferon (IFN)-γ, aminoglycoside antibiotics, thrombin, etc. also are designed [103]. Thrombin (or factor XII) is a zymogen of a trypsin-like serine protease that activates fibrinogen to fibrin and initiates clotting. It can be used for the detection of neurodegenerative diseases, Parkinson's disease, stroke, hemorrhage, multiple sclerosis, Alzheimer's and many more, when circulated in high concentrations in blood [104].

A strip biosensor made of aptamer-gold nanoparticles for the detection of Ramos cells, spiked into human blood, was also designed. When the target binds DNA receptors the aptamer is displaced off the DNA helix, the DNA backfiling hybridization occurs on the electrochemically-encoded surface [103].

Liu et al. (2015) in their review article, reported about aptamer-based devices developed by their own group and others which can target IFN-γ. An aptamer of hairpin-like structure (thiolated and MB conjugated) was anchored on the electrode made of gold. On approaching the target cytokine, the aptamer changes its shape, changing the redox current.

Researchers explored the computational field to generate databases of real-time observation values such as plasma drug levels, though it will take time to commercialize such a data system, due to lack of expert programmers [105].

Most recently, the era of CRISPR/Cas 9 technology plays a great role in various research fields [106]. E-CRISPR-Cas12a technology, based on electrochemical biosensors is considered to be the most recent and advanced biosensor device. The transducing system for clustered regularly interspaced short palindromic repeats (CRISPR) comprises type III, V, VI nucleases along with common disposable, micro-fabricated gold-based three-electrode sensor associated with thiolation and a redox reporter tagged non-specific DNA (Fig. 12.3). The reporter ssDNA strand is cleaved at a specific site of PAM-protospacer adjacent motif sequence after the conjugation of the target and the crRNA, resulting in a signal. A 5′ TTTN sequence of recognition strand detects the PAM [107].

12.4.2.2 Optical biosensors

Optical biosensors cover a vast range of applications in detecting biological elements, and they help to develop clinical diagnostics, treatments, food quality control, and environmental monitoring. Optical biosensors are highly sensitive, durable, and dependable, and can be integrated into a single chip without difficulty in pre-treatment or interference with the biological entity's nature. The recognition from target molecules is converted to a distinguishable optical signal by means of transducing units. In contrast to other physical signals, optical signals are highly sensitive, stable, and resistant to external disturbance. Hence, these biosensors take advantage of the traditional sensors with labels. The following configurations

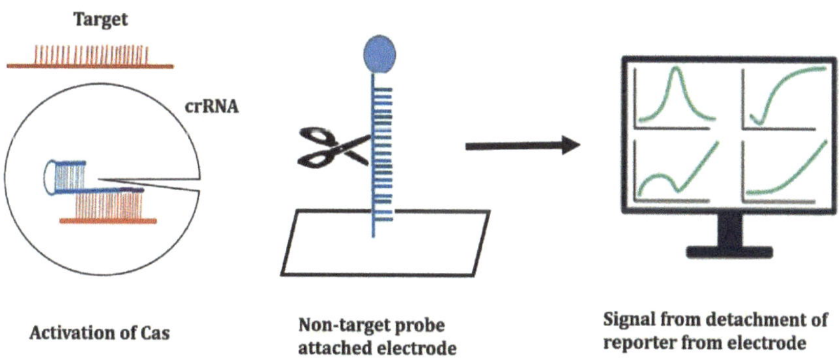

Figure 12.3 Working mechanism of the (CRISPR) nucleic acid-based electrochemical biosensors.

which characterize the optical biosensor are (1) surface plasmon resonance- (SPR) based biosensor, (2) optical waveguide-based biosensor, (3) optical resonator-based biosensor, (4) photonic crystal-based biosensor, and (5) optical fiber-based biosensor. Diverse optical biosensors are based on an evanescent field to detect the target analytes in the ambient medium. For NA analytes, evanescent-wave biosensors have made significant progress. The sample recognizes the receptors that have previously been placed onto the surface of a waveguide inside the evanescent field (see Fig. 12.4B) and so it influences the waveguide's characteristics and adjusts the effective mode index. The optical parameters of the altered waveguide, such as phase, amplitude, and resonant momentum, are used to determine the difference in the effective mode index, and the deviation is proportional to the object's concentration and conformation [108]. Optical transducers are generally used for NA detection [109]. Detection is based on the light properties measurement such as wavelength, intensity refractive index as well as polarization [110]. Optical biosensors can be labeled or non-labeled. Label-dependent sensors detect changes in color intensity or the presence of photons of a specific wavelength caused by labels such as DNA intercalator dyes or fluorescent compounds. The label serves as an indicator in the presence of the analyte. By inserting

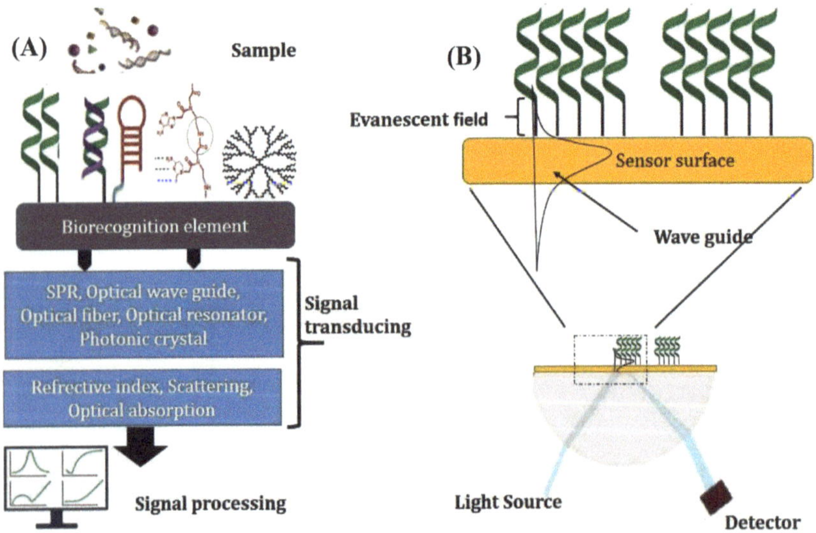

Figure 12.4 (A) Nucleic acid-based optical biosensor and (B) evanescent field principle and SPR biosensor. *SPR*, Surface plasmon resonance.

chromophore molecules into the hybridized target and the NA probe, the interaction between the two can be detected [111]. NA detection has also been done using fluorescence emission from molecular beacons [112], fluorescent-labeled signal probes [113], and quantum dots [114]. The label-free optical biosensors make it easier to determine the physical properties of analytes while maintaining their stability, since they require fewer assay conditions [115]. Label-free optical biosensors can detect a signal from an interaction and turn it into a calculable signal using the principles of light scattering or an evanescent wave.

SPR is the most widespread and renowned technology, which detects a shift in the refractive index produced by a molecular interaction at a metal surface via surface plasmon wave [116]. When photons from incident waves interact with electrons on the metal surface, surface plasmon occurs. The plasmon mode has a solid evanescent field just above the metal surface, which makes biosensing easier. For plasmon excitation, the light source must be efficiently linked to a thin gold film 45–50 nm thick. In the field of biosensing, modified equipment based on traditional SPR has emerged [2]. SPR-based devices are useful in diagnostic and laboratory applications, including the detection of SNPs and DNA methylation patterns (Table 12.3).

The DNA methylation profile of any gene can be measured by SPR biosensors as antibodies and proteins are used as receptors. A microfluidic device was prepared to determine the DNA methylation level [117]. The authors used fragmented DNA hybridized with a streptavidin-biotin interaction probe. Another biofunctionalized gold surface with a specific PNA-based LSPR biosensor was developed to detect the methylated p53 gene promoter [133]. SPR-based biosensors are being used to detect a variety of molecules like mRNA, long noncoding RNA, rRNA, short noncoding RNA/miRNA, etc. (Table 12.2). In 2017, SPRi biosensors were designed exclusively for the detection of 16S rRNAs pathogenic bacterial strains: *Legionella pneumophila, Pseudomonas aeruginosa, and Salmonella typhimurium* [121]. To detect a very low level or small size target like mi-RNA in a label-free system, amplification schemes were introduced to produce a momentous alteration in the refractive index. Integration of different amplification methods within the same assay was developed to achieve accuracy. For detection of mi-RNA accurately, a dual amplification system with two layers of GO-AuNPs, AuNPs-MoS2 composites and a combination of AuNP- antibody in SPR sensors was introduced. To achieve a sensitive detection of mi-RNA, a new-fangled DNA tail-clamp purine track was

Table 12.3 List of surface plasmon resonance (SPR)-based optical biosensors reported in literature for detection of nucleic acids.

Sensor	DNA/RNA	Sample	Amplification	References
SPR	Synthesized ODN	HCT116 human colon cancer cells and genomic DNA	NA	[117]
SPR	Synthesized *PAX-5* gene	Buffer	NA	[118]
SPR	*MGMT* promoter gene	Genomic lambda DNA		[119]
SPR	Fas57/Fas567 mRNA alternative splicing	Total RNA from HeLA cell	NA	[120]
SPRi	16S rRNAs	Total RNA *Pseudomonas aeruginosa* and *Legionella pneumophila, Salmonella typhimurium*	AuNPs	[121]
SPR	miR-145	Buffer	NA	[122]
SPR	miR-93	Buffer	Antibody	[123]
SPR	miR-141	Extraction of total RNA from cancer cell lines	GO-AuNPs	[124]
SPR	miR-21	Hepatocarcinoma cell line	AuNPs	[125]
SPR	miR-141	Extraction of total RNA from cancer cell lines	AuNPs-MoS2	[126]
SPR	miR-21	Extraction of total RNA from MCF-7 cells	CHA	[127]
SPRi	miR-21 and miR-192	Buffer	Nuclease + AuNPs	[128]
SPRi	miR-422; miR-223; miR-126; miR-23a	Extraction of total RNA from blood	AuNP + antibody	[129]

(Continued)

Table 12.3 (Continued)

Sensor	DNA/RNA	Sample	Amplification	References
SPRi	miR-29–3p	Extraction of total RNA from throat swab	RNase H	[130]
LSPR	E545K and E542K (SNPs: PIK3CA)	ctDNA in plasma	Immunogold colloids	[131]
LSPR	KRAS codon 12	Buffer	NA	[132]
LSPR	p53 gene promoter	DNA extraction from cancer cell line	NA	[133]

constructed [122]. This probe stimulated a triple helix with the RNA target, increasing the sensitivity 2.4 times then duplex helix forming probes. The biosensor based on Mirroring resonator, Mach-Zehnder Interferometer (MZI), Toroidal resonant cavity, and Bimodal wave guide (BiMW) recently advanced for the detection of NAs. The numerous optical biosensor applications are still at the preclinical stage and accomplished in very controlled laboratory environments.

12.4.3 Multiplex biosensors

Multiplexing is the need of the hour as it deals with multiple specimens diagnosis with reduced cost in a short time. It can be performed in a single go even with a large number of new pathogens or related molecules. All the pathogenic agents should be correctly and simultaneously detected with a single attempt to reduce the cost of the test (Fig. 12.5).

Nowadays, emerging viral diseases or even cancer can be identified using genetic testing for molecular biomarker detection for the diagnosis. Hot spot mutations such as PIK3CA/PTEN/EGFR are prevalent in solid tumors [134–136] and diagnostic kits are available in the market for the same. Recent metabolomics studies have shown that acetamide can be a diagnostic biomarker and propionic acid as a prognostic molecular biomarker for advanced stages of oral squamous cell carcinoma (OSCC) [137]. One more recent article has shown HPV as a diagnostic biomarker for the advanced stage OSCC patients [138].

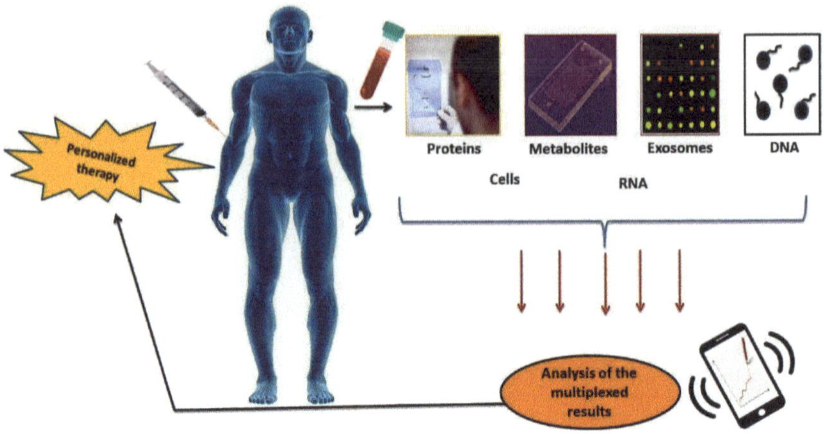

Figure 12.5 Overview of multiplexed POC methods. *POC*, Point-of-care.

A number of laboratories have already developed the ELISA, PCR and microarray-based diagnosis. Although these are very efficient and sensitive, they are not able to be explored in remote areas due to their high cost. Hence, there is a strong need to develop the POC platform emphasizing the assured criteria such as economic, specific, with higher sensitivity, easy to handle, rapid and mostly deliverable to end users anywhere (Fig. 12.3) [139].

A smartphone diagnostic tool based on POC multiplex immunoassay was developed by Laksanasopin and coworkers with comparable properties to ELISA, where power for optical, mechanical and electric functions can be replaced by the smartphone. The triplexed immunoassays were done by the dongle which includes HIV, syphilis treponemal and syphilis nontreponemal antibodies. It only takes 15 min to get the results with 92%–100% sensitivity and of 79%–100%. specificity [140].

One more multiplex platform using magnetic focusing cell sorting based on the principle of density has used smartphone for fluorescence imaging [72]. This 3D technology-based smartphone POC tool has shown that it is portable, compact and cost effective in nature and may be used for the identification and analysis of cell type and cell activity. Moreover, high resolution cameras of smartphones have overcome the use of the expensive microscope. Magnetic cameras with high resolution can separate the high density cells from blood and urine [141]. The "Slipdisc" multiplex low cost, portable platform was developed by researchers based on slipchip technology which can be moved precisely using a hand-wound clock and perform diagnostic assays. The colorimetric detection and image analysis can be performed using a smartphone. Diagnostic assays were be performed using the smartphone camera, which is sensitive to red, blue and green values. All the images were processed using image analysis software and the smartphone can be used in LAMP assay for colorimetric detection. LDH values were precisely measured using smartphone and slip disk technology using only 10 μL blood within 10 min [142].

Dubbed rapid amplification, (RAMP) a multiplex, rapid, two-stage, with high sensitivity and specificity, isothermal amplification assay, using a microfluidic chip was reported recently. In this device the amplicons are detected with a smartphone and colorimetric/ fluorescent dye. Apart from the NA from bacteria or viruses, multiple analytes of pathogens are detected in a single sample [143]. Shu et al. have developed a flow genetic analysis system (FGAS) for real time amplification and multiplexing of NA quantification. This POC platform assembly is powered by lithium batteries and is made up of thermal-gradient micro reactor, which has temperature and power management and a fluorescence imaging smartphone is attached, which is able to detect 10 copies/μL genomic DNA of *Salmonella enteric* in 15 min [144]. Furthermore, D4 assay was developed by Joh et al., by conversion of sandwich immunoassay to POC platform, where multiple analytes from a drop of blood could be investigated, comparable with the laboratory-based ELISA. The chips for D4 assays have shown durability as well stability against heat and prolonged storage [145].

12.5 Challenges and opportunities

12.5.1 Challenges

The POC in healthcare is not new in medicine as in the early days physicians also came to patients' houses for treatment with whatever device was available with them. Healthcare then shifted to specialized hospitals as technologies and medical discoveries developed along with a huge increase in the demands of patients, with an emphasis to curative medicine. Cost savings were realized through automated systems and large centralized laboratories for analyzing patient samples, which limited the POC devices in the intensive care unit of hospitals for urgent analysis or simple home testing kits used to test pregnancy. When POC testing (POCT) was available at the doorstep for a patient's rapid detection of analytes, for bettering the patient's diagnosis as well as monitoring health. Early stage diagnosis of any disease helps to make better and quick medical decisions which improve health outcomes, treatment modalities, and patient profiles. The economic outcomes offered by POC devices are many, like fewer unnecessary hospital visits, admissions and length of hospital stays, less inappropriate drug usage and early discharge from hospital which are the from the patients perspective. The benefits from the healthcare provider's perspective are reduced use of staff, equipment, estate and blood products, etc., In short, they can offer an improved quality of life [146].

Low-cost diagnostic imaging devices are in high demand for the point-of-patient care for disadvantaged and underserved populations in the developing world. The challenge is to develop reasonably-priced imaging devices and make them accessible and easily available, especially to remote communities and small hospitals that do not have accessibility to these facilities. Biosensors are used in clinical studies from the very familiar glucose self-testing kit, for toxicology and drug screenings, detection of blood related disorders, bedside diagnosis of heart disease through the levels of cardiac markers in blood. Commercialization of a newly developed devices has its generic problems. Like any other industry, there are always going to be advances and innovations at different points in time, and for some people, the cycle of development feels like a confusing leap that can be an obstacle when deciding if it's time to commercialize a developed biosensor. The degree of deviations in values are the major obstacle for appropriate decisions including biosensors as the first criterion. Though the random errors are quantified by the coefficient of variation (CV) as resulted from the analytical precision of a given measurement, the systematic error due to bias is a major challenge. Unlike random errors, it cannot be assessed statistically and can only be reduced through timely calibration of instruments with standardized reference materials. It is one of the queries that is still having disputes in TDM as standardized samples obtainable are insufficient for manufacturers to evaluate expected systematic errors in their products. The two well-known sources of bias are matrix effects and probe degradation. For the same analyte, it is possible to observe a bias across different certified laboratories, even when similar LC−MS configurations are utilized [147].

Sensor-based approaches are yet to achieve their true potential for multifarious clinical applications, despite conducting meticulous studies for a long time. We need to change the current perception on POC/TDM/personalized medicine by analyzing how it is acknowledged by different practitioners and what prevents/obstructs the advancement of these new technologies to the routine practice. It is said by [148], that we should not wait for the ultimate solution for the betterment of the patients, due to only the low sensitivity of these sensor based solutions in comparison to the other POC methods [149].

To achieve personalized healthcare monitoring system the POC devices should be equipped with smart mobile healthcare, thus concreting the path for next-generation POC testing [150]. Elgendi [151] introduced the two event-related moving averages (TERMA) framework associated with wearable and POC devices for biomedical signal analysis. However, commercialization clearance challenges in wearable sensing technologies are many such as durability, calibration, new biomarker identification, storage and data processing, translational problems on site, multiplexing, and security and privacy issues [147].

One of the most common challenges among the many encountered during the development of electrochemical biosensors, is coupling the target molecule and the electrode. The ability of the electron flow through the transducer as a signal is limited. Graphene, is an ideal electrode surface whose production still remains a challenge for scientists [152]. Graphene biosensors have some drawbacks even though they have extensive application prospects. Firstly, it is difficult to ensure the homogeneity of graphene, in terms of the number of surface functional groups, biocompatibility, size, thickness etc. The surface functionalization efficiency of graphene is affected by this inhomogeneity. An elaborate processing of the samples is essential to avoid interference from other substrates [153].

Nanocarbons or CNT (carbon nanotubes) playing a big role in various fields [154]. CNT and its role have emerged nowadays for drug delivery, invasion/metastasis, and cancer imaging. Its use in the E-DNA biosensor has been evidenced [155], but its acceptability is being challenged due to the biocompatibility and cytotoxicity of CNTs. Most of the new projects face many hurdles during their initial phase such as, short supply of funding, strict regulations for acquisition of medical devices, and a long transition time taken for shifting from the research bench to a patient bedside [19].

Investigations for biomolecules collected from patients' body fluids themselves is a challenging part in this field. Though blood sampling is done frequently, the method for getting blood samples still hurts patients and remains a challenge when continuous monitoring is required. New noninvasive techniques are sought to be implemented to get the analytes when required. Discovery of biomarkers from tear, sweat, or other body waste samples must also be investigated. Some biomarkers present in the blood, can also be found in these samples but correlation with diseases will be different from the concentration in the blood. Thus more study of their correlation is necessary [25]. A search on Pubmed evidenced that a diversified group of small molecules such as toxins, food and environmental contaminants (residual plastics), antibiotics, explosives, drugs, and protein cofactors can be targeted by simple aptamer biosensors. Nonetheless, to the best of our knowledge, only two kits are available for the quantification of thrombin and activated protein C (APC), which are commercially available protein targets [156]. Furthermore, difficulties are faced in the establishment of proper security in online data management exclusively for personal data, which is a precarious concern for mobile healthcare [155].

12.5.2 Opportunities

With such progressively advancing knowledge in the biosensing field in near future, bedside monitoring will help clinicians to design a proper regulation of diet in infants with inborn metabolism issues which currently requires complex testing in a hospital setting. A simple device of a chemical sensor can detect metabolic irregularities in the form of changed color, using a small sample of blood from a finger stick, are in the developing stage. The adverse effects like mental retardation can be prevented immediately, by adjusting the diet of the infants when such abnormalities are found. Most of the new generation POC equipment are designed to provide mobile healthcare (mH), hence it would be easy to manage at the personal level to next-generation POCT [155]. Microfluidic devices are being popularized for providing a single platform for sample processing and signal generation. Although using ELISA or LFIA for sample analysis mostly we get similar results as chip associated microfluidic assays, the later provide reduced complexity and enables mutiplexicity with high throughput screening [149]. Miniaturization and integration of the various assay steps, including (1) the lysis (2) the purification (3) amplification and, (4) on-chip detection of reaction products in the on-chip NA analysis technique are particularly promising [149].

Development of commercial POCs are triggered by the progress of lab-on-a-chip devices which in turn is realized through MEMS microelectromechanical systems (MEMS) technologies coupled with microfluidics, not forgetting the fact that it was invented decade ago [19].

Globalization of data is now possible due to open source module compatibility on a real-time mode with better tools and techniques being developed [105].

The developed countries will obviously promote home testing and personalized medicine as a result of which the number of patients requiring treatments at hospitals and the length of their stay is likely to decline due to theses new trends in personalized medicine and telemedicine [25]. The day is not far off where most people will be able to use POC devices, which would empower them to monitor and manage their own health [155].

In the present time, with COVID-19 getting a lot of attention for research exploration, there is a strong need for the drug repurposing using molecular biomarkers [157−159] it is suggested that efficient diagnosis of COVID-19 from intracellular biomarkers sensing apart from conventional method (RT-PCR) is no more an impossible task. Across the globe, a rapid test is widely accepted. IIT Delhi inventors developed a method which is testified 100% sensitive and specific and authenticated by the Indian Council of Medical Research (ICMR) [35]. However, the next-generation POC devices could have the answer to the advancement in complementary technologies and could lead to the improvement of diagnosis as well as treatment quality.

12.6 Conclusions and outlook

Original electrochemical biosensing devices along with optical and multiplexed devices are getting considerable attention due to their prodigious promise for generating sequence-specific information in a rapid, easier, and low-cost process compared to conventional assays (NA and biomolecular). Careful surveillance on the major challenges will help in resolving problems in NA isolation/amplification, sample collection, signal amplification, mismatch discrimination, etc. There will be ample opportunities for multiplex observations for multiple functions in a single microchip with microfluidic network.

Engineered cells and cellular components of biomolecules such as enzymes, antibodies, proteins, aptamers, oligonucleotides along with nanomaterials and other molecular tools are being implemented for successful detection of DNA and other target molecules on an electrode surface using signal amplification approaches. Recently introduced CRISPR technology will definitely occupy a significant position in both invasive and noninvasive devices for analysis of samples. In addition, for fast epidemic containment, in the absence of trained personnel, POC detection will be easy and simple to take observations of the signal and store the generated data (e.g., airports personnel). in silico molecular docking experiments are also of great help for virtual screening of appropriately interacting molecules before designing a sensor.

References

[1] A.A. Lubin, K.W. Plaxco, Folding-based electrochemical biosensors: the case for responsive nucleic acid architectures, Acc. Chem. Res. 43 (4) (2010) 496–505.

[2] W. Chen, J. Li, X. Wei, Y. Fan, H. Qian, S. Li, et al., Surface plasmon resonance biosensor using hydrogel-AuNP supramolecular spheres for determination of prostate cancer-derived exosomes, Microchim. Acta 187 (11) (2020) 1–10.

[3] D. Zhang, Y. Peng, H. Qi, Q. Gao, C. Zhang, Label-free electrochemical DNA biosensor array for simultaneous detection of the HIV-1 and HIV-2 oligonucleotides incorporating different hairpin-DNA probes and redox indicator, Biosens. Bioelectron. 25 (5) (2010) 1088–1094.

[4] M.A. Pellitero, A. Shaver, N. Arroyo-Currás, Critical review—approaches for the electrochemical interrogation of DNA-based sensors: a critical review, J. Electrochem. Soc. 167 (3) (2019) 037529.

[5] B.R. Eggins, Chemical Sensors and Biosensors, vol. 2, John Wiley & Sons, 2002.

[6] P.S. Ong, N.A. Yusof, I.B. Bwatanglang, J.I.B.A. Rashid, N. Nordin, I.A. Azmi, Impact of nanotechnology on diagnosis and therapy in biomedical industry, Handbook of Nanomaterials for Industrial Applications, Elsevier, 2018, pp. 662–695.

[7] L.C. Clark, L. Bargeron, Lyons c, Ann. N.Y. Acad. Sci. 102 (1962) 29–45.

[8] M.A.A. Valones, R.L. Guimarães, L.A.C. Brandão, P.R.E. de Souza, A. de Albuquerque Tavares Carvalho, S. Crovela, Principles and applications of polymerase chain reaction in medical diagnostic fields: a review, Braz. J. Microbiol. 40 (1) (2009) 1–11.

[9] P. Hornbeck, S.E. Winston, S.A. Fuller, Enzyme-linked immunosorbent assays (ELISA), Curr. Protoc. Mol. Biol. 15 (1991). Available from: https://doi.org/10.1002/0471142727.mb1102s15.

[10] I.J. Jahn, A.I. Radu, K. Weber, D. Cialla-May, J. Popp, Surface enhanced Raman spectroscopy for medical diagnostics, Nanotechnology Characterization Tools for Biosensing and Medical Diagnosis, Springer, Berlin, Heidelberg, 2018, pp. 1–66.

[11] E.Y. Su, A.B. Bocarsly, Electroanalytical chemistry in clinical analysis, Encycl. Anal. Chemistry: Applications, Theory Instrum. (2006).

[12] K. Ratajczak, M. Stobiecka, High-performance modified cellulose paper-based biosensors for medical diagnostics and early cancer screening: a concise review, Carbohydr. Polym. 229 (2020) 115463.

[13] C. Parolo, M. Medina-Sánchez, H. Montón, A. de la Escosura-Muñiz, A. Merkoçi, Paper-based electrodes for nanoparticles detection, Part. & Part. Syst. Charact. 30 (8) (2013) 662–666.

[14] A.W. Martinez, S.T. Phillips, G.M. Whitesides, E. Carrilho, Diagnostics for the developing world: microfluidic paper-based analytical devices, Anal. Chem. 82 (1) (2010) 3–10.

[15] D.M. Cate, J.A. Adkins, J. Mettakoonpitak, C.S. Henry, Recent developments in paper-based microfluidic devices, Anal. Chem. 87 (1) (2015) 19–41.

[16] B.K. Gale, A.R. Jafek, C.J. Lambert, B.L. Goenner, H. Moghimifam, U.C. Nze, S. K. Kamarapu, A review of current methods in microfluidic device fabrication and future commercialization prospects, Inventions 3 (3) (2018) 60.
[17] K. Mahesh, S. Vaidya, Microfluidics: a boon for biological research, Curr. Sci. (2017) 2021–2028.
[18] K. Iwai, K.C. Shih, X. Lin, T.A. Brubaker, R.D. Sochol, L. Lin, Finger-Powered Microfluidic Systems Using Multilayer Soft Lithography and Injection Molding Processes, Lab Chip 14 (2014) 3790–3799.
[19] S.L. Karsten, M.C. Tarhan, L.C. Kudo, D. Collard, H. Fujita, Point-of-care (POC) devices by means of advanced MEMS, Talanta 145 (2015) 55–59.
[20] K.M. Balss, W.N. Vreeland, K.W. Phinney, D. Ross, Simultaneous concentration and separation of enantiomers with chiral temperature gradient focusing, Anal. Chem. 76 (24) (2004) 7243–7249.
[21] M. Ludwig, D. Belder, Coated microfluidic devices for improved chiral separations in microchip electrophoresis, Electrophoresis 24 (15) (2003) 2481–2486.
[22] D.C. Christodouleas, B. Kaur, P. Chorti, From point-of-care testing to eHealth diagnostic devices (eDiagnostics), ACS Cent. Sci. 4 (12) (2018) 1600–1616.
[23] J.R. Mejía-Salazar, K.R. Cruz, E.M. MateronVasques, Microfluidic point-of-care devices: new trends and future prospects for eHealth diagnostics, Sensors 20 (7) (2020) 1951.
[24] M.W. Woo, J.W. Lee, K.H. Park, A reliable IoT system for personal healthcare devices, Future Gener. Computer Syst. 78 (2018) 626–640.
[25] Y. Manmana, T. Kubo, K. Otsuka, Recent developments of point-of-care (POC) testing platform for biomolecules, TrAC. Trends Anal. Chem. (2020) 116160.
[26] T. Nguyen, S.Z. Andreasen, A. Wolff, D.D. Bang, From lab on a chip to point of care devices: the role of open source microcontrollers, Micromachines 9 (8) (2018) 403.
[27] S. Shrivastava, T.Q. Trung, N.-E. Lee, Recent progress, challenges, and prospects of fully integrated mobile and wearable point-of-care testing systems for self-testing, Chem. Soc. Rev. 49 (6) (2020) 1812–1866.
[28] S. Malik, H. Parikh, N. Shah, S. Anand, S. Gupta, Non-invasive platform to estimate fasting blood glucose levels from salivary electrochemical parameters, Healthc. Technol. Lett. 6 (4) (2019) 87–91.
[29] J. Lu, S. Ge, L. Ge, M. Yan, J. Yu, Electrochemical DNA sensor based on three-dimensional folding paper device for specific and sensitive point-of-care testing, Electrochim. Acta 80 (2012) 334–341. Available from: https://doi.org/10.1016/j.electacta.2012.07.024.
[30] Y. Wang, S. Wang, S. Ge, S. Wang, M. Yan, D. Zang, J. Yu, Ultrasensitive chemiluminescence detection of DNA on a microfluidic paper-based analytical device, Monatsh. Chem. 145 (2014) 129–135. Available from: https://doi.org/10.1007/s00706-013-0971-1.
[31] Y.T. Chen, J.T. Yang, Detection of an amphiphilic biosample in a paper microchannel based on length, Biomed. Microdevices 17 (2015). Available from: https://doi.org/10.1007/s10544-015-9954-9.
[32] D. Zhang, D. Broyles, E.A. Hunt, E. Dikici, S. Daunert, S.K. Deo, A paper-based platform for detection of viral RNA, Analyst 142 (2017) 815–823. Available from: https://doi.org/10.1039/c6an02452a.
[33] W.Q. Ye, Y.X. Wei, Y.Z. Zhang, C.G. Yang, Z.R. Xu, Multiplexed detection of micro-RNAs based on microfluidic multi-color fluorescence droplets, Anal. Bioanal. Chem. 412 (2020) 647–655. Available from: https://doi.org/10.1007/s00216-019-02266-3.

[34] F. Dinter, M. Burdukiewicz, P. Schierack, W. Lehmann, J. Nestler, G. Dame, S. Ro€diger, Simultaneous detection and quantification of DNA and protein biomarkers in spectrum of cardiovascular diseases in a microfluidic microbead chip, Anal. Bioanal. Chem. 411 (2019) 7725−7735. Available from: https://doi.org/10.1007/s00216-019-02199-x, https://doi.org/10.1007/s00216-019-02199-x.
[35] F. Luo, Z. Li, G. Dai, Y. Lu, P. He, Q. Wang, Simultaneous detection of different bacteria by microchip electrophoresis combined with universal primer-duplex polymerase chain reaction, J. Chromatogr. A 1615 (2020) 460734.
[36] D. Chatterjee, D.S. Mansfield, A.T. Woolley, Microfluidic devices for label-free and non-instrumented quantitation of unamplified nucleic acids by flow distance measurement, Anal. Methods 6 (2014) 8173−8179. Available from: https://doi.org/10.1039/c4ay01845a.
[37] B. Yang, J. Kong, X. Fang, Bandage-like wearable flexible microfluidic recombinase polymerase amplification sensor for the rapid visual detection of nucleic acids, Talanta 204 (2019) 685−692. Available from: https://doi.org/10.1016/j.talanta.2019.06.031.
[38] M. Kong, Z. Li, J. Wu, J. Hu, Y. Sheng, D. Wu, Y. Lin, M. Li, X. Wang, S. Wang, A wearable microfluidic device for rapid detection of HIV-1 DNA using recombinase polymerase amplification, Talanta 205 (2019) 1−8. Available from: https://doi.org/10.1016/j.talanta.2019.120155.
[39] M. McKeague, E.M. McConnell, J. Cruz-Toledo, E.D. Bernard, A. Pach, E. Mastronardi, et al., Analysis of in vitro aptamer selection parameters, J. Mol. Evol. 81 (5) (2015) 150−161.
[40] N.J. Ronkainen, H.B. Halsall, W.R. Heineman, Electrochemical biosensors, Chem. Soc. Rev. 39 (5) (2010) 1747−1763.
[41] L. Farzin, S. Sadjadi, M. Shamsipur, S. Sheibani, Electrochemical genosensor based on carbon nanotube/amine-ionic liquid functionalized reduced graphene oxide nanoplatform for detection of human papillomavirus (HPV16)-related head and neck cancer, Journal of Pharmaceutical and Biomedical Analysis 179 (2020).
[42] F. Li, Q. Li, X. Zuo, C. Fan, DNA framework-engineered electrochemical biosensors, Sci. China Life Sci. (2020) 1−12.
[43] N. Jaiswal, C.M. Pandey, S. Solanki, I. Tiwari, B.D. Malhotra, An impedimetric biosensor based on electrophoretically assembled ZnO nanorods and carboxylated graphene nanoflakes on an indium tin oxide electrode for detection of the DNA of *Escherichia coli* O157: H7, Microchimica Acta 187 (1) (2020) 1−18.
[44] Q. Gong, H. Han, H. Yang, M. Zhang, X. Sun, Y. Liang, Z. Liu, W. Zhang, J. Qiao, Sensitive electrochemical DNA sensor for the detection of HIV based on a polyaniline/graphene nanocomposite, Journal of Materiomics 5 (2) (2019) 313−319.
[45] X. Niu, W. Chen, X. Wang, Y. Men, Q. Wang, W. Sun, G. Li, A graphene modified carbon ionic liquid electrode for voltammetric analysis of the sequence of the Staphylococcus aureus nuc gene, Microchimica Acta 185 (3) (2018) 1−9.
[46] N.K. Mogha, V. Sahu, R.K. Sharma, D.T. Masram, Reduced graphene oxide nanoribbon immobilized gold nanoparticle based electrochemical DNA biosensor for the detection of Mycobacterium tuberculosis, Journal of Materials Chemistry B 6 (31) (2018) 5181−5187.
[47] C. Chan, J. Shi, Y. Fan, M. Yang, A microfluidic flow-through chip integrated with reduced graphene oxide transistor for influenza virus gene detection, Sensors and Actuators B: Chemical 251 (2017) 927−933.
[48] Q. Gong, Y. Wang, H. Yang, A sensitive impedimetric DNA biosensor for the determination of the HIV gene based on graphene-Nafion composite film, Biosensors and Bioelectronics 89 (2017) 565−569.

[49] Z. Izadi, M. Sheikh-Zeinoddin, A.A. Ensafi, S. Soleimanian-Zad, Fabrication of an electrochemical DNA-based biosensor for Bacillus cereus detection in milk and infant formula, Biosensors and Bioelectronics 80 (2016) 582–589.
[50] J. Wang, A. Shi, X. Fang, X. Han, Y. Zhang, An ultrasensitive supersandwich electrochemical DNA biosensor based on gold nanoparticles decorated reduced graphene oxide, Analytical Biochemistry 469 (2015) 71–75.
[51] H. Huang, W. Bai, C. Dong, R. Guo, Z. Liu, An ultrasensitive electrochemical DNA biosensor based on graphene/Au nanorod/polythionine for human papillomavirus DNA detection, Biosensors and Bioelectronics 68 (2015) 442–446.
[52] M.A. Tabrizi, M. Shamsipur, A label-free electrochemical DNA biosensor based on covalent immobilization of salmonella DNA sequences on the nanoporous glassy carbon electrode, Biosensors and Bioelectronics 69 (2015) 100–105.
[53] A. Singh, M. Choudhary, M.P. Singh, H.N. Verma, S.P. Singh, K. Arora, DNA functionalized direct electro-deposited gold nanoaggregates for efficient detection of Salmonella typhi, Bioelectrochemistry 105 (2015) 7–15.
[54] M.H. Abdalhai, A.M. Fernandes, X. Xia, A. Musa, J. Ji, X. Sun, Electrochemical genosensor to detect pathogenic bacteria (*Escherichia coli* O157: H7) as applied in real food samples (fresh beef) to improve food safety and quality control, Journal of agricultural and food chemistry 63 (20) (2015) 5017–5025.
[55] Y. Hu, X. Xu, Q. Liu, L. Wang, Z. Lin, G. Chen, Ultrasensitive electrochemical biosensor for detection of DNA from Bacillus subtilis by coupling target-induced strand displacement and nicking endonuclease signal amplification, Analytical chemistry 86 (17) (2014) 8785–8790.
[56] A. Pandey, Y. Gurbuz, V. Ozguz, J.H. Niazi, A. Qureshi, Graphene-interfaced electrical biosensor for label-free and sensitive detection of foodborne pathogenic *E. coli* O157: H7, Biosensors and Bioelectronics 91 (2017) 225–231.
[57] M. Ligaj, M. Tichoniuk, D. Gwiazdowska, M. Filipiak, Electrochemical DNA biosensor for the detection of pathogenic bacteria Aeromonas hydrophila, Electrochimica Acta 128 (2014) 67–74.
[58] F. Jia, N. Duan, S. Wu, X. Ma, Y. Xia, Z. Wang, X. Wei, Impedimetric aptasensor for Staphylococcus aureus based on nanocomposite prepared from reduced graphene oxide and gold nanoparticles, Microchimica Acta 181 (9) (2014) 967–974.
[59] A. Singh, G. Sinsinbar, M. Choudhary, V. Kumar, R. Pasricha, H.N. Verma, S.P. Singh, K. Arora, Graphene oxide-chitosan nanocomposite based electrochemical DNA biosensor for detection of typhoid, Sensors and Actuators B: Chemical 185 (2013) 675–684.
[60] C. Luo, H. Tang, W. Cheng, L. Yan, D. Zhang, H. Ju, S. Ding, A sensitive electrochemical DNA biosensor for specific detection of Enterobacteriaceae bacteria by Exonuclease III-assisted signal amplification, Biosensors and Bioelectronics 48 (2013) 132–137.
[61] W. Sun, X. Qi, Y. Zhang, H. Yang, H. Gao, Y. Chen, Z. Sun, Electrochemical DNA biosensor for the detection of Listeria monocytogenes with dendritic nanogold and electrochemical reduced graphene modified carbon ionic liquid electrode, Electrochimica Acta 85 (2012) 145–151.
[62] Y.H. Bai, J.Y. Li, J.J. Xu, H.Y. Chen, Ultrasensitive electrochemical detection of DNA hybridization using Au/Fe_3O_4 magnetic composites combined with silver enhancement, Analyst 135 (7) (2010) 1672–1679.
[63] J.N. Appaturi, T. Pulingam, K.L. Thong, S. Muniandy, N. Ahmad, B.F. Leo, Rapid and sensitive detection of Salmonella with reduced graphene oxide-carbon nanotube based electrochemical aptasensor, Analytical biochemistry 589 (2020) 113489.
[64] Y. Jiang, W. Ma, W. Ji, H. Wei, L. Mao, Aptamer superstructure-based electrochemical biosensor for sensitive detection of ATP in rat brain with in vivo microdialysis, Analyst 144 (5) (2019) 1711–1717.

[65] S. Shahrokhian, S. Ranjbar, Development of a sensitive diagnostic device based on zeolitic imidazolate frameworks-8 using ferrocene−graphene oxide as electroactive indicator for Pseudomonas aeruginosa detection, ACS Sustainable Chemistry & Engineering 7 (15) (2019) 12760−12769.
[66] M. Pourmadadi, J.S. Shayeh, M. Omidi, F. Yazdian, M. Alebouyeh, L. Tayebi, A glassy carbon electrode modified with reduced graphene oxide and gold nanoparticles for electrochemical aptasensing of lipopolysaccharides from *Escherichia coli* bacteria, Microchimica Acta 186 (12) (2019) 1−8.
[67] G. Dai, Z. Li, F. Luo, S. Ai, B. Chen, Q. Wang, Electrochemical determination of Salmonella typhimurium by using aptamer-loaded gold nanoparticles and a composite prepared from a metal-organic framework (type UiO-67) and graphene, Microchimica Acta 186 (9) (2019) 1−9.
[68] F. Chekin, K. Bagga, P. Subramanian, R. Jijie, S.K. Singh, S. Kurungot, R. Boukherroub, S. Szunerits, Nucleic aptamer modified porous reduced graphene oxide/MoS_2 based electrodes for viral detection: Application to human papillomavirus (HPV), Sensors and Actuators B: Chemical 262 (2018) 991−1000.
[69] V. Mazzaracchio, D. Neagu, A. Porchetta, E. Marcoccio, A. Pomponi, G. Faggioni, N. D'Amore, A. Notargiacomo, M. Pea, D. Moscone, G. Palleschi, A label-free impedimetric aptasensor for the detection of Bacillus anthracis spore simulant, Biosensors and Bioelectronics 126 (2019) 640−646.
[70] M.R. Hasan, T. Pulingam, J.N. Appaturi, A.N. Zifruddin, S.J. Teh, T.W. Lim, F. Ibrahim, B.F. Leo, K.L. Thong, Carbon nanotube-based aptasensor for sensitive electrochemical detection of whole-cell Salmonella, Analytical biochemistry 554 (2018) 34−43.
[71] R. Chand, S. Neethirajan, Microfluidic platform integrated with graphene-gold nano-composite aptasensor for one-step detection of norovirus, Biosensors and Bioelectronics 98 (2017) 47−53.
[72] S. Muniandy, I.J. Dinshaw, S.J. Teh, C.W. Lai, F. Ibrahim, K.L. Thong, B.F. Leo, Graphene-based label-free electrochemical aptasensor for rapid and sensitive detection of foodborne pathogen, Analytical and Bioanalytical Chemistry 409 (29) (2017) 6893−6905.
[73] G. Wu, X. Tang, Z. Lin, M. Meyyappan, K.W.C. Lai. The effect of ionic strength on the sensing performance of liquid-gated biosensors. In 2017 IEEE 17th International Conference on Nanotechnology (IEEE-NANO) (pp. 242−245).
[74] H. Kaur, M. Shorie, M. Sharma, A.K. Ganguli, P. Sabherwal, Bridged Rebar Graphene functionalized aptasensor for pathogenic *E. coli* O78: K80: H11 detection, Biosensors and Bioelectronics 98 (2017) 486−493.
[75] Q. Pei, Y. Wang, S. Liu, Y. Qin, X. Leng, X. Cui, J. Huang, Exonuclease III-aided autonomous cascade signal amplification: A facile and universal DNA biosensing platform for ultrasensitive electrochemical detection of S. typhimurium, New Journal of Chemistry 41 (15) (2017) 7613−7620.
[76] E. Sheikhzadeh, M. Chamsaz, A.P.F. Turner, E.W.H. Jager, V. Beni, Label-free impedimetric biosensor for *Salmonella Typhimurium* detection based on poly [pyrrole-co-3-carboxyl-pyrrole] copolymer supported aptamer, Biosensors and bioelectronics 80 (2016) 194−200.
[77] X. Li, H. Fu, Y. He, Q. Zhai, J. Guo, K. Qing, G. Yi, Electrochemical aptasensor for rapid and sensitive determination of Salmonella based on target-induced strand displacement and gold nanoparticle amplification, Analytical Letters 49 (15) (2016) 2405−2417.
[78] F. Jia, N. Duan, S. Wu, R. Dai, Z. Wang, X. Li, Impedimetric Salmonella aptasensor using a glassy carbon electrode modified with an electrodeposited composite consisting of reduced graphene oxide and carbon nanotubes, Microchimica Acta 183 (1) (2016) 337−344.

[79] D.C. Vanegas, G. Clark, A.E. Cannon, S. Roux, P. Chaturvedi, E.S. McLamore, A self-referencing biosensor for real-time monitoring of physiological ATP transport in plant systems, Biosensors and Bioelectronics 74 (2015) 37–44.
[80] M.A. Tabrizi, M. Shamsipur, L. Farzin, A high sensitive electrochemical aptasensor for the determination of VEGF165 in serum of lung cancer patient, Biosensors and Bioelectronics 74 (2015) 764–769.
[81] M.H.M. Zaid, J. Abdullah, N.A. Yusof, H. Wasoh, Y. Sulaiman, M.F.M. Noh, R. Issa, Reduced graphene oxide/tempo-nanocellulose nanohybrid-based electrochemical biosensor for the determination of mycobacterium tuberculosis, Journal of Sensors (2020).
[82] P. Teengam, W. Siangproh, A. Tuantranont, C.S. Henry, T. Vilaivan, O. Chailapakul, Electrochemical paper-based peptide nucleic acid biosensor for detecting human papillomavirus, Analytica Chimica Acta 952 (2017) 32–40.
[83] M.H.M. Zaid, J. Abdullah, N.A. Yusof, Y. Sulaiman, H. Wasoh, M.F.M. Noh, R. Issa, PNA biosensor based on reduced graphene oxide/water soluble quantum dots for the detection of Mycobacterium tuberculosis, Sensors and Actuators B: Chemical 241 (2017) 1024–1034.
[84] C.A. Andrade, J.M. Nascimento, I.S. Oliveira, C.V. de Oliveira, C.P. de Melo, O.L. Franco, M.D. Oliveira, Nanostructured sensor based on carbon nanotubes and clavanin A for bacterial detection, Colloids and Surfaces B: Biointerfaces 135 (2015) 833–839.
[85] D. Feng, J. Su, G. He, Y. Xu, C. Wang, M. Zheng, Q. Qian, X. Mi, Electrochemical DNA Sensor for Sensitive BRCA1 Detection Based on DNA Tetrahedral-Structured Probe and Poly-Adenine Mediated Gold Nanoparticles, Biosensors 10 (7) (2020) 78.
[86] X. Wang, S. Niu, M. Wei, S. Liu, R. Liu, C. Shi, C. Ma, Ultrasensitive electrochemical DNA biosensor based on a tetrahedral structure and proximity-dependent surface hybridization, Analyst 145 (1) (2020) 150–156.
[87] L. Peng, P. Zhang, Y. Chai, R. Yuan, Bi-directional DNA walking machine and its application in an enzyme-free electrochemiluminescence biosensor for sensitive detection of microRNAs, Analytical chemistry 89 (9) (2017) 5036–5042.
[88] Y. Wen, H. Pei, Y.E. Shen, J. Xi, M. Lin, N. Lu, X. Shen, J. Li, C. Fan, DNA nanostructure-based interfacial engineering for PCR-free ultrasensitive electrochemical analysis of microRNA, Scientific reports 2 (1) (2012) 1–7.
[89] Z. Ge, M. Lin, P. Wang, H. Pei, J. Yan, J. Shi, Q. Huang, D. He, C. Fan, X. Zuo, Hybridization chain reaction amplification of microRNA detection with a tetrahedral DNA nanostructure-based electrochemical biosensor, Analytical chemistry 86 (4) (2014) 2124–2130.
[90] P. Miao, B. Wang, Z. Yu, J. Zhao, Y. Tang, Ultrasensitive electrochemical detection of microRNA with star trigon structure and endonuclease mediated signal amplification, Biosensors and Bioelectronics 63 (2015) 365–370.
[91] Y. Liu, N. Tuleouva, E. Ramanculov, A. Revzin, Aptamer-based electrochemical biosensor for interferon gamma detection, Anal. Chem. 82 (19) (2010) 8131–8136.
[92] X. Chen, G. Zhou, P. Song, J. Wang, J. Gao, J. Lu, C. Fan, X. Zuo, Ultrasensitive electrochemical detection of prostate-specific antigen by using antibodies anchored on a DNA nanostructural scaffold, Analytical chemistry 86 (15) (2014) 7337–7342.
[93] Y.L. Huang, S. Mo, Z.F. Gao, J.R. Chen, J.L. Lei, H.Q. Luo, N.B. Li, Amperometric biosensor for microRNA based on the use of tetrahedral DNA nanostructure probes and guanine nanowire amplification, Microchimica Acta 184 (8) (2017) 2597–2604.
[94] G. Zhou, M. Lin, P. Song, X. Chen, J. Chao, L. Wang, Q. Huang, W. Huang, C. Fan, X. Zuo, Multivalent capture and detection of cancer cells with DNA nanostructured biosensors and multibranched hybridization chain reaction amplification, Analytical chemistry 86 (15) (2014) 7843–7848.

[95] Y. Wen, H. Pei, Y. Wan, Y. Su, Q. Huang, S. Song, C. Fan, DNA nanostructure-decorated surfaces for enhanced aptamer-target binding and electrochemical cocaine sensors, Analytical chemistry 83 (19) (2011) 7418–7423.
[96] J. Wang, P. Cieplak, P.A. Kollman, How well does a restrained electrostatic potential (RESP) model perform in calculating conformational energies of organic and biological molecules? J. Comput. Chem. 21 (12) (2000) 1049–1074.
[97] J. Rainbow, E. Sedlackova, S. Jiang, G. Maxted, D. Moschou, L. Richtera, P. Estrela, Integrated electrochemical biosensors for detection of waterborne pathogens in low-resource settings, Biosensors 10 (4) (2020) 36.
[98] J. Wang, Electrochemical biosensors: towards point-of-care cancer diagnostics, Biosens. Bioelectron. 21 (10) (2006) 1887–1892.
[99] M. Senel, M. Dervisevic, F. Kokkokoğlu, Electrochemical DNA biosensors for label-free breast cancer gene marker detection, Anal. Bioanal. Chem. 411 (13) (2019) 2925–2935.
[100] P. Dauphin-Ducharme, K. Yang, N. Arroyo-Currás, K.L. Ploense, Y. Zhang, J. Gerson, M. Kurnik, T.E. Kippin, M.N. Stojanovic, K.W. Plaxco, Electrochemical aptamer-based sensors for improved therapeutic drug monitoring and high-precision, feedback-controlled drug delivery, ACS Sens. 4 (10) (2019) 2832–2837.
[101] A. Idili, J. Gerson, C. Parolo, T. Kippin, K.W. Plaxco, An electrochemical aptamer-based sensor for the rapid and convenient measurement of l-tryptophan, Anal. Bioanal. Chem. 411 (19) (2019) 4629–4635.
[102] E. Palecek, M. Bartosik, Electrochemistry of nucleic acids, Chem. Rev. 112 (6) (2012) 3427–3481.
[103] A.B. Iliuk, L. Hu, W. Andy Tao, Aptamer in bioanalytical applications, Anal. Chem. 83 (12) (2011) 4440–4452.
[104] A. Trapaidze, Integration of Thrombin-Binding Aptamers in Point-of-Care Devices for Continuous Monitoring of Thrombin in Plasma (PhD diss.), UT3, 2015.
[105] S.D. Curtis, K.L. Ploense, M. Kurnik, G. Ortega, C. Parolo, T.E. Kippin, K.W. Plaxco, N. Arroyo-Currás, Open source software for the real-time control, processing, and visualization of high-volume electrochemical data, Anal. Chem. 91 (19) (2019) 12321–12328.
[106] Y. Nalla, S. Shah, A new era in molecular biology clustered regularly interspaced short palindromic repeats/cas9 technology: a brief understanding, Adv. Hum. Biol. 11 (2) (2021) 152.
[107] O. Mukama, J. Wu, Z. Li, Q. Liang, Z. Yi, X. Lu, et al., An ultrasensitive and specific point-of-care CRISPR/Cas12 based lateral flow biosensor for the rapid detection of nucleic acids, Biosens. Bioelectron. 159 (2020) 112143.
[108] M.-C. Estevez, M. Alvarez, L.M. Lechuga, Integrated optical devices for lab-on-a-chip biosensing applications, Laser Photonics Rev. 6 (4) (2012) 463–487.
[109] U. Bora, A. Sett, D. Singh, Nucleic acid based biosensors for clinical applications, Biosens. J. 2 (1) (2013) 1–8.
[110] S. Sang, Y. Wang, Q. Feng, Y. Wei, J. Ji, W. Zhang, Progress of new label-free techniques for biosensors: a review, Crit. Rev. Biotechnol. 36 (3) (2016) 465–481.
[111] A. Szukalski, M. Moffa, A. Camposeo, D. Pisignano, J. Mysliwiec, All-optical switching in dye-doped DNA nanofibers, Journal of Materials Chemistry C 7 (1) (2019) 170–176.
[112] T. Yao, K. Kotegawa, A flow injection biosensor system for highly sensitive detection of 2, 4, 6-trichlorophenol based on preoxidation by ceric sulfate, Analytical sciences 19 (6) (2003) 829–833.
[113] X. Zhu, R. Wang, X. Zhou, H. Shi, Free-energy-driven lock/open assembly-based optical DNA sensor for cancer-related microRNA detection

with a shortened time-to-result, ACS Appl. Mater. Interfaces 9 (31) (2017) 25789–25795.

[114] N. Chen, A.A. Zinchenko, Y. Yamazaki, Y. Yoshikawa, S. Murata, K. Yoshikawa, Quantum dot probes for observation of single molecule DNA and a synthetic polyelectrolyte higher-order structure, Soft Matter 6 (12) (2010) 2834–2841.

[115] X. Chen, Y. Liu, J. Huang, W. Liu, J. Huang, Y. Zhang, W. Fu, Label-free techniques for laboratory medicine applications, Front. Laboratory Med. 1 (2) (2017) 82–85.

[116] M. Malmqvist, Biospecific interaction analysis using biosensor technology, Nature 361 (6408) (1993) 186–187.

[117] R. Kurita, H. Yanagisawa, K. Yoshioka, O. Niwa, On-chip sequence-specific immunochemical epigenomic analysis utilizing outward-turned cytosine in a DNA bulge with handheld surface plasmon resonance equipment, Anal. Chem. 87 (22) (2015) 11581–11586.

[118] C.S. Huertas, A. Aviñó, C. Kurachi, A. Piqué, J. Sandoval, R. Eritja, M. Esteller, L.M. Lechuga, Label-free DNA-methylation detection by direct ds-DNA fragment screening using poly-purine hairpins, Biosensors and Bioelectronics 120 (2018) 47–54.

[119] T. Kurinomaru, N. Kojima, R. Kurita, An alkylating immobilization linker for immunochemical epigenetic assessment, Chem. Commun. 53 (59) (2017) 8308–8311.

[120] C.S. Huertas, D. Fariña, L.M. Lechuga, Direct and label-free quantification of micro-RNA-181a at attomolar level in complex media using a nanophotonic biosensor. ACS, Sensors 1 (6) (2016) 748–756.

[121] F. Melaine, M. Saad, S. Faucher, M. Tabrizian, Selective and high dynamic range assay format for multiplex detection of pathogenic *Pseudomonas aeruginosa*, *Salmonella typhimurium*, and *Legionella pneumophila* RNAs using surface plasmon resonance imaging, Anal. Chem. 89 (14) (2017) 7802–7807.

[122] A. Aviñó, C.S. Huertas, L.M. Lechuga, R. Eritja, Sensitive and label-free detection of miRNA-145 by triplex formation, Anal. Bioanal. Chem. 408 (3) (2016) 885–893.

[123] S. Schmieder, J. Weißpflog, N. Danz, U. Klotzbach, F. Sonntag, Detection of miRNA using a surface plasmon resonance biosensor and antibody amplification, Curr. Directions Biomed. Eng. 2 (1) (2016) 135–138.

[124] Q. Li, Q. Wang, X. Yang, K. Wang, H. Zhang, W. Nie, High sensitivity surface plasmon resonance biosensor for detection of microRNA and small molecule based on graphene oxide-gold nanoparticles composites, Talanta 174 (2017) 521–526.

[125] R. Liu, Q. Wang, Q. Li, X. Yang, K. Wang, W. Nie, Surface plasmon resonance biosensor for sensitive detection of microRNA and cancer cell using multiple signal amplification strategy, Biosens. Bioelectron. 87 (2017) 433–438.

[126] W. Nie, Q. Wang, X. Yang, H. Zhang, Z. Li, L. Gao, Y. Zheng, X. Liu, K. Wang, High sensitivity surface plasmon resonance biosensor for detection of microRNA based on gold nanoparticles-decorated molybdenum sulfide, Anal. Chim. Acta 993 (2017) 55–62.

[127] D. Li, W. Cheng, Y. Yan, Y. Zhang, Y. Yin, H. Ju, S. Ding, A colorimetric biosensor for detection of attomolar microRNA with a functional nucleic acid-based amplification machine, Talanta 146 (2016) 470–476.

[128] X. Wei, X. Duan, X. Zhou, J. Wu, H. Xu, X. Min, et al., A highly sensitive SPRi biosensing strategy for simultaneous detection of multiplex miRNAs based on strand displacement amplification and AuNP signal enhancement, Analyst 143 (13) (2018) 3134–3140.

[129] A. Sguassero, Á. Artiga, C. Morasso, R.R. Jimenez, R.M. Rapún, R. Mancuso, et al., A simple and universal enzyme-free approach for the detection of multiple

microRNAs using a single nanostructured enhancer of surface plasmon resonance imaging, Anal. Bioanal. Chem. 411 (9) (2019) 1873–1885.

[130] H. Ho, F.C. Pui, S.Y. Loo, D. Wu, K.-T. Gu, Yong, S.K. Kong, MicroRNA biosensing with two-dimensional surface plasmon resonance imaging, Biosensors and Biodetection, Humana Press, New York, 2017, pp. 117–127.

[131] A.H. Nguyen, S.J. Sim, Nanoplasmonic biosensor: detection and amplification of dual bio-signatures of circulating tumor DNA, Biosens. Bioelectron. 67 (2015) 443–449.

[132] A. Rapisarda, N. Giamblanco, G. Marletta, Kinetic discrimination of DNA single-base mutations by localized surface plasmon resonance, J. Colloid Interface Sci. 487 (2017) 141–148.

[133] A.H. Nguyen, X. Ma, S.J. Sim, Gold nanostar based biosensor detects epigenetic alterations on promoter of real cells, Biosens. Bioelectron. 66 (2015) 497–503.

[134] S. Shah, S. Shah, H. Padh, K. Kalia, Genetic alterations of the PIK3CA oncogene in human oral squamous cell carcinoma in an Indian population, Oral. Surgery Oral Med. Oral Pathol. Oral Radiol. 120 (5) (2015) 628–635.

[135] S. Shah, D. Jajal, G. Mishra, K. Kalia, Genetic profile of PTEN gene in Indian oral squamous cell carcinoma primary tumors, J. Oral. Pathol. Med. 46 (2) (2017) 106–111.

[136] S. Shah, G. Mishra, K. Kalia, Single nucleotide polymorphism rs17849071 G/T in the PIK3CA gene is inversely associated with oral cancer, Oral. Cancer 2 (3) (2018) 83–89.

[137] A. Patel, V. Gupta, S. Shah, Role of Human Papillomavirus and its Association with an Indian Oral Squamous Cell Carcinoma Subjects, 2021.

[138] B. Turakhia, P. Turakhia, S. Shah, Green synthesis of zero valent iron nanoparticles from Spinacia oleracea (spinach) and its application in waste water treatment, J. Adv. Res. Appl. Sci 5 (1) (2018) 46–51.

[139] J. Li, J. Macdonald, Multiplexed lateral flow biosensors: technological advances for radically improving point-of-care diagnoses, Biosens. Bioelectron. 83 (2016) 177–192.

[140] T. Laksanasopin, T.W. Guo, S. Nayak, A.A. Sridhara, S. Xie, O.O. Olowookere, P. Cadinu, et al., A smartphone dongle for diagnosis of infectious diseases at the point of care, Sci. Transl. Med. 7 (273) (2015) 273re1.

[141] S. Knowlton, A. Joshi, P. Syrrist, A.F. Coskun, S. Tasoglu, 3D-printed smartphone-based point of care tool for fluorescence-and magnetophoresis-based cytometry, Lab Chip 17 (16) (2017) 2839–2851.

[142] I. Banerjee, T. Salih, H. Ramachandraiah, J. Erlandsson, T. Pettersson, A.C. Araújo, M. Karlsson, A. Russom, Slipdisc: a versatile sample preparation platform for point of care diagnostics, RSC Adv. 7 (56) (2017) 35048–35054.

[143] J. Song, C. Liu, M.G. Mauk, S.C. Rankin, J.B. Lok, R.M. Greenberg, H.H. Bau, Two-stage isothermal enzymatic amplification for concurrent multiplex molecular detection, Clin. Chem. 63 (3) (2017) 714–722.

[144] B. Shu, C. Zhang, D. Xing, A handheld flow genetic analysis system (FGAS): towards rapid, sensitive, quantitative and multiplex molecular diagnosis at the point-of-care level, Lab Chip 15 (12) (2015) 2597–2605.

[145] D.Y. Joh, A.M. Hucknall, Q. Wei, K.A. Mason, M.L. Lund, C.M. Fontes, et al., Inkjet-printed point-of-care immunoassay on a nanoscale polymer brush enables subpicomolar detection of analytes in blood, Proc. Natl. Acad. Sci. U. S. A. 114 (34) (2017) E7054–E7062.

[146] C.P. Price, Point of care testing: Potential for tracking disease management outcomes, Disease Management and Health Outcomes 10 (12) (2002) 749–761.

[147] H.C. Ates, J.A. Roberts, J. Lipman, A.E. Cass, G.A. Urban, C. Dincer, On-site therapeutic drug monitoring, Trends in biotechnology 38 (11) (2020) 1262–1277.

[148] M. Urdea, L.A. Penny, S.S. Olmsted, M.Y. Giovanni, P. Kaspar, A. Shepherd, P. Wilson, et al., Requirements for high impact diagnostics in the developing world, Nature 444 (1) (2006) 73–79.
[149] T.R. Kozel, A.R. Burnham-Marusich, Point-of-care testing for infectious diseases: past, present, and future, J. Clin. Microbiol. 55 (8) (2017) 2313–2320.
[150] S.K. Vashist, J.H.T. Luong, Trends in in vitro diagnostics and mobile healthcare, Biotechnol Adv 34 (3) (2016) 137–138.
[151] M. Elgendi, A. Mohamed, R. Ward, Efficient ECG compression and QRS detection for e-health applications, Scientific reports 7 (1) (2017) 1–16.
[152] H.A. Abdulbari, E.A.M. Basheer, Electrochemical biosensors: electrode development, materials, design, and fabrication, ChemBioEng Rev. 4 (2) (2017) 92–105.
[153] Z. Jiang, B. Feng, J. Xu, T. Qing, P. Zhang, Z. Qing, Graphene biosensors for bacterial and viral pathogens, Biosensors and Bioelectronics 166 (2020) 112471.
[154] V. Prasen, S. Sejal, Nanocarbons-Mediated water purification: an application towards wastewater treatment, Environmental Nanotechnology for Water Purification (2020) 57–99.
[155] S.K. Vashist, Point-of-care diagnostics: recent advances and trends, Biosensors 7 (4) (2017) 62.
[156] F. Pfeiffer, G. Mayer, Selection and biosensor application of aptamers for small molecules, Frontiers in chemistry 4 (2016) 25.
[157] T. Juneja, et al., Molecular Landscape and Computational Screening of the Natural inhibitors against HPV16 E6 Oncoprotein, Asian Pac J Cancer Prev 22 (8) (2021) 2461–2469. Available from: https://doi.org/10.31557/APJCP.2021.22.8.2461.
[158] M. Pandya, et al., Unravelling Vitamin B12 as a potential inhibitor against SARS-CoV-2: A computational approach, Informatics in Medicine Unlocked 30 (2022) 100951. Available from: https://doi.org/10.1016/j.imu.2022.100951.
[159] D. Menamadathil, et al., Artificial neural network based study predicting GS-441524 As potential inhibitor of SARS COV-2 activator protein furin: A Polypharmacology Approach, Applied Biochemistry and Biotechnology (2021).

CHAPTER 13

Strategies of detection and challenges for hepatitis C infectious disease

Maheswata Moharana, Subrat Kumar Pattanayak and Fahmida Khan
Department of Chemistry, National Institute of Technology, Raipur, Chhattisgarh, India

13.1 Introduction

Infectious diseases are defined as infections which have newly appeared in a community or already existed but are rapidly increasing in certain a geographic range. Infectious diseases are the second leading cause of death and primary cause of disability-adjusted life years worldwide (disability-adjusted life means a loss of 1 year from a healthy life) and the third cause of death in the United States [1,2], among the different infectious diseases which cause death worldwide such as HIV/AIDS, diarrheal diseases, lower respiratory tract infections, malaria and tuberculosis. In 1981, a new disease was first recognized: AIDS (Acquired Immune deficiency syndrome) which became a global killer. Before this in 1918–1920, an influenza pandemic killed around 50 million peoples [3,4]. The rate of morbidity (ill health) and mortality associated with the infectious diseases is more prevalent for people in developing countries particularly, infants and children (about three million children die every year from malaria and diarrheal diseases especially) [5,6]. In addition to HIV/AIDS and the influenza pandemic, there is a continual evolution of a wide range of emerging and reemerging infections throughout the world in recent years. Among them, Ebola virus and Nipah virus have been recognized as highly fatal but involved very small number of peoples. Dengue and West Nile fever are some of the reemerging infections that are causing considerable concern in the United States. Around 53,000 cases Dengue were reported in Brazil in 1998 [7].

Hepatitis A is the acute inflammation of the liver caused by infections with hepatitis A virus. Globally, around 126 million cases and 35,000 deaths were reported annually due to Hepatitis A infections [8]. This infection

Biosensors for Emerging and Re-emerging Infectious Diseases
DOI: https://doi.org/10.1016/B978-0-323-88464-8.00015-4

spread via person to person contact or ingestion of contaminated food or water. In some cases, the infection can be transmitted through infected blood [9].

Hepatitis B virus is another major public health threat which infects about 257 million people globally and more than 887,000 deaths are reported every year [10]. Approximately 40% of the hepatocellular carcinoma cases are detected due to chronic hepatitis B, which is the second leading cause of cancer-related mortality worldwide [11]. Hepatitis B virus infections are transmitted through contact with the infected blood or semen. It was reported that there are three major types of transmission. In the area of high endemicity, the transmission occurs from infected mothers to neonates. In low endemic regions, sexual transmission is the premier cause. The third cause of infection is unsafe injection and blood transfusion [12].

Hepatitis C is a major health problem affecting people of all age groups. It is a major cause of chronic liver diseases, liver cirrhosis and hepatocellular carcinoma [13]. Around 170 million people are infected with this virus worldwide [14]. It was reported that the hepatitis C cases are 3.25%–5% more in the people born during 1945–65 as compared to the adults born in other years [15]. The virus induces cell damages via a different mechanism. The generation of reactive oxygen species and oxidative stress plays a major role in the pathogenesis of chronic hepatitis C infections. The infection is also associated with increasing levels of many oxidative stress markers, including mitochondrial injury, iron overload and chronic inflammation [16]. The hepatitis C virus is an enveloped virus belonging to the Hepacivirus genus and the Flaviviridae family [17,18]. The hepatitis C virus is a single-stranded RNA virus with a positive sense strand. This positive sense strand has a 5' noncoding region and a long open reading frame followed by a 3' noncoding region. The hepatitis C virus contains a positive single-stranded RNA with ~9600 nucleotides. The nucleotides encode a single polypeptide of ~3000 amino acids which is further divided into structural (core, E1 and E2) and nonstructural (P7, nonstructural proteins (NS)2, NS3, NS4A, NS5A and NS5B) proteins through proteolysis [19,20]. According to phylogenetic classification, Hepatitis C virus is categorized into seven major groups and further subcategorized into 67 subtypes [21]. Among all of these variations, genotype 3 is dominant in south Asia and less prevalent in some European countries [22]. Besides these genotype variations, hepatitis C exists as a heterogeneous population of several viruses called quasispecies, within each infected individual [23].

13.1.1 Detection strategy for hepatitis C virus

A significant percentage of people having hepatitis C viral infections do not know that they are infected, so, they cannot benefit from the available treatments which is the main cause for the transmission of the disease. For this reason, the detection of the hepatitis C infections in asymptomatic people is very important. Llibre et al. reported in their study, a PoC assay for the qualitative detection of the hepatitis C virus-RNA on the PCR Genedrive instrument. The PoC assay identifies all the six major genotypes of hepatitis C virus with a detection limit of 2362 IU/mL, 98.6% sensitivity and with 100% specificity [24]. Caballeria et al. reported three methods for the detection of hepatitis C infection in their study. In the first method, they provide an explanatory letter to the adults assigned to two primary care teams (PCTs), inviting them to have a blood test. In the second method, posters and leaflets of important information were placed in highly visible and strategic locations in and around each PCT offices so that patients can easily read the information and the third method was to reexamine hepatitis C virus antibody test results in patients with hypertransaminasemia [25]. Chapko et al. reported that the identification of hepatitis C virus involved the detection in peripheral blood of antibody to hepatitis C virus (HCVAb). The serological presence of HCVAb signifies exposure to the hepatitis C virus but doesn't allow for discrimination between resolved and current hepatitis c infection. Current hepatitis C infection is identified by the detection of viremia, a marker for which is hepatitis C virus-RNA in peripheral blood [26]. Saab et al. reported three strategies i.e. biochemical, serological and viral-based screening for the detection of hepatitis C virus in hemodialysis patients [27].

13.1.2 Treatment options for hepatitis C infection

Without an effective vaccine, hepatitis C virus became a global challenge inflicting 170–300 million carriers throughout the globe causing cirrhosis and hepatocellular carcinoma. Today there is still no preventive hepatitis C vaccine available, so the application of medicinal plants has long served for the treatment of hepatitis C-like infectious and chronic diseases [28]. A number of studies were reported about the antioxidants such as glycyrrhizin, catechin, phytosterols, *N*-acetylcysteine and phytochemicals that are capable of decreasing hepatitis C virus replication and cell damage [29]. Acetylsalicylic acid reduces the level of hepatitis C virus-RNA and viral protein by decreasing the cellular oxidative stress and modifying the Cu/Zn superoxide dismutase expression. Carotene, vitamin D2 and linolinic acid also inhibited hepatitis C virus-RNA expression [30,31]. Govea-Salas et al. reported on the activity of gallic acid against hepatitis C viral infection. The effects of gallic acid were examined using a subgenomic hepatitis C virus replicon cell culture system that expressed hepatitis C virus nonstructural proteins. The cytotoxicity of gallic acid was evaluated at a concentration level of 100–600 mg/ mL using an MTT assay. Huh-7 replicon cells were incubated with 300 mg/mL gallic acid for different times and the hepatitis C virus-RNA and the protein levels were measured by reverse transcription quantitative polymerase chain reaction (PCR) and western blot analysis [32]. Wu reported 3-hydroxy caruilignan C isolated from Swietenia macrophylla stems which exhibited high anti-hepatitis C viral activity at both protein and RNA levels at nontoxic concentrations with an EC50 value of $10.5 \pm 1.2\ \mu M$ [33]. Rehman et al. reported *Nymphaea alba* and their phytochemicals as a valuable and alternative regimen against hepatitis C virus [34]. Kyeongjin Kim and his group reported a polyphenolic compound from the curry spice turmeric and curcumin, which shows antiviral activity against the influenza virus, adenovirus, coxsackievirus and human immunodeficiency virus. Curcumin also decreases hepatitis C virus gene expression via suppression of the Akt-SREBP-1 activation [35].

13.2 Computational details

The three-dimensional structure of the NS3 protease protein of hepacivirus C was obtained from the RCSB protein data bank with PDB ID: 2F9U. Similarly, from the Pubchem data base with Pubchem ID: 5280934 (alpha-linolenic acid), 102146586 (azadirachtanin), 14492795 (nimbaflavone) and 5280805 (rutin), three-dimensional structures of phytochemicals were retrieved. The three-dimensional protein structure was used as a receptor in this segment, with ligands including alpha-linolenic acid, azadirachtanin, nimbaflavone, and rutin. We cleaned all of the bad contacts from the protein before beginning the docking operation. Using the AutoDock method, we measured the Kollman charges and inserted the hydrogen atoms into the receptor's polar contacts. Meanwhile, we charged the ligands with Gasteiger charge. So all of the protein and ligand three-dimensional structures are converted to PDBQT format. On the protein, we created number of points of 126, 126, 126 in the x, y, and z directions, with core x, y, z of 197.744, 1.114, and 59.223, respectively, and spacing of 0.314 Å. We used a blind docking procedure to locate the studied protein's binding site. The gradient optimization algorithm was used in the docking process. Discovery studio visualizing software was used for structural visualization and outcome analysis.

13.3 Results and discussion

The main goal of this research is to determine the binding energy of the hepacivirus C NS3 protease protein (PDB ID: 2F9U) with alpha-linolenic acid, azadirachtanin, nimbaflavone, and rutin. The binding affinity between the compounds and proteins was determined using molecular docking. The binding energy of the receptor and the ligand was analyzed using molecular docking [36–52] or molecular interaction analysis. The main features of binding performance and types of bonds between the NS3 protease protein of hepacivirus C and the studied molecules were determined using molecular docking. The binding energy of hepacivirus C's NS3 protein with alpha-linolenic acid is −4.1 kcal/mol. With bond lengths of 3.00203 and 2.90941 Å, respectively, both ARG130 and ASP112 residues are involved in hydrogen bonding. Hydrophobic interaction involves the residues VAL113, PRO115, and TYR105. During this interaction, there are two hydrogen bonding interactions and three hydrophobic interactions. Fig. 13.1 shows a pictorial illustration of the interaction between the hepacivirus C NS3 protease protein and Alpha-linolenic acid.

Hepacivirus C's NS3 protease protein (PDB ID: 2F9U) interacts with azadirachtanin in a total of six hydrogen bonding and hydrophobic interactions. The residues LYS136, CYS159, THR42, and SER139 are involved in hydrogen bonding. Hydrophobic activity was found in the residues ILE132, LYS136, CYS159, and HIS57. Fig. 13.2 illustrates a diagrammatic picture of the relationship between hepacivirus C's NS3 protein (PDB ID: 2F9U) and azadirachtanin.

During the interaction of hepacivirus C's NS3 protease protein (PDB ID: 2F9U) with nimbaflavone, the binding energy was −9.6 kcal/mol. In comparison to hydrogen bonding, hydrophobic interactions play a significant role. In hydrogen bonding, only the GLN8 residue is involved. Fig. 13.3 provides a diagrammatic view of the relationship between hepacivirus C's NS3 protease protein (PDB ID: 2F9U) and nimbaflavone.

During in the interaction of hepacivirus C's NS3 protease protein (PDB ID: 2F9U) with rutin, the binding energy was −7.0 kcal/mol. In comparison to hydrogen bonding and electrostatic interaction, hydrophobic interaction plays a key role. With a bond length of 3.44 Å, only the ASP168 residue is engaged in electrostatic interaction. Fig. 13.4 demonstrates a diagrammatic representation of the interaction between hepacivirus C's NS3 protease protein (PDB ID: 2F9U) and rutin. Table 13.1 provides the detailed molecular interactions of phytochemicals with the NS3 protease protein (2F9U).

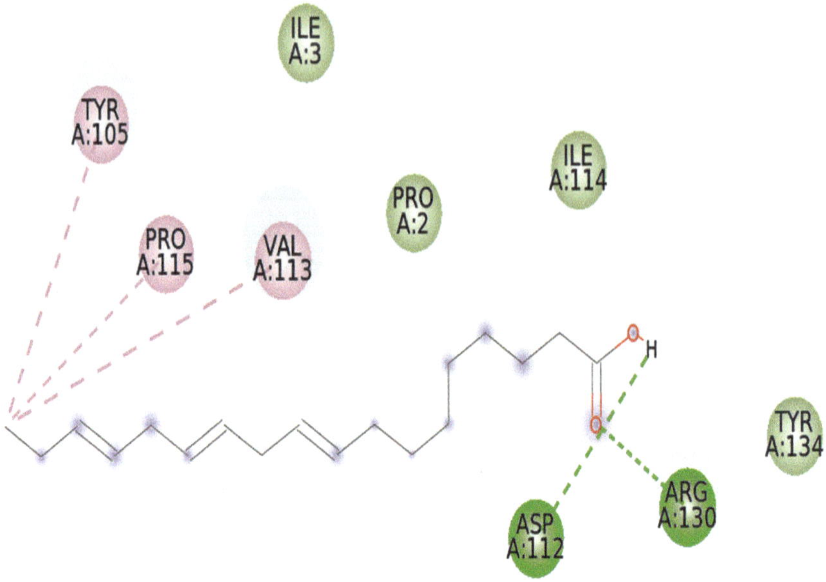

Figure 13.1 Molecular interaction of NS3 protease protein and alpha-linolenic acid.

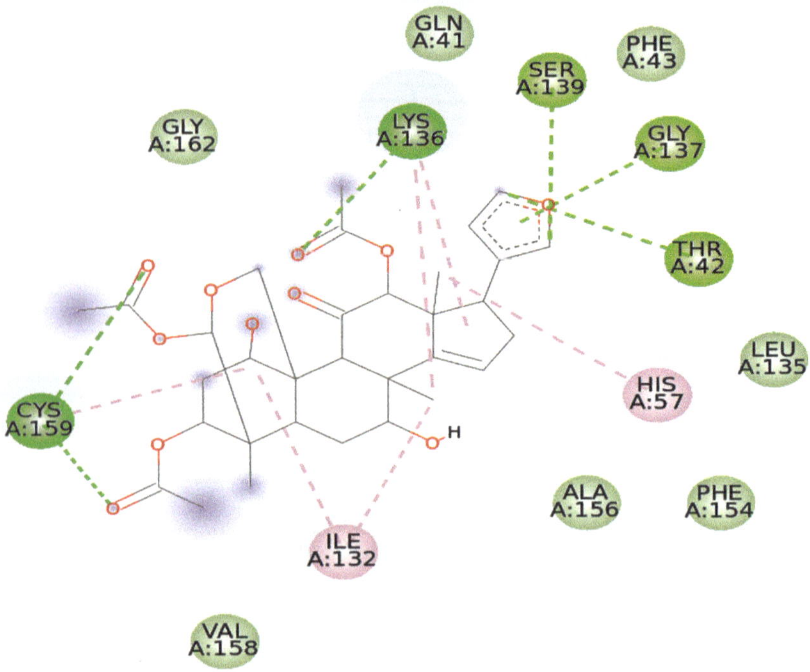

Figure 13.2 Molecular interaction of NS3 protease protein with azadirachtanin.

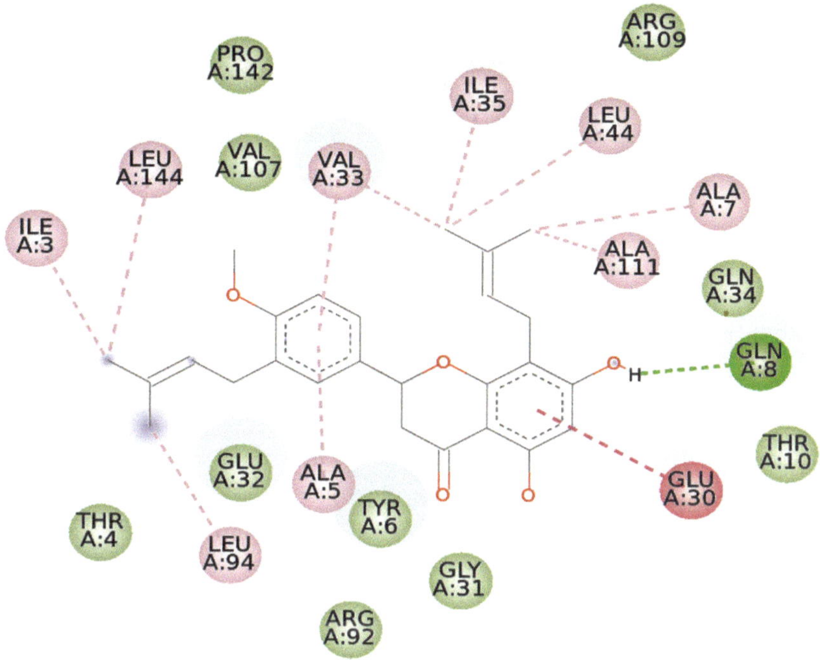

Figure 13.3 Molecular interaction of NS3 protease protein with nimbaflavone.

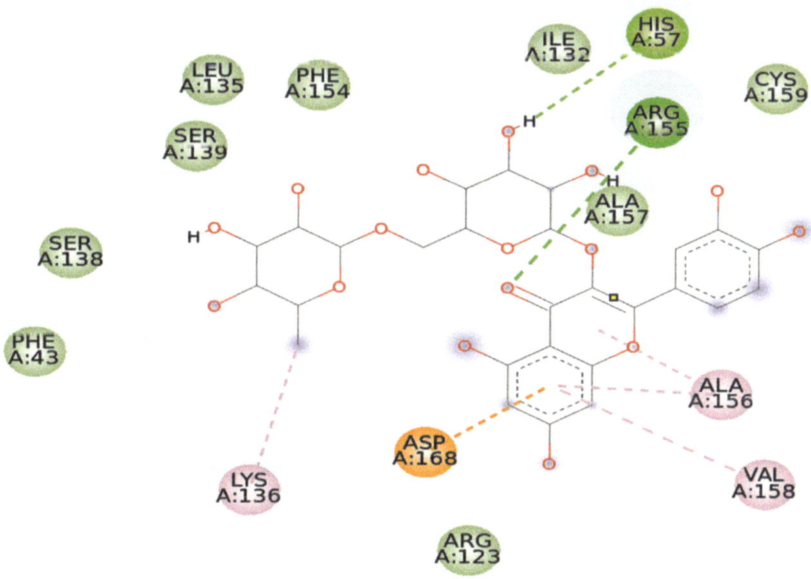

Figure 13.4 Molecular interaction of NS3 protease protein with rutin.

Table 13.1 Molecular interactions compounds with NS3 protease protein (2F9U).

Chemical compounds	Binding energy (kcal/mol)	Binding residues	Different interactions	Bond length (Å)
Alpha-linolenic acid	− 4.1	ARG130:NH1 -:UNK0:O	Hydrogen Bond	3.00203
		UNK0:H - A:ASP112:OD2	Hydrogen Bond	2.90941
		UNK0:C - A:VAL113	Hydrophobic	4.80651
		UNK0:C - A:PRO115	Hydrophobic	3.94842
		TYR105 -:UNK0:C	Hydrophobic	4.71551
Azadirachtanin	− 6.4	LYS136:NZ -:UNK0:O	Hydrogen Bond	2.80551
		CYS159:SG -:UNK0:O	Hydrogen Bond	3.64815
		CYS159:SG -:UNK0:O	Hydrogen Bond	3.55917
		UNK0:C - A:THR42:O	Hydrogen Bond	3.4059
		UNK0:C - A:SER139:OG	Hydrogen Bon	3.05325
		GLY137:N -:UNK0	Hydrogen Bond	4.04938
		ILE132 -:UNK0	Hydrophobi	5.43654
		LYS136 -:UNK0	Hydrophobic	4.39627
		CYS159 -:UNK0	Hydrophobic	5.37298
		UNK0:C - A:ILE132	Hydrophobic	4.21492
		UNK0:C - A:LYS136	Hydrophobic	4.1214
		HIS57 -:UNK0:C	Hydrophobic	4.73035
Nimbaflavone	− 9.6	UNK0:H - A:GLN8:O	Hydrogen Bond	2.22934
		GLU30:C,O -:UNK0	Hydrophobic	4.49715
		GLY31:N -:UNK0	Hydrophobic	4.49715
		ALA7 -:UNK0:C	Hydrophobic	3.46749
		ALA111 -:UNK0:C	Hydrophobic	3.78131
		UNK0:C - A:LEU94	Hydrophobic	5.10451
		UNK0:C - A:ILE3	Hydrophobic	4.98462

Rutin	− 7.0	UNK0:C - A:LEU144	Hydrophobic	4.97608
		UNK0:C - A:VAL33	Hydrophobic	4.04812
		UNK0:C - A:ILE35	Hydrophobic	4.68875
		UNK0:C - A:LEU44	Hydrophobic	4.82399
		UNK0 - A:ALA5	Hydrophobic	4.17891
		UNK0 - A:VAL33	Hydrophobic	4.85823
		ARG155:NH2 -:UNK0:O	Hydrogen Bond	3.27044
		ASP168:OD2 -:UNK0	Electrostatic	3.44734
		UNK0:H - A:HIS57	Hydrogen Bond	2.87699
		UNK0:C - A:LYS136	Hydrophobic	4.12988
		UNK0 - A:ALA156	Hydrophobic	4.13533
		UNK0 - A:ALA156	Hydrophobic	5.21664
		UNK0 - A:VAL158	Hydrophobic	5.35086

13.4 Conclusion

The primary objective of this research is to specify the binding energy of the hepacivirus C (PDB ID: 2F9U) NS3 protease protein with alpha-linolenic acid, azadirachtanin, nimbaflavone, and rutin. Hepacivirus C NS3 protease proteins had a binding performance of −4.1 kcal/mol with alpha-linolenic acid. The hydrophobic effect includes the residues VAL113, PRO115, and TYR105, and during this interaction, there were two hydrogen bonding interactions and three hydrophobic interactions. Hepacivirus C's NS3 protease protein (PDB ID: 2F9U) interacted with azadirachtanin in a total of six hydrogen bonding and hydrophobic interactions. The residues LYS136, CYS159, THR42, and SER139 were found in hydrogen bonding. Hydrophobic activity was found in the residues ILE132, LYS136, CYS159, and HIS57. During the interaction of hepacivirus C's NS3 protease protein (PDB ID: 2F9U) with nimbaflavone, the binding energy was −9.6 kcal/mol. In comparison to hydrogen bonding, hydrophobic interactions play a major role. In hydrogen bonding, only the GLN8 residue is involved. Hydrophobic interaction was observed in GLU30, GLY31, ALA7, ALA111, LEU94, ILE3, LEU144, VAL33, ILE35, ILE35, LEU44, ALA5, and VAL33 residues. During the interaction of hepacivirus C's NS3 protease protein (PDB ID: 2F9U) with rutin, the binding energy was −7.0 kcal/mol. In comparison to hydrogen bonding and electrostatic interaction, hydrophobic interaction plays a crucial role. With a bond length of 3.44 Å, only the ASP168 residue is involved in electrostatic interaction. Hydrogen bonding is present in both ARG155 and HIS57 residues. Hydrophobic interaction was observed in the residues LYS136, ALA156, and VAL158. The length of the bonds ranges from 4.12 to 5.35 Å. During the docking process, nimbaflavone has the best binding efficiency of the compounds studied.

References

[1] A.S. Fauci, Infectious diseases: considerations for the 21st century, Clin. Infect. Dis. 32 (5) (2001) 675−685.

[2] R.W. Pinner, S.M. Teutsch, L. Simonsen, L.A. Klug, J.M. Graber, M.J. Clarke, et al., Trends in infectious diseases mortality in the United States, Jama 275 (3) (1996) 189−193.

[3] N.P. Johnson, J. Mueller, Updating the accounts: global mortality of the 1918−1920 "Spanish" influenza pandemic, Bull. History Med. (2002) 105−115.

[4] Centers for Disease Control and Prevention (CDC), Multistate outbreak of monkeypox−Illinois, Indiana, and Wisconsin, 2003, MMWR. Morbidity Mortal. Wkly. Rep. 52 (23) (2003) 537−540.

[5] World Health Organization, The World Health Report 2004 (World Health Organization, Genève), 2004.
[6] R.L. Guerrant, B.L. Blackwood, Threats to global health and survival: the growing crises of tropical infectious diseases—our "unfinished agenda.", Clin. Infect. Dis. 28 (5) (1999) 966–986.
[7] R.E. Istúriz, D.J. Gubler, J.B. del Castillo, Dengue and dengue hemorrhagic fever in Latin America and the Caribbean, Infect. Dis. Clin. 14 (1) (2000) 121–140.
[8] R. Aggarwal, A. Goel, Hepatitis A: epidemiology in resource-poor countries, Curr. Opin. Infect. Dis. 28 (5) (2015) 488–496.
[9] C.M. Gossner, E. Severi, N. Danielsson, Y. Hutin, D. Coulombier, Changing hepatitis A epidemiology in the European Union: new challenges and opportunities, Eurosurveillance 20 (16) (2015) 21101.
[10] P.A. Revill, F.V. Chisari, J.M. Block, M. Dandri, A.J. Gehring, H. Guo, et al., A global scientific strategy to cure hepatitis B, Lancet Gastroenterol. Hepatol. 4 (7) (2019) 545–558.
[11] J.D. Stanaway, A.D. Flaxman, M. Naghavi, C. Fitzmaurice, T. Vos, I. Abubakar, et al., The global burden of viral hepatitis from 1990 to 2013: findings from the Global Burden of Disease Study 2013, Lancet 388 (10049) (2016) 1081–1088.
[12] C. Trépo, H.L. Chan, A. Lok, Hepatitis B virus infection, Lancet 384 (9959) (2014) 2053–2063.
[13] J.M. Pawlotsky, Therapy of hepatitis C: from empiricism to eradication, Hepatology 43 (S1) (2006) S207–S220.
[14] U.A. Ashfaq, T. Javed, S. Rehman, Z. Nawaz, S. Riazuddin, An overview of HCV molecular biology, replication and immune responses, Virology J. 8 (1) (2011) 1–10.
[15] M.K. Chapko, D.R. Dufour, R.I. Hatia, J. Drobeniuc, J.W. Ward, C.G. Teo, Cost-effectiveness of strategies for testing current hepatitis C virus infection, Hepatology 62 (5) (2015) 1396–1404.
[16] M. Govea-Salas, A.M. Rivas-Estilla, R. Rodríguez-Herrera, S.A. Lozano-Sepúlveda, C.N. Aguilar-Gonzalez, A. Zugasti-Cruz, et al., Gallic acid decreases hepatitis C virus expression through its antioxidant capacity, Exp. Ther. Med. 11 (2) (2016) 619–624.
[17] S.F. Wu, C.K. Lin, Y.S. Chuang, F.R. Chang, C.K. Tseng, Y.C. Wu, et al., Anti-hepatitis C virus activity of 3-hydroxy caruilignan C from *Swietenia macrophylla* stems, J. Viral Hepat. 19 (5) (2012) 364–370.
[18] B.D. Lindenbach, C.M. Rice, Unravelling hepatitis C virus replication from genome to function, Nature 436 (7053) (2005) 933–938.
[19] G. Waris, A. Siddiqui, Hepatitis C virus stimulates the expression of cyclooxygenase-2 via oxidative stress: role of prostaglandin E2 in RNA replication, J. Virol. 79 (15) (2005) 9725–9734.
[20] H. Huang, Y. Chen, J. Ye, Inhibition of hepatitis C virus replication by peroxidation of arachidonate and restoration by vitamin E, Proc. Natl Acad. Sci. U.S.A. 104 (47) (2007) 18666–18670.
[21] D.B. Smith, J. Bukh, C. Kuiken, A.S. Muerhoff, C.M. Rice, J.T. Stapleton, et al., Expanded classification of hepatitis C virus into 7 genotypes and 67 subtypes: updated criteria and genotype assignment web resource, Hepatology 59 (1) (2014) 318–327.
[22] J. Ampuero, M. Romero-Gomez, K.R. Reddy, HCV genotype 3—the new treatment challenge, Aliment. Pharmacol. Ther. 39 (7) (2014) 686–698.
[23] K. Murakawa, M. Esumi, T. Kato, H. Kambara, T. Shikata, Heterogeneity within the nonstructural protein 5-encoding region of hepatitis C viruses from a single patient, Gene 117 (2) (1992) 229–231.

[24] A. Llibre, Y. Shimakawa, E. Mottez, S. Ainsworth, T.P. Buivan, R. Firth, et al., Development and clinical validation of the Genedrive point-of-care test for qualitative detection of hepatitis C virus, Gut 67 (11) (2018) 2017–2024.
[25] L. Caballeria, G. Pera, J. Bernad, S. Canut, E. Navarro, M. Bruguera, Strategies for the detection of hepatitis C viral infection in the general population, Rev. Clín. Española (Engl. Ed.) 214 (5) (2014) 242–246.
[26] M.K. Chapko, D.R. Dufour, R.I. Hatia, J. Drobeniuc, J.W. Ward, C.G. Teo, Cost-effectiveness of strategies for testing current hepatitis C virus infection, Hepatology 62 (5) (2015) 1396–1404.
[27] S. Saab, M. Brezina, G. Gitnick, P. Martin, H.F. Yee Jr, Hepatitis C screening strategies in hemodialysis patients, Am. J. Kidney Dis. 38 (1) (2001) 91–97.
[28] S. Rehman, U.A. Ashfaq, B. Ijaz, S. Riazuddin, Anti-hepatitis C virus activity and synergistic effect of Nymphaea alba extracts and bioactive constituents in liver infected cells, Microb. Pathog. 121 (2018) 198–209.
[29] L. Patrick, Hepatitis C: epidemiology and review of complementary/alternative medicine treatments, Alternative Med. Rev. 4 (4) (1999) 220–238.
[30] K. Trujillo-Murillo, A.R. Rincón-Sánchez, H. Martínez-Rodríguez, F. Bosques-Padilla, J. Ramos-Jiménez, H.A. Barrera-Saldaña, et al., Acetylsalicylic acid inhibits hepatitis C virus RNA and protein expression through cyclooxygenase 2 signaling pathways, Hepatology 47 (5) (2008) 1462–1472.
[31] M. Yano, M. Ikeda, K.I. Abe, H. Dansako, S. Ohkoshi, Y. Aoyagi, et al., Comprehensive analysis of the effects of ordinary nutrients on hepatitis C virus RNA replication in cell culture, Antimicrob. Agents Chemother. 51 (6) (2007) 2016–2027.
[32] M. Govea-Salas, A.M. Rivas-Estilla, R. Rodríguez-Herrera, S.A. Lozano-Sepúlveda, C.N. Aguilar-Gonzalez, A. Zugasti-Cruz, et al., Gallic acid decreases hepatitis C virus expression through its antioxidant capacity, Exp. Ther. Med. 11 (2) (2016) 619–624.
[33] S.F. Wu, C.K. Lin, Y.S. Chuang, F.R. Chang, C.K. Tseng, Y.C. Wu, et al., Anti-hepatitis C virus activity of 3-hydroxy caruilignan C from *Swietenia macrophylla* stems, J. Viral Hepat. 19 (5) (2012) 364–370.
[34] S. Rehman, U.A. Ashfaq, B. Ijaz, S. Riazuddin, Anti-hepatitis C virus activity and synergistic effect of Nymphaea alba extracts and bioactive constituents in liver infected cells, Microb. Pathog. 121 (2018) 198–209.
[35] K. Kim, K.H. Kim, H.Y. Kim, H.K. Cho, N. Sakamoto, J. Cheong, Curcumin inhibits hepatitis C virus replication via suppressing the Akt-SREBP-1 pathway, FEBS Lett. 584 (4) (2010) 707–712.
[36] J. Panda, S.N. Sahu, J.K. Sahoo, S.P. Biswal, S.K. Pattanayak, R. Samantaray, et al., Efficient removal of two anionic dyes by a highly robust zirconium based metal organic framework from aqueous medium: experimental findings with molecular docking study. Environmental, Nanotechnol. Monit. Manage. 14 (2020) 100340.
[37] S.N. Sahu, B. Mishra, R. Sahu, S.K. Pattanayak, Molecular dynamics simulation perception study of the binding affinity performance for main protease of SARS-CoV-2, J. Biomol. Struct. Dyn. (2020) 1–16.
[38] S. N. Sahu, R. Sahu, S. K. Pattanayak, Molecular interaction study of phytochemicals with native and mutant protein related to nephrotic syndrome. In AIP Conference Proceedings (Vol. 2270, No. 1, p. 020002). AIP Publishing LLC, 2020, November.
[39] R. Panda, S. N. Sahu, F. Khan, & S. K. Pattanayak, Binding performance of phytochemicals with mutant threonine-protein kinase Chk2 protein: An in silico study. In AIP Conference Proceedings (Vol. 2270, No. 1, p. 020003). AIP Publishing LLC, 2020, November.

[40] J. Panda, S.N. Sahu, R. Pati, P.K. Panda, B.C. Tripathy, S.K. Pattanayak, et al., Role of pore volume and surface area of Cu-BTC and MIL-100 (Fe) metal-organic frameworks on the loading of rifampicin: collective experimental and docking study, ChemistrySelect 5 (40) (2020) 12398–12406.
[41] S. Sahoo, S.N. Sahu, S. kumar Pattanayak, N. Misra, M. Suar, Biosensor and its implementation in diagnosis of infectious diseases, Smart Biosensors in Medical Care, Academic Press, 2020, pp. 29–47.
[42] S.N. Sahu, M. Moharana, P.C. Prusti, S. Chakrabarty, F. Khan, S.K. Pattanayak, Real-time data analytics in healthcare using the Internet of Things, Real-Time Data Analytics for Large Scale Sensor Data, Academic Press, 2020, pp. 37–50.
[43] M. Moharana, S.K. Pattanayak, Biosensors: a better biomarker for diseases diagnosis, Smart Biosensors in Medical Care, Academic Press, 2020, pp. 49–64.
[44] S.N. Sahu, J. Panda, R. Sahu, T. Sahoo, S. Chakrabarty, S.K. Pattanayak, Healthcare information technology for rural healthcare development: insight into bioinformatics techniques, Internet of Things, Smart Computing and Technology: A Roadmap Ahead, Springer, Cham, 2020, pp. 151–169.
[45] S.N. Sahu, S.K. Pattanayak, Molecular docking and molecular dynamics simulation studies on PLCE1 encoded protein, J. Mol. Struct. 1198 (2019) 126936.
[46] S. N. Sahu, M. Moharana, R. Sahu, S. K. Pattanayak, Molecular docking approach study of binding performance of antifungal proteins. In AIP Conference Proceedings (Vol. 2142, No. 1, p. 060001). AIP Publishing LLC, 2019, August.
[47] S.N. Sahu, M. Moharana, R. Sahu, S.K. Pattanayak, Impact of mutation on podocin protein involved in type 2 nephrotic syndrome: insights into docking and molecular dynamics simulation study, J. Mol. Liq. 281 (2019) 549–562.
[48] S.M. Hiremath, A. Suvitha, N.R. Patil, C.S. Hiremath, S.S. Khemalapure, S.K. Pattanayak, et al., Synthesis of 5-(5-methyl-benzofuran-3-ylmethyl)-3*H*-[1,3,4] oxadiazole-2-thione and investigation of its spectroscopic, reactivity, optoelectronic and drug likeness properties by combined computational and experimental approach, Spectr. Acta. A: Mol. Biomol. Spectrosc. 205 (2018) 95–110.
[49] S.N. Sahu, M. Moharana, S.R. Martha, A. Bissoyi, P.K. Maharana, S.K. Pattanayak, Computational biology approach in management of big data of healthcare sector, Big Data Analytics for Intelligent Healthcare Management, Academic Press, 2019, pp. 247–267.
[50] A. Chand, P. Chettiyankandy, M. Moharana, S.N. Sahu, S.K. Pradhan, S.K. Pattanayak, et al., Computational methods for developing novel antiaging interventions, Molecular Basis and Emerging Strategies for Anti-aging Interventions, Springer, Singapore, 2018, pp. 175–193.
[51] A. Bissoyi, S.K. Pattanayak, A. Bit, A. Patel, A.K. Singh, S.S. Behera, et al., Alphavirus nonstructural proteases and their inhibitors, Viral Proteases and Their Inhibitors, Academic Press, 2017, pp. 77–104.
[52] S.N. Sahu, T. Sahoo, B. Mishra, S.K. Pattanayak, R. Sahu, Interaction of Ru-4′-(2-pyridyl)-2, 2′: 6′, 2″-Terpyridine with Diseased HER2 Protein, Advances in Mechanical Processing and Design, Springer, Singapore, 2021, pp. 413–419.

CHAPTER 14

Point-of-care electrochemical biosensors using CRISPR/Cas for RNA analysis

Rohan Panwar[1], Hardik Churi[2] and Sushma Dave[3]
[1]Seth GS Medical College and KEM Hospital, Mumbai, Maharashtra, India
[2]Department of Biotechnology SDSM College, Mumbai, Maharashtra, India
[3]Department of Applied Science, JIET Group of Institutions, Jodhpur, Rajasthan, India

14.1 Introduction

Pathogenic infections are a major concern in the healthcare sector, accounting for roughly a quarter of human deaths globally and when these findings are normalized to developing countries, the proportion grows exponentially [1].

Pandemics, are epidemics having worldwide diaspora; causing an excessive number of sicknesses, deaths in the world cause disruption in the socioeconomic situation affected countries. Globalization, lifestyle changes, and improvements on the social and economic fronts have caused these emerging infections and accelerated occurrence and circulation of new microbial agents. Globalization has also facilitated sharing information and experiences on diseases transmission and treatment [2].

Over the past 40 years, there are recurrent large-scale epidemics from emerging RNA viruses and pandemics of influenza (Spanish flu, swine flu, bird flu), outbreaks of Ebola and Zika virus, deadly and wide-spread epidemics of MERS and SARS, and the ongoing pandemic of COVID-19, originating in China in 2019, have swept across continents and emerged as the most widespread infections reported during this century.

These biomolecular agents (nucleic acids) are spread by an animal-to-human transmission event, with either clinically apparent or occult spread into vulnerable human populations. Each time, a scarcity of rapid, accessible and accurate molecular diagnostic testing has hindered the general public health response to the emerging viral threat [3].

According to the World Health Organization (WHO), the ideal approach for detecting foreign pathogens should be rapid, specific, sensitive, instrument-free and cost-effective [4,5]. Setting up motions capable of rapid and sensitive

Biosensors for Emerging and Re-emerging Infectious Diseases
DOI: https://doi.org/10.1016/B978-0-323-88464-8.00003-8

diagnosis is the need of hour. Although current diagnostics are both specific and sensitive, they happen to be time consuming and expenive; thus, they are not available in resource-limited areas, for the purposes of large-scale screenings and in the case of outbreaks and epidemics (Fig. 14.1).

Focus has shifted in research towards the development of molecular POC devices adhering to the ASSURED criteria (Affordable, Sensitive, Specific, User-friendly, Rapid and robust, Equipment-free and Deliverable to end-users), whose performances should be comparable to traditional molecular techniques [6].

The rapid and accurate detection of nucleic acids with high sensitivity plays a crucial role in different fields such as biotechnology, environmental monitoring, life science, and especially in the medical field. An early diagnosis of pathogens can be achieved by nucleic acid detection, particularly RNA, which is beneficial for better treatment of life-threatening diseases [7].

Small non-coding RNAs contain 18–25 nucleotides which are also termed microRNAs (miRNAs). The mature form of miRNAs bind with specific messenger RNAs which help in gene expression regulations in many biological processes [8,9]. The mRNA plays a key role between genome and

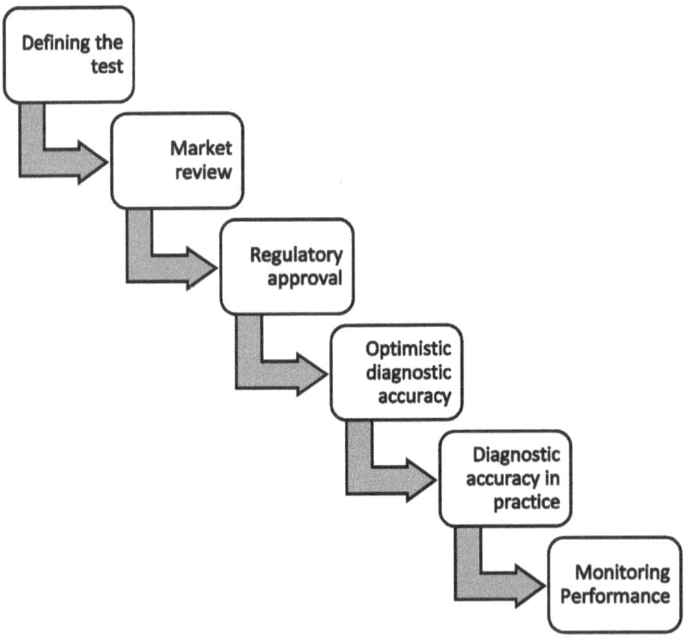

Figure 14.1 Ideal six steps that must be addressed before selecting a diagnostic test.

proteome processes. Sometimes miRNA can affect the translation and stability of mRNA by altering its structure and interaction with proteins [10]. Dysregulation of miRNAs can modify human body fluids such as saliva, plasma, serum, seminal, urine cerebrospinal fluid, and amniotic pleural effusion [11]. Some of the body fluids were enriched with specific circulating miRNAs. Different types of miRNAs are present in body fluids and their quantity varies according to the type of disease. In cancer diagnosis, the quality and quantity of circulating extracellular miRNAs are associated with stages of cancer [12]. Therefore such miRNAs are considered as potential diagnostic biomarkers. Different types of circulating miRNAs have been found in body fluids of cancer patients which indicates that these circulating miRNAs are linked with cancer initiation and progression [13]. Therefore, circulating miRNA have been widely used as noninvasive biomarkers for cancer diagnosis.

The ideal biomarker should be inexpensive, noninvasive, specific to the disease, and should be accurate in early detection [11]. The biomarker should give a response before clinical symptoms appear in the patient. For accurate on-site miRNA detection, highly sensitive electrochemical microfluidic biosensors are necessary. Lab-on chip (LOC) devices are used for on-site detection, which contains multiple immobilization areas in a single channel. Nowadays, novel biosensors used for RNA detection include nucleic acid-based, aptamer-based, optical biosensors, electrochemical biosensors using silver/gold nanoparticles, and Surface Plasmon Resonance [14]. Health professionals have used different types of electrochemical biosensors for early diagnosis of diseases such as enzymatic electrochemical biosensors, bioaffinity-based electrochemical biosensors, and microfluidic LOC methods [15]. In most of the cases, the sample obtained for detection contains minute amounts of miRNAs so traditionally, different methods were used such as polymerase chain reaction (PCR), microarrays, and RNA sequencing methods. These techniques are tedious and also require expensive instruments with well-equipped laboratories and skilled personnel. An advantageous, affordable, and alternative technique for genomic diagnosis is CRISPR/Cas (clustered regularly interspaced short palindromic repeats and associated nuclease proteins) system [16]. Many of the miRNAs require hundreds of transcriptional processes, therefore, it is very difficult to diagnose, but CRISPR-based biosensors fulfill the demand for ultrasensitive nucleotide detection [17]. CRISPR/Cas system is considered a powerful tool for transcription regulation, genome editing, and molecular diagnostics [18,19].

CRISPR was first discovered in 1987 [20]. CRISPR is a natural process found in bacteria and archaea as an adaptive immune defense system protecting against viral invasion [16]. It is mainly composed of two components. First, are short DNA repeats called Clustered Regulatory Interspaced Short Palindromic Repeats (CRISPR), and second is CRISPR-associated (Cas) systems [21]. CRISPR repeaters and spacers are attached with CRISPR-associated (Cas) genes and previously, viral or bacterial plasmids could present in between the spacer segments. The CRISPR system gives protection to bacteria from repeated viral attacks. When a virus infects bacteria for the first time then the viral genome gets incorporated as a new spacer DNA amongst existing spacers. These spacers formed from the viral genome act as a genetic memory in the bacterial CRISPR genome. The CRISPR sequence is transcribed to form crRNA molecules which are associated with bacterial molecular machinery (Cas complex) [22,23]. Transactivating CRISPR RNA also called TracrRNA combines with crRNA to form guide RNA (gRNA). When a virus attacks bacteria again, then gRNA guides the bacterial Cas complex to invade the viral genome and when the match is found, then the Cas complex triggered the degradation of invading phage genome. Some CRISPR types require Protospacer Adjacent Motifs (PAM) sequences which are very short sequences present in target nucleic acids. These PAM sequences help guide RNA to find the target sequence and then Cas enzymes cut the target sequence. In this way, CRISPR protects bacteria from viral attacks. This CRISPR-Cas system plays a crucial role in a variety of technologies like genetics and biomedical research [24].

Electrochemical detection of RNAs is an inexpensive and quite easy method to fabricate biosensor chips [16]. Thus early, rapid, and ultrasensitive RNA-based diagnosis involves the use of electrochemical CRISPR biosensors.

14.2 CRISPR/Cas classification

The most recent classification of CRISPR/Cas is based on Cas effector enzymes. These effector Cas proteins provide immunity by cleaving foreign genetic material [25]. CRISPR-Cas is divided into two classes, class 1 and class 2, which are further divided into six types and several subtypes. The effector molecule of class 1 is characterized by a multi-subunit effector complex while the effector molecule of class 2 consists of one single multi-domain effector protein.

The class 1 system of CRISPR includes type I, type III, and type IV systems and is further subdivided into twelve subtypes. Most types belong to the class 1 CRISPR system and are found in various bacteria and archaea. The class 1 system is composed of 4–7 Cas protein subunits [26]. In the class 1 system, type I and type III are most common because they are responsible for various activities such as crRNA, site binding, target cleavage, and spacer regulation. Type I system uses Cas3 enzymes which recognize double-stranded DNA with the help of PAM sequences and introduce a single strand break. Cas3 enzymes are signature enzymes of the CRISPR class 1 system. Type III system is slightly different in its mechanism than type I. The type III system uses Cas10 to introduce single-stranded breaks in DNA and RNA which are specifically involved in the transcriptional process. The type IV system is new and functionally uncharacterized. Type IV system lacks genetic material of Cas nucleases, integrases, and other important enzymes which are commonly observed in other CRISPR systems, therefore it is difficult to predict the mechanism and function of a type IV system.

The class 2 system of CRISPR consists of a single multi-domain effector protein and it includes type II, type V and type VI systems. The class 2 system of CRISPR is identified in various bacteria but is not yet found in archaea. Type II CRISPR system uses Cas9 enzymes that require both crRNA and tracrRNA which combine to form gRNA. This gRNA helps Cas9 to target and produce a blunt end in target double-stranded DNA with the help of PAM sequences. The type V CRISPR system uses Cas12 enzymes which mainly depend on its subtypes. Cas12a requires PAM sequences to create staggered double-stranded breaks in DNA. The difference between Cas9 and Cas12a is that Cas9 recognizes the G-rich PAM sequence while Cas12 recognizes the T-rich PAM sequence [27]. Cas12b requires crRNA and tracrRNA for its activity. The type VI system of CRISPR uses C2c2 enzyme which is also termed as Cas13 enzymes. Cas13 is completely different from the rest of the CRISPR/Cas enzymes because it targets RNA. CRISPR/Cas13 recognizes a Protospacer Flanking Sequence (PFS) in target single-stranded RNA and then the crRNA-Cas13 complex introduces single-stranded breaks in target RNA sequences [26].

CRISPR/Cas13 targets viral RNA sequences as well as RNAs found in plants. Cas13 can detect foreign genomes without PAM sequences. Most viruses and pathogens contain RNA as genetic material. For cancer diagnosis, miRNAs play a key role. Therefore CRISPR/Cas13 can be used for the early diagnosis of cancer and different diseases [28].

14.3 Electrochemical biosensor

Sensors are device operations that convert physical inputs into a readable output. Chemical sensors are gaining a leading position among the presently commercially available ones with a wide array of clinical, industrial, environmental, and agricultural applications. Biosensors have gone viral as analytical and diagnostic tools for widespread use as they outperform any other device presently in use. Thanks to their operational simplicity, low cost, and no skills requirements, they have become the standard man's tools of everyday use. These advantages have won them increasingly wide applications in such varied fields as diabetic and cardiac self-monitoring, forensic investigations, drug discovery, agricultural and environmental detection systems, the food industry, and in biodefense. No doubt, further commercialization of biosensors relies on such improved features as enhanced selectivity, sensitivity, stability, reproducibility, and portability, all at lower costs [29]. Electrochemical biosensors, especially with reference to advancing point-of-care (POC) medical devices, includes low fabrication costs, minimal power requirements, basic designs, facile user interfaces, simple miniaturization, robust measurements, low detection limits and small operating volumes. However, they also reduce turbidity-based biases which are common in conventional optical systems [30] (Fig. 14.2).

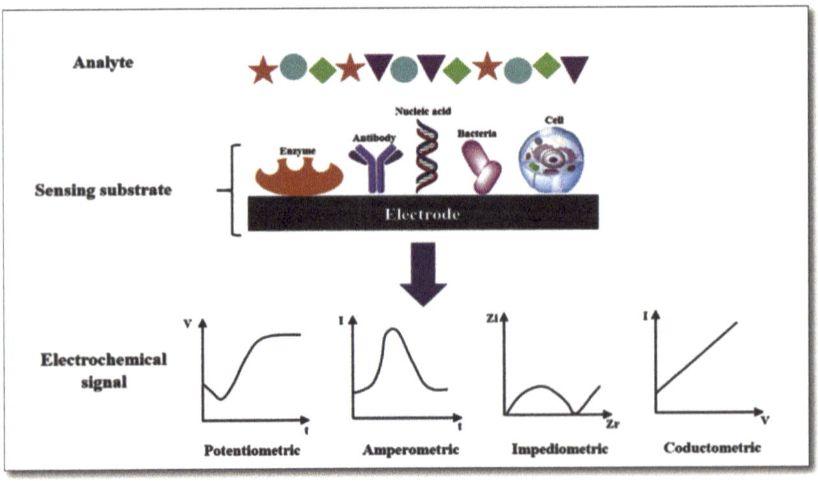

Figure 14.2 Schematic representation of an electrochemical biosensor [29].

14.4 CRISPR/Cas

CRISPR is an inherent bacterial defense scrutiny that assists in the identification of unknown genetic signatures, such as bacteriophages. Using CRISPR-associated proteins (Cas), the bacteria fragmentizes foreign nucleic acid and thus has rendered protection from the bioparticle.

CRISPR-associated protein is an emerging technology for gene editing and diagnostics development. CRISPR/Cas system has been differentiated into types and subclasses on the basis of its loci and signatures of protein. Although employed for genome altering operations, the CRISPR/Cas effectors have entered the biosensing forefront.

14.5 Crispr/Cas biosensing systems

It is an established fact that electrochemical biosensing of viruses is not an easy climb and is therefore considered ambitious. However, electrochemistry provides several well-known advances for biosensor design, such as innate high sensitivity, relatively low-cost sensors and equipment, user-friendly manipulation, fast analysis, and suitability for miniaturization, hence creating POC devices [1,31].

Nucleic acids are important targets for molecular diagnostic techniques, detecting infectious diseases, antimicrobial resistance genes, mutations associated with cancer or genetic diseases (e.g., Huntington, CF and Duchenne muscular dystrophy), single nucleotide polymorphism (SNPs) and biomarkers (miRNA) [32]. Several efficient and effective nucleic acid diagnostic kits have been formulated and authenticated using Cas9, Cas12, and Cas13 proteins [33].

CRISPR/Cas systems have displayed remarkable potential in developing novel biosensing applications for macromolecule detection due to the collateral cleavage activity of Cas effector proteins (Cas12, Cas13, etc.).

Owing to their sequence programmability, much research has focused on the use of CRISPR-Cas technologies for improved accuracy in the development of disease treatment and diagnostic platforms. Despite the potential risk of nonspecific RNA targeting effects, it is believed that the CRISPR-Cas13 system presents a better level of specificity over the available RNA-targeting approaches. Continued development of Cas variants with increased specificities and activities is critical for the application of these technologies to viral infection control [34]. Diagnostics that require processing of samples in a well-equipped centralized laboratory in resource-limited settings, where a lack of infrastructure and skilled medical personnel preclude effective and fast diagnosis. Hence, attempts for engineering of reliable and rapid diagnostic tests that can be conducted outside the clinical laboratory is vital for effective disease treatment and management. Development of rapid initial screening tests that can be used at the POC will minimize the delay and expense of testing in centralized laboratories [1] (Fig. 14.3).

The latest finding of the collateral RNA cleavage property of the Cas13a effector has ignited an unparalleled interest in developing novel biosensing technologies for nucleic acid detection, promising significant advances in CRISPR diagnostics. Now, alongside the invention of Cas12 collateral cleavage activities on single-stranded DNA (ssDNA), several CRISPR/Cas systems have been established for detecting various targets, including bacteria, viruses, cancer mutations, etc. Based on the key Cas effectors, we offer an in-depth classification of CRISPR/Cas biosensing systems and propose their future utility. As the field continues to mature, CRISPR/Cas systems have the potential to become promising candidates for next-generation diagnostic biosensing platforms [35].

A study of the sort VI CRISPR/Cas system demonstrated the collateral cleavage enzymatic activity of Cas13a after its target-specific binding. This is evident in the study of Bruch et al. [18] to develop an electrochemical macromolecule (miRNA) biosensor. Collateral cleavage-based CRISPR/Cas technology has the potential to significantly affect the sector of biosensors by offering a more rapid and precise method for ultrasensitive macromolecule detection. Talk of creating CRISPR/Cas-based deployable POC paper devices are in the works. Crispr /Cas has gauged

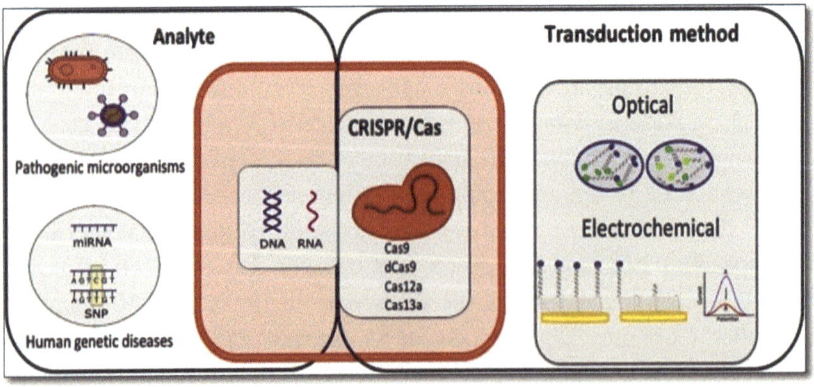

Figure 14.3 Electrochemical biosensor of CRISPR/Cas [32]. *CRISPR/Cas*, Clustered regularly interspaced short palindromic repeats and associated nuclease proteins.

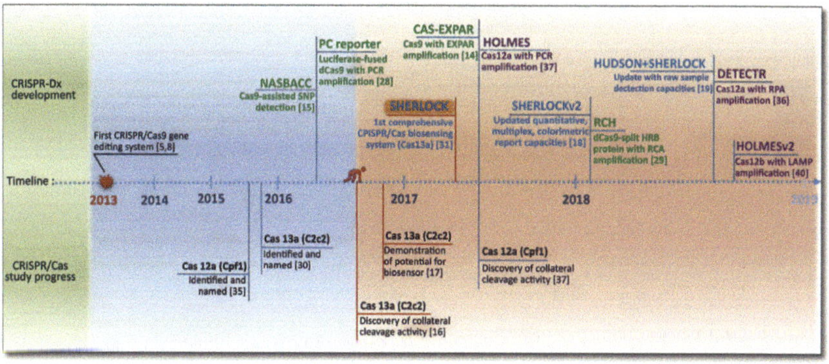

Figure 14.4 The development of CRISPR/Cas biosensing technology [35]. *CRISPR/Cas*, Clustered regularly interspaced short palindromic repeats and associated nuclease proteins.

the interest of many scientists from all walks of life implicated through its compressive publication in the recent years [31] (Fig. 14.4).

Despite significant progress in recent years, the prevailing CRISPR/Cas-based biosensing platforms have several limitations, including dependence on proper amplification methods, overpriced fluorescence detection equipment, or lateral flow biosensor (LFBMolecular methods for detecting nucleic acids based on CRISPR-Cas systems appear to be highly sensitive, specific, and capable of one-step detection of both RNA and DNA) [4]. CRISPR-Cas, a CRISPR-associated nuclease that can promiscuously cleaves RNAs, enables a rapid and inexpensive test for single-molecule detection and single-base discrimination of nucleic acids. This technique can also detect bacteria, microRNAs and cancer mutations, in a simple and easily scalable manner, by merely changing target-specific crRNA/sgRNA. The CRISPR–Cas12-based lateral flow assay technique is easy to implement and is an accurate and good replacement for real-time diagnosis by reverse transcription PCR (RT-PCR) [36,37]. More importantly, they have been successfully integrated in POC devices with performances like routine techniques [32]. The CRISPR-Cas system has recently shown promise in identifying nucleic acids during a POC testing format, which might remove the necessity for expensive laboratory equipment and would offer rapid and cost-effective detection with high sensitivity and specificity in various RNA and DNA samples. Fig. 14.5 summarizes various CRISPR-Cas diagnostic biosensing systems and their properties of such including SHERLOCK, SHERLOCKv2, HOLMES, HOLMESv2, DETECTR, CAS-EXPAR, NASBACC, STOPCovid, ctPCR, and AIOD-CRISPR, are developed [38,39].

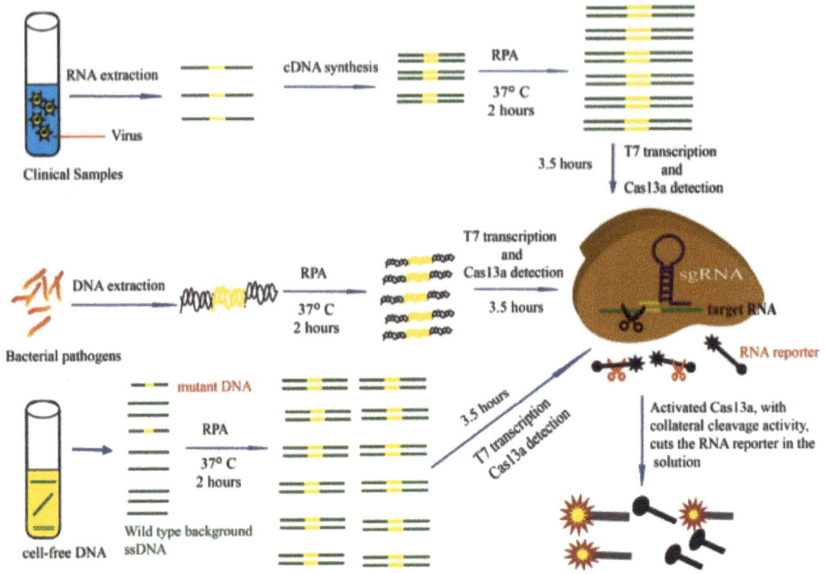

Figure 14.5 Schematic for SHERLOCK system. Target sequence detection involves the following steps: (1) isothermal amplification of target sequence with RPA or RT-RPA, (2) transcription of amplified DNA to RNA with T7 RNA polymerase, (3) activation of Cas13a by identifying and cutting target sequence, and (4) activated Cas13a, with collateral cleavage activity cuts the RNA reporter in the solution of nucleic acid biosensing systems developed using CRISPR/Cas [38].

14.6 Mechanism of RNA targeting by CRISPR/Cas13

CRISPR/Cas class II type VI system targets RNA elements with help of gRNA and is also termed as Cas13. Cas13 enzymes protect prokaryotes from RNA elements like RNA phages. The size of Cas13 family enzymes is between 900–1300 amino acids [40]. Cas13 family is furthered divided into 4 types such as Cas13a (C2c2), Cas13b (C2c6), Cas13c (C2c7), and Cas13d. CRISPR/Cas13 targets and cleaves particularly ssRNA.

Cas13 possesses RNA-guided RNase activity with the help of two Higher Eukaryotic and Prokaryotic Nucleotide-binding (HEPN) domains. Cas13 family contains HEPN domains that process pre-CRISPR RNA into mature crRNAs. The crRNA-Cas13 complex searches Protospacer by scanning the ssRNA. These Protospacer Flanking Sequences (PFS) help the Cas13 complex to find and cleave the target ssRNA. When the crRNA-Cas13 complex binds with ssRNA then the HEPN1 and HEPN2 domains of the Cas13 complex undergo conformational changes which bring both HEPN domains close to each other. This conformational change causes activation of the HEPN nuclease site. HEPN domains combine to form an RNA-guided RNA cleavage complex. This Cas13 complex cleaves target RNA under cis-cleavage activity and nonspecific RNA under trans-cleavage activity also called collateral cleavage activity. Cas13 cleaves specifically at the Uridine (U) and Adenosine (A) rich regions. Sometimes Cas13 cleaves host RNAs. Collateral cleavage activity helps to develop specific and sensitive pathogen detection methods [28]. In this catalytic cleavage, RNA which are complementary to the target sequence is cleaved by Cas13 complex and also triggers the catalytic cleavage of nonspecific RNA molecules in trans-cleavage [41]. The cis-cleavage mechanism of target RNA is still unclear.

The activation of the Cas13 complex after binding with target RNA sequence causes a collateral cleavage mechanism which is a promising characteristic of Cas 13 enzymes. With the help of this mechanism, the Cas13 complex can be used for the diagnosis of miRNA. A cas13 complex is a tool used for the optical detection of viral RNAs. The CRISPR/Cas13 technology can be successfully used to create electrochemical biosensors which help in measuring miRNA levels in serum samples of patients.

14.7 Electrochemical biosensor for RNA detection

For rapid and early diagnosis of pathogenic diseases, different types of samples are used. Detection of pathogens based on genetic material such as DNA/RNA plays a crucial role in the diagnosis and accurate treatment of diseases. Scientists created different types of biosensors for the detection of nucleic acids which involves various detection techniques such as fluorescent detection, colorimetric detection, and electrochemical detection. Some electrochemical biosensors use the CRISPR/Cas13 mechanism for miRNA detection.

An electrochemical biosensor for miRNA-21 was developed by Cui et al. [12] combining CRISPR/Cas13 system and catalytic hairpin assembly (CHA) [12]. CHA reaction used two different DNA strands for the preparation of stable hairpin DNA structures. Target miRNA hybridizes with the spacer region of Cas13-crRNA complex. This hybridization activates trans-cleavage activity of CRISPR-Cas13 complex and cleaves hairpin DNA which produces amplified electrochemical signals with the help of a CHA reaction [12]. This electrochemical biosensor shows a good linear relationship of miRNAs with logarithmic concentrations from 10 fM to 1 nM with a detection limit of 2.6 fM. This ultrasensitive biosensor shows better results than previously reported analytical methods [28]. Cui et al. used this CRISPR-Cas13 and CHA cascade signal amplification method to develop an ultrasensitive electrochemical biosensor for miRNA-21 detection [12].

Bruch et al. (2019) were the first scientists to successfully develop a microfluidic electrochemical biosensor for the detection of miRNA with the help of CRISPR-Cas13. This CRISPR-Cas 13 based microfluidic electrochemical biosensor chip can be used for on-site miRNAs detection. In this method, streptavidin (SA) molecules are applied on-chip to functionalize it and a vacuum is used to remove unbound molecules from the chip while Bovine Serum Albumin (BSA) is used to block remaining surface channels to prevent adsorption of unnecessary biomolecules. For CRISPR-Cas13, a complex mechanism and collateral cleavage of reporter RNA (reRNA), CRISPR-Cas 13 effector molecule with target-specific crRNA, 6-FAM-biotin labeled reRNA, and sample with target miRNA are mixed separately. Then, this mixture is applied on a microfluidic chip. If the sample does not contain target miRNAs then the CRISPR-Cas13 complex fails to cleave 6-FAM-biotin-labeled reRNA and bound to SA molecules on the surface channels. Then GOx (glucose oxidase)—conjugated anti-6-FAM-antibody binds with 6-FAM-biotin labeled reRNA which triggers an enzymatic reaction in which glucose is converted into hydrogen peroxide (H_2O_2). This H_2O_2 is amperometric and is detected with the help of an electrochemical cell. The amount of amperometric signals produced by the enzymatic reaction is directly proportional to the amount of GOx-conjugated anti-6-FAM-antibody bound to uncleaved 6-FAM-biotin-labeled reRNA and hence inversely proportional to the number of target miRNAs present in the sample. If the sample contains target miRNAs, then CRISPR-Cas13 complex triggers the cleavage of 6-FAM-biotin-labeled reRNA due to which 6-FAM-antibody will not bind and no enzymatic reaction will occur. This method helps in the detection of miRNAs levels in brain tumor patients.

These CRISPR/Cas13-based RNA detection technologies provide high sensitivity, are inexpensive, and require less time. CRISPR-Cas13-based electrochemical biosensors were successfully used for onsite miRNA detection which helps in the detection of early-stage cancers and other life-threatening pathogenic diseases.

14.8 Applications

CRISPR/Cas13-based RNA targeting tools show promising results in medical research, biotechnology, and therapeutic industries. These RNA targeting tools are especially used as diagnostic tools.

Electrochemical biosensors based on CRISPR/Cas13 systems are used to detect and target RNA viruses. Bruch et al. (2019) developed a CRISPR/Cas13a-based low-cost electrochemical microfluidic biosensor for the detection of tumor markers such as miR-19b and miR-20 from serum samples of children [42].

Other applications of the CRISPR/Cas13 system show promising results in diagnosis and as a defense mechanism Lu et al. [42] and Cox et al. [43] developed a tool with the help of Cas13b termed REPAIRv2 (RNA Editing for Programmable A to I Replacement version 2). This REPAIRv2 is useful to study G to A disease-relevant mutations and also used in the treatment. RESCUE is an RNA editing tool used for the conversion of cytidine to uridine. Live imaging of RNAs in living cells is beneficial to understand their function. For such live imaging, RNA-guided RNA targeting Cas13 RNases (dCas13) are used by Yang and co-workers (2019). Yang et al. [44] combined dCas13 and Ms2-MCP for dual-color imaging of RNAs in single cells. Wang et al. [45] developed a new tool termed Live-cell Fluorescent in situ hybridization (LiveFISH) by combining Cas9 and Cas13 systems for real-time live imaging of both DNA and RNA in living cells. Mahas et al. studied the CRISPR/Cas13 system to target RNA viruses in plants [46]. They constructed CRISPR/Cas13d (CasRx) which can target one RNA virus at a time or else can target two RNA viruses simultaneously and made this helpful to study plant virology easily. Freije et al. developed Cs13 assisted restriction of viral expression and readout (CARVER) [47]. CARVER is used to detect RNA viruses that can infect humans and cleave ssRNA with the help of collateral activity of Cas13 [47]. Abbott et al. [48] demonstrated a CRISPR/Cas13 based system that helps to degrade RNA from live influenza virus (IAV) and SARS-CoV-2 in epithelial cells of the human lung. This helps to decrease the viral load of IAV from the respiratory epithelial cells of human lungs. Tng et al. discovered *Prevotella* sp. *P5–125* CRISPR/Cas13b system used to target RNA viruses of chikungunya (CHIKV) in mosquitoes [49].

CRISPR/Cas13 system requires a lot of studies so that this mechanism can be used to target various life-threatening RNA viruses. New research and study of CRISPR/Cas13 will help for RNA manipulation, transcriptome engineering, RNA imaging, and nucleic acid detection.

14.9 Conclusion

CRISPR/Cas system is divided into different types and subtypes from which some are used as diagnostic tools. Collateral cleavage activity of

CRISPR/Cas13 specifically cleaves ssRNA transcripts. The electrochemical biosensor based on a combination of CRISPR/Cas13 with enzymatic reaction is used for miRNA detection in patients' body fluids. For such diagnoses, miRNAs serve as biomarkers in the clinical field. The new technologies based on CRISPR/Cas13 electrochemical biosensors can be used for early, on-site, accurate, and sensitive diagnosis to provide effective treatment to patients. Different types of CRISPR/ Cas13 electrochemical biosensors are successfully used for the detection of RNA viruses and tumors with the help of enzymatic reactions such as CHA, Glucose-oxidase (GOx)—H_2O_2 microfluidic system, and catalytic hairpin DNA circuit. Furthermore, the discoveries in CRISPR/Cas technology will transform clinical and biotechnological research. CRISPR is an emerging technology that will undoubtedly help to combat viral outbreaks and antibiotic-resistant pathogenic species in the near future.

In a nutshell, CRISPR/Cas systems provide biosensing ease with its diagnostic biodata. Scientists are hopeful that with time and more research on the CRISPR/Cas systems, many insights are promised and minor shortcomings can be resolved. These trying times of COVID-19 serve as a reminder to the healthcare community for the necessity in maturation of diagnostic setups.

References

[1] R. Aman, A. Mahas, M. Mahfouz, Nucleic acid detection using CRISPR/Cas biosensing technologies, ACS Synth. Biol. 9 (6) (2020). Available from: https://doi.org/10.1021/acssynbio.9b00507.

[2] L. Akin, M.G. Gözel, Understanding dynamics of pandemics, Turkish J. Med. Sci. 50 (SI-1) (2020). Available from: https://doi.org/10.3906/sag-2004-133.

[3] J.P. Broughton, X. Deng, G. Yu, C.L. Fasching, V. Servellita, J. Singh, et al., CRISPR–Cas12-based detection of SARS-CoV-2, Nat. Biotechnol. 38 (7) (2020). Available from: https://doi.org/10.1038/s41587-020-0513-4.

[4] A. Kostyusheva, S. Brezgin, Y. Babin, I. Vasilyeva, D. Glebe, D. Kostyushev, et al., CRISPR-Cas systems for diagnosing infectious diseases, Methods (2021). Available from: https://doi.org/10.1016/j.ymeth.2021.04.007.

[5] M.I. Mustafa, A.M. Makhawi, Sherlock and detectr: CRISPR-cas systems as potential rapid diagnostic tools for emerging infectious diseases, J. Clin. Microbiol. 59 (3) (2021) 1–30. Available from: https://doi.org/10.1128/JCM.00745-20.

[6] C.S. Kosack, A.L. Page, P.R. Klatser, A guide to aid the selection of diagnostic tests, Bull. World Health Organ. 95 (9) (2017). Available from: https://doi.org/10.2471/BLT.16.187468.

[7] S.I. van Dongen, K. de Nooijer, J.M. Cramm, A.L. Francke, W.H. Oldenmenger, I.J. Korfage, et al., Self-management of patients with advanced cancer: a systematic review of experiences and attitudes, Palliat. Med. 34 (2) (2020) 160–178. Available from: https://doi.org/10.1177/0269216319883976.

[8] C.N. Correia, N.C. Nalpas, K.E. McLoughlin, J.A. Browne, S.V. Gordon, D.E. MacHugh, et al., Circulating microRNAs as potential biomarkers of infectious disease, Front. Immunol. 8 (2017). Available from: https://doi.org/10.3389/fimmu.2017.00118.
[9] J. Hausser, M. Zavolan, Identification and consequences of miRNA-target interactions-beyond repression of gene expression, Nat. Rev. Genet. 15 (9) (2014). Available from: https://doi.org/10.1038/nrg3765.
[10] W. Filipowicz, S.N. Bhattacharyya, N. Sonenberg, Mechanisms of post-transcriptional regulation by microRNAs: Are the answers in sight? Nat. Rev. Genet. 9 (2) (2008). Available from: https://doi.org/10.1038/nrg2290.
[11] J.A. Weber, D.H. Baxter, S. Zhang, D.Y. Huang, K.H. Huang, M.J. Lee, et al., The microRNA spectrum in 12 body fluids, Clin. Chem. 56 (11) (2010). Available from: https://doi.org/10.1373/clinchem.2010.147405.
[12] Y. Cui, S. Fan, Z. Yuan, M. Song, J. Hu, D. Qian, et al., Ultrasensitive electrochemical assay for microRNA-21 based on CRISPR/Cas13a-assisted catalytic hairpin assembly, Talanta 224 (2021) 121878. Available from: https://doi.org/10.1016/j.talanta.2020.121878.
[13] G. Cheng, Circulating miRNAs: roles in cancer diagnosis, prognosis and therapy, Adv. Drug Deliv. Rev. 81 (2015). Available from: https://doi.org/10.1016/j.addr.2014.09.001.
[14] R. Samson, G.R. Navale, M.S. Dharne, Biosensors: frontiers in rapid detection of COVID-19, 3 Biotech 10 (9) (2020). Available from: https://doi.org/10.1007/s13205-020-02369-0.
[15] M.L. Kovarik, D.M. Ornoff, A.T. Melvin, N.C. Dobes, Y. Wang, A.J. Dickinson, et al., Micro total analysis systems: fundamental advances and applications in the laboratory, clinic, and field, Anal. Chem. 85 (2) (2013). Available from: https://doi.org/10.1021/ac3031543.
[16] Y. Sheng, T. Zhang, S. Zhang, M. Johnston, X. Zheng, Y. Shan, et al., A CRISPR/Cas13a-powered catalytic electrochemical biosensor for successive and highly sensitive RNA diagnostics, Biosens. Bioelectron. 178 (2021). Available from: https://doi.org/10.1016/j.bios.2021.113027.
[17] J.E. van Dongen, J.T.W. Berendsen, R.D.M. Steenbergen, R.M.F. Wolthuis, J.C.T. Eijkel, L.I. Segerink, Point-of-care CRISPR/Cas nucleic acid detection: recent advances, challenges and opportunities, Biosens. Bioelectron. 166 (2020). Available from: https://doi.org/10.1016/j.bios.2020.112445.
[18] R. Bruch, M. Johnston, A. Kling, T. Mattmüller, J. Baaske, S. Partel, et al., CRISPR-powered electrochemical microfluidic multiplexed biosensor for target amplification-free miRNA diagnostics, Biosens. Bioelectron. 177 (2021). Available from: https://doi.org/10.1016/j.bios.2020.112887.
[19] S. Konermann, M.D. Brigham, A.E. Trevino, J. Joung, O.O. Abudayyeh, C. Barcena, et al., Genome-scale transcriptional activation by an engineered CRISPR-Cas9 complex, Nature 517 (7536) (2015). Available from: https://doi.org/10.1038/nature14136.
[20] Y. Ishino, H. Shinagawa, K. Makino, M. Amemura, A. Nakatura, Nucleotide sequence of the IAP gene, responsible for alkaline phosphatase isoenzyme conversion in *Escherichia coli*, and identification of the gene product, J. Bacteriol. 169 (12) (1987). Available from: https://doi.org/10.1128/jb.169.12.5429-5433.1987.
[21] K.G. Andersen, A. Rambaut, W.I. Lipkin, E.C. Holmes, R.F. Garry, The proximal origin of SARS-CoV-2, Nat. Med. 26 (4) (2020). Available from: https://doi.org/10.1038/s41591-020-0820-9.
[22] L. Chou-Zheng, A. Hatoum-Aslan, A type iii-a CRISPR-Cas system employs degradosome nucleases to ensure robust immunity, ELife 8 (2019). Available from: https://doi.org/10.7554/eLife.45393.
[23] M. Hryhorowicz, D. Lipiński, J. Zeyland, R. Słomski, CRISPR/Cas9 immune system as a tool for genome engineering, Archivum Immunologiae et. Therapiae Experimentalis 65 (3) (2017). Available from: https://doi.org/10.1007/s00005-016-0427-5.

[24] F. Hille, E. Charpentier, CRISPR-cas: biology, mechanisms and relevance, Philos. Trans. R. Soc. B: Biol. Sci. 371 (1707) (2016). Available from: https://doi.org/10.1098/rstb.2015.0496.
[25] D. Zhang, Y. Yan, H. Que, T. Yang, X. Cheng, S. Ding, et al., CRISPR/Cas12a-mediated interfacial cleaving of hairpin DNA reporter for electrochemical nucleic acid sensing, ACS Sens. 5 (2) (2020). Available from: https://doi.org/10.1021/acssensors.9b02461.
[26] O.O. Abudayyeh, J.S. Gootenberg, P. Essletzbichler, S. Han, J. Joung, J.J. Belanto, et al., RNA targeting with CRISPR-Cas13, Nature 550 (7675) (2017). Available from: https://doi.org/10.1038/nature24049.
[27] I. Fonfara, H. Richter, M. BratoviÄ, A. Le Rhun, E. Charpentier, The CRISPR-associated DNA-cleaving enzyme Cpf1 also processes precursor CRISPR RNA, Nature 532 (7600) (2016). Available from: https://doi.org/10.1038/nature17945.
[28] J.T. Granados-Riveron, G. Aquino-Jarquin, Crispr/cas13-based approaches for ultra-sensitive and specific detection of micrornas, Cells 10 (7) (2021). Available from: https://doi.org/10.3390/cells10071655.
[29] A.A. Ensafi, Chapter 1 – An introduction to sensors and biosensors, Electrochemical Biosensors, Elsevier Inc, 2019. Available from: https://doi.org/10.1016/B978-0-12-816491-4.00001-2.
[30] A.J. Veloso, X.R. Cheng, Electrochemical biosensors for medical applications, in: *Biosensors for Medical Applications*, Woodhead Publishing, 2012, pp. 3–40.
[31] J. Kudr, P. Michalek, L. Ilieva, V. Adam, O. Zitka, COVID-19: a challenge for electrochemical biosensors, TrAC. – Trends Anal. Chem. 136 (2021) 116192. Available from: https://doi.org/10.1016/j.trac.2021.116192.
[32] A. Bonini, N. Poma, F. Vivaldi, A. Kirchhain, P. Salvo, D. Bottai, et al., Advances in biosensing: the CRISPR/Cas system as a new powerful tool for the detection of nucleic acids, J. Pharm. Biomed. Anal. 192 (2021) 113645. Available from: https://doi.org/10.1016/j.jpba.2020.113645.
[33] S. Srivastava, D.J. Upadhyay, A. Srivastava, Next-generation molecular diagnostics development by CRISPR/Cas tool: rapid detection and surveillance of viral disease outbreaks, Front. Mol. Biosci. 7 (2020). Available from: https://doi.org/10.3389/fmolb.2020.582499.
[34] A. Ghorbani, S. Hadifar, R. Salari, K. Izadpanah, M. Burmistrz, A. Afsharifar, et al., A short overview of CRISPR-Cas technology and its application in viral disease control, Transgenic Res. 30 (3) (2021). Available from: https://doi.org/10.1007/s11248-021-00247-w.
[35] Y. Li, S. Li, J. Wang, G. Liu, CRISPR/Cas systems towards next-generation biosensing, Trends Biotechnol. 37 (7) (2019) 730–743. Available from: https://doi.org/10.1016/j.tibtech.2018.12.005.
[36] O. Mukama, J. Wu, Z. Li, Q. Liang, Z. Yi, X. Lu, et al., An ultrasensitive and specific point-of-care CRISPR/Cas12 based lateral flow biosensor for the rapid detection of nucleic acids, Biosens. Bioelectron. 159 (2020). Available from: https://doi.org/10.1016/j.bios.2020.112143.
[37] P. Ranjan, A. Singhal, S. Yadav, N. Kumar, S. Murali, S.K. Sanghi, et al., Rapid diagnosis of SARS-CoV-2 using potential point-of-care electrochemical immunosensor: toward the future prospects. In, Int. Rev. Immunology 40 (1–2) (2021). Available from: https://doi.org/10.1080/08830185.2021.1872566.
[38] M. Vatankhah, A. Azizi, A. Sanajouyan Langeroudi, S. Ataei Azimi, I. Khorsand, M. A. Kerachian, et al., CRISPR-based biosensing systems: a way to rapidly diagnose COVID-19, Crit. Rev. Clin. Lab. Sci. 0 (0) (2020) 1–27. Available from: https://doi.org/10.1080/10408363.2020.1849010.

[39] M. Vatankhah, A. Azizi, A. Sanajouyan Langeroudi, S. Ataei Azimi, I. Khorsand, M. A. Kerachian, et al., CRISPR-based biosensing systems: a way to rapidly diagnose COVID-19, Crit. Rev. Clin. Lab. Sci. 58 (4) (2021) 225–241. Available from: https://doi.org/10.1080/10408363.2020.1849010.

[40] M.M. Kaminski, O.O. Abudayyeh, J.S. Gootenberg, F. Zhang, J.J. Collins, CRISPR-based diagnostics, Nat. Biomed. Eng. 5 (7) (2021). Available from: https://doi.org/10.1038/s41551-021-00760-7.

[41] A. Varble, L.A. Marraffini, Three new Cs for CRISPR: collateral, communicate, cooperate, Trends Genet. 35 (6) (2019). Available from: https://doi.org/10.1016/j.tig.2019.03.009.

[42] L. Lu, L. Zhou, J. Chen, F. Yan, J. Liu, X. Dong, et al., Nanochannel-confined graphene quantum dots for ultrasensitive electrochemical analysis of complex samples, ACS Nano (2018). Available from: https://doi.org/10.1021/acsnano.8b07564.

[43] D.B.T. Cox, J.S. Gootenberg, O.O. Abudayyeh, B. Franklin, M.J. Kellner, J. Joung, F. Zhang, RNA editing with CRISPR-Cas13, Science 358 (6366) (2017) 1019–1027.

[44] L.Z. Yang, Y. Wang, S.Q. Li, R.W. Yao, P.F. Luan, H. Wu, G.G. Carmichael, L.L. Chen, Dynamic imaging of RNA in living cells by CRISPR-Cas13 systems, Mol. Cell 76 (6) (2019) 981–997.

[45] H. Wang, M. Nakamura, T.R. Abbott, D. Zhao, K. Luo, C. Yu, C.M. Nguyen, et al., CRISPR-mediated live imaging of genome editing and transcription, Science 365 (6459) (2019) 1301–1305.

[46] A. Mahas, R. Aman, M. Mahfouz, CRISPR-Cas13d mediates robust RNA virus interference in plants, Genome Biol. 20 (1) (2019). Available from: https://doi.org/10.1186/s13059-019-1881-2.

[47] C.A. Freije, C. Myhrvold, C.K. Boehm, A.E. Lin, N.L. Welch, A. Carter, et al., Programmable inhibition and detection of RNA viruses using Cas13, Mol. Cell 76 (5) (2019). Available from: https://doi.org/10.1016/j.molcel.2019.09.013.

[48] T.R. Abbott, G. Dhamdhere, Y. Liu, X. Lin, L. Goudy, L. Zeng, A. Chemparathy, et al., Development of CRISPR as an antiviral strategy to combat SARS-CoV-2 and influenza, Cell 181 (4) (2020) 865–876.

[49] P.Y.L. Tng, L. Carabajal Paladino, S.A.N. Verkuijl, J. Purcell, A. Merits, P.T. Leftwich, et al., Cas13b-dependent and Cas13b-independent RNA knockdown of viral sequences in mosquito cells following guide RNA expression, Commun. Biol. 3 (1) (2020). Available from: https://doi.org/10.1038/s42003-020-01142-6.

CHAPTER 15

Biosensing technologies applied in virus detection as rapid tools during pandemics: past lessons and recent trends

Hichem Moulahoum[1,*], Faezeh Ghorbanizamani[1,*], Figen Zihnioglu[1], Tuncay Goksel[2], Kutsal Turhan[3] and Suna Timur[1,4]
[1]Biochemistry Department, Faculty of Sciences, Ege University, Izmir, Turkey
[2]Department of Pulmonary Medicine, Faculty of Medicine, Ege University, Izmir, Turkey
[3]Department of Thoracic Surgery, Faculty of Medicine, Ege University, Izmir, Turkey
[4]Central Research Test and Analysis Laboratory Application and Research Center, Ege University, Izmir, Turkey

15.1 Introduction

Throughout history, humans have faced various pandemics that have caused millions of deaths and significant financial losses [1]. The destruction from such infectious diseases (e.g., tuberculosis, malaria, and human immunodeficiency virus [HIV]) is more pronounced in developing countries [2]. A large amount of yearly research funding, specifically in developed countries, is allocated to find diagnosis tools and control methods for infectious diseases [3]. Many infectious diseases are caused by a pathogen (virus or bacteria) invading host cells. Viruses invade the host cell and use its environment to spread and establish genetic material replication. The host cell response is via immune system activation, however, most viruses are able to circumvent the protection mechanisms by demolishing and manipulating the host's immune reactions. One of the successful methods for virus detection is viral isolation methods. Just like the other detection methods, the viral cultivation method has its advantages and disadvantages. High sensitivity and accuracy are the positive features of this method, while the long waiting period (several days) for results is one of its limitations [4–6]. More recently, virus and bacteria detection techniques such as

* Authors Hichem Moulahoum and Faezeh Ghorbanizamani contributed equally to this chapter.

Biosensors for Emerging and Re-emerging Infectious Diseases
DOI: https://doi.org/10.1016/B978-0-323-88464-8.00004-X

polymerase chain reaction (PCR) [7] and Enzyme-Linked Immunosorbent Assay (ELISA) [8] can be utilized for the direct recognition of pathogen-specific genetic material or proteins. However, both of PCR and ELISA methods require trained individuals, pricey instruments, and long turnarounds [9–11]. As a result, the need for fast, precise, reliable, and small-sized tools for virus recognition is understandable [12].

Among the various available methods, biosensors have shown their high potential due to their pronounced sensitivity and selectivity in the detection of pathogens [13,14]. The biosensor concept was first introduced in the 1960s via a glucose detector using an oxidase enzyme electrode [15]. Later, with scientific advancements and the use of nanotechnology, biosensors have become more renowned and their utilization has increased in various fields including biomedicine, drug delivery, environmental and food sciences [16–20]. The success of using a DNA-based piezoelectric biosensor for the diagnosis of human papillomavirus (HPV) in 2007 has attracted the attention of the international scientific community in using DNA biosensors for clinical studies [21]. The use of biosensors for clinical viral and bacterial pathogen detection has demonstrated the superiority of these devices compared to previous methods (portability, speed, and accuracy) [22–24]. Biosensors have been effectively employed in the diagnosis of human-infecting viruses (Ebola, HIV, and the newest SARS-CoV-2) and bacteria (*Escherichia coli* and *Salmonella*) [25–28]. In this chapter, a comprehensive and integrative update of the use of biosensor techniques to diagnose infectious diseases is presented. The chapter aims to introduce biosensors-related novel technologies employed for microbe detection with special consideration given to nanotechnological materials applied such as nanoparticles (gold, magnetic, polymeric) and electrospun nanofibers (carbon, or graphene). A special interest will be given to the COVID-19 pandemic and the current state of diagnostics.

15.2 Nanomaterials-enabled biosensors

Nanotechnology and nanoscience developments have been accompanied with improved methods for identifying biomolecules as well as biosensing devices. Nanomaterials such as nanoparticles, nanotubes, nanorods, and nanowires with superior features like excellent electrical conductivity, tunable optical, electrical, and magnetic properties have been applied for faster responses and higher sensitivity of devices to the target molecules. However, some important challenges in biosensor structures are the detection difficulty of a small amount of analytes as well as transmitting the signal between bioreceptor and analyte samples. The application of labeled nanomaterials has significantly resolved this problem and enhanced signal intensity for the detection of low concentrated analytes. Metallic nanoparticles (from gold or silver) and quantum dots (QDs), like cadmium and lead-based, have been proposed as labels for DNA probing [29,30]. The electrochemical signal amplification resulting from using nanolabels increases the sensitivity and selectivity of the systems and by taking advantage of nanomaterials, biosensors with the ability to detect virus-mediated infections are advancing [30]. Depending on the targeting mechanism, these types of sensing apparatuses are divided into nucleic acids, antibodies, aptamers, and antigen-based biosensors [31].

15.2.1 Gold nanostructures-enabled biosensing

Gold nanoparticles (AuNPs) possessing great biocompatibility, optical and electrochemical features, and simple fabrication and modification methods are a logical choice for sensing applications. The higher surface energy of AuNPs compared to other nanoparticles (NP) makes them an attractive candidate to immobilize different kinds of biomolecules. The conductivity of AuNPs also leads to rapid and direct electron transfer across the electroactive species and electrode constituents. Moreover, the magnetic and light scattering properties of AuNPs can promote the signal intensity and as a result, the sensitivity of the system (Fig. 15.1) [32]. Consequently, AuNPs can be used to enhance the performance of optical, electrochemical, and piezoelectric biosensors. Table 15.1 shows some recent applications of AuNPs in virus detection.

15.2.2 Lanthanide-based biosensors

The conversion of long wavelengths into short-wavelength emissions can be performed by upconversion nanoparticles (UCNPs) that absorb photons and transform energy [53]. Lanthanide (Ln^{3+}) with its superior electronic structure in crystal form, can effectively emit photons from near-infrared (NIR) photons (low-energy) into the visible spectrum. As a result, Ln^{3+}-doped UC materials that have both efficient energy-transfer features and nanotechnology aspects have gained a growing interest in the fabrication of solid-state lasers, flat panel displays, optical fiber-based communications, photonic devices, and bioapplications [54]. Among all these applications, Ln^{3+}-doped NPs are most remarkable because of their long luminescence lifetimes, and visible and pure color which makes them interesting luminescence-based sensors [55]. More

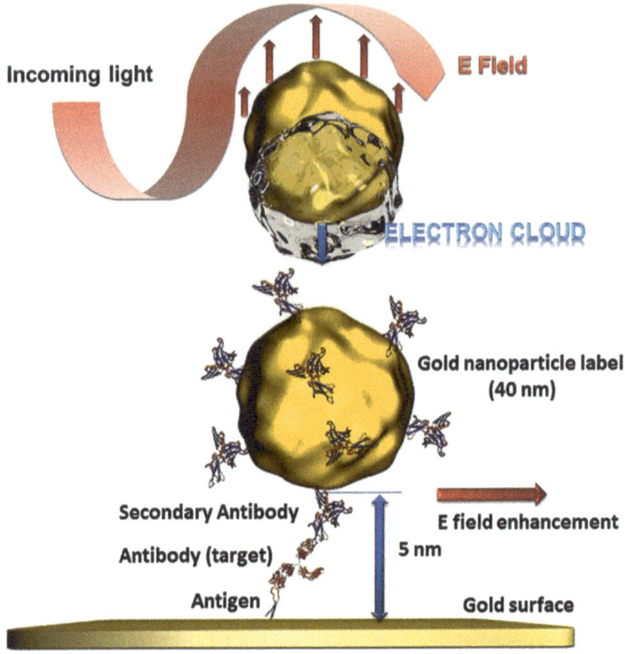

Figure 15.1 Illustration of the polarization of gold nanoparticles and their application in Au-based biosensing. *Reused from M. Holzinger, A. Le Goff, S. Cosnier, Nanomaterials for biosensing applications: a review, Front. Chem. 2 (2014) 63. © 2014 Holzinger, Le Goff and Cosnier under open access license.*

recently, Ln^{3+}-doped NPs have been employed for designing an LFA-based detector for infectious agents [56]. Chen et al. proposed an LFA-based SARS-CoV-2 detection system by using modified lanthanide-doped polystyrene NPs as the florescent probe. The comparison of the LFA results with the RT-PCR data showing close similarity, which confirms the accuracy of this method for COVID-19 diagnosis [57]. In another study, a luminescence resonance energy transfer (LRET) biosensor was constructed to detect avian influenza virus (AIV) Subtype H7. The system used amine-functionalized lanthanide-doped NaYF 4 UCNPs and demonstrated a detection ranging between 10 pM and 10 nM with an LOD of 7.0 pM [58].

15.2.3 Graphene-enabled biosensing

Since the discovery of graphene (G) in 2004 and its superior features, it has been widely employed in different applications including the structure of sensors [59]. Graphene's unique electrochemical characteristics (high conductivity,

Table 15.1 Some recent development of gold nanoparticles based biosensors in virus detection.

Virus	Target	Biomolecule	Assay	References
Bunyaviridae HTNV	NC protein	DNA	IPCR	[33]
Bunyaviridae RVFV	Capsid antigen	Antibody	SERS	[34]
Filoviridae EBOV	NP gene	DNA	Scanometric	[35]
Coronaviridae SARS	PP1ab gene	None	Colorimetric	[36]
	NC protein	DNA	Electrochemical	[37]
Flaviviridae HCV	5' UTR gene	DNA	Electrochemical	[38]
Flaviviridae DENV	E gene	DNA	QCM	[39]
Flaviviridae WNV	HBeAg	Antibody	SERS	[34]
Hepadnaviridae HBV	HBsAg	Antibody	Electrochemical	[40]
	HBV DNA	DNA	Fluorometric	[41]
Herpesviridae HEV	ORF1gene	DNA	Colorimetric	[42]
H5N1	HA gene	Antibody	Colorimetric	[43]
	HA gene	DNA	Electrochemical	[44]
Papillomaviridae HPV	L1 gene	Antibody	Electrochemical	[45]
Retroviridae HIV	p24 antigen	Avidin DNA	DLS	[46]
	pol gene	DNA	Electrochemical	[47]
SARS-CoV-2	E genes	DNA	PPT and LSPR	[48]
	N-gene	DNA	Colorimetric	[49]
	RdRp gene	Oligo probe	Colorimetric	[50]
	Recombinant antigen	Antibody	Colorimetric	[51]
	Monoclonal antibody	Antigen	Amperometric	[52]

5 UTR, Five prime untranslated region; *DENV*, dengue virus; *DLS*, dynamic light scattering; *E gene*, Envelope gene; *EBOV*, Ebola virus; *H5N1*, influenza A virus subtype H5N1; *HA gene*, hemagglutinin gene; *HBeAg*, hepatitis B e-antigen; *HBsAg*, hepatitis B surface antigen; *HBV DNA*, hepatitis B virus DNA; *HBV*, hepatitis B virus; *HCV*, hepatitis C virus; *HEV*, hepatitis E virus; *HIV*, human immunodeficiency virus; *HPV*, human papillomavirus; *HTNV*, Hantaan virus; *ICPR*, immuno-polymerase chain reaction; *L1 gene*, late expressed gene1; *LSPR*, localized surface plasmon resonance; *NC protein*, nucleocapsid protein; *N-gene*, nucleocapsid phosphoprotein; *NP gene*, nucleoprotein gene; *ORF1*, open reading frame1; *P24 antigen*, P24 capsid protein antigen; *PP1ab*, ployprotein1ab; *PPT*, plasmonic photothermal; *QCM*, quartz crystal microbalance; *RdRp*, RNA-dependent RNA polymerase; *RVFV*, Rift Valley fever virus; *SARS-CoV-2*, severe acute respiratory syndrome-related coronavirus; *SERS*, surface-enhanced Raman scattering; *WNV*, West Nile virus.

linear current-voltage, and capacity to immobilize different molecules) are considered great assets in sensor application and several construction methods for graphene preparation have been described. There are nonhomogenous G (with complex assemblies, sizes, oxidation states, layers, and colloidal stability), impure G (consisting of other carbon and metal-based materials, and minerals), and G nanocomposites (consisting organic and inorganic compounds), which are all being used for graphene-based biosensing application [60]. Studies have confirmed that graphene in its normal form (G), modified (oxide (GO), reduced (rGO), functionalized (fG)), or coupled with some derivatives (QD (GQD), N-doped G aerogel) are effective materials in biosensing applications [61]. Functionalized graphene (with nucleic acids or antibodies/proteins/enzymes) has become an important player in biosensing research [62–64]. Biofunctionalized graphene compounds are stable and protected against any interference from the proximate environment which enhances the detection sensitivity and analysis efficiency. Table 15.2 summarizes some of the recently reported functionalized graphene-based biosensors used in the diagnosis of pathogens.

15.2.4 Magnetic nanoparticles-based biosensing

Magnetic nanoparticles (MNPs) with both electrical and magnetic properties are being used in biotechnology, biomedicine, material science, engineering, and environmental fields [92–100]. MNPs are divided into metals (such as iron, cobalt, and nickel) [101], metal oxides (such as iron oxides and ferrites) [102], and composite MNPs (such as FeCo and FePt) [103]. The properties of MNPs (small size, good monodispersity, superparamagnetic behavior, high coercivity, low Curie temperature, and high magnetic susceptibility) make them attractive candidates for the field of biosensing. Recently, biofunctionalized MNPs (including immobilized biomolecules such as antibodies, proteins, enzymes, etc.) have been used in the detection of pathogens such as bacteria and viruses [103–107]. For instance, MNPs had been used in the detection of hemagglutinin (HA) which is a surface glycoprotein of the H5N1 Influenza virus [108]. The polyaniline coated MNP platform used in this study could recognize the antigen at 1.4 mM in serum. Krejcova et al. reported the use of paramagnetic NPs to detect viral protein-labeled QDs [109]. The proposed platform was able to diagnose many influenza sub-types with QDs which is important to treat severe human influenza cases in the early stages to avoid the virus spreading. MNPs-based biosensors have shown promising results

Table 15.2 Recent graphene-based biosensors developed for the diagnosis of pathogens.

Bioreceptors	Biosensor design	Target	Biosensor type	References
Specific cDNA	GO	*Bacillus anthracis* pagA gene	Fluorescence	[65]
	Graphene-Nafion	HIV gene	Electrochemistry	[66]
	Polyaniline-graphene	HIV gene	Electrochemistry	[67]
	MWCNT-NH_2-IL-rGO	HIV gene	Electrochemistry	[68]
	Magnetic rGOCuNCs	HCV subtype gene	Electrochemistry	[69]
	3D Graphene	*Staphylococcus aureus* N gene seq.	Electrochemistry	[70]
	Graphene-AuNPs	*S. aureus* gene seq.	SAW	[71]
Aptamers	rGO-CNT	*Salmonella typhimurium*	Electrochemistry	[72]
	GO	*Pseudomonas aeruginosa*	Electrochemistry	[73]
	rGO-AuNPs	LPS from *Escherichia coli* 055: B5	Electrochemistry	[74]
	GO	LPS	Fluorescence	[75]
	rGO	LPS from *E. coli* 055: B5	Fluorescence	[76]
Peptide/nucleic acids	Graphene-polyaniline	HPV type 16 DNA	Electrochemistry	[77]
	GO-CdS QDs	*Mycobacterium tuberculosis* DNA	Electrochemistry	[78]
	rGO-TEMPONCC	*M. tuberculosis* DNA	Electrochemistry	[79]
Anti-EBOV antibody	rGO	EBOV glycoprotein	Electrochemistry	[80]
PEDV-2C11 Antibody	AuNP–MoS_2-rGO	PEDV	Electrochemistry	[69]
Anti-HIV p24 antibody	Graphene	HIV antigen	Electrochemistry	[81]
Goat anti-*E. coli* O157: H7	AuNPs-Graphene	*E. coli* O157: H7	Electrochemistry	[82]
Rabbit anti-*E. coli* antibody	AuNPs-Graphene	*E. coli* K12 ER2925	Electrochemistry	[83]
DENV E-protein monoclonal antibody	CdS–NH_2GO	DENV E protein	SPR	[84]

(*Continued*)

Table 15.2 (Continued)

Bioreceptors	Biosensor design	Target	Biosensor type	References
McAb against *Salmonella enteritidis* flagellin	Graphene	*S. enteritidis*	Colorimetry	[85]
Spike protein antibody	Graphene	SARS-CoV-2 and MERS-CoV antigen proteins	FET	[86]
Peptide sequence	AgNPs-GO	LPS	Electrochemistry	[87]
	GO	TEV	Fluorescence	[88]
	GO	HIV-1 protease	Fluorescence	[89]
	AgNPs-rGO	*E. coli* O157:H7	SPR	[90]
Bacitracin A	Au@Ag-GO	*E. coli*, *S. aureus* and *P. aeruginosa*	SERS	[91]

AgNPs, Silver nanoparticles; *AuNPs*, gold nanoparticles; *CdS QDs*, cadmium selenide cadmium sulfide quantum dots; *CNTs*, carbon nanotubes; *CuNCs*, copper nanoclusters; *DENV*, Dengue virus; *EBOV*, Ebola virus; *FET*, field-effect transistor-based amperometric biosensing; *GO*, graphene oxide; *HCV*, hepatitis C virus; *HIV*, human immunodeficiency virus; *HPV*, human papillomavirus; *IL*, ionic liquid; *LPS*, lipopolysaccharide; MoS_2, molybdenum disulphide; *MWCNT*, multi walled carbon nanotubes; *NCC*, nanocrystalline cellulose; *PEDV*, porcine epidemic diarrhea virus; *RGO*, reduced graphene oxide; *SAW*, surface acoustic wave; *SERS*, surface-enhanced Raman scattering; *SPR*, surface plasmon resonance; *TEMPO*, 2,2,6,6-tetramethylpiperidine-1-oxyl; *TEV*, tobacco etch virus.

in identifying deadly respiratory viruses [100,106]. Tian et al. suggested the isothermal circle to amplify the magnetic signals of iron oxide NPs used for nucleic acid quantification [110,111]. The suggested platform was effectively applied for the specific detection of COVID-19 with the capability of differentiating between COVID-19 and other coronaviruses [111]. Carboxyl polymer-coated MNPs (pcMNPs) have been proposed for efficient extraction of COVID-19 RNA [107]. The prepared pcMNPs-enabled RT-PCR-based biosensor had a linearity between 10 and 10^5 SARS-CoV-2 virus copies. Moreover, the application of MNPs is also seen in LFA and microfluidic platforms (lab-on-a-chip (LOC) devices). Promising results have been reported which proves the high potential of these materials for developing portable and easy-to-use devices [112].

15.3 Biosensing technologies applied in virus diagnosis

A biosensor is a measurement device that can detect a specific analyte by a biological component and a physicochemical detector combination [113]. In clinics, biosensors are used to detect disease-related biomolecules via monitoring the biological markers in body liquids (blood, saliva, and urine) [114,115]. For example, Zhang et al. successfully developed a real-time salivary glucose tracking sensor. The accuracy of this biosensor has been proved superior to the results from the UV spectrophotometer [116]. There are numerous examples of using biosensors in diagnostics including the detection of cancers, heart diseases, allergies, bacteria, viruses, etc. [117–120]. Regardless of the type of application, selectivity, sensitivity, and fast detection are crucial criteria for a biosensing device [121]. Their performance is defined by their design and purpose. The reliability, accuracy, and robustness of these tools should be taken into consideration during the design and planning of the diagnostic systems [122].

15.3.1 Point-of-care biosensors

Point-of-care testing (POCT) is a combination of effective diagnostic tools for fast and early detection of clinical biomarkers near the patients which predicts potential diseases at an early stage and helping in the management of further treatment steps. Along with healthcare management personalization, POC tools especially biosensors, have attracted tremendous attention for the ability to detect any changes in biological markers or any chemical reaction and converting them into signals. Utilizing nanomaterials with superior features such as optical, magnetic, catalytic, and electric activities has led to the creation of POC biosensors with properties like early detection, high sensitivity, portability, robustness, and affordability [123–126]. Zhang et al. employed a simple and effective dot-blot immunoassay for hepatitis B virus detection. In the proposed system, QD nanobeads (QDNBs) were used as fluorescent amplified signal indicators. The application of the system under UV lamp brought promising results which confirmed the ability of this simple POCT system to diagnose hepatitis B [127]. Numerous research endeavors have considered the great potential of gold and silver nanoparticles in the development of nanodiagnostic tools due to their ability to derive strong absorption following contact with electromagnetic radiation [128]. Darbha et al. developed a POCT system by taking advantage of gold nanorods' optical features to detect HIV. The single-base mismatch in the DNA of the HIV-1 virus was detected efficiently by an intensity change under hyper-Rayleigh scattering (HRS) spectroscopy [129]. Similarly, hepatitis C virus (HCV) was diagnosed using a POCT system based on AuNPs and HRS applications. In this work, rhodamine 6G was used to tag HCV ssRNA and the complex was conjugated onto AuNPs. The system had a HCV ssRNA LOD = 80 pM and was able to selectively detect a single-base mismatch [130].

Chung et al. manufactured a dual probe-nanoparticles system with the ability to detect human pathogens via sandwich hybridization involving two oligonucleotide probes. The 16S rRNA and the constructed probe had been used for amplified DNA detection while the signals had been measured by a micronuclear magnetic resonance (micro-NMR) device. This magneto-DNA platform can be used for various clinically-relevant bacteria-specific detection [131]. A portable diagnostic magnetic resonance (DMR) device was used by Lee et al. to form a platform for unprocessed biological sample measurement. The DMR device showed applicability in sensing and characterizing different types of infections induced by fungi, bacteria, and viruses [132]. This study proved the potential of DMR-based POC nanodiagnostic systems for large-scale screening of future pandemics due to their low-cost and high-throughput. Liong et al. utilized an MNP-based barcode platform for *Mycobacterium tuberculosis* detection [133]. The mycobacterial genes were amplified and captured in an MNP-labeled platform and the portable system boasted high sensitivity, high-throughput, and low-cost [133].

Inci et al. established a nanoplasmonic-based instrument for HIV diagnosis using specific AuNPs and antibodies that can selectively detect viral subtypes. In this platform, the captured viruses induced a shift in signals transmitted by the AuNPs-coated surface. They confirmed the performance of the platform by using different samples including a spiked blood sample or HIV-infected blood from each subtype (HIV A to G). The ability of this platform to quickly capture and isolate viruses for detection and quantification is an important leap in the creation of POCT that are usable at hospital levels [134]. In another study, a sandwich immunoassay was constructed using AuNPs' optoplasmonics features for the detection of capsid antigen p24 of HIV-1 in human serums (Fig. 15.2) [136]. Their

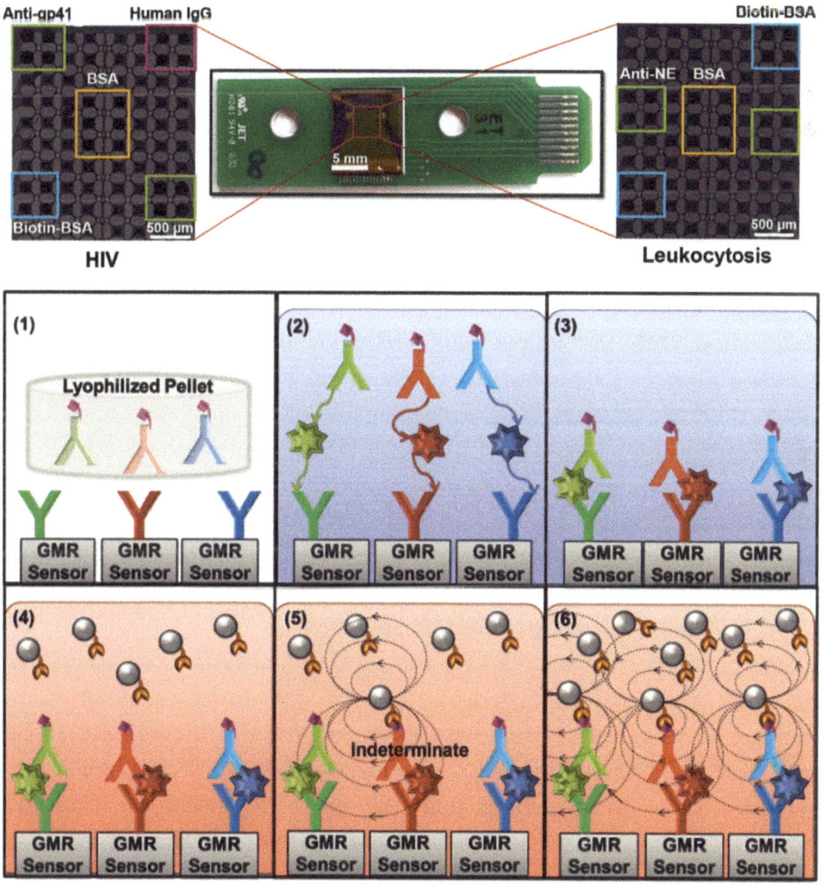

Figure 15.2 Magneto-biosensor POCT system using multiplexed PCB cards for the detection of HIV and leukocytosis. The upper panel demonstrates a photograph of the proposed system with the lower diagram describes the sensing system employed for the detection of the infection/disease. *POCT, Point-of-care testing. Reproduced from E. Ng, et al., Magneto-nanosensor smartphone platform for the detection of HIV and leukocytosis at point-of-care, Nanomedicine 16 (2019) 10–19. ©2019 Elsevier Inc.*

data showed an LOD = 10^{-17} g/mL that corresponds to one virion per 10 mL of plasma. These values are five times more sensitive than the clinically-approved previous generation of immunoassays.

Ng et al. created a magneto-nanosensor POCT platform for the detection of HIV and leukocytosis in saliva and blood, respectively. The nanosensor contains a chip with 80 separate clusters for the detection of multiple markers at the same time. The proposed system is still in the development process and was tested on a low number of samples. However, the addition of a smartphone app attached to the system demonstrates the full potential of the device as a portable diagnostic system. Additionally, on-site detection can be performed in around 15 min with an accuracy ranging between 80% and 90% [135].

15.3.2 Nucleic acid-based biosensors

Nucleic acid-based electrochemical biosensors (NAEBs) have been widely utilized in biological studies [137], diagnosis [138] and therapeutic applications [139]. For a given target pathogen, it is always important to detect the proper biorecognition element [4,121]. Hence, the electrode surface modification with the proper oligonucleotide probes is the key step to successful design of a nucleic-based biosensor. The various types of electroactive compounds such as carbon, graphene, gold, etc., have been applied and examined for NAEBs [140,141]. Several methods have been employed to improve the sensitivity, selectivity, and fast response of the nucleic-based biosensors including hybridization of nucleic probes combined with optical or electrochemical-based sensing platforms [142].

15.3.2.1 Optical nucleic acid-based biosensor

Surface plasmon resonance (SPR) and localized SPR (LSPR)-based biosensors are the most used optical biosensors for nucleic acid detection. The plasma wave interference with the recognition elements through the movement over the surface causes a reduction of the reflected light intensity [143]. The distinguishing point of the SPR biosensors from the LSPRs is the use of metallic nanostructures that function as active in LSPR compared to the large metallic structures in SPR [144]. The light incidence into the LSPR metallic nanostructure causes changes in the electromagnetic field and consequently, the light absorbance within the ultraviolet-visible band. As a result, capturing targets in LSPR leads to a shift in the absorbance wavelength. Choi et al. created an LSPR-based label-free immunosensor for HIV-1 detection [145]. They produced a uniform gold nanoparticle layer on a glass substrate coated with indium tin oxide (ITO) (via electrochemical deposition technique). The HIV-1 monoclonal antibody gold modified surface was employed for the detection of several amounts of HIV-1 particles by measuring the UV–Vis wavelength shift (LOD = 200 fg/mL). The simple construction, high sensitivity, and selectivity of the LSPR biosensor makes it a favorable methodology for the detection of viruses in biological samples.

Besides SPR and LSPR, fluorescent biosensors have been widely used as optical biosensors that recognize the electromagnetic emission produced by exposing a fluorescent dye to radiation. The fluorescent biosensor can directly detect fluorescent-labeled targets or indirectly by employing fluorescence resonance energy transfer (FRET) [146,147]. Zhu et al. developed a molybdenum disulfide (MoS_2) sensing platform for DNA detection [148]. The ability of MoS_2 to adsorb dye-labeled targets makes it possible to recognize dye-labeled DNA probes after being hybridized by its complementary target DNA. In another study by the same group, they considered the effect of applying a single-layer transmembrane domain (TMD) nanosheet made of MoS_2, titanium disulfide (TiS_2), or tantalum disulfide (TaS_2) for the detection of target DNA [149]. Despite the positive results, the fluorescent optical biosensor has some shortcomings. For instance, labeling the nucleic acid of the real virus samples with expensive organic dyes is not an applicable procedure. Moreover, the nonuniformity of the sensing platform could cause poor reproducibility.

15.3.3 Electrochemical sensors

In electrochemical sensors, any electrochemical agitation and change due to a chemical reaction is measured and converted to analytical information of an analyte. The passing electrochemical current through the electrode is composed of two parts: (1) capacitive or changing current, (2) faradic current [150]. The capacitive or changing current is used for semiquantitative determination while the faradic current is applied for the quantitative determination of ions and molecules in solution. Several analytical methods use faradic current for their analysis including cyclic voltammetry, coulometry, amperometry, square wave voltammetry, differential pulse voltammetry, chronocoulometry, chronopotentiometry, and impedance spectroscopy [150]. The application of sensors for pathogens is a new subject that has emerged during the last two decades [150]. In 2001, the first report of using an electrochemical biosensor for virus detection was reported [14,151].

The developed electrochemical biosensors for virus detection generally use current, voltage, conductance, and impedance measurements. Table 15.3 shows recent developments in virus detection by electrochemical biosensors. In the electrochemical biosensors using faradic current as a measuring tool, the antibody is immobilized on the working electrodes such as gold or platinum followed by an attachment to a protein-formed membrane. The cyclic voltammetric peak of the activated electrode changes when the electrode is in touch with the virus solution, therefore, the current difference is proportional to the virus concentration [150]. In impedance electrochemical biosensors, the working electrode is covered with a protein-membrane after and before virus interaction. The protein intervention and virus modification could cause some changes in the electrochemical impedance of the system which can be used to provide information about virus concentration.

The conductometric sensors, which are known as immunosensors, contain metallic electrodes as electronic conductors, which causes magnetic separation. The antibody-immobilized electrodes can be used for pathogen detection while the measured resistance change by the system is related to virus concentration.

The field-effect transistor (FET) sensors are electrochemical biosensors that have recently attracted increasing attention for biomolecular recognition. In general, optical lithography and extra modification for virus detection have been used for developing FET biosensing systems. In other words, FET biosensors are solid-state devices based on the electroconductivity of the semiconductor between the source and drain terminals which is regulated by a third gate electrode through an insulator. FET biosensors are surface-charge-sensitive tools that can be used to detect various biomolecules including pathogens, label-free [187].

A pop-up system for label-free HBV DNA diagnosis was recently developed using pyrrolidinyl peptide nucleic acid attached on an electrode surface to detect target DNA through covalent bonding [188]. Differential pulse voltammetry (DPV) showed decreased signal in the presence of the target HBV DNA and the detection interval of the platform ranged between 50 pM and 100 nM (LOD = 1.45 pM).

15.3.4 Chiral biosensors

The enantiomeric feature of chemical substances is defined as the preferential response of the only one chiral selector of an enantiomer to external molecules which can be used specifically for stereo-specific synthesis,

Table 15.3 Recently developed electrochemical biosensors for pathogen detection.

Target pathogen	Working electrode	Biorecognition element	Electrochemical method	References
Escherichia coli	AuNP/Chitosan/CNT and SiO_2/thionine NP composite on Au electrode	Antibody	CV	[152]
E. coli	Au electrode	Antibody	CV	[153]
Rotavirus	Graphene microelectrode	Antibody	CV	[154]
Enterobacter cloacae	Au electrode	Concanavalin A lectin, ricinus	CV	[155]
Salmonella typhimurium	AuNP-functionalized poly (amidoamine)-CNT-chitosan	Antibody	CV	[156]
Dengue virus (types 1–4)	AuNPs on Au electrode	Antibody	CV	[157]
JEV	Carbon NPs on carbon electrode	Antibody	CV	[158]
E. coli	Silica NPs on polyelectrolyte multilayer on Au electrode	Antibody	CV	[159]
Norovirus	Au microelectrode	Concanavalin A lectin, polyclonal anti-norovirus	CV	[160]
AIV H5N1	Au electrode	Anti-AIVH5N1, concanavalin A lectin	CV	[161]
Marine pathogenic SRB	AuNPs on nickel foam electrode	Antibody	EIS	[162]
E. coli	Au interdigitated microelectrode array	Magainin I peptide	EIS	[163]
Staphylococcus aureus	Pt wire electrode	Antibody	EIS	[164]
S. typhimurium	AuNPs on graphene oxide on carbon electrode	Nucleic acid	EIS	[165]

(Continued)

Table 15.3 (Continued)

Target pathogen	Working electrode	Biorecognition element	Electrochemical method	References
Klebsiella pneumoniae	CNTs on Au electrode	Clavanin A peptide	EIS	[166]
AIV H5N1	Au interdigitated microelectrode array	Antibody	EIS	[167]
E. coli	4-(3-pyrrol) butryic acid electrode	Concanavalin A lectin, *Arachis hypogaea* lectin	EIS	[168]
Listeria monocytogenes	Au interdigitated microelectrode array	Antibody	EIS	[169]
S. aureus	Ag interdigitated electrode array	Melittin peptide	EIS	[170]
Bacillus *anthracis*	Ag electrode	Antibody	Conductometry	[171]
E. coli	Ag nanofiber array electrode	Antibody	Conductometry	[172]
SIV H1N1	PDDA/CNT composite on Au microelectrode	Antibody	Conductometry	[173]
Human influenza A viruses H1N1 and H3N2	Silicon nanowire electrode array	Antibody	Conductometry	[174]
Helicobacter pylori	Graphene interdigitated microelectrode array	Odoranin-HP peptide	Conductometry	[175]
B. subtilis	CNTs on Au interdigitated microelectrode array	Antibody	Conductometry	[176]
Influenza A, H1N1	SiO_2/glycan modified FET gate	Virus particle	FET	[177]
Influenza A, H3N2, H5N2, H5N1	Silicon nanowire	Virus particle	FET	[174,178]

Influenza A	Silicon nanowire	Virus particle	FET	[179]
Influenza A	Silicon nanowire	Antibody	FET	[180]
Dengue	Poly-silicon nanowire	Nucleic acid	FET	[181]
HIV-1	Coplanar-gate graphene	Virus particle	FET	[182]
Rotavirus	Reduced graphene oxide	Virus particle	FET	[183]
SARS-CoV-2	Graphene	Virus particle	FET	[86]
SARS-CoV-2	Graphene	Viral antigen	FET	[86]
Hepatitis B	Silicon nanowire	Nucleic acid	FET	[184]
Zika	Graphene	Virus particle	FET	[185]
Plum pox	Au electrode	Virus particle	FET	[186]

AIV, Avian influenza virus; *JEV*, Japanese encephalitis virus; *NPs*, nanoparticles; *SIV*, Swine influenza virus; *SRB*, sulphate-reducing bacteria.
Source: Readapted in part with permission from E. Cesewski, B.N. Johnson, Electrochemical biosensors for pathogen detection, Biosens. Bioelectron. 159 (2020) 112214.

pharmaceutical production, pesticides, food additives, biomonitoring, and in the development of various chiral biosensors [189−191]. The chiral nanoparticles (CNPs) among other substances, have been able to express their great features and applications as chirality-based biosensors. These CNPs are made from a metal, semiconductor, or UCNPs [192]. At the moment, different CNPs such as Au-Au NP dimers [193,194], Au−Ag NP dimers [195], Au−QD NP dimers [196], Au nanorod (NR) dimers [197], and twisted superstructures [198−200] have been manufactured and utilized in biosensors. These chirality-based devices are used for the recognition of various small-sized molecules such as antibiotics [201], toxins [202], macromolecules such as DNA [203,204] or proteins [195,205], and biomarkers in living cells [197,205]. Chirality-based technology has been used for virus detection. Ahmed et al. developed a bottom-up, time-effective molybdenum disulfide (MoS_2)-based biosensor using QDs to detect avian influenza virus proteins or virus clusters in complex blood samples [206]. This nanohybrid platform was able to detect influenza virus A (H5N1) with a limit of detection (LOD) of 0.35 and 80.92 pg/mL, respectively. In another study, a synthetic way of chiral zirconium QDs (Zr QDs) preparation was performed to detect coronavirus [207]. The synthesized Zr QDs were conjugated with anti-infectious bronchitis virus (IBV) antibodies of coronavirus to form an immuno-link in the presence of the target analyte. The chiro-immunosensor using chiral AuNPs and QDs was demonstrated to measure picomolar amounts of H5N1 influenza [208]. This platform has illustrated that the circular dichroism (CD) coupling with QD may help to recognize the target analyte in lower concentrations (high limit of detection) and with higher sensitivity.

More recently, a sensor using commercial DVD disks as a metasurface were used to measures the layer-by-layer functionalization of biomolecules and target HIV-1 capture [209]. The resulting metasurface-based biosensor was covered with a multilayer of titanium (Ti), silver (Ag), and gold (Au) and then attached with an anti-gp120 HIV antibody (Fig. 15.3). As a result, the DVD-based metasurface had a wide active area and, dielectric-metal layers which could stimulate plasmonic Fano resonance for real-time measurements.

15.4 Conclusions and prospects

The recent pandemic of COVID-19 has been a wake up call demonstrating the destructiveness of viral infections as well as the importance of rapid diagnostic method development to prevent the spread of the disease on a large scale. Among the available methods, the ones with features such as

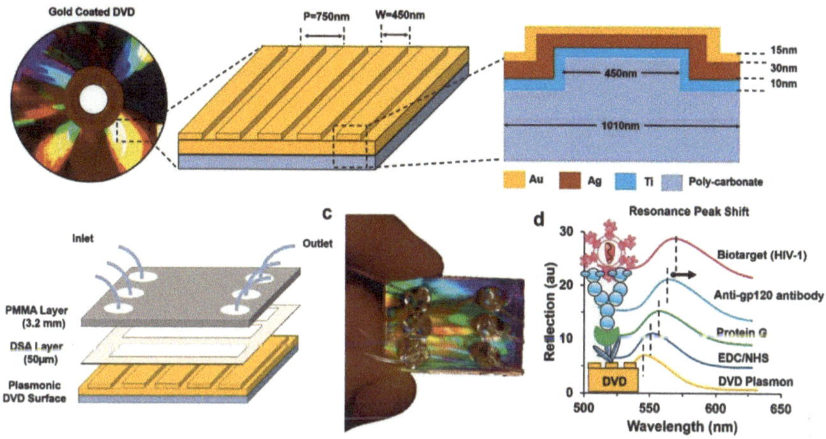

Figure 15.3 A Fano-resonant plastic-templated metasurface and a portable microfluidic sensor for HIV-1 detection. *Reproduced with permission from R. Ahmed, et al., Tunable fano-resonant metasurfaces on a disposable plastic-template for multimodal and multiplex biosensing, Adv. Mater. 32(19) (2020) e1907160. ©2020 WILEY-VCH Verlag GmbH & Co. KGaA, Weinheim.*

high sensitivity, short-time analysis, high result reliability, and low cost in the event of viral infections are needed. In this regard, by analyzing the available scientific literature resources and experiences gained during the spread of past contagious diseases, nanomaterials-enabled biosensors have proven their potential towards the diagnosis of numerous pathogens and distinguished themselves as reliable analytical devices.

The function of nanomaterials-based biosensors is mainly relying on bioreceptors such as DNA, proteins, antibodies and peptide sequences. Moreover, nonrepeatable results or falsely negative/positive results in the event of a pandemic can be misleading and irreparable. Thus, the implication of some alternative methods such as gene-editing CRISPR [210], and an aerosol-mediated diagnosis approach [211] could provide promising results. In addition, the main components of the biosensor structure, like nanoparticles such as metal, magnetic, or graphene-based nanoparticles, are always subject to the possibility of sedimentation. Therefore, fundamental studies on their molecular structure and physicochemical properties, as well as experimental testing for their optimal conditions and the methods for their synthesis, application, storage, and development, are of great importance. Finally, it is crucial that novel biosensors are adaptable and multiplex to accommodate different matrixes (urine, blood, saliva, and nasopharyngeal swabs) and are able to detect various targets of the desired virus.

References

[1] A.A. Malik, C. Nantasenamat, T. Piacham, Molecularly imprinted polymer for human viral pathogen detection, Mater. Sci. Eng. C. 77 (2017) 1341–1348.
[2] P.J. Hotez, Blue marble health and "the big three diseases": HIV/AIDS, tuberculosis, and malaria, Microbes Infect. 17 (8) (2015) 539–541.
[3] L.G. Carrascosa, et al., Nanomechanical biosensors: a new sensing tool, TrAC. Trends Anal. Chem. 25 (3) (2006) 196–206.
[4] R. Esfandyarpour, et al., Simulation and fabrication of a new novel 3D injectable biosensor for high throughput genomics and proteomics in a lab-on-a-chip device, Nanotechnology 24 (46) (2013) 465301.
[5] A. Hu, et al., Simultaneous detection, subgrouping, and quantitation of respiratory syncytial virus A and B by real-time PCR, J. Clin. Microbiol. 41 (1) (2003) 149–154.
[6] M. Gueudin, et al., Quantitation of respiratory syncytial virus RNA in nasal aspirates of children by real-time RT-PCR assay, J. Virol. Methods 109 (1) (2003) 39–45.
[7] O.I. Guliy, et al., Virus detection methods and biosensor technologies, Biophysics 64 (6) (2020) 890–897.
[8] M.S. Cheng, et al., Development of an electrochemical membrane-based nanobiosensor for ultrasensitive detection of dengue virus, Anal. Chim. Acta 725 (2012) 74–80.
[9] M. Qasim, et al., Nanotechnology for diagnosis and treatment of infectious diseases, J. Nanosci. Nanotechnol. 14 (10) (2014) 7374–7387.
[10] W. Sun, Electrochemical ELISA for the detection of cucumber mosaic virus using o-pheneylenediamine as substrate, Talanta 55 (6) (2001) 1211–1218.
[11] L.N. Cella, et al., Single-walled carbon nanotube-based chemiresistive affinity biosensors for small molecules: ultrasensitive glucose detection, J. Am. Chem. Soc. 132 (14) (2010) 5024–5026.
[12] N. Chartuprayoon, et al., Polypyrrole nanoribbon based chemiresistive immunosensors for viral plant pathogen detection, Anal. Methods 5 (14) (2013) 3497.
[13] A. Pandey, et al., Graphene-interfaced electrical biosensor for label-free and sensitive detection of foodborne pathogenic E. coli O157:H7, Biosens. Bioelectron. 91 (2017) 225–231.
[14] E. Cesewski, B.N. Johnson, Electrochemical biosensors for pathogen detection, Biosens. Bioelectron. 159 (2020) 112214.
[15] L.C. Clark Jr., C. Lyons, Electrode systems for continuous monitoring in cardiovascular surgery, Ann. N. Y. Acad. Sci. 102 (1) (1962) 29–45.
[16] S.K. Metkar, K. Girigoswami, Diagnostic biosensors in medicine – a review, Biocatal. Agric. Biotechnol. 17 (2019) 271–283.
[17] T. Lakshmipriya, S.C.B. Gopinath, An introduction to biosensors and biomolecules, Nanobiosensors for Biomolecular Targeting, Elsevier, 2019, pp. 1–21.
[18] K. Scholten, E. Meng, A review of implantable biosensors for closed-loop glucose control and other drug delivery applications, Int. J. Pharm. 544 (2) (2018) 319–334.
[19] M.K. Yazdi, et al., Nanotechnology-based biosensors in drug delivery, Nanoengineered Biomaterials for Advanced Drug Delivery, Elsevier, 2020, pp. 767–779.
[20] C. Griesche, A.J. Baeumner, Biosensors to support sustainable agriculture and food safety, TrAC. Trends Anal. Chem. 128 (2020) 115906.
[21] D. Dell'Atti, et al., Development of combined DNA-based piezoelectric biosensors for the simultaneous detection and genotyping of high risk Human Papilloma Virus strains, Clin. Chim. Acta 383 (1–2) (2007) 140–146.
[22] S.E. Seo, et al., Smartphone with optical, physical, and electrochemical nanobiosensors, J. Ind. Eng. Chem. 77 (2019) 1–11.

[23] R. Monosik, M. Stred'ansky, E. Sturdik, Application of electrochemical biosensors in clinical diagnosis, J. Clin. Lab. Anal. 26 (1) (2012) 22–34.
[24] A. Mobed, et al., Advances in detection of fastidious bacteria: from microscopic observation to molecular biosensors, TrAC. Trends Anal. Chem. 113 (2019) 157–171.
[25] A. Petrosova, et al., Development of a highly sensitive, field operable biosensor for serological studies of Ebola virus in central Africa, Sens. Actuators B Chem. 122 (2) (2007) 578–586.
[26] J.M. Encarnacao, et al., Piezoelectric biosensors for biorecognition analysis: application to the kinetic study of HIV-1 Vif protein binding to recombinant antibodies, J. Biotechnol. 132 (2) (2007) 142–148.
[27] R.E. Ionescu, Biosensor Platforms for Rapid Detection of E. coli Bacteria, in: A. Samie (Ed.), Escherichia coli – Recent Advances on Physiology, Pathogenesis and Biotechnological Applications, InTech, 2017.
[28] F. Malvano, R. Pilloton, D. Albanese, A novel impedimetric biosensor based on the antimicrobial activity of the peptide nisin for the detection of Salmonella spp, Food Chem. 325 (2020) 126868.
[29] Y. Saylan, et al., An alternative medical diagnosis method: biosensors for virus detection, Biosensors (Basel) 9 (2) (2019) 65.
[30] T. Ozer, B.J. Geiss, C.S. Henry, Review-chemical and biological sensors for viral detection, J. Electrochem. Soc. 167 (3) (2020) 037523.
[31] Y. Xiao, et al., "Plugging into Enzymes": nanowiring of redox enzymes by a gold nanoparticle, Science 299 (5614) (2003) 1877–1881.
[32] M. Holzinger, A. Le Goff, S. Cosnier, Nanomaterials for biosensing applications: a review, Front. Chem. 2 (2014) 63.
[33] L. Chen, et al., Gold nanoparticle enhanced immuno-PCR for ultrasensitive detection of Hantaan virus nucleocapsid protein, J. Immunol. Methods 346 (1–2) (2009) 64–70.
[34] J. Neng, et al., Surface-enhanced Raman scattering (SERS) detection of multiple viral antigens using magnetic capture of SERS-active nanoparticles, Biosens. Bioelectron. 41 (2013) 316–321.
[35] Y.C. Cao, R. Jin, C.A. Mirkin, Nanoparticles with Raman spectroscopic fingerprints for DNA and RNA detection, Science 297 (5586) (2002) 1536–1540.
[36] H. Li, L. Rothberg, Colorimetric detection of DNA sequences based on electrostatic interactions with unmodified gold nanoparticles, Proc. Natl. Acad. Sci. USA 101 (39) (2004) 14036–14039.
[37] G. Martínez-Paredes, M.B. González-García, A. Costa-García, Genosensor for SARS virus detection based on gold nanostructured screen-printed carbon electrodes, Electroanalysis 21 (3–5) (2009) 379–385.
[38] W. Li, et al., Catalytic signal amplification of gold nanoparticles combining with conformation-switched hairpin DNA probe for hepatitis C virus quantification, Chem. Commun. (Camb.) 48 (63) (2012) 7877–7879.
[39] S.H. Chen, et al., A method of layer-by-layer gold nanoparticle hybridization in a quartz crystal microbalance DNA sensing system used to detect dengue virus, Nanotechnology 20 (21) (2009) 215501.
[40] S. Wu, et al., Gold nanoparticle-labeled detection antibodies for use in an enhanced electrochemical immunoassay of hepatitis B surface antigen in human serum, Microchim. Acta 166 (3–4) (2009) 269–275.
[41] X. Lu, et al., A gold nanorods-based fluorescent biosensor for the detection of hepatitis B virus DNA based on fluorescence resonance energy transfer, Analyst 138 (2) (2013) 642–650.

[42] H.H. Liu, et al., Array-based nano-amplification technique was applied in detection of hepatitis E virus, J. Biochem. Mol. Biol. 39 (3) (2006) 247–252.
[43] J.C. Wu, et al., Electrophoresis-enhanced detection of deoxyribonucleic acids on a membrane-based lateral flow strip using avian influenza H5 genetic sequence as the model, Sensors (Basel) 14 (3) (2014) 4399–4415.
[44] X. Liu, et al., Electrochemical detection of avian influenza virus H5N1 gene sequence using a DNA aptamer immobilized onto a hybrid nanomaterial-modified electrode, Electrochim. Acta 56 (18) (2011) 6266–6270.
[45] T.J. Baek, et al., Development of a photodiode array biochip using a bipolar semiconductor and its application to detection of human papilloma virus, Anal. Bioanal. Chem. 390 (5) (2008) 1373–1378.
[46] E.Y. Kim, et al., Detection of HIV-1 p24 Gag in plasma by a nanoparticle-based bio-barcode-amplification method, Nanomedicine (Lond.) 3 (3) (2008) 293–303.
[47] Y. Hu, et al., Green-synthesized gold nanoparticles decorated graphene sheets for label-free electrochemical impedance DNA hybridization biosensing, Biosens. Bioelectron. 26 (11) (2011) 4355–4361.
[48] G. Qiu, et al., Dual-functional plasmonic photothermal biosensors for highly accurate severe acute respiratory syndrome coronavirus 2 detection, ACS Nano 14 (5) (2020) 5268–5277.
[49] P. Moitra, et al., Selective naked-eye detection of SARS-CoV-2 mediated by N gene targeted antisense oligonucleotide capped plasmonic nanoparticles, ACS Nano 14 (6) (2020) 7617–7627.
[50] Kumar, V., et al., *Development of RNA-based assay for rapid detection of SARS-CoV-2 in clinical samples*. bioRxiv, 2020.
[51] Z. Li, et al., Development and clinical application of a rapid IgM-IgG combined antibody test for SARS-CoV-2 infection diagnosis, J. Med. Virol. 92 (9) (2020) 1518–1524.
[52] Mahari, S., et al., *eCovSens-Ultrasensitive Novel In-House Built Printed Circuit Board Based Electrochemical Device for Rapid Detection of nCovid-19 antigen, a spike protein domain 1 of SARS-CoV-2*. bioRxiv, 2020.
[53] L. Yao, et al., Iridium-complex-modified upconversion nanophosphors for effective LRET detection of cyanide anions in pure water, Adv. Funct. Mater. 22 (13) (2012) 2667–2672.
[54] W. Zheng, et al., Lanthanide-doped upconversion nano-bioprobes: electronic structures, optical properties, and biodetection, Chem. Soc. Rev. 44 (6) (2015) 1379–1415.
[55] J. Zhao, et al., A robust luminescent Tb(III)-MOF with Lewis basic pyridyl sites for the highly sensitive detection of metal ions and small molecules, Inorg. Chem. 55 (7) (2016) 3265–3271.
[56] R. Banerjee, A. Jaiswal, Recent advances in nanoparticle-based lateral flow immunoassay as a point-of-care diagnostic tool for infectious agents and diseases, Analyst 143 (9) (2018) 1970–1996.
[57] Z. Chen, et al., Rapid and sensitive detection of anti-SARS-CoV-2 IgG, using lanthanide-doped nanoparticles-based lateral flow immunoassay, Anal. Chem. 92 (10) (2020) 7226–7231.
[58] W.W. Ye, et al., Upconversion luminescence resonance energy transfer (LRET)-based biosensor for rapid and ultrasensitive detection of avian influenza virus H7 subtype, Small 10 (12) (2014) 2390–2397.
[59] K.S. Novoselov, et al., Electric field effect in atomically thin carbon films, Science 306 (5696) (2004) 666–669.
[60] H. Gu, et al., Biomarkers-based biosensing and bioimaging with graphene for cancer diagnosis, Nanomaterials (Basel) 9 (1) (2019) 130.

[61] A. Cordaro, et al., Graphene-based strategies in liquid biopsy and in viral diseases diagnosis, Nanomaterials (Basel) 10 (6) (2020) 1014.
[62] S.-Y. Wang, et al., Fabrication of fluorescent biosensing platform based on graphene oxide-DNA and their application in biomolecule detection, TrAC. Trends Anal. Chem. 106 (2018) 53–61.
[63] M. Adeel, et al., Graphene and graphene oxide: functionalization and nano-bio-catalytic system for enzyme immobilization and biotechnological perspective, Int. J. Biol. Macromol. 120 (Pt B) (2018) 1430–1440.
[64] J. Pena-Bahamonde, et al., Recent advances in graphene-based biosensor technology with applications in life sciences, J. Nanobiotechnol. 16 (1) (2018) 75.
[65] R. Ziółkowski, et al., Toward fluorimetric-paired-emitter-detector-diode test for Bacillus anthracis DNA based on graphene oxide, Microchem. J. 154 (2020) 104592.
[66] Q. Gong, Y. Wang, H. Yang, A sensitive impedimetric DNA biosensor for the determination of the HIV gene based on graphene-Nafion composite film, Biosens. Bioelectron. 89 (Pt 1) (2017) 565–569.
[67] Q. Gong, et al., Sensitive electrochemical DNA sensor for the detection of HIV based on a polyaniline/graphene nanocomposite, J. Mater. 5 (2) (2019) 313–319.
[68] L. Farzin, et al., Electrochemical genosensor based on carbon nanotube/amine-ionic liquid functionalized reduced graphene oxide nanoplatform for detection of human papillomavirus (HPV16)-related head and neck cancer, J. Pharm. Biomed. Anal. 179 (2020) 112989.
[69] X. Li, et al., An impedimetric immunosensor for determination of porcine epidemic diarrhea virus based on the nanocomposite consisting of molybdenum disulfide/reduced graphene oxide decorated with gold nanoparticles, Mikrochim. Acta 187 (4) (2020) 217.
[70] X. Niu, et al., A graphene modified carbon ionic liquid electrode for voltammetric analysis of the sequence of the Staphylococcus aureus nuc gene, Mikrochim. Acta 185 (3) (2018) 167.
[71] J. Ji, et al., Single-layered graphene/Au-nanoparticles-based love wave biosensor for highly sensitive and specific detection of Staphylococcus aureus gene sequences, ACS Appl. Mater. Interfaces 12 (11) (2020) 12417–12425.
[72] J.N. Appaturi, et al., Rapid and sensitive detection of Salmonella with reduced graphene oxide-carbon nanotube based electrochemical aptasensor, Anal. Biochem. 589 (2020) 113489.
[73] S. Shahrokhian, S. Ranjbar, Development of a sensitive diagnostic device based on zeolitic imidazolate frameworks-8 using ferrocene–graphene oxide as electroactive indicator for Pseudomonas aeruginosa detection, ACS Sustain. Chem. Eng. 7 (15) (2019) 12760–12769.
[74] M. Pourmadadi, et al., A glassy carbon electrode modified with reduced graphene oxide and gold nanoparticles for electrochemical aptasensing of lipopolysaccharides from Escherichia coli bacteria, Mikrochim. Acta 186 (12) (2019) 787.
[75] L.X. Wen, et al., A fluorescent probe composed of quantum dot labeled aptamer and graphene oxide for the determination of the lipopolysaccharide endotoxin, Mikrochim. Acta 186 (2) (2019) 122.
[76] J. Niu, et al., Femtomolar detection of lipopolysaccharide in injectables and serum samples using aptamer-coupled reduced graphene oxide in a continuous injection-electrostacking biochip, Anal. Chem. 91 (3) (2019) 2360–2367.
[77] P. Teengam, et al., Electrochemical paper-based peptide nucleic acid biosensor for detecting human papillomavirus, Anal. Chim. Acta 952 (2017) 32–40.
[78] M.H. Mat Zaid, et al., PNA biosensor based on reduced graphene oxide/water soluble quantum dots for the detection of *Mycobacterium tuberculosis*, Sens. Actuators B: Chem. 241 (2017) 1024–1034.

[79] M.H.M. Zaid, et al., Reduced graphene oxide/TEMPO-nanocellulose nanohybrid-based electrochemical biosensor for the determination of Mycobacterium tuberculosis, J. Sens. 2020 (2020) 1–11.
[80] X. Jin, et al., A field effect transistor modified with reduced graphene oxide for immunodetection of Ebola virus, Mikrochim. Acta 186 (4) (2019) 223.
[81] S. Islam, et al., A smart nanosensor for the detection of human immunodeficiency virus and associated cardiovascular and arthritis diseases using functionalized graphene-based transistors, Biosens. Bioelectron. 126 (2019) 792–799.
[82] Z. You, et al., Laser-induced noble metal nanoparticle-graphene composites enabled flexible biosensor for pathogen detection, Biosens. Bioelectron. 150 (2020) 111896.
[83] W. Zhao, et al., Monolayer graphene chemiresistive biosensor for rapid bacteria detection in a microchannel, Sens. Actuators Rep. 2 (1) (2020) 100004.
[84] N.A.S. Omar, et al., Sensitive surface plasmon resonance performance of cadmium sulfide quantum dots-amine functionalized graphene oxide based thin film towards dengue virus E-protein, Opt. Laser Technol. 114 (2019) 204–208.
[85] T. Bu, et al., New functional tracer-two-dimensional nanosheet-based immunochromatographic assay for Salmonella enteritidis detection, J. Agric. Food Chem. 67 (23) (2019) 6642–6649.
[86] G. Seo, et al., Rapid detection of COVID-19 causative virus (SARS-CoV-2) in human nasopharyngeal swab specimens using field-effect transistor-based biosensor, ACS Nano 14 (4) (2020) 5135–5142.
[87] N. Yu, et al., Highly sensitive endotoxin assay combining peptide/graphene oxide and DNA-modified gold nanoparticles, ACS Omega 4 (10) (2019) 14312–14316.
[88] J. Chen, S.R. Nugen, Detection of protease and engineered phage-infected bacteria using peptide-graphene oxide nanosensors, Anal. Bioanal. Chem. 411 (12) (2019) 2487–2492.
[89] Y. Zhang, et al., Graphene oxide-based biosensing platform for rapid and sensitive detection of HIV-1 protease, Anal. Bioanal. Chem. 410 (24) (2018) 6177–6185.
[90] C. Zhou, et al., Fiber optic surface plasmon resonance sensor for detection of E. coli O157:H7 based on antimicrobial peptides and AgNPs-rGO, Biosens. Bioelectron. 117 (2018) 347–353.
[91] K. Yuan, et al., Antimicrobial peptide based magnetic recognition elements and Au@Ag-GO SERS tags with stable internal standards: a three in one biosensor for isolation, discrimination and killing of multiple bacteria in whole blood, Chem. Sci. 9 (47) (2018) 8781–8795.
[92] H.B. Na, I.C. Song, T. Hyeon, Inorganic nanoparticles for MRI contrast agents, Adv. Mater. 21 (21) (2009) 2133–2148.
[93] A.K. Gupta, M. Gupta, Synthesis and surface engineering of iron oxide nanoparticles for biomedical applications, Biomaterials 26 (18) (2005) 3995–4021.
[94] M. Takafuji, et al., Preparation of poly(1-vinylimidazole)-grafted magnetic nanoparticles and their application for removal of metal ions, Chem. Mater. 16 (10) (2004) 1977–1983.
[95] S. Sun, et al., Monodisperse FePt nanoparticles and ferromagnetic FePt nanocrystal superlattices, Science 287 (5460) (2000) 1989–1992.
[96] T. Guo, et al., The recent advances of magnetic nanoparticles in medicine, J. Nanomater. 2018 (2018) 1–8.
[97] M.R. Dudek, et al., Magnetocaloric materials with ultra-small magnetic nanoparticles working at room temperature, Sci. Rep. 9 (1) (2019) 17607.
[98] S. Ale Ebrahim, et al., Publication trends in drug delivery and magnetic nanoparticles, Nanoscale Res. Lett. 14 (1) (2019) 164.
[99] W. Yang, et al., Transition metal (Fe, Co and Ni) carbide and nitride nanomaterials: structure, chemical synthesis and applications, ChemNanoMat 1 (6) (2015) 376–398.

[100] M. Aminul Islam, M. Ziaul Ahsan, Plausible approach for rapid detection of SARS-CoV-2 virus by magnetic nanoparticle based biosensors, Am. J. Nanosci. 6 (2) (2020) 6.
[101] S. Fathi Karkan, et al., Magnetic nanoparticles in cancer diagnosis and treatment: a review, Artif. Cell Nanomed. Biotechnol. 45 (1) (2017) 1–5.
[102] S.M. Dadfar, et al., Iron oxide nanoparticles: diagnostic, therapeutic and theranostic applications, Adv. Drug Deliv. Rev. 138 (2019) 302–325.
[103] D. Li, et al., Recent developments of rare-earth-free hard-magnetic materials, Sci. China Phys. Mech. Astron. 59 (1) (2015).
[104] L. Lu, et al., Recent advances in biological detection with magnetic nanoparticles as a useful tool, Sci. China Chem. 58 (5) (2015) 793–809.
[105] T.C. Chou, et al., Rapid and specific influenza virus detection by functionalized magnetic nanoparticles and mass spectrometry, J. Nanobiotechnol. 9 (2011) 52.
[106] J.M. Barnett, et al., Initial trail results of a magnetic biosensor for the rapid detection of Porcine Reproductive and Respiratory Virus (PRRSV) infection, Sens. Bio-Sensing Res. 27 (2020) 100315.
[107] Zhao, Z., et al., *A Simple Magnetic Nanoparticles-Based Viral RNA Extraction Method for Efficient Detection of SARS-CoV-2*. BioRxiv, 2020.
[108] T.L. Kamikawa, et al., Nanoparticle-based biosensor for the detection of emerging pandemic influenza strains, Biosens. Bioelectron. 26 (4) (2010) 1346–1352.
[109] L. Krejcova, et al., Paramagnetic particles coupled with an automated flow injection analysis as a tool for influenza viral protein detection, Electrophoresis 33 (21) (2012) 3195–3204.
[110] B. Tian, et al., Nicking-assisted on-loop and off-loop enzymatic cascade amplification for optomagnetic detection of a highly conserved dengue virus sequence, Biosens. Bioelectron. 160 (2020) 112219.
[111] B. Tian, et al., Homogeneous circle-to-circle amplification for real-time optomagnetic detection of SARS-CoV-2 RdRp coding sequence, Biosens. Bioelectron. 165 (2020) 112356.
[112] Q.H. Nguyen, M.I. Kim, Nanomaterial-mediated paper-based biosensors for colorimetric pathogen detection, Trends Anal. Chem. 132 (2020) 116038.
[113] J. Chao, et al., DNA nanotechnology-enabled biosensors, Biosens. Bioelectron. 76 (2016) 68–79.
[114] R. Kizek, et al., Nanoscale virus biosensors: state of the art, Nanobiosens. Dis. Diagnosis (2015) 47.
[115] R.S. Malon, et al., Saliva-based biosensors: noninvasive monitoring tool for clinical diagnostics, Biomed. Res. Int. 2014 (2014) 962903.
[116] W. Zhang, Y. Du, M.L. Wang, Noninvasive glucose monitoring using saliva nano-biosensor, Sens. Bio-Sensing Res. 4 (2015) 23–29.
[117] R.K. Mishra, R. Rajakumari, Nanobiosensors for biomedical application, Characterization and Biology of Nanomaterials for Drug Delivery, Elsevier, 2019, pp. 1–23.
[118] V. Aiello, et al., Fabrication and characterization of the sensing element for glucose biosensor applications, Sensors and Microsystems, WORLD SCIENTIFIC, 2008, pp. 3–13.
[119] S. Selvarajan, et al., Unconventional active biosensor made of piezoelectric $BaTiO_3$ nanoparticles for biomolecule detection. Sens. Actuators B: Chem. 253 (2017) 1180–1187.
[120] I. Bhatnagar, et al., Chitosan stabilized gold nanoparticle mediated self-assembled gliP nanobiosensor for diagnosis of Invasive Aspergillosis, Int. J. Biol. Macromol. 110 (2018) 449–456.
[121] L.D. Tran, et al., Electrochemical detection of short HIV sequences on chitosan/Fe3O4 nanoparticle based screen printed electrodes, Mater. Sci. Eng.: C. 31 (2) (2011) 477–485.

[122] N. Bhalla, et al., Introduction to biosensors, Essays Biochem. 60 (1) (2016) 1–8.
[123] T. Laksanasopin, et al., A smartphone dongle for diagnosis of infectious diseases at the point of care, Sci. Transl. Med. 7 (273) (2015) 273re1.
[124] Y. Wang, et al., Charge-selective fractions of naturally occurring nanoparticles as bioactive nanocarriers for cancer therapy, Acta Biomater. 10 (10) (2014) 4269–4284.
[125] L. Sun, et al., Sundew-inspired adhesive hydrogels combined with adipose-derived stem cells for wound healing, ACS Appl. Mater. Interfaces 8 (3) (2016) 2423–2434.
[126] F. Chai, et al., A solid self-nanoemulsifying system of the BCS class IIb drug dabigatran etexilate to improve oral bioavailability, Nanomedicine (Lond.) 11 (14) (2016) 1801–1816.
[127] P. Zhang, et al., Simple and sensitive detection of HBsAg by using a quantum dots nanobeads based dot-blot immunoassay, Theranostics 4 (3) (2014) 307–315.
[128] P. Tallury, et al., Nanobioimaging and sensing of infectious diseases, Adv. Drug Deliv. Rev. 62 (4–5) (2010) 424–437.
[129] G.K. Darbha, et al., Gold-nanorod-based sensing of sequence specific HIV-1 virus DNA by using hyper-Rayleigh scattering spectroscopy, Chemistry 14 (13) (2008) 3896–3903.
[130] J. Griffin, et al., Sequence-specific HCV RNA quantification using the size-dependent nonlinear optical properties of gold nanoparticles, Small 5 (7) (2009) 839–845.
[131] H.J. Chung, et al., A magneto-DNA nanoparticle system for rapid detection and phenotyping of bacteria, Nat. Nanotechnol. 8 (5) (2013) 369–375.
[132] H. Lee, et al., Chip-NMR biosensor for detection and molecular analysis of cells, Nat. Med. 14 (8) (2008) 869–874.
[133] M. Liong, et al., Magnetic barcode assay for genetic detection of pathogens, Nat. Commun. 4 (2013) 1752.
[134] F. Inci, et al., Nanoplasmonic quantitative detection of intact viruses from unprocessed whole blood, ACS Nano 7 (6) (2013) 4733–4745.
[135] E. Ng, et al., Magneto-nanosensor smartphone platform for the detection of HIV and leukocytosis at point-of-care, Nanomedicine 16 (2019) 10–19.
[136] P.M. Kosaka, et al., Ultrasensitive detection of HIV-1 p24 antigen by a hybrid nanomechanical-optoplasmonic platform with potential for detecting HIV-1 at first week after infection, PLoS One 12 (2) (2017) e0171899.
[137] P. Mehrotra, Biosensors and their applications – a review, J. Oral. Biol. Craniofac Res. 6 (2) (2016) 153–159.
[138] S. Atay, et al., Quartz crystal microbalance based biosensors for detecting highly metastatic breast cancer cells via their transferrin receptors, Anal. Methods 8 (1) (2016) 153–161.
[139] E.B. Bahadir, M.K. Sezginturk, Applications of commercial biosensors in clinical, food, environmental, and biothreat/biowarfare analyses, Anal. Biochem. 478 (2015) 107–120.
[140] J. Vlachova, et al., A 3D microfluidic chip for electrochemical detection of hydrolysed nucleic bases by a modified glassy carbon electrode, Sensors (Basel) 15 (2) (2015) 2438–2452.
[141] M. Trefulka, et al., Voltammetry of osmium end-labeled oligodeoxynucleotides at carbon, mercury, and gold electrodes, Electroanalysis 19 (12) (2007) 1334–1338.
[142] V. Jayanthi, A.B. Das, U. Saxena, Recent advances in biosensor development for the detection of cancer biomarkers, Biosens. Bioelectron. 91 (2017) 15–23.
[143] N.A.S. Omar, et al., Development of an optical sensor based on surface plasmon resonance phenomenon for diagnosis of dengue virus E-protein, Sens. Bio-Sensing Res. 20 (2018) 16–21.

[144] S. Klinghammer, et al., Plasmonic biosensor based on vertical arrays of gold nanoantennas, ACS Sens. 3 (7) (2018) 1392–1400.
[145] J.H. Lee, et al., Highly sensitive localized surface plasmon resonance immunosensor for label-free detection of HIV-1, Nanomedicine 9 (7) (2013) 1018–1026.
[146] Y. Shen, et al., A novel and sensitive turn-on fluorescent biosensor for the DNA detection using Sm3 + -modulated glutathione-capped CdTe quantum dots, Sens. Actuators B Chem. 199 (2014) 389–397.
[147] S. Takalkar, K. Baryeh, G. Liu, Fluorescent carbon nanoparticle-based lateral flow biosensor for ultrasensitive detection of DNA, Biosens. Bioelectron. 98 (2017) 147–154.
[148] C. Zhu, et al., Single-layer MoS_2-based nanoprobes for homogeneous detection of biomolecules. J. Am. Chem. Soc. 135 (16) (2013) 5998–6001.
[149] Y. Zhang, et al., Single-layer transition metal dichalcogenide nanosheet-based nanosensors for rapid, sensitive, and multiplexed detection of DNA, Adv. Mater. 27 (5) (2015) 935–939.
[150] K.S.V. Santhanam, Electrochemical approaches towards sensing viruses: a mini review, Med. Dev. Sens. 4 (1) (2020).
[151] D.R. Thévenot, et al., Electrochemical biosensors: recommended definitions and classification1International Union of Pure and Applied Chemistry: Physical Chemistry Division, Commission I.7 (Biophysical Chemistry); Analytical Chemistry Division, Commission V.5 (Electroanalytical Chemistry)0.1, Biosens. Bioelectron. 16 (1–2) (2001) 121–131.
[152] Y. Li, et al., Amperometric immunosensor for the detection of Escherichia coli O157:H7 in food specimens, Anal. Biochem. 421 (1) (2012) 227–233.
[153] V. Escamilla-Gomez, et al., Gold screen-printed-based impedimetric immunobiosensors for direct and sensitive Escherichia coli quantisation, Biosens. Bioelectron. 24 (11) (2009) 3365–3371.
[154] F. Liu, et al., Graphene-based electrochemical biosensor for pathogenic virus detection, BioChip J. 5 (2) (2011) 123–128.
[155] F. Xi, et al., Discrimination and detection of bacteria with a label-free impedimetric biosensor based on self-assembled lectin monolayer, J. Electroanalytical Chem. 656 (1–2) (2011) 252–257.
[156] J. Dong, et al., A label-free electrochemical impedance immunosensor based on AuNPs/PAMAM-MWCNT-Chi nanocomposite modified glassy carbon electrode for detection of Salmonella typhimurium in milk, Food Chem. 141 (3) (2013) 1980–1986.
[157] D.M.N. Luna, et al., Electrochemical immunosensor for dengue virus serotypes based on 4-mercaptobenzoic acid modified gold nanoparticles on self-assembled cysteine monolayers, Sens. Actuators B: Chem. 220 (2015) 565–572.
[158] S.F. Chin, et al., Carbon nanoparticle modified screen printed carbon electrode as a disposable electrochemical immunosensor strip for the detection of Japanese encephalitis virus, Microchim. Acta 184 (2) (2016) 491–497.
[159] M. Mathelié-Guinlet, et al., Silica nanoparticles-assisted electrochemical biosensor for the rapid, sensitive and specific detection of Escherichia coli, Sens. Actuators B: Chem. 292 (2019) 314–320.
[160] S.A. Hong, et al., A rapid, sensitive and selective electrochemical biosensor with concanavalin A for the preemptive detection of norovirus, Biosens. Bioelectron. 64 (2015) 338–344.
[161] Q. Zhang, et al., Electrochemical conversion of Fe_3O_4 magnetic nanoparticles to electroactive prussian blue analogues for self-sacrificial label biosensing of avian influenza virus H5N1. Anal. Chem. 89 (22) (2017) 12145–12151.
[162] Y. Wan, et al., A 3D-impedimetric immunosensor based on foam Ni for detection of sulfate-reducing bacteria, Electrochem. Commun. 12 (2) (2010) 288–291.

[163] M.S. Mannoor, et al., Electrical detection of pathogenic bacteria via immobilized antimicrobial peptides, Proc. Natl. Acad. Sci. USA 107 (45) (2010) 19207–19212.
[164] F. Tan, et al., A PDMS microfluidic impedance immunosensor for E. coli O157: H7 and Staphylococcus aureus detection via antibody-immobilized nanoporous membrane, Sens. Actuators B Chem. 159 (1) (2011) 328–335.
[165] X. Ma, et al., An aptamer-based electrochemical biosensor for the detection of Salmonella, J. Microbiol. Methods 98 (2014) 94–98.
[166] C.A.S. Andrade, et al., Nanostructured sensor based on carbon nanotubes and clavanin A for bacterial detection, Colloids Surf. B Biointerfaces 135 (2015) 833–839.
[167] J. Lin, et al., An impedance immunosensor based on low-cost microelectrodes and specific monoclonal antibodies for rapid detection of avian influenza virus H5N1 in chicken swabs, Biosens. Bioelectron. 67 (2015) 546–552.
[168] N.M. Saucedo, S. Srinives, A. Mulchandani, Electrochemical biosensor for rapid detection of viable bacteria and antibiotic screening, J. Anal. Test. 3 (1) (2019) 117–122.
[169] Q. Chen, et al., Fast and sensitive detection of foodborne pathogen using electrochemical impedance analysis, urease catalysis and microfluidics, Biosens. Bioelectron. 86 (2016) 770–776.
[170] D. Wilson, et al., Erratum to "Electrical detection of pathogenic bacteria in food samples using information visualization methods with a sensor based on magnetic nanoparticles functionalized with antimicrobial peptides" [Talanta 194 (2019) 611–618], Talanta, 200, 2019, p. 562.
[171] S. Pal, E.C. Alocilja, Electrically active polyaniline coated magnetic (EAPM) nanoparticle as novel transducer in biosensor for detection of Bacillus anthracis spores in food samples, Biosens. Bioelectron. 24 (5) (2009) 1437–1444.
[172] Y. Luo, et al., Surface functionalization of electrospun nanofibers for detecting E. coli O157:H7 and BVDV cells in a direct-charge transfer biosensor, Biosens. Bioelectron. 26 (4) (2010) 1612–1617.
[173] D. Lee, et al., Carbon nanotube electric immunoassay for the detection of swine influenza virus H1N1, Biosens. Bioelectron. 26 (8) (2011) 3482–3487.
[174] F. Shen, et al., Rapid flu diagnosis using silicon nanowire sensor, Nano Lett. 12 (7) (2012) 3722–3730.
[175] M.S. Mannoor, et al., Graphene-based wireless bacteria detection on tooth enamel, Nat. Commun. 3 (1) (2012) 763.
[176] M.S. Yoo, et al., Development of electrochemical biosensor for detection of pathogenic microorganism in Asian dust events, Chemosphere 175 (2017) 269–274.
[177] S. Hideshima, et al., Glycan-immobilized dual-channel field effect transistor biosensor for the rapid identification of pandemic influenza viral particles, Sci. Rep. 9 (1) (2019) 11616.
[178] P.L. Chiang, et al., Nanowire transistor-based ultrasensitive virus detection with reversible surface functionalization, Chem. Asian J. 7 (9) (2012) 2073–2079.
[179] F. Patolsky, et al., Electrical detection of single viruses, Proc. Natl. Acad. Sci. USA 101 (39) (2004) 14017–14022.
[180] J.Y. Kim, et al., Multiplex electrical detection of avian influenza and human immunodeficiency virus with an underlap-embedded silicon nanowire field-effect transistor, Biosens. Bioelectron. 55 (2014) 162–167.
[181] M.N.M. N, et al., Electrical responses of dengue virus (DENV) using poly-Si nanowire array biosensor, AIP Conf. Proc. 2045 (1) (2018) 020034.
[182] J.W. Kim, et al., Attomolar detection of virus by liquid coplanar-gate graphene transistor on plastic, Nanotechnology 30 (34) (2019) 345502.
[183] F. Liu, et al., Micropatterned reduced graphene oxide based field-effect transistor for real-time virus detection, Sens. Actuators B Chem. 186 (2013) 252–257.

[184] C.-Y. Wu, et al., Real-time sensing of hepatitis B virus X gene using an ultrasensitive nanowire field effect transistor, Nano Online, De Gruyter., 2016.
[185] S. Afsahi, et al., Novel graphene-based biosensor for early detection of Zika virus infection, Biosens. Bioelectron. 100 (2018) 85–88.
[186] M. Berto, et al., Label free detection of plant viruses with organic transistor biosensors, Sens. Actuators B Chem. 281 (2019) 150–156.
[187] A. Poghossian, et al., Field-effect sensors for virus detection: from Ebola to SARS-CoV-2 and plant viral enhancers, Front. Plant. Sci. 11 (2020) 598103.
[188] C. Srisomwat, et al., Pop-up paper electrochemical device for label-free hepatitis B virus DNA detection, Sens. Actuators B: Chem. 316 (2020) 128077.
[189] A. Kuzyk, et al., Reconfigurable 3D plasmonic metamolecules, Nat. Mater. 13 (9) (2014) 862–866.
[190] C. Zhou, X. Duan, N. Liu, A plasmonic nanorod that walks on DNA origami, Nat. Commun. 6 (2015) 8102.
[191] M. Sun, et al., Intracellular localization of nanoparticle dimers by chirality reversal, Nat. Commun. 8 (1) (2017) 1847.
[192] H.-E. Lee, et al., Biomolecule-enabled chiral assembly of plasmonic nanostructures, ChemNanoMat 3 (10) (2017) 685–697.
[193] Z. Xu, et al., Chirality based sensor for bisphenol A detection, Chem. Commun. (Camb.) 48 (46) (2012) 5760–5762.
[194] L. Xu, et al., Highly selective recognition and ultrasensitive quantification of enantiomers, J. Mater. Chem. B 1 (35) (2013) 4478–4483.
[195] X. Wu, et al., Unexpected chirality of nanoparticle dimers and ultrasensitive chiroplasmonic bioanalysis, J. Am. Chem. Soc. 135 (49) (2013) 18629–18636.
[196] M. Sun, et al., Chirality of self-assembled metal–semiconductor nanostructures, J. Mater. Chem. C. 2 (15) (2014) 2702–2706.
[197] W. Ma, et al., Chiral plasmonics of self-assembled nanorod dimers, Sci. Rep. 3 (2013) 1934.
[198] W. Ma, et al., Attomolar DNA detection with chiral nanorod assemblies, Nat. Commun. 4 (2013) 2689.
[199] W. Ma, et al., A chiral-nanoassemblies-enabled strategy for simultaneously profiling surface glycoprotein and MicroRNA in living cells, Adv. Mater. 29 (42) (2017) 1703410.
[200] L. Xu, et al., 2D chiroptical nanostructures for high-performance photooxidants, Adv. Funct. Mater. 28 (18) (2018) 1707237.
[201] H. Kuang, et al., Immuno-driven plasmonic oligomer sensor for the ultrasensitive detection of antibiotics, RSC Adv. 3 (38) (2013) 17294.
[202] J. Cai, et al., Chiral shell core-satellite nanostructures for ultrasensitive detection of mycotoxin, Small 14 (13) (2018) e1703931.
[203] W. Yan, et al., Plasmonic chirogenesis from gold nanoparticles superstructures, J. Phys. Chem. C. 117 (34) (2013) 17757–17765.
[204] Y. Zhao, et al., Shell-engineered chiroplasmonic assemblies of nanoparticles for zeptomolar DNA detection, Nano Lett. 14 (7) (2014) 3908–3913.
[205] L. Tang, et al., Chirality-based Au@Ag nanorod dimers sensor for ultrasensitive PSA detection, ACS Appl. Mater. Interfaces 7 (23) (2015) 12708–12712.
[206] S.R. Ahmed, S. Neethirajan, Chiral MoS2 quantum dots: dual-mode detection approaches for avian influenza viruses, Glob. Chall. 2 (4) (2018) 1700071.
[207] S.R. Ahmed, et al., Chiral zirconium quantum dots: a new class of nanocrystals for optical detection of coronavirus, Heliyon 4 (8) (2018) e00766.
[208] S.R. Ahmed, É. Nagy, S. Neethirajan, Self-assembled star-shaped chiroplasmonic gold nanoparticles for an ultrasensitive chiro-immunosensor for viruses, RSC Adv. 7 (65) (2017) 40849–40857.

[209] R. Ahmed, et al., Tunable fano-resonant metasurfaces on a disposable plastic-template for multimodal and multiplex biosensing, Adv. Mater. 32 (19) (2020) e1907160.
[210] C. Lucia, P.-B. Federico, G.C. Alejandra, *An Ultrasensitive, Rapid, and Portable Coronavirus SARS-CoV-2 Sequence Detection Method Based on CRISPR-Cas12*. bioRxiv, 2020.
[211] J.A. Huffman, et al., Real-time sensing of bioaerosols: review and current perspectives, Aerosol Sci. Technol. 54 (5) (2019) 465–495.

CHAPTER 16

Biosensors for healthcare: an artificial intelligence approach

Shivani Dave[1], Aditya Dave[2], S. Radhakrishnan[3], Jayashankar Das[4] and Sushma Dave[5]

[1]Department of Computer Science, MBM Engineering College, Jodhpur, Rajasthan, India
[2]IIT Madras (B.Sc. Data Science), Chennai, Tamil Nadu, India
[3]Department of Organic Materials and Fiber Engineering, Jeonbuk National University, Jeonju-si, Jeollabuk-do, South Korea
[4]Valnizen Healthcare PVT Limited, Mumbai, Maharashtra, India
[5]Department of Applied Science, JIET Group of Institutions, Jodhpur, Rajasthan, India

16.1 Introduction

Biosensors are analytical instruments used to detect and quantify specific molecules in the presence of a mixture of various molecules [1]. The development of biosensors for the screening of biologically-important molecules is highly desirable in various sectors and can have a serious impact on human health. Different biomolecules (glucose, uric acid, ascorbic acid, vitamins, dopamine, DNA, enzymes, etc.) are present in human body and play key roles in the proper activity of human life [2]. Changes, either an increase or decrease, of their concentration will have several health complications. Currently, many analytical techniques (spectroscopy, chromatography, capillary electrophoresis, etc.) are available to identify these biomolecules. These methods have major drawbacks including high expense, tedious process, lengthy time, requiring high volumes of sample for analysis, and a trained person to handle the instrument. In order to solve these issues, electrochemical methods [88] are the most powerful and key alternative technique due to their easy handling, low-cost, rapid response with good precision, portability and miniaturization [3–8].

Generally, the electrochemical biosensor analytical tool contains a bioreceptor, transducer, and analytical output (Fig. 16.1). Here, the bioreceptors are usually joined with transducers, and an analytical signal is received from the transducer. The signal varies depending on the type of interaction and concentration of biomolecules with bioreceptors. The simple construction of biosensor devices creates for huge interest in many

Biosensors for Emerging and Re-emerging Infectious Diseases
DOI: https://doi.org/10.1016/B978-0-323-88464-8.00008-7

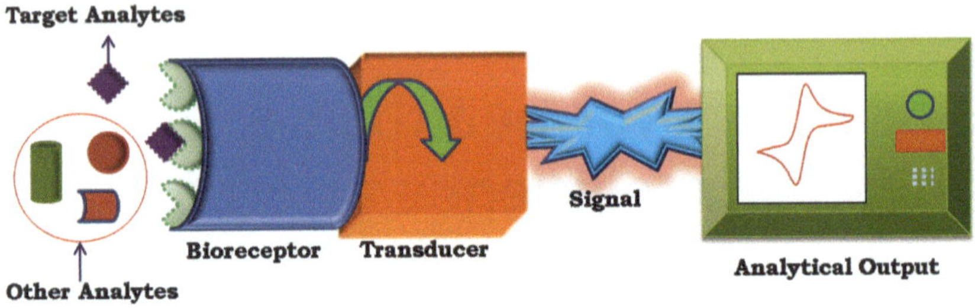

Figure 16.1 Schematic illustration of the basic structure of a sensor.

applications due to their easy operation, portability, rapid detection, good selectivity, sensitivity, reproducibility, low-cost, etc. [9,10].

It is well known that biosensors are broadly classified into two categories based on the transducer and nature of bioreceptors. For example, if an optical transducer is used in the biosensor device, they are called optical sensors. Similarly, where electrical transducers are used, it would be called an electrochemical biosensor. Likewise, we can classify different biosensors as piezoelectric, thermal, or colorimetric. The second category is based on the type of bioreceptor used in the biosensor device. For example, if an enzyme is used as a bio-receptor in the device, they are called enzymatic biosensors. Similarly, if using DNA, antibodies, microbial cells, then it would be called a DNA biosensor, immunosensor, and microbial sensor, respectively. Electrochemical biosensors have unique properties compared to other types of biosensors. Hence, in recent years many research groups have largely focused on the electrochemical sensors for the detection of various applications, including environmental monitoring, pharmaceutical, and food industries. The electrochemical biosensors are subdivided into three classifications based on the nature of the working principle and their method of output measurements including: potentiometric, amperometric, and impedimetric [11–17]. This chapter also covers recently developed biosensors and their mechanisms of action and technological strategies involved in their application to health care. In addition, it provides the key to the ontogeny of effective and supportive hierarchical protocols for clinical decision-making in personalized medicine through an interdisciplinary AI approach integrated with information analysis. It includes benefits of AI to patients and society as a whole and how it will bring new precision to health care. Key issues include: proper access to data, the right regulatory environment, measures to support innovation, large and small research institutions and industries, facilitating the incorporation of innovation by healthcare organizations, and protecting patients and their rights. A number of basic legal and ethical issues are also outlined [18–22] (Fig. 16.2).

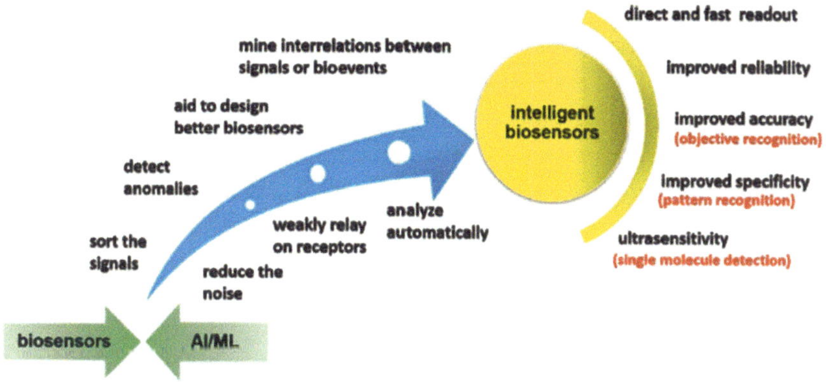

Figure 16.2 Benefits of ML in biosensors. *ML*, Machine learning.

No one could guess how radically the world would change after steam power began, and the beneficial transformation of industry and transportation years ago. Similarly, the influence of ancient research was not spoken of when the then scientists and pioneers pushed the frontier of innovation, leading to another era of comfort and connection. As the 21st century gains momentum, the world is at stake in another similar progressive transformation and that is in the field of artificial intelligence. Artificial intelligence (AI) is re-characterizing how civilizations order themselves. From space exploration to life science advances, and from building independent vehicles to managing illegal tax avoidance, AI creates a path to a phenomenal new perspective [23]. The healthcare sector is one of the most important areas for using AI. Instead of just an idea, coordinated and accurate medical care may become a reality, significantly improving patient outcomes, re-energizing health professionals, and increasing efficiency across the welfare area [24–26]. AI is advanced in core areas. Due to the historical background of mankind, a framework with some degree of independence will emerge and the most popular missconception about artificial intelligence is that it is related with natural intelligence. This misconception is that intelligence is a single dimension. This is quite different from the traditional standard, like based programming that requires explicit instructions to complete a task. The adaptive numerical model at the core of AI uses a large number of calculations to distinguish between complex indirect connections in large datasets. The easiest way is for AI to consistently interpret dialects, subtitle recordings, and prevent email spam [27].

The use of AI in various areas of medicine has the on-the-spot advantage of easing and accelerating workflows for the general analysis of images; however, it's also a route to, give particular information of the ailment. Predicting the route of most cancers from imaging information may grow to be a part of the science recurring within the next few years. According to reports, the growing use of AI in radiology holds the possibility of basically converting the translation of clinical images. It may also cause extra specific information of an ailment—as an example through the analysis of ailment development primarily based on previously unused imaging information [28,29]. Image reconstructions from imaging information and radiological reporting are facilitated radically through AI. Besides this, key values and parameters may be mounted fast and with great precision. For example, lung nodules and tumor foci may be exactly measured and lung volumes routinely quantified. When AI answers remotely, it verifies an affected person's symptoms and supplies indicators to clinicians most effectively while the care for the affected person is needed. Thus it reduces needless health facility visits and can also reduce the burden on clinical professionals. AI can save a massive percentage of clinical caretakers' time through warding off needless visits. The enhancements could be felt in the extent of healthcare systems, in addition to the individual patients [30,31]. Innovative answers using virtual technology can offer assistance for reforms within healthcare and long-term care systems. The world now speaks of how "Artificial Intelligence and supercomputing provide new possibilities to convert healthcare systems." AI in hospital treatment is now a reality. Gadgets depending on AI are being worn which have new complexity and even picture processing [84]. Algorithms and calculations are currently being applied in mammography for breast malignancy location, in CT for colon ailment analysis, in chest radiographs for the identity of pulmonary nodules, in MRI for cerebrum increase division, and for locating neurologic problems, like Alzheimer's disease. Calculations can help dermatologists with enhancing analyses, for example figuring out more than 90% of pores and skin malignant growths through substantial preparations of scientific pictures. The improvements could be felt at the level of clinical finding, simply through individual patients [32–36] (Fig. 16.3).

AI in health care is now a reality. Medical care suppliers are beginning to insert this innovation into their work processes and the dynamic cycles at all phases of medical services, bringing enhancements for patients, suppliers, payers, and other medical care partners. The benefits incorporate

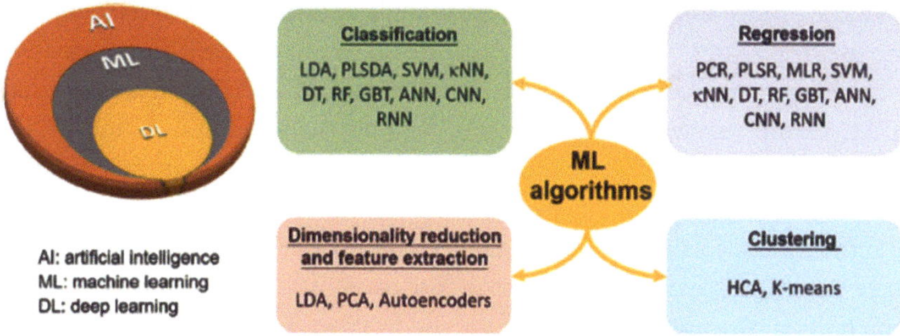

Figure 16.3 Relationship between AI, ML, and DL (Deep learning), and various ML algorithms involved in this review. *AI*, Artificial intelligence; *CNN*, convolutional neural network; *DT*, decision tree; *Feedforward ANN*, feedforward artificial neural network; *GBT*, gradient-boosted trees; *HCA*, hierarchical clustering analysis; *κNN*, κ-nearest neighbor; *LDA*, linear discriminant analysis; *ML*, machine learning; *MLR*, multiple linear regression; *NB*, Naive Bayes; *PCA or PCR*, principal component analysis or regression; *PLSDA or PLSR*, partial least-squares discriminant analysis or regression; *RF*, random forest; *RNN*, recurrent neural network; *SVM*, support vector machine. *With permission from Advancing Biosensors with Machine Learning Feiyun Cui et al., 2020.*

new treatment openings, for instance, in perceiving such wounds that could have ordinarily been dismissed, to back-office persons, for instance, leaders and other equipment support system engineers [37,38].

With AI, patients will receive more quick and precise judgments than is currently available and AI will aid in dynamics, taking into consideration custom-made treatment mediations with higher first-time achievement rates. Medical service conveyances can be improved, and unfavorable outcomes decreased. Customizing care creates for more compelling consideration, less exercising in futility and assets, and increased patient fulfillment. There are massive expense reserve funds to be had from conveying AI as well—especially because of the better-designated care it grants. It can even facilitate a portion of the developing strains in staff deficiencies across medical care communities [39].

The most thrilling developments are in cutting-edge care, where patients will, after some time, have an advantage. There is an enormous scope of patient-announced side effects or biometrics and therefore AI can recommend expected analyses and follow-up stages for these patients.

AI permits the checking in on patients to further develop results. Patients and doctors approach persistent appraisal and backing in areas like consistency, through applications such as a cell phone interfaces that guarantees patients take their drugs accurately; the scope of AI chatbots are intended to offer constant help and guidance to the patient getting treated [40,41].

Exploration and development in analysis and treatment are likewise in line with gigantic advances due to joining new volumes of information with the preparation limit of supercomputers and the insightful resources of AI. Computer-based intelligence opens up the way to a switch in disclosure techniques for drug improvement. For example, another battling item now in clinical preliminaries is from AI-based checking of large sets of information focused on both dangerous and nonharmful cells. This has been in front of the conventional disclosure strategy for a long time, which is at the core of exemplary medication revelation—a technique with an under 10% achievement rate.

AI-based models are now helping with clinical preliminary plans and patient enlistment, and AI-based checking frameworks are being utilized to support adherence and lessening dropout rates. "Artificial intelligence is anything but an enchantment shot and is a lot of a work in progress, yet it holds a lot of guarantees for the fate of medical care and medication advancement," says PC researcher Stefan Harrer in Trends in Pharmacological Sciences [42].

The AI process is state-of-the-art diagnostics and is being driven to provide multiple types of information, including patient-published symptoms, biometric and biosensor information, imaging, and biomarker information, for continuous treatment from the onset of an emergency. Sensors detect symptoms early on, predict potential well-being episodes, and enable preventative measures [43]. AI, which submits applications related to oncology, nervous system science, and analysis of cardiovascular disease, has shown powerful results. According to a study at the Belgian University of Ruben Catholic, AI is twice as accurate as a pulmonologist in decoding breath test results; with 1430 Belgian patients, compared to 45% of specialists, 82% of the AI effectively diagnosed the essential illness [44].

AI as advanced personalized care segment empowers the advancement of clinical pathways for individual patients. In complex infections, like disease, new difficulties are arising with new science, and AI offers a possible pathway to the goal. Where successful medication mixes are needed for a critical restorative effect, as is progressively the case today, it turns out to be harder to plan viable medication blends and to pick the right medication mix for the proper understanding as the medications become more explicit to subatomic targets [45,46].

Medical care utilizes AI to anticipate which therapies would be most advantageous to individual patients, and applications can give wellbeing counsel dependent on a client's particular hereditary make-up. Computerized advancements would themselves be able to empower residents to get to data about wellbeing hazard factors and prosperity measures, and can assist them with taking part in the appropriate wellbeing conduct and infection avoidance [47–49].

Chemometrics assume a basic part in biosensors-based recognition, investigation, and analysis. These days, as a part of man-made reasoning, machine learning (ML) has accomplished great advances. Nonetheless, novel progressed ML techniques, particularly profound realizing, which is popular for picture investigation, facial recognition, and discourse acknowledgment, has remained moderately tricky to the biosensor local area. In this chapter, how ML can be advantageous to biosensors is methodically examined. The benefits and drawbacks of the most famous ML calculations are summarized based on a detecting information examination. Exceptionally, profound learning techniques, for example, convolutional neural organization (CNN) and repetitive neural organization (RNN) are stressed. Various ML-helped electrochemical biosensors, wearable hardware, SERS and other spectra-based biosensors, fluorescence biosensors, and colorimetric biosensors are completely explained. Moreover, biosensor networks and multi-biosensor information combinations are presented [21].

Biosensors-based gadgets are changing the clinical determination of illnesses through observing patient signs. The advancement of savvy and computerized molecular diagnostic tools outfitted with enormous biomedical examination information, distributed computing, and clinical man-made brainpower can be an optimal methodology for the recognition and checking of infections, treatment, and for strong medical choices. This survey zeroed in on the utilization of AI approaches for the improvement of modern CRISPR biosensors dependent on CPUs and the utilization of the Internet of Things for remote transmission of signs over the cloud for continous help. The current survey additionally examined the revelation of CRISPR, its use as a quality-altering apparatus, and CRISPR-based biosensors with a high affectability of Attomolar (10^{-18} M), Femtomolar (10^{-15} M), and Picomolar (10^{-12} M) in contrast to traditional biosensors with an affectability of nanomolar 10^{-9} M and micromolar 10^{-3} M. Furthermore, the survey showed limits and openly explored issues with the current state of CRISPR-based biosensing applications [50].

The improvement in the field of AI and ML approaches offers another measurement in the field of medication and medical services. These strategies have been effectively utilized for managing issues in biomedical sciences by utilizing displaying and prescient techniques. Clinicians alongside PC masters have been cooperating to develop new strategies for finding and observing illnesses like a malignant growth, pathogenic infections, hereditary illnesses, cardiovascular infections, etc. [51]. The utilization of information mining in medication contributes to the comprehension of biomedical information, determination, and expectation of illnesses. The shrinking of computational expenses and expanding amount of information has permitted diverse ML calculations to assume an indispensable part in the biosensing framework for clinical findings [52]. General procedures utilized for SML incorporate support vector machine (SVM), neural networks (NNs) random forest, decision tree, etc. [53–56]. Solo ML utilizes an unlabeled dataset and models determine how to anticipate results dependent on designs gained from input information. Fundamental Unsupervised ML calculations incorporate bunching and rule mining. The most used ML approach is supported where learning measures depend on the utilization of programming to consequently decide the best exhibition inside a particular setting. The advances in amalgamations of biosensors with AI have prompted the improvement of keen biosensors. Chang et al. [57] created XPRIZE DeepQ Tricorder biosensor that can analyze 12 distinct problems like tuberculosis, diabetes, sickliness, pneumonia, atrial fibrillation, hepatitis A, urinary parcel contamination, otitis media, stroke, leukocytosis, persistent obstructive aspiratory infection (COPD) and rest apnea, and catch five ongoing fundamental signs like respiratory rate, internal heat level, oxygen immersion, ECG and pulse. A scientist has also reported the applied CNN model for expectation of bacterial pollutions utilizing information acquired from photonic precious stone-based optical biosensor [58]. Connecting biosensors with ML approaches will improve expectation, analysis and clinical dynamics [59]. A man-made AI approach was created for the expectation of hypoglycemia (low blood glucose level). The approach utilizes long-transient memory (LSM) and the outcome has shown superior organization in expectation of hypoglycemia after thirty minutes [60]. AI was applied to further develop identification of low blood glucose level from information procured from noninvasive sensors. Dataset incorporate pulse, skin air temperature and galvanic skin reaction. The approach used SVM to prepare and characterize information into hypoglycemic and hyperglycemic [85].

Cardiovascular infections (CVD) and stroke are the leading causes of death worldwide. Early and rapid detection is essential for a fruitful assessment of cardiovascular disease and stroke. Over the last decade, many biosensors have been developed to detect a wide range of cardiac marker to reduce the costs for healthcare. One of the major challenges is to find a way of predicting the risk that an individual can suffer from CVD. There has been considerable interest in finding diagnostic and prognostic biomarkers that can be detected in blood and predict CVD risk. Of these, C-reactive protein (CRP) is the best known biomarker followed by cardiac troponin I or T (cTnI/T), myoglobin, lipoprotein-associated phospholipase A(2), interlukin-6 (IL-6), interlukin-1 (IL-1), low-density lipoprotein (LDL), myeloperoxidase (MPO) and tumor necrosis factor alpha (TNF-) has been used to predict cardiovascular events [61] Cardiovascular diseases (CVDs)—diseases of the heart and blood vessels—remain the most prevalent causes of death worldwide accounting for 31% of recorded annual mortalities [1]. According to the World Health Organization, cardiovascular diseases include heart failure, rheumatic heart disease, arrhythmia, coronary heart disease, atherosclerosis, atrial fibrillation and cerebrovascular disease [62]. There have been great advances in the era of biosensors, but in reality there are some limitations. Most of the biosensors manufactured follow the traditional approach, are tested in a central research center, and it takes hours or days to manufacture the final product. In addition, in cardiovascular assessments, the patient must meet two of the three conditions: adjustment of blood biomarker levels, characteristic chest pain, and symptomatic electrocardiogram (ECG). In this sense, there is general interest in the finer, more robust and practical symptomatic stages that also help in the continuous identification and monitoring of the severity of CVD patients. With continued progress in the field of artificial consciousness, AI and its effective use in biomedical sciences can be used to develop new displays and predictive technologies for clinical practice, including heart disease. New areas and equipment have been created [63].

AI was termed by John McCarthy in 1956 through a convention delivered on this subject [64,65]. However, the opportunity of machines being capable of simulating human conduct was raised in advance with by Alan Turing who created the Turing check to be able to differentiate people from machines. Since then, computational energy has grown to the factors of spontaneous calculations and high potential to compare new data, consistent with formerly assessed data, in actual time [66,67]. Today, AI is incorporated into our day-to-day lives in many forms, inclusive of private assistants (Siri, Alexa, Google assistant, etc.), computerized mass transportation, aviation and PC gaming. More recently, AI has additionally started to be integrated into medication to enhance the affected person's care with the aid of using dashing up approaches and accomplishing more accuracy; establishing the direction to offering improved healthcare overall. Radiological images, pathology slides, and the patient's digital scientific records (EMR) are being evaluated with the aid of using system learning, helping with the method of analysis to remedy sufferers and augmenting the physicians' capabilities.

As the world fights the COVID-19 pandemic, man-made brainpower, epidemiological investigation and novel detecting advances can assume key parts in alleviating the effects of this extraordinary worldwide emergency. The areas of profound learning, mathematical demonstration and bio detecting have quickly reacted with answers for this pandemic situation through the advancement of new techniques for early identification and gauging of the effects of the COVID-19 infection [86,87]. New applications in the space of profound learning-based PC vision and image processing instruments are talked about which have arisen with promising early COVID-19 identification techniques, utilizing clinical trials. What's more, an assortment of prescient models created in the current year for assessing the transmission, contamination and death rates due to the novel COVID-19 virus are depicted which give urgent data for deciding social and legislative measures for containing the viral episode. Lastly, the best in class clinical test techniques is utilizing nanolevel sensors for behind-care testing which could empower quick and accurate findings of the COVID-19 diseases [68–73].

To deal with the COVID-19 pandemic, advancement of fast, specific, delicate symptomatic frameworks for beginning phase β-Covid extreme intense respiratory condition (SARS-CoV-2) infection protein recognition is arising as a vital reaction to create the bioinformatics required for proficient diagnostics, enhancement of treatment, and examination of treatments of higher quality. The requirement for such symptomatic frameworks is prescribed by specialists to accomplish the mass and designated SARS-CoV-2 recognition needed to manage the COVID-19 pandemic through the comprehension of disease movement and ideal treatment choices. To accomplish these, there is a need for creating shrewd sensors to quickly and specifically distinguish SARS-CoV-2 protein at the picomolar level. Coronavirus disease, because of human-to-human transmission, requests diagnostics at the point-of-care (POC) without the need for experienced workers and modern research centers. Therefore, we propose to investigate the compartmentalization approach by planning and creating nanoenabled scaled-down electrochemical biosensors to recognize SARS-CoV-2 infection at the site of the scourge as the most ideal way to manage the pandemic. Such a diagnostics approach is dependent on a POC detection and can be communicated with the Internet of things and computerized reasoning (AI) procedures, (e.g., AI and deep learning for diagnostics) for exploring helpful informatics through information stockpiling, sharing, and examination [74,75]. In light of COVID-19 administrative-related difficulties and viewpoints, our work in this survey presents an aggregate methodology including electrochemical SARS-CoV-2 biosensing upheld by AI to create the bioinformatics required for the beginning phase of COVID-19 analysis, connection of viral burden with pathogenesis, comprehension of pandemic movement, treatment advancement, POC diagnostics, and infections in a customized way.

Significant advances over the last decade have made the sensor easier to use. This white paper defines the usefulness of sensors as a function of systems that provide intelligence. This is done by collecting data, extracting information (e.g., from data analysis), and interpreting that information into actionable information. Many analytical tools are available, including ML, AI, and MatLab's toolbox that have changed the way information and intelligence are extracted and realized. Therefore, you need to be cautious that you are getting the correct information. It has been reported that the system strategy for data analytics and examples of intelligence are extracted specifically from an application perspective. To illustrate, this example SensorComm's emission monitoring solution, WiNOx (https://www.sensorcommtech.com/) has been introduced. The system was developed to monitor emissions from vehicles, and more recently modified to monitor individuals remotely for healthcare applications, and now to monitor natural gas for oil and gas applications.

While wearable and versatile synthetic sensors have encountered enormous development over the previous decade, their potential for monitoring and directing treatment has arisen in recent years. As of now, rules from specialists and dieticians address the most well known methodology for maintaining an ideal sustenance status. In any case, such proposals depend on populace midpoints and don't consider individual inconstancy in reacting to supplements. Accurate nourishment has arisen to address the enormous heterogeneity in people's reactions to eat less carbs, by noting sustenance is dependent on the particular necessities of every individual. It targets forestalling and overseeing sicknesses by defining customized dietary protocols to people based on their metabolic profile, foundation, and willingness. Recent advances in computerized nourishment innovation, including calorie-counting versatile applications and wearable GPS movement beacons, do not have the capacity to monitor sustenance at the atomic level. The acknowledgment of successful nourishment requires collaboration from various sensor modalities to make convenient expectations and effective input.

Work done by a group of scientists surveyed key opportunities and difficulties in the successful acceptance of viable wearable and portable sustenance checking platforms. Painless wearable and portable electrochemical sensors, fit for checking transient substance varieties upon the admission of food and enhancements, are a fantastic contender to overcome issues among computerized and biochemical examinations for a successful customized sustenance approach. By giving opportune (beforehand inaccessible) dietary data, such wearable and versatile sensors create the direction essential for supporting dietary conduct change to reach nourishment equilibrium. Combining the quickly arising wearable substance-detecting gadgets which produce unique scientific information with information mining techniques that recognize examples and make forecasts, are relied upon to interrupt dietary decision-making towards more viable sustenance [76].

One of the reports [77] show that to foster a compact remote keen nanosensor (PWIN) for fast, practical on-location assurance of terbutaline (TRA) buildup in meat items, it must be combined with customary nanosensors and the superior execution of fluid chromatography (HPLC). The layer-by-layer sandwiched nanohybrid created by platinum-palladium nanoparticles, carboxylatedgraphene and graphene-like molybdenum disulfide, showed a wide and detection of 0.55–14.9 μmol/L, utilizing the versatile potentiostat with a cell phone. The outcome was near the direct reach (0.4–14 μmol/L), using the customary potentiostat with a work station for TRA. The constraint of recognitions were distinguished as 0.44 and 0.18 μmol/L, respectively. PWIN showed palatable recuperation (91%–98.43%) of TRA in examples by the standard expansion strategy and in examination with both a conventional sensor (93.79%–98%) and HPLC (93.4%–98.6%), uncovering that PWIN for fast, nearby investigation in the food handling field is possible.

Mechano-computational strategies related to man-made brainpower (AI) are upsetting the translations of pivotal data from clinical information and changing it into enhanced and coordinated data for diagnostics. It is conceivable, because of important flawlessness in man-made brainpower, PC-supported diagnostics, remote helpers, automated medical procedures, expanded reality and genome altering (in light of AI) innovations. Such procedures are developing as the items for diagnosing new microbial or nonmicrobial infections. This article addresses a procedure for using such methodologies and giving helpful arrangements for using these strategies in infection diagnostics [78].

Biosensor-based gadgets are spearheading the advanced biomedical applications and will be the eventual fate of cardiac medical services. The coupling of AI for heart checking-based biosensors for POC diagnostics is conspicuously assessed here. This assessment interprets the main AI calculations for the modern biosensor alongside the Internet of Things, computational methods and central processor-based cardiovascular biomarkers for constant wellbeing checking. This review also examines the current cardiovascular biosensors along with specialized methodologies associated with their instruments and their applications in medical services. Moreover, it gives a key to the ontogeny of a powerful and strong progressive convention for clinical decision-making about customized medication through data investigation, and coordinated multidisciplinary AI.

A significant test for Education 4.0 is to utilize wearable gadgets to aid in observing their learning conduct and exercises (steps, pulse fluctuation, and pulse) continuously. The main point of this paper is to introduce our execution of adaptivity and AI strategies inside the Education 4.0 interaction. In this work, we research implanted biosensors (harmless, minimal expense, and interruption-free) utilized in cell phones and smartwatches. The goal is to empower higher education. For example, a clever framework helped by AI that considers inserted biosensor information and natural information to assess the understudies' wellbeing. In such a manner, we propose a system that utilizes wearable gadgets to gather information with biofeedback techniques to help with scholastic achievement [20,79,80].

There is an outbreak of COVID-19 and its variants around the world. Execution of customary wellbeing measures, for example, segregation and quarantine stays, is the best way to control the episode. At the point when an asymptomatic person with COVID-19 is being isolated, it is important to perform temperature and side effect observation. As such investigation is discontinuous in nature and exceptionally reliant upon self-restraint, it has restricted adequacy. Biosensor advancements have made it conceivable to persistently screen physiological boundaries using wearable biosensors with an assortment of structural factors. It is necessary to investigate the capacity of involving wearable biosensors for constantly screening multifaceted physiological boundaries for early recognition of COVID-19 clinical movement. This randomized, controlled open-named preliminary includes 200−1000 asymptomatic subjects with close COVID-19 contact under required quarantine at assigned offices in Hong Kong. Subjects were randomized to get a remote observation methodology (mediation gathering) or standard system (control bunch) in a 1:1 proportion during the multi day-quarantine period. Notwithstanding fever and indication inspection in the benchmark group, subjects in the intercession gathering wore wearable biosensors on their arms to constantly screen skin temperature, respiratory rate, circulatory strain, heart rate, blood oxygen immersion and everyday exercises. These physiological boundaries are moved continuously to a cell phone application called Biovitals Sentinel. This information will then be handled utilizing a cloud-based multivariate physiology investigation motor called Biovitals to recognize simple physiological changes. The outcomes will be shown on an online dashboard for the clinicians' review. The primary result is the opportunity to end COVID-19 [81].

AI vows to give valuable data to clinicians representing considerable authority in hypertension. As of now, there are a few critical AI applications on enormous approved information collections. A report presents the utilization of AI to foresee clinical results in huge amounts of information. For example, information with high volume, assortment, veracity, speed and worth. Four models are remembered for this audit. In the principal model, profound learning and backing vector machine (SVM) anticipated the event of cardiovascular occasions with 56%–57% precision. In the subsequent model of an information base of 378,256 patients, a neural network calculation anticipated the event of cardiovascular incidents during long-term back to responsiveness (68%) and particularity (71%). In the third model, an AI calculation characterized 1,504,437 patients on the presence or nonappearance of hypertension with 51% responsiveness, almost 100% particularity and region under the bend 87%. In model four, wearable biosensors and convenient gadgets were used to evaluate an individual's danger of developing hypertension by using photo-plethysmography to isolate people who were at risk of hypertension, with responsiveness higher than 80% and positive prescient higher than 90%. The effects of the above examinations were adapted to socioeconomics and the expected risk factors for atherosclerotic sickness. These models depict the utilization of computerized reasoning strategies in the field of hypertension [82].

Worldwide, the pandemic is indicative of increased occurrences of infectious diseases brought about by microbes. The rapid spread of these illnesses is compromising the general population health and possibly, the economy. The lack of simple, demonstrative instruments, instead of a tedious and complex research center-based finding process, is a significant quest for clinical consideration. At present, the advancement of reasoning behind POCT is sought after for quick identification of illnesses alongside "on location" results that are useful in for early diagnosis and better treatment. What's more, POCT gadgets assume an essential part in forestalling the transmission of illnesses by offering constant testing and lab quality microbial findings in less time. Ideal conclusions and further treatment improvements work to control the reemergence of these sicknesses. By and by, endeavors are being made to help such POCT by the innovative advancement in the field of the Internet of Medical Things (IoMT). The IoMT offers remote-based activity and availability of POCT gadgets with wellbeing mastery and a clinical focus. In this survey, the new POC diagnostics coordinated or future prospects of incorporating with IoMT are examined, centered around new and recurring illnesses like intestinal sickness, dengue fever, influenza A (H1N1), human papilloma infection (HPV), Ebola infection illness (EVD), Zika infection (ZIKV), and (COVID-19). The IoMT-helped POCT frameworks are fit to fill the gap between bioinformatics age, fast examination, and clinical approval. An advanced IoMT-mediated POCT will be valuable in understanding the illness' movement, treatment choice, and assessing the viability of treatment [83].

16.2 Conclusion

AI in healthcare is a reality today. Health care providers are beginning to innovate their work processes and dynamic cycles at every stage of their delivery, bringing improvements to patients, healthcare providers, payers and other healthcare partners. Benefits include new treatments (e.g., identifying suspicious injuries that may have been overlooked) and the future in this field is yet to be explored.

References

[1] P. Mehrotra, Biosensors and their applications – a review, J. Oral. Biol. Craniofacial Res. (2016).

[2] A Mathur, S Dave, S. Dave, Predicting Chemical Reactions with Artificial Intelligence and Machine Learning. 2021.

[3] L. Lu, Recent advances in synthesis of three-dimensional porous graphene and its applications in construction of electrochemical (bio)sensors for small biomolecules detection, Biosens. Bioelectron. (2018).

[4] S. Kumar, R. Singh, Recent optical sensing technologies for the detection of various biomolecules: review, Opt. Laser Technol. (2021).

[5] E. Prabakaran, K. Pillay, Electrochemical detection of 4-nitrophenol by using graphene based nanocomposite modified glassy carbon electrodes: a mini review, Nanoarchitectonics (2021).

[6] S. Radhakrishnan, H.Y. Kim, B.S. Kim, A novel CuS microflower superstructure based sensitive and selective nonenzymatic glucose detection, Sens. Actuators, B Chem. (2016).

[7] A.M.V. Mohan, V. Rajendran, R.K. Mishra, M. Jayaraman, Recent advances and perspectives in sweat based wearable electrochemical sensors, TrAC. - Trends Anal. Chem. (2020).

[8] A. Venkadesh, J. Mathiyarasu, S. Dave, S. Radhakrishnan, Amine mediated synthesis of nickel oxide nanoparticles and their superior electrochemical sensing performance for glucose detection, Inorg. Chem. Commun. (2021).
[9] S. Dave, S. Dave, J. Das, Electrochemical monitoring as an emerging technology for detection of environmental pollutants with special reference to pesticides, Future Effl. Treat. Plants (2021).
[10] S. Dave, S. Kumbhat, Electrochemical and spectral characterization of silver nanoparticles synthesized employing root extract of curculigo orchioides, Indian J. Chem. Technol. (2018).
[11] C. Griesche, A.J. Baeumner, Biosensors to support sustainable agriculture and food safety, TrAC. - Trends Anal. Chem. (2020).
[12] G.K. Mishra, A. Barfidokht, F. Tehrani, R.K. Mishra, Food safety analysis using electrochemical biosensors, Foods (2018).
[13] L. Alvarado-Ramírez, M. Rostro-Alanis, J. Rodríguez-Rodríguez, J.E. Sosa-Hernández, E.M. Melchor-Martínez, H.M.N. Iqbal, et al., Enzyme (Single and multiple) and nanozyme biosensors: recent developments and their novel applications in the water-food-health nexus, Biosensors (2021).
[14] Z. Kotsiri, J. Vidic, A. Vantarakis, Applications of biosensors for bacteria and virus detection in food and water—A systematic review, J. Environ. Sci. (China) (2022).
[15] Y. Ye, J. Ji, Z. Sun, P. Shen, X. Sun, Recent advances in electrochemical biosensors for antioxidant analysis in foodstuff, TrAC. - Trends Anal. Chem. (2020).
[16] H. Kumar, K. Kuča, S.K. Bhatia, K. Saini, A. Kaushal, R. Verma, et al., Applications of nanotechnology in biosensor-based detection of foodborne pathogens, Sensors (Switz.) (2020).
[17] J. Zhou, Q. Qi, C. Wang, Y. Qian, G. Liu, Y. Wang, et al., Surface plasmon resonance (SPR) biosensors for food allergen detection in food matrices, Biosens. Bioelectron. (2019).
[18] X. Jin, C. Liu, T. Xu, L. Su, X. Zhang, Artificial intelligence biosensors: challenges and prospects, Biosens. Bioelectron. (2020).
[19] R. Vashistha, A.K. Dangi, A. Kumar, D. Chhabra, P. Shukla, Futuristic biosensors for cardiac health care: an artificial intelligence approach, 3 Biotech (2018).
[20] Ciolacu MI, Binder L, Popp H. Enabling IoT in Education 4.0 with BioSensors from wearables and artificial intelligence, in: Proceedings of the IEEE Twenty-fifth International Symposium for Design and Technology in Electronic Packaging, SIITME 2019. 2019.
[21] F. Cui, Y. Yue, Y. Zhang, Z. Zhang, H.S. Zhou, Advancing Biosensors with Machine Learning, ACS Sensors (2020).
[22] S. Balamurugan, Biosensors based artificial intelligence driven healthcare robots, J. Diagnostic Tech. Biomed. Anal. (2018).
[23] D. Horgan, M. Romao, S.A. Morré, D. Kalra, Artificial intelligence: Power for civilisation - and for better healthcare, Public. Health Genomics (2020).
[24] K.S. Lata, S. Kumar, V. Vaghasia, P. Sharma, S.B. Bhairappanvar, S. Soni, et al., Exploring Leptospiral proteomes to identify potential candidates for vaccine design against Leptospirosis using an immunoinformatics approach, Sci. Rep. (2018).
[25] S. Kumar, K.S. Lata, P. Sharma, S.B. Bhairappanavar, S. Soni, J. Das, Author Correction: Inferring pathogen-host interactions between *Leptospira interrogans* and *Homo sapiens* using network theory, Sci. Rep. 9 (1) (2019) 1434. Available from: https://doi.org/10.1038/s41598-018-38329-1.
[26] R. Khandia, S. Singhal, U. Kumar, A. Ansari, R. Tiwari, K. Dhama, et al., Analysis of nipah virus codon usage and adaptation to hosts, Front. Microbiol. (2019).
[27] C. Castelluccia, D. Le Métayer, Understanding algorithmic decision-making: opportunities and challenges, Eur. Parliam. Res. Serv. (2019).

[28] K.G. van Leeuwen, S. Schalekamp, M.J.C.M. Rutten, B. van Ginneken, M. de Rooij, Artificial intelligence in radiology: 100 commercially available products and their scientific evidence, Eur. Radiol. (2021).
[29] A. Hosny, C. Parmar, J. Quackenbush, L.H. Schwartz, H.J.W.L. Aerts, Artificial intelligence in radiology, Nat. Rev. Cancer (2018).
[30] A.P. Brady, E. Neri, Artificial intelligence in radiology-ethical considerations, Diagnostics (2020).
[31] A. Teramoto, Application of artificial intelligence in radiology, Japanese J. Cancer Chemother. (2019).
[32] T. Davenport, R. Kalakota, The potential for artificial intelligence in healthcare, Futur. Healthc. J. (2019).
[33] J.M. Puaschunder, The Potential for Artificial Intelligence in Healthcare, SSRN Electron. J. (2020).
[34] T. Davenport, R. Kalakota, DIGITAL TECHNOLOGY The potential for artificial intelligence in healthcare, Fut. Healthcare J. (2019).
[35] C. Hayhurst, Get ting smarter: the promise and potential of artificial intelligence in healthcare, Biomed. Instrum. Technol. (2019).
[36] L. Sun, R.K. Gupta, A. Sharma, Review and potential for artificial intelligence in healthcare, Int. J. Syst. Assur. Eng. Manag. (2021).
[37] N. Mehta, A. Pandit, S. Shukla, Transforming healthcare with big data analytics and artificial intelligence: a systematic mapping study, J. Biomed. Inform. (2019).
[38] J.J. Hopkins, P.A. Keane, K. Balaskas, Delivering personalized medicine in retinal care: from artificial intelligence algorithms to clinical application, Curr. Opin. Ophthalmol. (2020).
[39] D. Gruson, Artificial intelligence in healthcare: major potential for innovations in our health system, Soins (2019).
[40] M. Kyrarini, F. Lygerakis, A. Rajavenkatanarayanan, C. Sevastopoulos, H.R. Nambiappan, K.K. Chaitanya, et al., A survey of robots in healthcare, Technologies (2021).
[41] S. Romiti, M. Vinciguerra, W. Saade, I. Anso Cortajarena, E. Greco, Artificial intelligence (AI) and cardiovascular diseases: an unexpected alliance, Cardiology Res. Pract. (2020).
[42] S. Harrer, P. Shah, B. Antony, J. Hu, Artificial intelligence for clinical trial design, Trends Pharmacol. Sci. (2019).
[43] E.H. Weissler, T. Naumann, T. Andersson, R. Ranganath, O. Elemento, Y. Luo, et al., The role of machine learning in clinical research: transforming the future of evidence generation, Trials (2021).
[44] P. Venkatesan, Annual congress of the European respiratory society, Lancet. Respiratory Med. 2018 (2018).
[45] T. Habuza, A.N. Navaz, F. Hashim, F. Alnajjar, N. Zaki, M.A. Serhani, et al., AI applications in robotics, diagnostic image analysis and precision medicine: current limitations, future trends, guidelines on CAD systems for medicine, Inform. Med. Unlocked (2021).
[46] K.B. Johnson, W.Q. Wei, D. Weeraratne, M.E. Frisse, K. Misulis, K. Rhee, et al., Precision medicine, AI, and the future of personalized health care, Clin. Transl. Science. (2021).
[47] R. Sparrow, J. Hatherley, High hopes for "deep medicine"? AI, economics, and the future of care, Hastings Cent. Rep. (2020).
[48] A.T. Greenhill, B.R. Edmunds, A primer of AI in medicine, Tech. Gastrointest. Endoscopy (2019).
[49] A. Jean, A brief history of artificial intelligence, Medecine/Sciences (2020).
[50] A.U. Ibrahim, F. Al-Turjman, Z. Sa'id, M. Ozsoz, Futuristic CRISPR-based biosensing in the cloud and internet of things era: an overview, Multimed. Tools Appl. (2020).

[51] M. Milano, The digital skills gap is widening fast. Here's how to bridge it | World Economic Forum, World Econ. Forum (2019).
[52] Y. Juhn, H. Liu, Artificial intelligence approaches using natural language processing to advance EHR-based clinical research, J. Allergy Clin. Immunol. (2020).
[53] S. Huang, C.A.I. Nianguang, P. Penzuti Pacheco, S. Narandes, Y. Wang, X.U. Wayne, Applications of support vector machine (SVM) learning in cancer genomics, Cancer Genomics Proteom. (2018).
[54] F. Cumbo, E. Cappelli, E. Weitschek, A brain-inspired hyperdimensional computing approach for classifying massive DNA methylation data of cancer, Algorithms (2020).
[55] S. Patil, K. Habib Awan, G. Arakeri, C. Jayampath Seneviratne, N. Muddur, S. Malik, et al., Machine learning and its potential applications to the genomic study of head and neck cancer—a systematic review, J. Oral. Pathol. Med. (2019).
[56] R.K. Maurya, S.K. Yadav, P. Tiwari, Timely recognition of breast cancer (BC) via support vector machines (SVMS) algorithms-a study, Int. J. Adv. Sci. Technol. (2020).
[57] E.Y. Chang, M.H. Wu, K.F. Tang, H.C. Kao, C.N. Chou, Artificial intelligence in XPRIZE DeepQ tricorder, in: MMHealth 2017 - Proceedings of the Second International Workshop on Multimedia for Personal Health and Health Care, co-located with MM 2017. 2017.
[58] S.A. Nehal, D. Roy, M. Devi, T. Srinivas, Highly sensitive lab-on-chip with deep learning AI for detection of bacteria in water, Int. J. Inf. Technol (2020).
[59] T.M. Quan, T. Doike, D.C. Bui, K. Hayashi, S. Arata, A. Kobayashi, et al. AI-based edge-intelligent hypoglycemia prediction system using alternate learning and inference method for blood glucose level data with low-periodicity, in: Proceedings of the IEEE International Conference on Artificial Intelligence Circuits and Systems, AICAS 2019. 2019.
[60] Marling C, Lijie X, Razvan BC, Schwartz FL. Machine learning experiments with noninvasive sensors for hypoglycemia detection. Knowl. Discov. Healthc. Data co-located Int. Jt. Conf. Artif. Intell. 2016.
[61] A. Qureshi, Y. Gurbuz, J.H. Niazi, Biosensors for cardiac biomarkers detection: a review, Sens. Actuators, B: Chem. (2012).
[62] H. Alawieh, T.El Chemaly, S. Alam, M. Khraiche, Towards point-of-care heart failure diagnostic platforms: Bnp and nt-probnp biosensors, Sensors (Switz.) (2019).
[63] I. Kavakiotis, O. Tsave, A. Salifoglou, N. Maglaveras, I. Vlahavas, I. Chouvarda, Machine learning and data mining methods in diabetes research, Comput. Struct. Biotechnol. J. (2017).
[64] L. Morgenstern, S.A. McIlraith, John McCarthy's legacy, Artif. Intell. (2010).
[65] S.L. Andresen, John McCarthy: father of AI, IEEE Intell. Syst. (2002).
[66] G. Strawn, Alan turing, IT Prof. (2014).
[67] S. Muggleton, Alan turing and the development of artificial intelligence, AI Commun (2014).
[68] R. Vaishya, M. Javaid, I.H. Khan, A. Haleem, Artificial intelligence (AI) applications for COVID-19 pandemic, Diabetes Metab. Syndr. Clin. Res. Rev. (2020).
[69] D.C. Nguyen, M. Ding, P.N. Pathirana, A. Seneviratne, . Blockchain and AI-based solutions to combat coronavirus (COVID-19)-like epidemics: a survey, IEEE Access. (2021).
[70] K.C. Santosh, AI-driven tools for coronavirus outbreak: need of active learning and cross-population train/test models on multitudinal/multimodal data, J. Med. Syst. (2020).
[71] H. Greenspan, R. San José Estépar, W.J. Niessen, E. Siegel, M. Nielsen, Position paper on COVID-19 imaging and AI: from the clinical needs and technological challenges to initial AI solutions at the lab and national level towards a new era for AI in healthcare, Med. Image Anal. (2020).
[72] O. Dogan, S. Tiwari, M.A. Jabbar, S. Guggari, A systematic review on AI/ML approaches against COVID-19 outbreak, Complex. Intell. Syst. (2021).

[73] A.S. Ahuja, V.P. Reddy, O. Marques, Artificial intelligence and COVID-19: a multidisciplinary approach, Integr. Med. Res. (2020).

[74] A. Serag, A. Ion-Margineanu, H. Qureshi, R. McMillan, M.J. Saint Martin, J. Diamond, et al., Translational AI and deep learning in diagnostic pathology, Front. Med. (2019).

[75] P. Hunter, The advent of AI and deep learning in diagnostics and imaging, EMBO Rep. (2019).

[76] M. Parrilla, A. Vanhooydonck, R. Watts, K. De Wael, Wearable wristband-based electrochemical sensor for the detection of phenylalanine in biofluids, Biosens. Bioelectron. (2022).

[77] Y. Ge, P. Liu, L. Xu, M. Qu, W. Hao, H. Liang, et al., A portable wireless intelligent electrochemical sensor based on layer-by-layer sandwiched nanohybrid for terbutaline in meat products, Food Chem. (2022).

[78] M. Marino, G. Vairo, P. Wriggers, Mechano-chemo-biological computational models for arteries in health, disease and healing: from tissue remodelling to drug-eluting devices, Curr. Pharm. Des. (2020).

[79] Z. Chen, J. Zhang, X. Jiang, Z. Hu, X. Han, M. Xu, et al., Education 4.0 using artificial intelligence for students performance analysis, Intel. Artif. (2020).

[80] M. Ciolacu, A.F. Tehrani, L. Binder, P.M. Svasta. Education 4.0 - artificial intelligence assisted higher education: early recognition system with machine learning to support students' success, in: Proceedings of the IEEE Twenty-fourth International Symposium for Design and Technology in Electronic Packaging, SIITME 2018. 2019.

[81] C.K. Wong, D.T.Y. Ho, A.R. Tam, M. Zhou, Y.M. Lau, M.O.Y. Tang, et al., Artificial intelligence mobile health platform for early detection of COVID-19 in quarantine subjects using a wearable biosensor: protocol for a randomised controlled trial, BMJ Open (2020).

[82] D. Amaratunga, J. Cabrera, D. Sargsyan, J.B. Kostis, S. Zinonos, W.J. Kostis, Uses and opportunities for machine learning in hypertension research, Int. J. Cardiol. Hypertens. (2020).

[83] S. Jain, M. Nehra, R. Kumar, N. Dilbaghi, T.Y. Hu, S. Kumar, et al., Internet of medical things (IoMT)-integrated biosensors for point-of-care testing of infectious diseases, Biosens. Bioelectron. (2021).

[84] Junaid Bajwa, Usman Munir, Aditya Nori, Williams Bryan, Artificial intelligence in healthcare: transforming the practice of medicine, Future Healthcare Journal 8 (2) (2021) e188.

[85] Liu Tang, Shwu Jen Chang, Ching-Jung Chen, Liu Jen-Tsai, Non-invasive blood glucose monitoring technology: a review, Sensors 20 (23) (2020) 6925.

[86] M. Dhanalakshmi, Kajari Das, Medha Pandya, Sejal Shah, Ayushman Gadnayak, Sushma Dave, Das. Jayashankar, Artificial Neural Network-Based Study Predicts GS-441524 as a Potential Inhibitor of SARS-CoV-2 Activator Protein Furin: a Polypharmacology Approach, Applied Biochemistry and Biotechnology (2022) 1–19.

[87] Medha Pandya, Sejal Shah, M. Dhanalakshmi, Tanzil Juneja, Amisha Patel, Ayushman Gadnayak, Sushma Dave, Kajari Das, and Jayashankar Das. Unravelling Vitamin B12 as a potential inhibitor against SARS-CoV-2: A computational approach, Informatics in Medicine Unlocked 30 (2022) (n.d.) 100951.

[88] Sivaprakasam Radhakrishnan, Byoung-Suhk Kim, and Sushma Dave. Surface modification with nanomaterials for electrochemical biosensing application, in: *Advanced Nanomaterials for Point of Care Diagnosis and Therapy*, Elsevier, 2022, pp. 101–120.

CHAPTER 17

Development of biosensors based on biomarkers in body fluids for the diagnosis of emerging infectious diseases

Janak Paudyal
Department of Chemistry and Biochemistry, University of Colorado, Colorado Springs, CO, United States

17.1 Introduction

COVID-19 showed us that early diagnosis and treatment are critical to contain a pandemic of an infectious disease. In the past, we also saw many public health crises resulting from emerging diseases such as the Zika virus, Ebola, SARS, etc. One of the many reasons for these public health crises is the lack of diagnosis and/or treatment plans for emerging and re-emerging infectious diseases. Unfortunately, despite the advancement in science and medicine, we still lack early and rapid diagnostic methods for many existing infectious diseases. The early diagnosis and treatment can be enabled by utilizing a decentralized testing system, which is accessible to people in remote and resource-limited settings. The easy collection of the sample, possibly through a noninvasive way, is the critical step in the diagnosis of the diseases. These samples often are bodily fluids such as saliva, urine, nasopharyngeal fluid, blood, plasma, lymph, cerebrospinal fluids, pleuroperitoneal fluids, etc. A liquid biopsy involves the detection and analysis of various biomarkers that are naturally present in body fluids as a result of infection [1–3]. These biomarkers, often called circulating targets (CTs), are present in bodily fluids because of the presence of the disease-causing agent or the body's response to it.

The analysis of CTs in bodily fluids may provide a real-time picture of one's physiological and pathological condition. This can be done noninvasively or in minimally-invasive fashion, unlike tissue biopsies which require original tissues obtained via invasive surgery or fine-needle aspiration. The liquid biopsy also enables the measurement of the diseases in physically inaccessible tissues. It can also be used to monitor the disease condition

Biosensors for Emerging and Re-emerging Infectious Diseases
DOI: https://doi.org/10.1016/B978-0-323-88464-8.00020-8

over a time period of a treatment plan. More importantly, these CTs may be present at an early stage of the disease and sampling can be done in a convenient and noninvasive way. This type of sampling from the body fluids presents the opportunity to develop biosensor-like devices that can be used for the screening of infectious diseases in resource-limited areas.

To realize biosensing of emerging diseases, there are many challenges [4–6]. First, the complex nature of the bodily fluids and the extremely low concentration of CTs in biofluids make it very difficult, if not impossible to detect the biomarkers without sample preparation. Second, an extremely low concentration of biomarkers (in the Femtomolar range) requires highly-sensitive analytical methods. Target and/or signal amplification steps are usually required to achieve high sensitivity which makes the detection method complex and time-consuming. Third, the integration of the sample preparation and amplification steps along with the detector in a single portable format is challenging.

Despite these challenges, significant advancement has been made in the development of analytical methods that may lead to the true point-of-care (POC) devices for infectious disease diagnosis. In this chapter, we will be discussing the various types of CTs and their presence in biofluids, sample preparation methods that can be applied to biosensors, analytical methods based on amplification, and bioaffinity materials. The chapter also provides future remarks.

17.2 Biomarkers for infectious disease diagnosis

17.2.1 Antigen

The presence of the disease-causing agent (antigen) in the biofluid can be used for disease diagnosis. The detection of an antigen usually consists of the detection of either protein or polysaccharide molecules present on the surface of the pathogen, which are dispersed in patient tissues. Since these proteins are the integral component of pathogens, the detection of these molecules confirms the presence of pathogens. For example, most of the commercial antigen tests utilize detection of either Nucleocapsid Protein (N-protein) or Spiked Protein (S-Protein) present in the virus for the diagnosis of SARS-CoV-2 disease [5]. These antigens can be detected in various biofluids including saliva, urine, plasma, nasopharyngeal samples, etc.

17.2.2 Antibodies

Antibodies are the proteins produced by the body's immune system as a response to the presence of pathogens. The presence of antibodies in biofluids does not necessarily indicate the presence of pathogens, however it may be used with other symptoms to predict the current stage of the disease. The change in concentration of immunoglobulin M (IgM) or immunoglobulin G (IgG) antibodies or in some cases a change in the total antibody (IgG, IgM, IgE, IgD) in the biofluids, may indicate that the person is infected. The detection of antibodies is especially important in scenarios when there is no easy access to the sample containing an infectious organism. The presence of antibodies also offers a wider window of time for the indirect diagnosis of diseases.

17.2.3 Nucleic acids

The presence of pathogens can also be confirmed by detecting the presence of certain nucleic acids (RNA or DNA) in a patient's biofluid samples. For analytical purposes, the unique DNA/RNA sequence within the genomic material is first established through genomic sequencing methods. Once established, these unique DNA/RNA sequences can act as a molecular fingerprint for the detection of pathogens in biological samples. The diagnosis based on nucleic acids has been widely used to detect a large number of infectious diseases including common viral and bacterial infectious diseases.

17.2.4 Circulating cell-free DNA/RNA

Circulating cell-free DNA or RNA (cfDNA or cfRNA) are small fragments of DNA/RNA found in the acellular fraction of biological fluids. These small fragments of DNA/RNA (average size of 170 bp), most likely are the genetic material from dying human cells and microorganisms. The difference in the concentration of cfDNA/cfRNA of an individual under certain physiological disease conditions can be used as a biomarker for infectious disease detection.

17.2.5 MicroRNAs

MicroRNAs (miRNAs) are single-stranded, non-coding RNA molecules of ~22 nucleotides. While most miRNAs are present in the cells, a handful of miRNAs are also present in extracellular biofluids such as serum, plasma, urine, saliva, tears, breast milk, etc. [7]. These extracellular miRNAs, like intracellular miRNAs, may represent physiological and/or pathological conditions. The comparison of miRNA profiles with healthy controls can be used as a biomarker for the detection of infectious diseases.

Circulating miRNAs are present in the form of either miRNA – protein complexes or EV-shuttled complexes. These complex forms of miRNA make them more stable to the RNase degradation compared to the circulating DNA or RNA. These miRNA molecules may serve as robust CTs due to their remarkable stability towards multiple freeze-thaw cycles, pH, heat, etc. The use of miRNAs as a biomarker is limited due to their non-specificity. For example, diverse infections may have similar changes in the level of a particular miRNA or sets of miRNAs. Also, miRNAs from homologous members have similar sequence structures, which require the use of detectors able to differentiate a single-base mismatch.

17.3 Sample collection and preparation

17.3.1 Sample collection

Sample collection is the first step in the molecular diagnosis of disease. The sample should contain enough molecular biomarkers, diseased cell or viral load to avoid false-negative results. Various samples such as saliva, urine, nasal discharges, blood, cerebrospinal fluid, etc. can be used in the diagnosis of infectious diseases. Due to the complexity of bio-samples, the sample with a higher concentration of the analyte is preferable from the analytical point of view. While noninvasive samples such as urine or saliva that can be self-collected, are ideal for the POC diagnosis, it is ultimately determined by the sensitivity and selectivity of the analytical method chosen. For example, samples from the lower and higher respiratory tract (viral titer of $10^3 - 10^5$ copies/mL and $10^4 - 10^7$ copies/mL, respectively) have been used in the quantitative polymerase chain reaction (qPCR) method for diagnosis of COVID-19 [5,8].

It is important to collect the sample at a suitable point in time as the concentration of biomarkers may increase or decrease with the change in time following the onset of symptoms. That's why it is necessary to the diagnosis along with other disease symptoms. reverse transcription (RT)-qPCR is considered the gold standard method; the sample is usually stored at cold temperature (at 2°C–8°C for the first 72 h and −70°C after that) in case of delayed analysis to avoid the degradation of the nucleic acid and protein biomarkers present in the sample.

17.3.2 Sample preparation

The sample preparation method is unique to each sample and biomarker type. Here we have explained the commonly used sample preparation method in the routine RT-qPCR method [9]. The sample preparation usually involves virus inactivation followed by isolation and purification of nucleic acid material. The virus can be inactivated through the use of 70% alcohol or heating (at 56°C for 30 min or 65°C for 10 min). It should be noticed that the inactivation step could potentially damage the nucleic acid content, especially in a poor sample leading to false-negative results.

The sample treatment for the isolation and purification of nucleic acid and other biomarkers is necessary due to the complex nature of biofluids and their very low concentration. The presence of a large amount of extraneous material may interfere with the analytical method. For instance, in the regular RT-qPCR method, a large amount of the protein in the sample significantly interferes with the efficiency of the PCR reaction. The isolation and purification can be done through classical organic extraction to remove proteins and fats. In most cases, the isolation and purification are carried out using commercially available kits such as QIAGEN or Roche, which are either based on column separation, magnetic separation, or separation based on the polymer interaction. These separation methods are complex and require a long processing time. In some cases, it may require the use of either centrifuge or vacuum pumps [10] which makes them ill-suited for POC applications. Furthermore, these sample preparation steps use chaotropic reagents for cell lysis that will release the DNA/RNA from the cell and may interfere with the analysis of cf DNA or miRNA.

Ideally, analysis without any sample preparation methods are required for the POC applications. In practice, however, sample preparation is indispensable with very few exceptions. Furthermore, poor handling during extraction and/or the presence of PCR inhibitors may lead to false negative results. Scientists have tried to simplify the complex sample preparation steps into one or two steps so that they are adequate for POC analysis depending on their sample. Some examples include, extraction-free amplification such as the use of NaOH solution to directly release the target nucleic acid from the collected clinical nasopharyngeal swab samples [11], without any other sample preparation steps for the detection of SARS-Cov-2 (Fig. 17.1).

Choong Eun Jin et. al developed a unique sample preparation method that could potentially be used in POC microfluidic devices for the isolation and purification of cfDNA from liquid samples [12]. In their system, they have developed bis(sulfosuccinimidyl)suberate (BS3) and helix-shaped microchannels (BSH) which can be used for the enrichment and

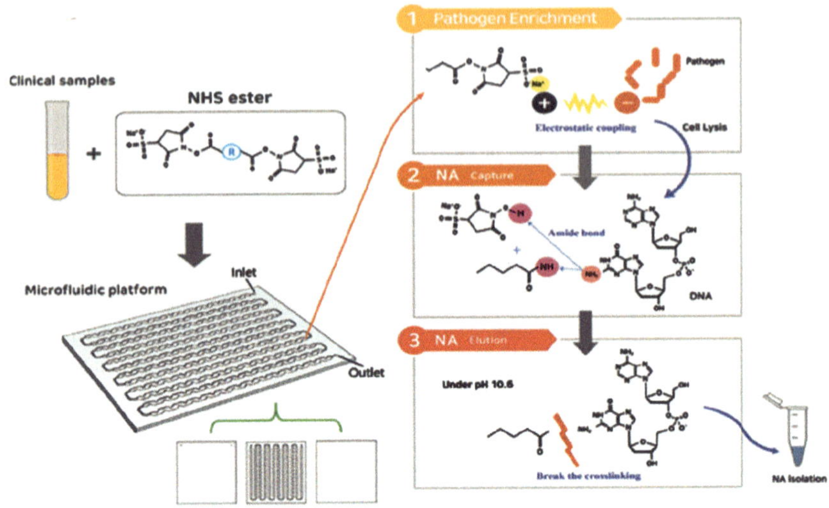

Figure 17.1 The preparation method that could potentially be applied for the isolation of cfDNA from blood samples for POC analysis of infectious diseases. The separation of *Mycobacterium tuberculosis* from the liquefied sputum sample takes ~ 40 min. However, the separation of cfDNA from the blood sample can be done within ~20 min without the use of an instrument or harsh chemicals. *POC*, Point-of-care. *Taken from C.E. Jin, B. Koo, H.J. Lee, I.J. Park, S.H. Kim, Y. Shin, Bis(sulfosuccinimidyl)suberate-based Helix-shaped microchannels as enhancers of biomolecule isolation from liquid biopsies, Anal. Chem. 92(17) (2020) 11994–12001 with permission.*

release of pathogenic genetic material from liquid biopsy samples. The bis-BS3 was assembled on one side with the microfluidic platform using APDMS-amino group while the other side contained a positively charged ester to bind the negatively charged pathogen by electrostatic interaction in the pathogen enrichment step 1. After cell lysis, the BS3 is released, the amine of ester forms an amide bond with the primary amine of nucleic acids in the nucleic acid capture step 2. Finally, the sample was washed with BPS buffer, for disruption of the cross-linking to elute the nucleic acids in step 3. This system is useful for the extraction of high-quality cfDNA from blood samples with low background due to the lack of a lysis step.

17.4 Biosensors based on isothermal nucleic acid amplification

Even after the sample preparation step, detection of biomarkers in complex biofluids is a needle-in-a-haystack problem. These biomarkers may be present in the femtomolar concentration and are surrounded by millions of different molecules present in the micro- to millimolar concentrations. The analysis of these biomarkers in the biofluids, therefore, demands analytical methods with very high levels of sensitivity and specificity. In practice, however, there is always a trade-off between the sensitivity and specificity of any analytical method. This is because the optimized detection strategies to detect low levels of nucleic acids may be susceptible to high background signals due to the presence of interferences. Furthermore, the transducer should be resistant to biofouling. Also, the detection may lead to false positive or negative results due to off-the-target events.

One of the best ways of overcoming these challenges is to make a copy of biomarkers present in the biofluids. PCR is the method of making the copies of DNA and reverse transcription-PCR (RT-PCR) is the method of making the copies of RNA. Hence, qPCR for DNA, and reverse transcription-qPCR (RT-qPCR) for RNA have been used as the gold standard methods for detecting several nucleic acid biomarkers for infectious diseases diagnosis in biofluids. Besides being an excellent technique for the amplification and detection of target nucleic acid, it is prone to false positives resulting from the amplification artifacts such as primer dimers. Furthermore, it is time consuming, requires specialized thermal cycling equipment, and is not applicable in resource-poor settings.

As alternatives to PCR, various isothermal nucleic acid amplification processes have been researched so that amplification can be carried out in a POC setting [13]. The avoidance of thermal cycling makes the reaction more compatible and accessible for nucleic acid analysis in biofluids.

17.4.1 Loop-mediated isothermal amplification

Ever since the discovery of loop-mediated isothermal amplification (LAMP) in 2000, it has become one of the most popular isothermal techniques for nucleic acid amplification. LAMP utilizes four primers, two inner primers (forward inner primer-FIP and backward inner primer-IP) and two outer primers (forward outer primer-F3 and backward outer primer-B3) that recognize the four different regions within the target DNA. Additional loop primers can be introduced to accelerate the reaction. The amplification is based on the strand displacement activity (unwinding of double-stranded DNA) of the Bst polymerase enzyme. The optimal temperature of the Bst polymerase is between 55°C and 65°C and the reaction can be completed in a water bath without the need of a thermocycler [14].

The performance of LAMP has been very comparable to those of the gold standard qPCR method. On top of this, LAMP-based assays can be performed on-site without the need for a centralized lab. For example, Jang et al. compared the LAMP assay, for the diagnosis of SARS-COV-2, to that of the commercial RT-qPCR method [15]. They utilized the clinical nasopharyngeal swab sample and detected the RNA polymerase (RdRP), Envelope (E), and Nucleocapsid protein (N) gene of SARS-COV-2 and compared their results with the commercial RT-qPCR method. The sample was extracted using a commercial viral RNA mini kit from Qiagen, Hilden Germany. They found that the developed assay had comparable and/or better results as compared to Allplex 2019-nCoV Assay or Power- Chek 2019-nCoV Real-time PCR assay.

One of the key advantages of LAMP is its superior amplification capacity. It is possible to achieve $\sim 10^9$-fold amplification within an hour. The high yield of LAMP can be used to facilitate the colorimetric detection of diseases via the naked eye. For example, Yu et al. detected COVID-19, in their RT-LAMP-based platform called iLACO, through visualization of the color change of the pH indicator with the naked eye [16]. The system can achieve a limit of detection (LOD) of 2000 copies per reaction in 20 min after the sample extraction step. The sensitivity of this method was about 89.9% in 248 clinical samples. Similarly, in another work, Thi et al. developed a colorimetric assay for the diagnosis of COVID-19 [17], by using the RT-LAMP method to amplify the clinical RNA samples from pharyngeal swabs. The color of the phenol-red dye changed within the first 30 min of the reaction for positive samples ($C_T < 30$). The clinical samples, with the RNA isolation step, yielded high sensitivity (97.5%) and specificity (99.7%). They can also applied the same method without the RNA extraction step, amid the lower sensitivity, showing the potential of POC diagnosis of COVID-19.

LAMP-based methods also have higher specificity as compared to the PCR-based methods. This is due to the need for four to six different primers, as compared to two primers in PCR, to bind to specific sites on the template strand. However, the use of four to six primers also makes it more complex as compared to the PCR. It is somewhat limited due to non-specific amplification as in other amplification-based methods. The non-specific amplification can be mitigated by using barcode-based technologies or combining it with the CRISPR-based method. Due to its potential, it has been widely used as an isothermal amplification strategy while constructing POC devices for infectious diseases.

17.4.1.1 LAMP in lateral flow assays

The robustness of LAMP makes it compatible with the lateral flow assays (LFA), which also makes it a strong candidate for POC diagnostics [18]. LFA's are POC devices that can be used to quickly establish the presence of target analytes in complex biological samples. These test strips are ideal for the resource-limited setting because of their simple operation, lower cost, and portability.

In a typical LFA, a liquid sample containing the analyte moves along various zones (sample pad, conjugate release pad, test zone, and adsorption pad) of the polymeric strip through capillary action. The sample pad acts as a sample preparation step where it interacts with the buffer, salts, surfactants, etc. which makes it suitable for binding with the recognition element in the conjugate release pad. The conjugate release pad contains the target-specific recognition element with the labeled detection probe. The analyte present in the sample interacts with the recognition element while migrating through the conjugate release pad. The sample, along with the recognition element-analyte complex, then migrates to the detection zone. The detection zone contains a porous membrane with two lines, one for the test and one for the control. The test line contains the analyte-specific components which react with the recognition element-analyte complex to generate the detectible signal. The response in the control line shows the proper flow of samples in the test strips.

The most common analytes for LFA are proteins that are present in relatively higher concentrations in biofluids that are detected based on the recognition ability of their antibodies. The detection of human chorionic gonadotropin (hCG) protein in urine samples for the pregnancy test is one of the most common POC devices. The format of LFA can also be applied to nucleic acid targets. However, nucleic acid samples require a pre-amplification process due to their lower concentration in biological samples compared to proteins used in immunoassays. Also, primer-dimer and correct amplified product may not be distinguishable in the usual primer tag strategy. This makes the direct detection of amplicons based on the primer tag strategy, difficult. Furthermore, cumbersome hybridization and sample-extraction processes are required before nucleic acid detection. Despite these challenges, recent works have shown the promise to make a true POC device for the detection of infectious diseases in biofluids (Fig. 17.2).

In their effort, Zhang et al. used the unique combination of the later flow assay and LAMP method to develop a rapid one-pot detection of SARS-CoV-2 in clinical samples [11]. They first released RNA from the

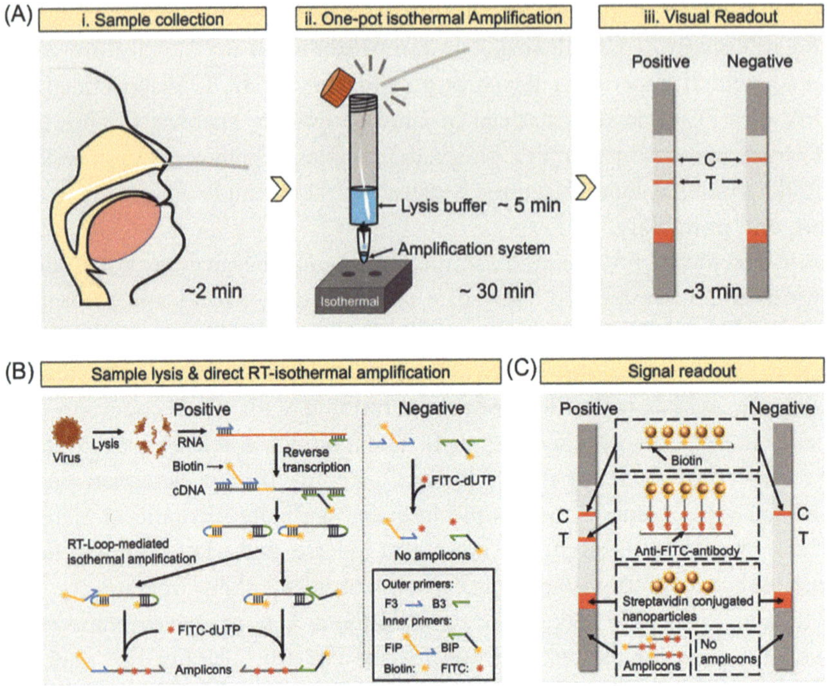

Figure 17.2 Lateral flow assay that detects SARS-CoV-2 in nasopharyngeal extracts. (A) Schematic diagram showing the sample collection, isothermal amplification, and visual readout in RT-LAMP-LFA system. (B) Schematic showing the working principle of RT-LAMP and (C) visual detection with LFA. *LAMP*, Loop-mediated isothermal amplification; *LFA*, lateral flow assays; *RT*, reverse transcription. *Taken from C. Zhang, T. Zheng, H. Wang, W. Chen, X. Huang, J. Liang, et al., Rapid one-pot detection of SARS-CoV-2 based on a lateral flow assay in clinical samples, Anal. Chem. 93(7) (2021) 3325–3330 with permission.*

collected clinical nasopharyngeal samples using 50 mM NaOH which does not interfere with the downstream amplification reaction. The released RNA was then amplified at 60°C using RT-LAMP. In their LAMP procedure, they utilized the fluorescein (FITC)-labeled dUTP to generate labeled amplicons. These label-amplified products migrated through the test strip where they interacted with the streptavidin-coated particles. The anti-FITC antibody immobilized at the test line then captured the amplicons-streptavidin-coated particle conjugates. The presence of FITC-labeled amplicons that originated from FITC-dUTP formed the red T line. The red line indicates the presence of the amplified product.

The unreacted streptavidin-coated particles continue moving in the membrane. The biotin immobilized at the control line ultimately captures the unreacted streptavidin-coated particles (C). This system can be applied for the diagnosis of a clinical sample of SARS-CoV-2 within 40 min of the sampling. The results also showed that the system is specified with no cross-reaction with other respiratory pathogenic microorganisms. The development of this assay is very close to the realization of achieving the ultimate on-site screening of pathogens.

17.4.2 Nucleic acid sequence-based amplification

NASBA (nucleic acid sequence-based amplification) is an isothermal amplification method in which the target RNA is transcribed into a double-stranded RNA:DNA hybrid using the reverse transcription. The cDNA is then generated through degradation of the RNA strand from the hybrid using RNase H. The generated cDNA is then used to create antisense copies of the original RNA target through the use of T7 RNA polymerase [19].

NASBA offers a higher amplification efficiency as compared to PCR (109 copies can be done in just 90 min). The reaction can be performed at a lower temperature 41°C than in LAMP. Unlike amplification methods, it yields a single-stranded RNA. These advantages have been used to develop a sensor for the detection of the Zika virus and SARS-Cov-2. In their work, Pardee et al. developed a paper-based sensor for the detection of Zika virus by combining NASBA with the toehold switch RNA sensors in synthetic RNA samples spiked in saliva [20]. By further incorporating the CRISPR/Cas9-based module, they were able to discriminate between viral strains with single-base resolution and detect the Zika virus from the plasma of a viremic macaque. In another work, Wu et al. developed a POC lateral flow assay for the detection of SARS-COV-2 RNA in saliva samples spiked with the synthetic RNA [19]. They used a crude saliva sample without any extraction process and utilized the complimentary molecular beacons labeled with the fluorophore for the amplified sequence for the visualization.

17.4.3 Recombinase polymerase amplification

Recombinase polymerase amplification (RPA) uses recombinase enzyme to join a pair of primers having homologous sequences in duplex DNA. A single-stranded DNA binding (SSB) protein is also used which not only binds the displaced DNA strand, but also stabilizes the resulting D loop. Bsu DNA polymerase then uses the primer to initiate the amplification process if the target sequence is present. The amplification reaction is very fast, and the DNA amplification can be done within minutes.

The high amplification efficiency of the RPA ($10^9 - 10^{11}$-fold) makes rolling circle amplification (RCA) an excellent method for the detection of trace nucleic acid biomarkers for infectious disease diagnosis. Zhang et al. developed an assay by combining high amplification RPA along with the sequence recognition ability of the Cas system to achieve high sensitivity and specificity [21]. They utilized a DNA probe that specifically targeted the ORF1ab and N regions of the SARS-CoV-2 genome. By attaching gold nanoparticles as a colorimetric tag in the DNA probe, they were able to detect the 4 copies/μL of virus in clinical samples. The avoidance of false-positive results from other beta coronaviruses demonstrated the specificity of the method. The method also showed the high potential of practical testing for SARS-CoV-2.

17.4.4 Rolling circle amplification

RCA utilizes circular DNA templates to hybridize target-specific sequences. The amplification is achieved by adding dNTPs to a primer in the presence of DNA or RNA polymerase. The RCA product contains a complementary sequence of circular templates in tandem. These tandem complementary sequences can be cleaved using enzymes which produce copies of the target DNA or RNA fragments. RCA provides the potential opportunity to amplify the nucleic acid targets at room temperature. In RCA, the original DNA target multiplies, which makes it less prone to errors compared to the other amplification methods. In one of the examples, Wang et al. developed an assay for the detection of severe acute respiratory syndrome (SARS) by utilizing the isothermal RCA method [22]. The method was useful to detect the SARS in clinical samples.

17.5 CRISPR–Cas for improved nucleic acid detection

The genome-editing tool of Clustered Regularly Interspaced Short Palindromic Repeats (CRISPR) and their associated proteins (CRISPR/Cas) have expanded in to the nucleic acid diagnosis. This system uses CRISPR RNA (crRNA) or a single-guide RNA (sgRNA) to direct the functionality of Cas proteins [21,23,24]. These Cas proteins can recognize and cleave the specific nucleic acid sequences. A guided RNA strand contains CrRNA and tracer RNA. A CrRNA is a sequence complementary to the target nucleic acid whereas the tracer RNA is a binding scaffold for the Cas nuclease. This CRISPR–Cas system can be programmed for various functions and activities which makes it a versatile tool for the cleavage of both single or double-stranded DNA or RNA at programmable sites. The Cas protein can cleavage site-specific double-strand breaks on target and non-target nucleic acid molecules, that produce single-stranded reporter nucleic acids. This cleavage activity is used to build a reporter system for optical readouts. These Cas proteins act at a moderate temperature of ~37°C, making them a suitable candidate for POC testing. The programmable functions of the CRISPR–Cas system can be used to improve nucleic acid detection for various purposes.

In one of the applications, the CRISPR–Cas system can be used to simplify the nucleic acid assay by facilitating the unwinding of dsDNA. The nucleic acid amplification techniques generally use multiple enzymes to unwind and separate dsDNA during amplification steps. In the CRISPR–Cas system (Cas9 and Cas12), various crRNA can be used, instead of restriction enzymes to unwind the dsDNA. The system does not require the use of single-stranded binding proteins. One of the works by Jiang et al. was able to develop a magnetic pull-down-assisted colorimetric method based on the RPA-CRISPR/Cas12a system for the amplification and detection of SARS-CoV-2 [25]. The developed method can be applied in various samples including cultured viruses with a detection limit of 50 RNA copies per reaction. More importantly, the developed method can be applied in the nasopharyngeal and throat swab samples. The method also produced consistent results (95.12%) as compared to the clinically approved RT-qPCR Diagnosis kit.

In another application, the CRISPR–Cas system can be used for sequence-specific detection to improve the specificity of analysis. The combination of CRISPR–Cas systems with other nucleic acid amplification strategies has improved both the specificity and sensitivity of the target analytes. Wang et al. utilized the sequence recognition ability of the Cas13a/crRNA to improve the specificity of the transcription amplification [26]. They were able to selectively detect SARS-CoV-2 from its mutated variant called D614G. The method can be applied for the detection of SARS-CoV-2 infection in various samples including throat swabs, serum, etc. This method can also detect and differentiate various respiratory coronavirus-related diseases such as SARS- CoV-2, H1V1, the Middle East respiratory syndrome (MERS), influenza virus, and SARS.

17.5.1 Potential of POC devices

The combination of specificity from the sequence-recognition ability of the CRISPR/Cas system along with isothermal amplification methods and ambient optical method show the great potential of the development of POC devices for the detection of infectious diseases [27]. Please note that nonspecific amplification is one of the major factors that generates false-positive signals in amplification-based assays. The combination of the CRISPR/Cas system along with isothermal amplification could potentially reduce false positive results while still maintaining sufficient sensitivity.

In the creative work of Pang et al., SARS-CoV-2 can be detected in a single pot from the clinical swab samples with 100% clinical specificity with 40 min assay time [28]. In their work, they collected the samples from respiratory swabs, and RNA was extracted from commercially available kits (Fig. 17.3).

The whole assay can be done in a PCR tube, the bottom of which contains 25 μL RT-LAMP reagent mixture and its cap contains 10 μL Cas12a-mediated detection reagent mixture. First, an aliquot of the RNA extract was mixed with RT-LAMP reagents which are present at the bottom of the tube. The RT-LAMP reaction was carried out at 62°C for 30 min and allowed to cool down at room temperature. After the completion of the RT-LAMP, Cas12a reagents were added by inverting the tube and mixing well by wrist-flicking. The reagent also contained the fluorophore which

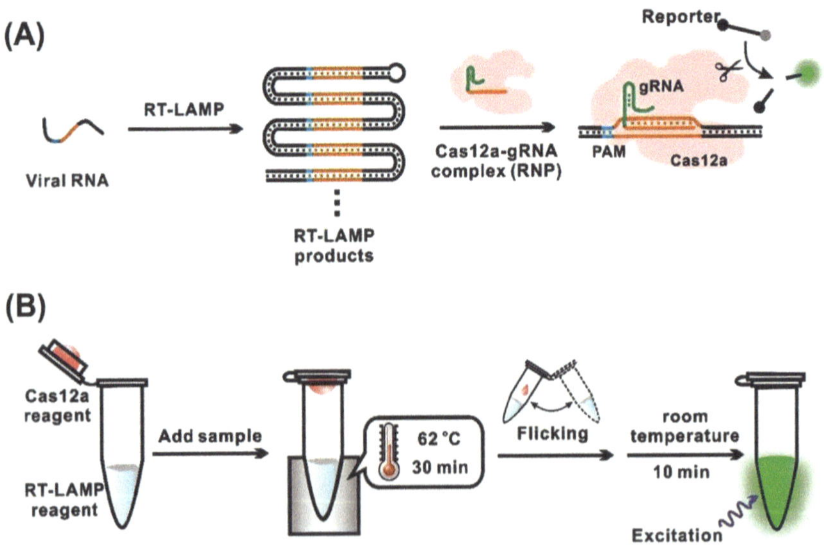

Figure 17.3 The combination of CRISPR/Cas system along with the isothermal amplification process has the great potential for diagnosis of infectious diseases: [28] Figure showing the working principle of the assay (A) and the schematic diagram of the procedure (B) for SARS-CoV-2 assay. *Taken from the cited reference with permission.*

generates the green fluorescence signal on completion of the reaction. The green fluorescence can simply be recorded using a smartphone camera. It was found that the use of the CRISPR/Cas system produced 100% clinical specificity while analyzing the 100 samples. The assay uses the respiratory swab samples for SARS-CoV-2 detection and targeted N and E genes.

17.6 Analysis without amplification of target or probe

The analysis of biomarkers without the need for target or probe amplification greatly reduces the complexity and analysis time. These methods often rely on superior sensitivity of the method achieved through the higher binding affinity of the probe, target recycling, signal amplification, sample concentration, or their combinations. Among a large number of detection and amplification methods, we discuss a few selected techniques here.

17.6.1 CRISPR/Cas-based methods for nucleic acid detection

While target/probe amplification is a general approach for the detection of the trace amount of nucleic acid target in biofluids, simpler analytical methods without the target or probe amplification have been developed. These simpler methods are usually free from the false-positive signals derived from nonspecific amplification and complexity of the primer design.

The versatile programming of the CRISPR/Cas system has been used for the detection of nucleic acids without target amplification. Yue et al. developed a droplet Cas12a method for the direct quantification of DNA without amplification steps [29]. The DNA quantification method can be used for the detection of various virus diseases such as Hepatitis B, African swine flu, etc. in clinical serum samples. Using their optimized reaction conditions, they were able to detect as low as 100 fM of DNA which shows a ~50-fold improvement as compared to the other existing Cas12a assays. This very low-level detection with a high specificity could potentially act as the basis for next-generation nucleic acid quantification methods. Similarly, in another Cas13a assay, Tian et al. [30] quantified the single RNA molecule without the need for reverse transcriptase, labeling, and amplification. The excellent performance of the assay can also be expanded for precise counting of other miRNAs or SARS-CoV-2 RNA, in both synthetic and clinical samples. Similarly, Ding et al. developed the assay called digital warm-start CRISPR (dWS-CRISPR) that can quantify SARS-CoV-2 genetic material from clinical saliva and swab samples without the need for RNA extraction

[31]. Furthermore, the assay can also eliminate the premature amplification of targets at temperatures higher than 50°C.

17.6.1.1 Chemiluminescence based methods

The chemiluminescence method utilizes the light emitted through the reaction of an analyte with the chemiluminescence probe to quantify the presence of an analyte. The combination of specificity of the recognition element along with the sensitivity of the chemiluminescence method, has shown great potential for the detection of biomarkers in body fluids (Fig. 17.4).

In one example, Yuhui et. al. developed the electro-chemiluminescence (ECL) sensor, in which the light generated through the electrochemical reaction is measured, to detect the Zika virus (ZIKV) [32]. The sensor is

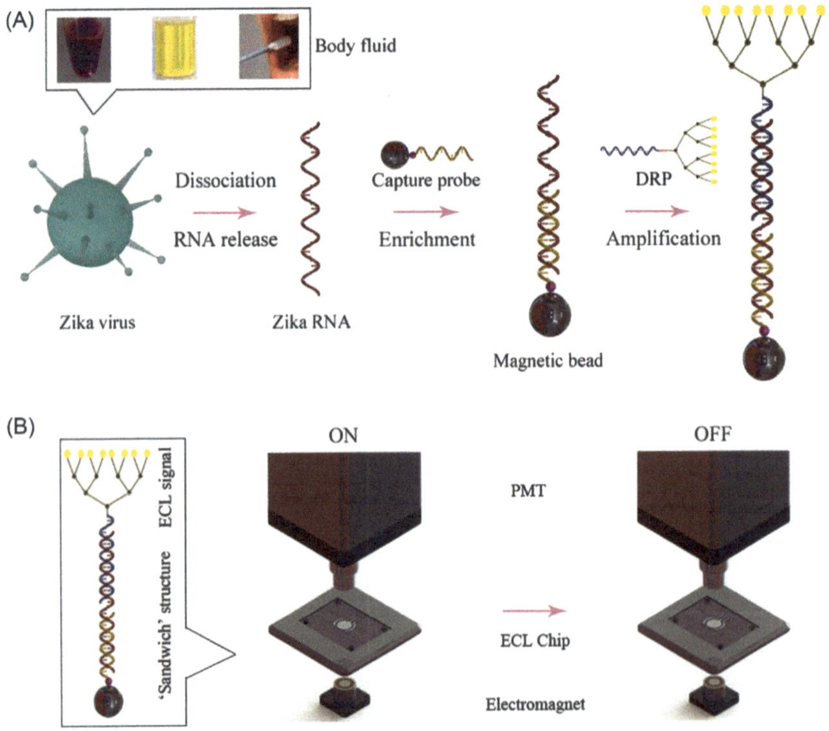

Figure 17.4 Detection of Zika virus without the target amplification in liquid biopsy system based on the dendritic Ru(bpy)32 + -polymer-amplified ECL method [32]. (A) Zika virus was released from the cells in body fluids through the use of SDS (Sodium dodecyl sulfate) lysis buffer for dissociation. Magnetic bead immobilized probes were used for enrichment (B) ECL signal-producing step. *ECL*, Electro-chemiluminescence. *Taken from the cited reference with permission.*

based on newly developed dendritic $Ru(bpy)_3^{2+}$-polymer (DRP) as an ECL probe. In this method, the ZIKV samples from body fluids were pretreated with the lysis buffer to release the viral RNA. The sample was then concentrated by using capture probes which are attached to the beads for magnetic separation. Once enriched, the ECL probe recognizes the Zika virus-specific RNA that are captured in the magnetic beads. This method utilizes only a drop of blood to detect as low as 500 copies of the virus without target amplification. In another work, Yao et al. detected SARS-Cov-2 by utilizing the ECL amplification probe called Au@Ti_3C_2@PEI-$Ru(dcbpy)_3^{2+}$nanocomposite [33]. This method targets the RdRp gene and can detect as low as 0.21 fM RNA in the spiked serum sample.

17.6.1.2 Colorimetric methods

Immunoassays in the form of Lateral flow rapid tests or chromatographic format with the colorimetric probes provide an excellent solution for the large-scale screening of infectious diseases [34]. These tests rely on the detection of proteins or antibodies specific to the pathogen through the utilization of recognition elements. The presence of pathogens can be detected by the naked eye without the use of any instruments. These easy-to-use tests provide a very quick "yes" or "no" answer on if the virus is present in the body fluids.

Several antigen tests for SARS-Cov-2 are commercially available and have received an emergency use authorization from the United States Food and Drug Administration (FDA) [35]. These commercially available tests are based on the recognition of the spike protein or nucleocapsid or other proteins present on the surface of the virus by an antibody specific to these proteins. The test can be done in test slides or strips following the lysis of the body fluid and are available in various assay formats. In one format, the test strips are coated with the capture antibody which recognizes and captures the targeted protein in the virus. Following the addition of labeled antibodies, the signal is generated in the form of color. Besides the success of these assays, it is still necessary to increase the sensitivity to improve the utility of these tests so that can be used for a larger window of infection.

In one of the examples, Loynachan et al. used platinum core − shell nanocatalysts (PtNCs) to amplify the colorimetric signal in lateral flow immunoassays [36]. They were able to detect P24, a protein biomarker for HIV, in the spiked serum sample via the naked eye within 20 minutes.

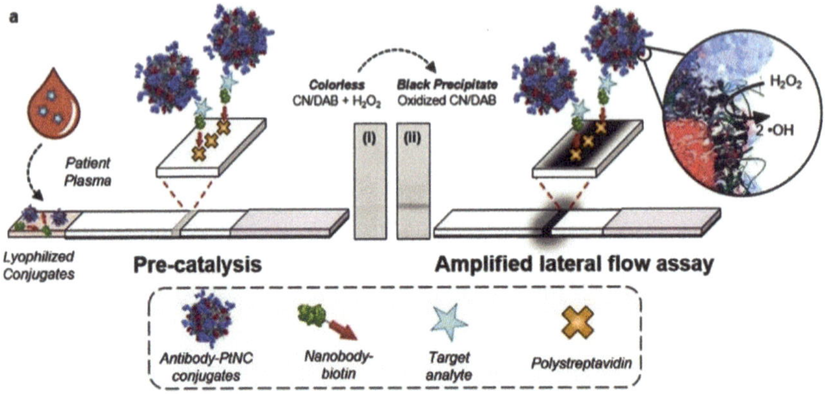

Figure 17.5 Schematic representation of the PtNC-integrated LFIA (later-flow immunoassay) for amplified detection of P24 biomarker in spiked plasma and serum samples. *Taken from the reference [36] with permission.*

They claimed that the sensitivity can be increased by a factor of twenty as compared to various commercially available rapid tests for p24 (Fig. 17.5).

In their work, they designed a biotinylated antibody fragment called nanobody-biotin. The PtNCs and nanobody-biotin were mixed with the samples containing HIV. When the target protein is present in the sample, it forms a complex with the PtNC and becomes biotinylated. The biotinylated PtNC then concentrates in the streptavidin-coated test line through biotin-streptavidin binding. The deposition of PtNC in the test line indirectly represents the concentration of the target protein in the sample. The PtNC in the test line can be detected by utilizing peroxidase mimicking activity as it can catalyze a peroxidase substrate (CN/DAB, in this case) in the presence of hydrogen peroxide to generate a black product.

17.7 Conclusion and future remarks

Liquid biopsy provides a noninvasive or semi-invasive alternative to traditional tissue biopsies. The analysis of the CT biomarkers in these biofluids provides the information essential for screening, diagnosis, treatment, and the management of infectious diseases. Biosensors represent the simple, portable, and cheap device that can be used for POC diagnosis of infectious diseases without the use of sophisticated instruments or manpower in remote areas or resource-limited settings. More importantly, the biosensor should reliably measure circular biomarkers, with sufficient sensitivity and selectivity, for the diagnosis of infectious diseases.

The complex nature of the body fluids is one of the major obstacles in developing biosensors to analyze biomarkers in liquid biopsy samples. While no sample preparation or minimal sample preparation is required for the detection of proteins in immunoassay format, sample preparation is very critical to analyze the nucleic acid biomarkers in liquid biopsy samples. The progress on microfluidics as well as assay format technologies has greatly reduced the sample preparation steps. The use of new affinity materials and/or nanomaterials may further minimize the sample preparation steps. An extremely lower concentration of biomarkers requires highly sensitive analytical methods for their detection. Due to their extremely lower concentration, detection of nucleic acid biomarkers generally require the target of probe amplification steps. The standard amplification processes require thermal recycling but the development in the isothermal amplification process and the CRISPR/Cas technologies has simplified the amplification process. Various signal amplification strategies have been used as an alternative to target/probe amplification with limited success for both protein and nucleic acid biomarkers. Recent developments in microfluidics, nanotechnology, and isothermal target amplification methods have provided ways of designing biosensors that can be applied in liquid biopsy samples.

Lateral flow immunoassays detect the antigen and represent biosensors that can be applied in the resource-limited setting. Research has shown that a similar format can be used for nucleic acid biomarkers as well. However, these tests are qualitative and prone to false positive or negative results. To make truly quantitative liquid-biopsy based biosensors, further work is needed especially in (1) reducing and/or eliminating the sample preparation steps, (2) developing new recognition elements with better binding affinity and stability in biofluids and, (3) testing the developed method with sufficiently large clinical samples.

References

[1] R. Khan, Z. Khurshid, F. Yahya Ibrahim Asiri, Advancing point-of-care (PoC) testing using human saliva as liquid biopsy, Diagnostics. 7 (3) (2017) 39.

[2] D. Han, R. Li, J. Shi, P. Tan, R. Zhang, J. Li, Liquid biopsy for infectious diseases: a focus on microbial cell-free DNA sequencing, Theranostics. 10 (12) (2020) 5501–5513.

[3] G. Wen, T. Zhou, W. Gu, The potential of using blood circular RNA as liquid biopsy biomarker for human diseases, Protein Cell. 12 (12) (2021) 911–946.

[4] J. Martín, N. Tena, A.G. Asuero, Current state of diagnostic, screening and surveillance testing methods for COVID-19 from an analytical chemistry point of view, Microchem. J. 167 (April) (2021) 106305.

[5] A. Hosseini, R. Pandey, E. Osman, A. Victorious, F. Li, T. Didar, et al., Roadmap to the bioanalytical testing of COVID-19: from sample collection to disease surveillance, ACS Sensors. 5 (11) (2020) 3328−3345.
[6] O. Pashchenko, T. Shelby, T. Banerjee, S. Santra, A comparison of optical, electrochemical, magnetic, and colorimetric point-of-care biosensors for infectious disease diagnosis, ACS Infect Dis. 4 (8) (2018) 1162−1178.
[7] L. Tribolet, E. Kerr, C. Cowled, A. G.D. Bean, C.R. Stewart, R.J. Farr, MicroRNA biomarkers for infectious diseases: from basic research to biosensing, Front Microbiol. 11 (1197) (2020).
[8] T.J. Johnson, R.T. Nishida, A.P. Sonpar, Y.-C.J. Lin, K.A. Watson, S.W. Smith, et al., Viral load of SARS-CoV-2 in droplets and bioaerosols directly captured during breathing, speaking and coughing, Sci Rep. 12 (1) (2022) 1−13.
[9] P. Kralik, M. Ricchi, A basic guide to real time PCR in microbial diagnostics: definitions, parameters, and everything, Front Microbiol. 8 (FEB) (2017) 1−9.
[10] C.E. Jin, B. Koo, T.Y. Lee, K. Han, S.B. Lim, I.J. Park, et al., Simple and low-cost sampling of cell-free nucleic acids from blood plasma for rapid and sensitive detection of circulating tumor DNA, Adv. Sci. 5 (10) (2018).
[11] C. Zhang, T. Zheng, H. Wang, W. Chen, X. Huang, J. Liang, et al., Rapid one-pot detection of SARS-CoV-2 based on a lateral flow assay in clinical samples, Anal. Chem. 93 (7) (2021) 3325−3330.
[12] C.E. Jin, B. Koo, H.J. Lee, I.J. Park, S.H. Kim, Y. Shin, Bis(sulfosuccinimidyl)suberate-based Helix-shaped microchannels as enhancers of biomolecule isolation from liquid biopsies, Anal. Chem. 92 (17) (2020) 11994−12001.
[13] G. Pooria, G. Amir, Nucleic acid isothermal amplification technologies—a review, Nucleosides Nucleotides Nucleic Acids. (2008) 224−243.
[14] L. Becherer, N. Borst, M. Bakheit, S. Frischmann, R. Zengerle, F. Von Stetten, Loop-mediated isothermal amplification (LAMP)-review and classification of methods for sequence-specific detection, Anal Methods. 12 (6) (2020) 717−746.
[15] W.S. Jang, D.H. Lim, J. Yoon, A. Kim, M. Lim, J. Nam, et al., Development of a multiplex loop-mediated isothermal amplification (LAMP) assay for onsite diagnosis of SARS CoV-2, PLoS One (2021).
[16] X. Yin, W. Chen, V. Pelechano, X. Hao, S. Wu, L. Yu, Rapid detection of COVID-19 coronavirus using a reverse transcriptional loop-mediated isothermal amplification (RT-LAMP) diagnostic platform, Clin. Chem. 66 (7) (2020) 975−977.
[17] V.L. Dao Thi, K. Herbst, K. Boerner, M. Meurer, L.P.M. Kremer, D. Kirrmaier, et al., A colorimetric RT-LAMP assay and LAMP-sequencing for detecting SARS-CoV-2 RNA in clinical samples, Sci Transl Med. 12 (556) (2020).
[18] M. Chaouch, Loop-mediated isothermal amplification (LAMP): an effective molecular point-of-care technique for the rapid diagnosis of coronavirus SARS-CoV-2, Rev. Med. Virol. (2021). October.
[19] Q. Wu, C. Suo, T. Brown, T. Wang, S.A. Teichmann, A.R. Bassett, INSIGHT: a population-scale COVID-19 testing strategy combining point-of-care diagnosis with centralized high-throughput sequencing, Sci. Adv. 7 (7) (2021) 1−19.
[20] K. Pardee, A.A. Green, M.K. Takahashi, D. Braff, G. Lambert, J.W. Lee, et al., Rapid, low-cost detection of zika virus using programmable biomolecular components, Cell. 165 (5) (2016) 1255−1266.
[21] Y. Yu, G. Su, W.S. Zhang, J. Pan, F. Li, M. Zhu, et al., Reverse transcription recombinase polymerase amplification coupled with CRISPR-Cas12a for facile and highly sensitive colorimetric SARS-CoV-2 detection, Anal. Chem. 93 (8) (2021) 4126−4133.
[22] B. Wang, S.J. Potter, Y. Lin, A.L. Cunningham, D.E. Dwyer, Y. Su, et al., Rapid and sensitive detection of severe acute respiratory syndrome coronavirus by rolling circle amplification, J. Clin. Microbiol. 43 (5) (2005) 2339−2344.

[23] H. Yue, M. Huang, T. Tian, E. Xiong, X. Zhou, Advances in clustered, regularly interspaced short palindromic repeats (CRISPR)-based diagnostic assays assisted by micro/nanotechnologies, ACS Nano. 15 (5) (2021) 7848–7859.
[24] W. Feng, A.M. Newbigging, J. Tao, Y. Cao, H. Peng, C. Le, et al., CRISPR technology incorporating amplification strategies: molecular assays for nucleic acids, proteins, and small molecules, Chem. Sci. 12 (13) (2021) 4683–4698.
[25] Y. Jiang, M. Hu, A.A. Liu, Y. Lin, L. Liu, B. Yu, et al., Detection of SARS-CoV-2 by CRISPR/Cas12a-enhanced colorimetry, ACS Sensors. 6 (3) (2021) 1086–1093.
[26] Y. Wang, Y. Zhang, J. Chen, M. Wang, T. Zhang, W. Luo, et al., Detection of SARS-CoV-2 and its mutated variants via CRISPR-Cas13-based transcription amplification, Anal. Chem. 93 (7) (2021) 3393–3402.
[27] Y. Zhang, M. Chen, C. Liu, J. Chen, X. Luo, Y. Xue, et al., Sensitive and rapid on-site detection of SARS-CoV-2 using a gold nanoparticle-based high-throughput platform coupled with CRISPR/Cas12-assisted RT-LAMP, Sensors Actuators, B Chem. 345 (June) (2021) 130411.
[28] B. Pang, J. Xu, Y. Liu, H. Peng, W. Feng, Y. Cao, et al., Isothermal amplification and ambient visualization in a single tube for the detection of SARS-CoV-2 using loop-mediated amplification and CRISPR technology, Anal. Chem. 92 (24) (2020) 16204–16212.
[29] H. Yue, B. Shu, T. Tian, E. Xiong, M. Huang, D. Zhu, et al., Droplet Cas12a assay enables DNA quantification from unamplified samples at the single-molecule level, Nano Lett. 21 (11) (2021) 4643–4653.
[30] T. Tian, B. Shu, Y. Jiang, M. Ye, L. Liu, Z. Guo, et al., An ultralocalized Cas13a assay enables universal and nucleic acid amplification-free single-molecule RNA diagnostics, ACS Nano. 15 (1) (2021) 1167–1178.
[31] X. Ding, K. Yin, Z. Li, M.M. Sfeir, C. Liu, Sensitive quantitative detection of SARS-CoV-2 in clinical samples using digital warm-start CRISPR assay, Biosens. Bioelectron. 184 (March) (2021) 113218.
[32] Y. Liao, Z. Fan, H. Deng, Y. Yang, J. Lin, Z. Zhao, et al., Zika virus liquid biopsy: a dendritic Ru(bpy)32 + -polymer-amplified ECL diagnosis strategy using a drop of blood, ACS Cent. Sci. 4 (10) (2018) 1403–1411.
[33] B. Yao, J. Zhang, Z. Fan, Y. Ding, B. Zhou, R. Yang, et al., Rational engineering of the DNA walker amplification strategy by using a Au@Ti3C2@PEI-Ru(dcbpy) 32 + nanocomposite biosensor for detection of the SARS-CoV-2 RdRp gene, ACS Appl. Mater. Interfaces. 13 (17) (2021) 19816–19824.
[34] H. Kim, D.R. Chung, M. Kang, A new point-of-care test for the diagnosis of infectious diseases based on multiplex lateral flow immunoassays, Analyst. 144 (8) (2019) 2460–2466.
[35] FDA.gov. https://www.fda.gov/medical-devices/coronavirus-disease-2019-covid-19-emergency-use-authorizations-medical-devices/eua-authorized-serology-test-performance.
[36] C.N. Loynachan, M.R. Thomas, E.R. Gray, D.A. Richards, J. Kim, B.S. Miller, et al., Platinum nanocatalyst amplification: redefining the gold standard for lateral flow immunoassays with ultrabroad dynamic range, ACS Nano. 12 (1) (2018) 279–288.

CHAPTER 18

Advantages of silicon nanowire-based biosensors as wireless technology for infectious disease diagnosis

Maheswata Sahoo[1], Ayushman Gadnayak[1], Ananya Nayak[1], Swayamprabha Sahoo[1], Sushma Dave[2], Padmaja Mohanty[3], Jatindra Nath Mohanty[1] and Jayashankar Das[4]

[1]Centre for Genomics and Molecular Therapeutics, IMS and SUMH, Siksha "O" Anusandhan University (Deemed to be), Bhubaneswar, Odisha, India
[2]Department of Applied Science, JIET Group of Institutions, Jodhpur, Rajasthan, India
[3]Department of Botany, Bioinformatics and Climate change, University School of Sciences, Gujarat University, Ahmedabad, Gujarat, India
[4]Valnizen Healthcare PVT Limited, Mumbai, Maharashtra, India

18.1 Introduction

Infectious disease are denoted as a special disorder which affects the surrounding environment or the whole body. Infectious disease is defined as "a disease caused by epiphytes, bacteria, parasites, and viruses that can spread throughout the population" [1−3]. Highly sensitive biomedical sensors could make it promising to diagnose or identify disease at an early stage, that dramatically increases the likelihood of detection and life-saving interventions. We find more research has been directed to expand the application of the biomedical sensor, where recent advances in nano-technology are going to provide the most potential solutions in biomedical sensor development area. Nanotechnology covers a wide scientific and technical research area for verification of materials and structures, but is around 100 nm [4]. In the nanometer range, the physical and chemical properties of materials are closely related to the surface/volume ratio and quantum size affects that, leading to fully diverse properties compared to at the macroscale [5−7]. As the surface-to-volume ratio in nanoscale range is very large, it affects a large area of the sensory structure, making the nanostructures too sensitive to changes in their surface. The movement of electrons in nanostructures is limited by the quantization effect,

Biosensors for Emerging and Re-emerging Infectious Diseases
DOI: https://doi.org/10.1016/B978-0-323-88464-8.00019-1

producing quantum size special effects where the individual energy level of the device relies on the structure size. Both the semiconductor energy is in the smallest state as well as volume standard oscillator quality is expanded with reducing the size [8], wherein nanostructures can have elevated energy conversion competence with relatively less thermal noise. Biomolecular sizes (approx 10–100 nm), that are in a similar range as synthetic nanostructures also support the progress of the biomedical sensor by means of nanotechnology. The most important goal of biomedical sensors is to identify and characterize chemical as well as biological species, from disease diagnosis to drug discovery [9].

In response to this limitation, silicon nanowires (SiNW) were used that proceed as field effect-transistor (FET) sensors. This electrically-mediated biosensor incorporates biomolecules through a supplementary metal oxide semiconductor (MOS)-attuned technology to facilitate detection of charge changes as the target analyte is bound to a fixed probe. SiNW is extremely sensitive, direct & definite for detecting a large number of species because of its one dimensional structure and high surface-to-volume ratio. Thus has huge biological as well as clinical applications. The obtainable device offers many significance roles, like the real time transmission of electrical signals and the ability to detect without labels. Despite this property, the sensitivity is very high in comparison with other methods. In addition, the superior profitability of the current top-down manufacturing strategy translates into lower prices for goods on the market. However, nanowire sensors can be dominated over the previously used gold standard methods like PCR and ELISA, only when they are sophisticated enough in terms of specificity, simplicity, reliability and importantly sensitivity. Especially with promising wearable biosensors, nanowires are expected to come to an end [10,11].

Over the years, SiNW devices have been studied extensively in many applications as biosensors. The commercialization of technology will be successful because handheld devices open tremendous opportunities for a revolution in COVID-19 diagnostics for selective detection of the coronavirus in minutes.

18.2 Overview of advents of silicon nanowires-based biosensor

The essential part of biomedical sensors is to distinguish and portray substances and organic species, from applications in illness analysis to mediate revelation. Utilizing nanomaterials (like nanoparticles, nanowires, and carbon nanotubes) with precise attractive electrical and optical properties, these essential biomedical activities are often carried out immediately. In addition, semiconductor nanowires allow for the detection of various types of species electrically and without labels. These nanowires are made from a semiconductor material and the surface is changed regularly to make them sensitive to chemicals and species. Carbon nanotubes that can even be utilized in biomedical sensors are still regulated by integrating semiconducting metals & materials, involving additional decontamination. In addition, the protocol for linking replicated analytes to nanotubes is not well developed. Therefore, the nanostructures created by nanowires are the best alternative to biomedical sensors, which have high impact, consistency, reproducibility and flexibility in general production.

18.2.1 The working principle

The SiNW sensors are a FET-mediated device that contains electrodes for the source, line and cover. The source and drainage electrodes connect the semiconductor conduit, whereas the gate electrodes modulate the conductivity of the conduit. The SiNW is fixed in-between the source and thus the conduit in semiconductor lines is the sensor part of the device. In order to perceive a precise target, the receptor molecules which distinguish the target, are immobilized on the surface of SiNW. Using the example of an immunologically-modified SiNW sensor of the n-type, the semiconductor channels have the same conductivity which is mainly determined via the electron concentration. At the point when surface receptors are presented to the target molecule, which holds a negative charge, the particular binding causes a diminishing in conductivity. Then the conductivity is collected through an electronic recording system. This standard can be extrapolated into p-type SiNW and charged biomolecules [12] (Fig. 18.1A and B).

18.2.2 Fabrication details of silicon nanowire biosensors

Two crucial nanoscale productions progressions, one is the top down or bottom up approach, typically do not produce nanowire (NW) semiconductor biosensors. The efficiency of the SiNW biosensors is influenced by a number of factors, like diameter, density and carrier mobility along with surface chemistry. In a different report, it has been obtained experimentally that a smaller diameter causes greater sensitivity [13–15]. In the same way, the simulation results also show that the smallest amount of analyte involved for noticeable signal changes, diminishing the sensor diameter (below 10 nm

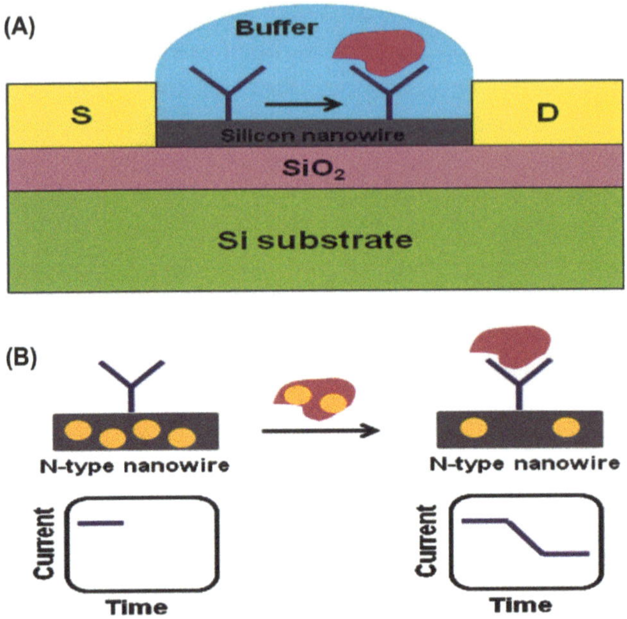

Figure 18.1 SiNW biosensor's functional principle. (A) SiNW as a semiconductor conduit that has connected to the source and conducts the electrodes. The SiNW surface is having immobilized receptor molecules that designed specifically for capturing the target. (B) Shows a negatively charged objective molecule attaches to a receptor that found repelling the −ve charge in the C3 regions, which is a basis of decrease in the conductivity for N doped C3. *SiNW*, silicon nanowire.

diameter). Besides, the theoretical study shows the doping concentration determines the biosensors sensitivity. An example is like a slightly touched S3 exhibits greater sensitivity than one that is deeply touched or untouched. These results demonstrate the significance of the NW fabrication procedure in achieving a small size and suitable doping concentration to attain high sensitivity levels. It has been widely reported that bottom- up techniques like Para-liquid–solid (VLS) grow SiNW along with its application as a biosensor. The 20 nm SiNW of n-type or p-type was supplied with the VLS technique via Au nanoclusters as boosters and phosphine or diborane as preservative precursors [16,17] (Fig. 18.2).

18.2.3 Surface functionalization

The SiNW exterior area must be customized with the interrogation particles, and as a result the biosensor will be suitable for receiving the precise target molecule. Generally, there are two methods to bind the probe

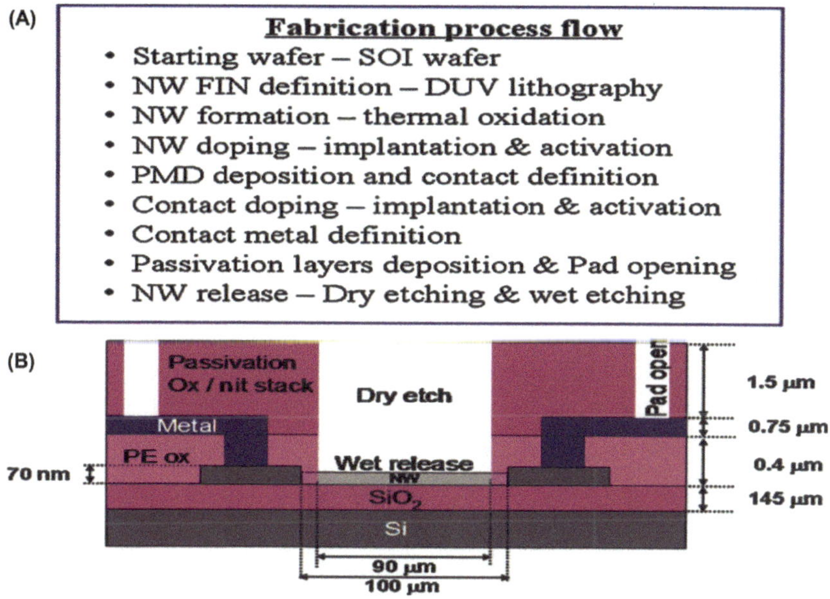

Figure 18.2 (A) Flow of the SiNW biosensor production process (B) SiNW sensor's cross-section area representation. *SiNW*, silicon nanowire.

molecules at the surface, covalent coupling and electrostatic adsorption, which were used in the SiNW biosensor. In electrostatic adsorption, attractive objects are used to adsorb ionic solutes on a reverse adsorbent. The demonstration of the detection of DNA in electrolyte solutions by a SiNW sensor were done by health group, where the main strands of DNA were adsorbed electrostatically onto an amine supplemented organic single layer on NW surface. Here in the covalent binding process, the probe molecules bind to the SiNW surface via covalent bonds. As we know, the oxide can also naturally raise on the SiNW surface area, and various methods are believed to be able to function with the oxide layer using silane chemistry [18].

18.2.4 Fluid exchange

The efficient fluid trading framework is extremely attractive for the fast transfer of analyte to the SiNW surface. Two important techniques for trading fluids are being developed, with the help of either open-space or closed microfluidic channels. Microfluidic channels made of polydimethylsiloxane (PDMS) are usually used and are set to the highest value of the SiNW sensor. This coordinates the appropriate response to SZ and allows the least volume of tests to be ensured. Nevertheless, the laminar microfluid flow halves the concentration of the analyte particles to lead the SiNW surface area. Here, the target biomolecules can be absorbed in the Polydimethylsiloxane sidewall. Therefore, the SiNW sensor does not achieve high applicability. We make macroscale elongated cameras, not micro channels for trading fluids. The camera is made cut-to size with a laser machine and acrylic material. The arrangement of the spaces on the SiNW surface area is basically confirmed via twice-sized bands. Even though the strategy does not depend on unbroken flow, a corresponding response is transmitted from the upper pipe to the main sensor below the NW, avoiding moderate spread problems but increasing susceptibility to detection. Often a higher level of fluid is also used or applied. This simple and smart design includes one reservoir that holds an adequate volume of buffer solution to make the most of binding competence while keeping away from the propagation resistance experienced by inherent microchannel frameworks [19].

18.2.5 The Debye length effect

Debye length is a significant constraint on device execution. The Debye length is the length from the SiNW surface at which the majority of the charge can show up. This achievement is legitimately identified with the separation that the NWs can obtain. Debye length is increased regularly by utilizing aqueous buffer solutions where electrolyte concentrations are low. In either case, extreme dilution resulting in a low salt center can have an effect on the biological action of the protein [20].

18.2.6 Electrical detection

To convert the attaching function of molecules into functional electrical marks, mainly two techniques are used: the identification of the current and the assessment of the state, respectively. Both assessments were carried out using the Alessi REL 6100 test station (Cascade Microtech; Beaverton, OR) connected to the HP Boundary Analyzer (HP-4155A). A framework for two tests has been found where a regulated voltage can be useful to the output electrodes of a SiNW machine even as the output terminals are grounded. For continuous identification, a short heart rate voltage of 0.1 V is applied to the SiNW sensor and the current through the SiNW is estimated. Information points are retrieved every 1 s to show the heating effect caused by the intensity dissipation by the SiNW sensor. Then the buffer solution is poured into SiNW machine position through the inlet channel. Standards are created by recording gifts via SiNW while immersing them in a buffer solution about 100–200 s. The biomolecular analytic solution is next added through the inlet tube to localize the SiNW device, consequencing in the conductivity correction due to the precise binding of the biomolecules to the operating surface. Direct current is achieved after 100–200 s and will not maintain SiNW conductivity after coupling function [21,22].

18.3 Detection of diverse biological marker or agents

18.3.1 Protein

Through proteomic and genomic studies, several new biomarkers have been discovered that could potentially be used to diagnose disease. Single marker tests do not provide an acceptable diagnosis for complex diseases such as cancer, which are characterized by significant heterogeneity. In order to analyze a confusing disease such as malignant growth, which is represented by significant heterogeneity, a single marker test does not provide an adequate diagnosis. The main application of the "p" type of silicon nanowire sensor is to electrically identify proteins during solution. From this point onwards, various stages based on the nanowires continue to be produced, while different marker proteins are identified [23].

18.3.2 DNA and RNA

With the nanowire sensor, specific DNA and RNA sequences can be distinguished. The surface of silicon nanowires contains single-stranded PNA sequences tuned to act as DNA receptors. The ability of silicon nanowires to recognize DNA at 10 femtomolar levels was demonstrated. It is regularly a much better idea for the DNA detection to occur when weigh against quartz crystal microbalances, SPR (surface plasmon resonance) and SPR enhanced by nanoparticles. In addition, it can be used to supplementary enhance the informative value and applicability of the nanowire sensors. Primary information suggests that the limit of detection (LOD) of the ssDNA site can be achieved at approximately 1 femtomolar level. Apart from the recognition of individual DNA strands, nanowires can also be used to recognize the bonds between proteins and DNA [24].

18.3.3 Viruses

Proper virus detection is a very significant task of the nanoscale sensor in protecting the social fabric, as most infections cause serious illness in humans and can be used as biological weapons. Currently, Si nanowire sensors do not typically identify many high-risk viruses, including dengue fever, H1N1, H3N2 influenza A and HIV. The sensors surface of the nanowires is only functional with antibodies that explicitly attach to the target virus and affect the conductivity of nanowires. Shen et al., built a nano conductors-based biosensor which detected more than 29 flu viruses/μL from exhaled condensate samples (EBC). Very recently, Ibarlucea et al., introduced the Si nanowires with the scaffold protein of Ebola VP40 award. The exact LOD after 30 min of hatching was approximately 6.25 nM, six orders before the ELISA strategy. With preferences such as portability, speed, accuracy and nanowire biosensors that play an unstoppable role in diagnosing epidemics [25,26].

18.4 Outlook

The significance of the SiNW cluster are that it has high reproducibility and consistency, high return, as well as phenomenal flexibility and manufacturing capability. Additionally, production of SiNW clusters according to the suggested top and down methodology by standard innovations to prepare semiconductors enables SiNW production for both low and high multiplex capacities and allows combination directly with electric reading circuits. Flow attempts in our lab are indebted to the belief in a platform, where the electric biosensor discovery module consists of the SiNW sensor cluster chip, a miniature electrical test station where biochips are often exposed to electronic hardware that measures opposition to nanowires and refers to verified information and information about the tin frame. This detection module is of exceptional logical value and should provide a compact and integrated framework to allow quality profiling and atomic diagnostics. Additionally, like extra nanowire biosensor devices, the logical intensity stays moderately low and superimposed on a elevated background. The capacity to develop signal intensity is critical to the confidence of a realistic SiNW sensor. Diverse surface chemicals are improved in our laboratory to form sensitive and robust SiNW sensors. However, the possible introduction of semiconductor nanowire detection schemes for diagnostic applications hinges on how this new scheme compares to the existing gold standard, i.e. PCR and ELISA in terms of specificity, simplicity, reliability and sensitivity [27,28].

Nanowire is an important technology that has been developed in current years to unravel problems inherent in customary biosensors, together with vivo sensors, they incorporate low cost moveable devices and novel strategies for processing metered signals. Through this technology, nanowire-based

5.

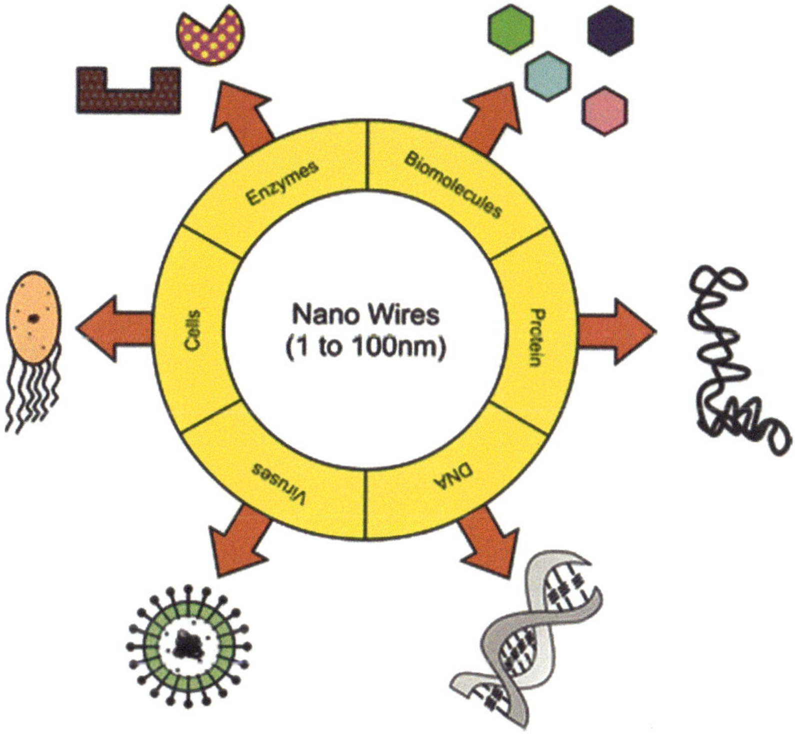

Figure 18.3 Applications of nanowire biosensor.

biosensors are additionally practical and suitable for a variety of biomedical appliances (Fig. 18.3) [29].

18.5 Summary and future prospective

With precise receptors, the sensor nanowires have revealed immense potential to be sensible detection platforms for biological and clinical applications. The accessible device has numerous advantages, like the real-time transmission of electrical signals with elevated sensitivity and the viability of unmarked discovery, as summarize in Table 18.1. Apart from these interesting features, a few improvements are still required for commercialization. Although the sensitivity is very elevated in comparison to other methods, the intensity of the analytic signal remains very low and can be contaminated with high surrounding noise, particularly in in vivo environments. Advancement in the receptor attaching process could solve the problem with superior sensitivity and the problem of simpler manufacturing processing. In addition, the better

Table 18.1 Summary of the salient features and the target application of nanowire-based biosensors.

Features	Application	References
Top-down fabrication process using single-crystalline silicon wafer	Photodiode and FET for the retinal prosthetic systems	[11,12,17]
High sensitivity using poly (ethylene glycol) cross-linker	Detection of protein and DNA	[11,18]
Long-term stability using Al_2O_3 shell coating	In vivo sensing	[8,18]
Integrating nanowires with disposable device	Glucose detection	[18,21]
Multivariable detection using machine learning	Multiple disease diagnosis	[12,18,19,25]

FET, field effect-transistor.

recovery rates of current top-down production methods allow lower costs for commercial products. However, the success of the nanowire sensor depends on how far it has progressed in terms of sensitivity, specificity, simplicity and reliability contrasted to current gold standards like ELISA and PCR processes. Especially with promising wearable biosensors, nanowires are expected to come to an end.

References

[1] J.A. Greene, Prescribing by Numbers: Drugs and the Definition of Disease, JHU Press, 2007.

[2] M.A. Woodbury, J. Clive, A. Garson Jr., Mathematical typology: a grade of membership technique for obtaining disease definition, Comput. Biomed. Res. 11, no. 3 (1978) 277–298.

[3] A. Venkadesh, J. Mathiyarasu, S. Dave, Amine mediated synthesis of nickel oxide nanoparticles and their superior electrochemical sensing performance for glucose detection, Inorg. Chem. Commun. 131 (2021) 108779.

[4] T. Kurkina, K. Balasubramanian, Towards in vitro molecular diagnostics using nanostructures, Cell Mol. Life Sci. 69 (2012) 373–388.

[5] S. Radhakrishnan, B.-S. Kim, S. Dave, Surface modification with nanomaterials for electrochemical biosensing application, in: Advanced Nanomaterials for Point of Care Diagnosis and Therapy, Elsevier, 2022, pp. 101–120.

[6] S. Sahoo, A. Nayak, A. Gadnayak, M. Sahoo, S. Dave, P. Mohanty, et al., Quantum dots enabled point-of-care diagnostics: a new dimension to the nanodiagnosis, in: Advanced Nanomaterials for Point of Care Diagnosis and Therapy, Elsevier, 2022, pp. 43–52.

[7] S. Dave, S. Dave, D. Jayashankar, Biological synthesis of platinum, palladium, copper, and zinc nanostructures, in: Nanobiotechnology, Elsevier, 2021, pp. 211–223.

[8] P. Alivisatos, The use of nanocrystals in biological detection, Nat. Biotechnol. 22 (2004) 47–52.
[9] B. Ballou, B.C.; Lagerholm, L.A. Ernst, M.P. Bruchez, A.S. Waggoner, Noninvasive imaging of quantum dots in mice, Bioconjugate Chem. 15 (2004) 79–86.
[10] F.A. Jaffer, R. Weissleder, SeeingWithin: molecular imaging of the cardiovascular system, Circ. Res. 94 (2004) 433–445.
[11] J.M. Perez, L. Josephson, R. Weissleder, Use of magnetic nanoparticles as nanosensors to probe for molecular interactions, ChemBioChem 5 (2004) 261–264.
[12] S. Sang, Y. Wang, Q. Feng, Y. Wei, J. Ji, W. Zhang, Progress of new label-free techniques for biosensors: a review, Crit. Rev. Biotechnol. 36 (2015) 465–481.
[13] G.-J. Zhang, N. Yong, Silicon nanowire biosensor and its applications in disease diagnostics: a review, Anal. Chim. Acta 749 (2012) 1–15.
[14] X.T. Vu, R. Stockmann, B. Wolfrum, A. Offenhäusser, S. Ingebrandt, Fabrication and application of a microfluidic-embedded silicon nanowire biosensor chip, Phys. Status Solidi (a) 207, no. 4 (2010) 850–857.
[15] P. Namdari, D. Hadis, E. Ali, Recent advances in silicon nanowire biosensors: synthesis methods, properties, and applications, Nanoscale Res. Lett. 11 (1) (2016) 1–16.
[16] Y. Cui, Q. Wei, H. Park, C.M. Lieber, Nanowire nanosensors for highly sensitive and selective detection of biological and chemical species, Science 293 (2001) 1289–1292.
[17] F. Seker, K. Meeker, T.F. Kuech, A.B. Ellis, Surface chemistry of prototypical bulk II-VI and III-V semiconductors and implications for chemical sensing, Chem. Rev. 100 (2000) 2505–2536.
[18] R.F. Taylor, J.S. Schultz, Handbook of Chemical and Biological Sensors, Institute of Physics Pub., Washington, DC, 1996.
[19] Y. Cui, Z.H. Zhong, D.L. Wang, W.U. Wang, C.M. Lieber, High performance silicon nanowire field effect transistors, Nano Lett. 3 (2003) 149–152.
[20] C.M. Lieber, The incredible shrinking circuit—researchers have built nanotransistors and nanowires. Now they just need to find a way to put them all together, Sci. Am. 285 (2001) 58–64.
[21] P.M. Ajayan, J.M. Tour, Nanotube composites, Nature 447 (2007) 1066–1068.
[22] G.J. Zhang, Y. Ning, Silicon nanowire biosensor and its applications in disease diagnostics: a review, Anal. Chim. Acta 749 (2012) 1–15.
[23] A. Kaushik, R. Kumar, E. Huey, S. Bhansali, N. Nair, M. Nair, et al., Fabrication and evaluation of silicon nanowire photodetectors on flexible substrate for retinal prosthetic system, Sens. Mater. 24 (2012) 205–220.
[24] S. Lee, S.W. Jung, J. Ahn, H.J. Yoo, S.J. Oh, Microelectrode array with integrated nanowire FET switches for high-resolution retinal prosthetic systems, J. Micromech. Microeng. 24 (2014) 075018.
[25] R. Janissen, P.K. Sahoo, C.A. Santos, A.M. Da Silva, A.A.G. Von Zuben, D.E.P. Souto, et al., InP nanowire biosensor with tailored biofunctionalization: ultrasensitive and highly selective disease biomarker detection, Nano Lett. 17 (2017) 5938–5949.
[26] W. Zhou, X. Dai, T. Fu, C. Xie, J. Liu, C.M. Lieber, Long term stability of nanowire nanoelectronics in physiological environments, Nano Lett. 14 (2014) 1614–1619.
[27] X. Li, C. Zhao, X. Liu, A paper-based microfluidic biosensor integrating zinc oxide nanowires for electrochemical glucose detection, Microsyst. Nanoeng. 1 (2015) 15014.
[28] N. Shehada, J.C. Cancilla, J.S. Torrecilla, E.S. Pariente, G. Brönstrup, S. Christiansen, et al., Silicon nanowire sensors enable diagnosis of patients via exhaled breath, ACS Nano 10 (2016) 7047–7057.
[29] M.K. Nakhleh, H. Amal, R. Jeries, Y.Y. Broza, M. Aboud, A. Gharra, et al., Diagnosis and classification of 17 diseases from 1404 subjects via pattern analysis of exhaled molecules, ACS Nano 11 (2017) 112–125.

CHAPTER 19

Computational biology and biosensors as surveillance tools for emerging and re-emerging infectious diseases

Swati Goswami, Vijay Kumar and Joydeep Chakraborty
Department of Microbiology, RK University, Rajkot, Gujarat, India

19.1 Introduction

Emerging infectious diseases (EIDs) are a significant economic and public health concern around the world [1,2]. Their emergence is assumed to be primarily influenced by socioeconomic, environmental, and ecological factors, but no comparison study has looked at these connections in order to better understand global EID temporal and spatial trends [1–10]. This chapter provides current information on emerging and reemerging infectious illnesses, as well as the epidemiology, computational biology, and biosensing applications used in infectious disease management, with a focus on the World Health Organization's (WHO) top priority for EIDs [10]. These diseases include Rift Valley fever (RVF), Crimean-Congo hemorrhagic fever, Zika, Lassa fever (LF), Ebola virus disease (EVD), Nipah and henipaviral diseases, and COVID-19.

19.2 Global burden of emerging and reemerging infectious diseases

19.2.1 Rift Valley fever

RVF is a zoonotic viral infection that affects both ruminants and humans [11]. The disease is named after the Rift Valley in East Africa, where the etiologic virus was discovered in 1930 during an inquiry into an outbreak of infection among sheep on a farm in Kenya [12]. As early as 1912, epizootic outbreaks caused by this virus could be identified retroactively [13].

Biosensors for Emerging and Re-emerging Infectious Diseases
DOI: https://doi.org/10.1016/B978-0-323-88464-8.00018-X

More than 30 outbreaks of this illness have occurred in Africa since 1930. RVF has spread over sub-Saharan Africa and the Arabian Peninsula. The RVF virus (RVFV), a member of the *Bunyaviridae* family of the genus *Phlebovirus*, causes the sickness. Mosquitoes transmit the virus, and virus replication in domestic ruminants causes substantial mortality and miscarriage rates. In humans, RVFV infection normally results in a self-limiting, acute, febrile illness; nevertheless, a small percentage of cases progress to neurological problems, partial or total blindness, hemorrhagic fever, or thrombosis.

The epidemiological, clinical, and laboratory aspects of the RVF pandemic in Saudi Arabia from August 26, 2000 to September 22, 2001 are summarized in this associated descriptive research. There were a total of 886 instances reported; 81.9% of the 834 reported cases with laboratory findings were laboratory verified, with 51.1% positive for only RVF immunoglobulin M, 35.7% positive for only RVF antigen, and 13.2% positive for both. The average age (standard deviation [19.4 years]) of the patients was 46.9 years, with a 4:1 male to female ratio. Fever (92.6% of patients), nausea (59.4%), vomiting (52.6%), abdominal pain (38.0%), diarrhea (22.1%), jaundice (18.1%), neurological manifestations (17.1%), hemorrhagic manifestations (7.1%), vision loss or scotomas (1.5%), elevated liver enzyme levels (98%) and elevated lactate levels (1.5%) were among the clinical and laboratory findings. There was a 13.9% fatality rate. The presence of bleeding, neurological symptoms, and jaundice were all linked to a high fatality rate. Patients with leukopenia showed a reduced mortality rate (2.3% vs 27.9%; odds ratio, 0.09; 95% confidence range, 0.01–0.63) than those with a normal or high leukocyte count [14].

Human cases of RVF have been documented in Kenya in the counties of Isiolo and Mandera, as well as in animals in the counties of Isiolo, Mandera, Murang'a, and Garissa. There were 32 human cases (14 confirmed positive) and 11 deaths as of February 4, 2021, a case fatality rate of 34% [15].

19.2.2 Crimean-Congo hemorrhagic fever

A virus of the genus *Nairovirus* causes Crimean-Congo hemorrhagic fever (CCHF), a severe form of hemorrhagic fever. Various mammal species that are asymptomatic serve as amplifying hosts. Tick bites or contact with animal blood can infect humans; CCHF is indigenous to Africa, Asia (particularly the Middle East), and South East Europe, with a wide geographic distribution

[16–18]. Two indigenous instances were recorded in Spain in 2016 and 2018, indicating that it has grown in this area in recent years. The incubation period is brief, with symptoms appearing in less than a week. Fever, headache, myalgia, and gastrointestinal problems are all common signs of various infectious disorders. During the second phase, the hemorrhagic syndrome develops, with severe bleeding in and out of the mucous membranes and skin [19,20]. To prevent secondary and nosocomial transmission, strict barrier preventative measures are essential. Polymerase chain reaction (PCR) detection of the virus genome within the first days following the onset of disease, followed by serological testing for IgM antibodies starting the second week after infection, can be used to confirm CCHF. Supportive care is the mainstay of patient management.

Despite a few optimistic retrospective reports, there is no conclusive proof that ribavirin can be used to cure CCHF. Despite this, the WHO continues to recommend ribavirin for the treatment of CCHF, citing the low medical risk associated with short-term medication. However, post-exposure ribavirin prescriptions for medical professionals should be encouraged to prevent secondary infection.

After Dengue virus, CCHF has the broadest geographical distribution of all tick-borne viruses that cause disease in humans, as well as the second-largest geographical distribution of all arboviruses [21].

19.2.3 Zika virus

Zika virus (ZIKV) is a flavivirus related to yellow fever, dengue fever, West Nile encephalitis, and Japanese encephalitis. On Yap Island in the southwestern Pacific Ocean, ZIKV caused an outbreak of very mild disease characterized by rash, arthralgia, and conjunctivitis in 2007. ZIKV was discovered for the first time outside of Africa and Asia. ZIKV disease's history, transmission dynamics, virology, and clinical manifestations are examined, as well as the possibility of misdiagnosing ZIKV illness as dengue fever. The discovery of ZIKV outside of its previously recognized geographic range should raise concerns about the virus's potential to spread to neighboring Pacific islands and the Americas. On Yap Island in the Federated States of Micronesia, an outbreak of the disease characterized by rash, arthralgia, and conjunctivitis was recorded in April 2007. RNA of the Zika virus (ZIKV), a flavivirus related to yellow fever, dengue, West Nile, and Japanese encephalitis viruses, was found in serum samples from individuals in the acute phase of disease [22].

These data indicate that ZIKV has migrated beyond its traditional geographic area [23,24]. Fever had developed in a rhesus monkey kept in a cage on a tree platform in Uganda's Zika Forest 60 years before, on April 18, 1947 [25–26]. The monkey, Rhesus 766, was a sentinel animal in the Rockefeller Foundation's jungle yellow fever research program [27].

In research conducted in 1968 and between 1971 and 1975 in Nigeria, ZIKV was isolated from humans; 40% of those examined exhibited neutralizing antibodies to ZIKV [25,26,28–31]. Human isolates were recovered from feverish toddlers aged 10 months, 2 years (2 cases), and 3 years, all of whom had no other clinical characteristics recorded, as well as a 10-year-old boy who had a fever, headache, and bodily pains [32–34]. Serologic evidence of human ZIKV infection was reported from Uganda, Tanzania, Egypt, the Central African Republic, Sierra Leone (10), and Gabon between 1951 and 1981, as well as from India, Malaysia, the Philippines, Thailand, Vietnam, and Indonesia [35–37]. A patient in Tahiti sought treatment for hematospermia in December 2013, amid a ZIKV outbreak in French Polynesia, and ZIKV was isolated from his sperm. ZIKV infection through sexual contact has been suspected in the past. This finding lends credence to the idea that ZIKV can be transmitted sexually [38].

19.2.4 Lassa fever

LF is a rare viral hemorrhagic fever that recurs in endemic West African countries such as Sierra Leone, Guinea, Liberia, and Nigeria. The name comes from Lassa, a Nigerian hamlet where the first cases of the disease were reported in 1969 [39–41]. Lassa virus (LASV) belongs to the *Arenaviridae* family, which includes both Old and New World species [42,43]. LASV is an Old World arenavirus, and the multimammate rodent, *Mastomys nataliensis*, which lives in close proximity to people, is its principal reservoir host [44]. The African wood mouse, *Hylomyscus pamfi* and the Guinea multimammate mouse, *Mastomys erythroleucus* are secondary rodent reservoirs [45]. Infections in humans are spread mostly through touch or ingestion of contaminated foodstuffs contaminated with infected rat excrement, urine, or blood. Infections can also be contracted by inhaling aerosols containing dried excreta or urine from infected mice. During the care of the sick or deceased, human-to-human transmission can occur through contact with blood or body fluids.

According to recent estimates, LF causes 2 million infections and 5000–10,000 deaths per year [46,47]. The high prevalence of LASV-specific antibodies in LF endemic areas suggests that many infections are moderate or asymptomatic [48]. Acute LF manifests itself with a variety of symptoms that, in the early stages, can be mistaken for other endemic diseases.

Clinical diagnosis and laboratory confirmation have long been significant obstacles to successful disease management and control in West African afflicted areas. The development of new diagnostic and surveillance methods for LF virus has been aided by recent advances in molecular biology, recombinant DNA technologies, and genomic sequencing. A few examples of the multiplex, magnetic bead-based immunodiagnostics for both LASV antigens and antibodies include: molecular probe-based quantitative real-time PCR for genomic signatures, rapid diagnostics tests that detect the most prevalent West African lineages, and the successful use of next-generation sequencing technology to diagnose and characterize LASV in West Africa. Disease treatment, control, and prevention will continue to improve as a result of these advancements [49–52].

19.2.5 Ebola virus disease

During 2014–16, West Africa had experienced the greatest Ebola virus disease (EVD) outbreak in history since its first reported outbreak in 1976. As of April 2022, the democratic republic of Congo is experiencing sixth outbreak since 2018. As per the World Health Organization (WHO) records, previous outbreaks in Equateur Province were in 2018 and 2020, with 54 and 130 recorded cases, respectively. The Zaire Ebolavirus (EBOV), which is responsible for this outbreak, belongs to the genus *Ebolavirus*, which, along with the genus *Marburgvirus*, makes up the Filoviridae family. EBOV is one of the most dangerous viruses with case fatality rates as high as 90%. Multiorgan failure and severe bleeding complications are the leading causes of death among the Ebola patient. The WHO named the EVD epidemic in West Africa a public health emergency of international concern on August 8, 2014, emphasizing the importance of international attention and collaboration in order to contain the outbreak. The imported EVD case in Nigeria, which resulted in a small outbreak, and similar imported cases in the United States and Spain, which appeared to be well contained at first but eventually led to infection of health-care workers, demonstrate the importance of adequate isolation methods, personnel training, and proper use of personal protective equipment [53,54]. Due to continuing categorization, retroactive investigation, and the availability of test findings, the total number of cases in the West Africa outbreak is subject to change. In the Democratic Republic of Congo, a second, unrelated EVD outbreak has been reported, with a total of 62 confirmed and suspected cases [55].

19.2.6 Nipah and Henipaviral diseases

Nipha virus is of zoonotic origin that spreads from animals to human. It was first recognized during outbreak in Malaysia in the year 1999. Many outbreaks of NiV have been reported during 2001 and 2011 from Bangladesh. In India, NiV infection was reported for the first time in 2018, followed by 2019 and now its incidences have been reported recently from Kerala in September 2021 where a 12-year-old boy was found Nipah virus positive in Kozhikode district, Kerala and he died after 4 days of hospitalization. When Nipah and Henipaviral illnesses spread rapidly, they offer a significant risk of an international public health emergency [56]. Following the current emergency circumstances caused by Ebola and Zika virus during the last 5 years [57,58], the public is now concerned about Nipah virus disease epidemics. EVD outbreaks and epidemics (2014–16) prompted a massive mobilization of researchers to develop new technologies in the areas of efficient and rapid diagnostics, vaccines, therapies, and drug targets to combat EVD and save the lives of a large human population around the world. Scientists are working to combat the Nipah virus, just as they do with Zika.

Nipah virus (NiV), a paramyxovirus belonging to the genus *Henipavirus* in the family Paramyxoviridae, causes a zoonotic infection known as Nipah viral sickness. It's a biosafety level-4 pathogen that is spread by certain types of fruit bats, primarily *Pteropus* spp., which serve as natural reservoir hosts. In 1998, the sickness was initially reported from the Malaysian town of Kampung Sungai Nipah. There is also human-to-human transfer. Outbreaks have also been detected in various South and Southeast Asian nations. Based on currently available complete N and G gene sequences, phylogenetic analysis confirmed the circulation of two major clades of NiV. In the NiV-MY clade, NiV isolates from Malaysia and Cambodia clustered together, while isolates from Bangladesh and India clustered together in the NiV-BD clade. Thailand's NiV isolates had a mixed population of sequences.

In humans, the virus is responsible for a fast advancing severe sickness that can include severe respiratory illness and/or fatal encephalitis. Respiratory sickness, as well as anxious signs, may develop in pigs under the age of 6 months. For diagnostic reasons, many enzyme-linked immunosorbent tests and molecular approaches based on polymerase chain reaction have been developed. Due to the high cost of antibody medicines, finding broad-spectrum antivirals is critical, as is concentrating on tiny interfering RNAs (siRNAs). The high pathogenicity of NiV in humans, as well as the lack of vaccinations or medicines to combat the disease, has piqued the interest of researchers all over the world who are working to produce effective NiV vaccines and treatment regimens [59–62].

19.2.7 COVID-19

SARS-CoV-2 (formerly known as 2019-nCoV) is an enveloped, positive-sense single-stranded genomic RNA virus (+ssRNA) that caused the coronavirus disease of 2019 (COVID-19). SARS-CoV-2, which was first identified in the Chinese city of Wuhan, is contagious in humans and has spread fast over the world due to close human contact or available respiratory material (cough, sneeze) from sick persons. As a result of the rising infection rate outside of China, WHO Director-General Dr. Margaret Chan designated the COVID-19 outbreak a "pandemic" on March 12, 2020 [63]. SARS-Cov-2 is a member of the *coronavirus* family and the *Sarbecovirus* subgenus, which includes numerous other species that cause mild to severe human infections. After 229E, NL63, OC43, HKU1, MERS-CoV, and the previous SARS-CoV, SARS-CoV-2 is the eighth coronavirus to infect people [64–66]. Between December 2019 and early April 2020, Khailany et al. published a complete genome characterization study based on 95 SARS-CoV-2 sequences that were available on GenBank, the National Microbiology Data Center (NMDC), and the NGDC Genome Warehouse, with the NC 045512 genome sequence serving as the reference for alignment [67]. According to the report, the SARS-CoV-2 complete genome is around 30 kb, with orf1ab encoding orf1ab polyproteins accounting for two-thirds of the 5′ and structural proteins accounting for one-third of the 3′. Surface glycoprotein (S), envelope protein (E), membrane protein (M), and nucleocapsid (N) proteins are currently known [67]. Furthermore, the genome comparison analysis discovered 116 mutations, the most common of which are 8782C > T in the ORF1ab gene, 28144T > C in the ORF8 gene, and 29095C > T in the N gene. According to the study, the alterations may have an impact on the severity and spread of SARS-CoV-2. A quick and accurate self-testing method for COVID-19 diagnosis has become essential for determining the actual number of cases worldwide and taking appropriate medical and governmental measures. The SARS-CoV-2 (previously, 2019-nCoV) virus was first recorded in Wuhan (China) in December 2019, and it quickly spread throughout the world, resulting in 14 million active cases and 582,000 deaths as of July 2020. The diagnosis tools that have been provided so far are based on viral gene detection, human antibody detection, and viral antigen detection, with viral gene detection by RT-PCR being the most reliable methodology. Yüce et al. [68–70].

19.3 Computational biology and biosensors as surveillance tools for emerging and reemerging infectious diseases

Surveillance of infectious diseases is one of the most important tools for monitoring the emergence of new pathogenic outbreaks as well as the reemergence of various infectious diseases. Surveillance helps not only in monitoring the status of an infectious disease, but also helps decide measures to control it well in time once there has been an outbreak. Infectious disease surveillance is composed of three major aspects, that is, current epidemiological status of diseases, monitoring trends in spread of infection and the identification of new pathogenic outbreaks [71]. There are primarily two types of surveillance systems. One is the active one, where each and every case related to an infectious disease is identified and recorded by active engagement of public health staff. The cumulative data is then submitted to a public health agency, which plays the key role in data management and analysis. In passive surveillance, on the other hand, relevant cases are reported to the public health agency directly by medical professionals based on case studies. However, in both cases, data handling becomes labor-intensive and requires high-end analysis.

19.3.1 Alert and response systems

With the advent of information technology (IT) in late 20th century, and its rapid evolution over the last couple of decades, application of IT tools in large data management has become indispensable. The Monitoring

Emerging Diseases (ProMED) program (http://www.promedmail.org), run by the International Society for Infectious Diseases (ISID), was launched in 1994 with an aim to identify unusual health events related to emerging and reemerging infectious diseases (RID). Outbreak analyses obtained from ProMED have not only been used in the past to raise alarms during major outbreaks like Zika and SARS, but it also came up with the first detailed report of early cases of COVID-19 pandemic [72]. Another such tool is HealthMap (http://www.healthmap.org) which displays the live global status of diseases. Data from both these tools are used by the United States Center for Disease Control and Prevention (US CDC) and the WHO.

In April 2000, the WHO formalized an infrastructure called the "Global Outbreak Alert and Response Network" (GOARN) (https://extranet.who.int/goarn/) for responding to the heightened need for early awareness of outbreaks and preparedness to respond. The Network, which unites 110 existing networks, is supported by several new mechanisms and a computerized tool for real-time gathering of disease intelligence. The procedure for outbreak alert and response has four phases: systematic detection, outbreak verification, real time alerts, and rapid response. With an aim to identify a flu pandemic early and respond quickly, Google launched a collective intelligence-based web service called "Google Flue Trends" in 2008 [73]. The algorithm calculated predictions based on Google search queries and thus, monitored millions of users' health tracking behaviors online. Privacy of users' was maintained by only considering anonymous search queries, without identifying individuals that performed the search. The tool was able to predict influenza disease with some accuracy, close to the CDC influenza reports based on laboratory-confirmed influenza surveillance. In fact, in 2010, Google's data of search queries about flu symptoms predicted a spike in flu cases in the mid-Atlantic region of the United States 2 weeks prior to the CDC report being released. Despite all of this, Google Flue Trends had to be discontinued in 2015 following instances of overprediction of the number of influenza cases, especially during 2011–13. The system is no longer active, but is available for research and development purposes.

19.3.2 Surveillance through social networking and outbreak modeling

Currently, surveillance of infectious diseases has gone beyond traditional active and passive systems and includes data retrieval through electronic health records, mobile phones and social media. Members of the International Society for Disease Surveillance (ISDS) established a social media workgroup to develop research and innovations in electronic public health surveillance by reviewing various ways on how social media data (e.g., those obtained from Twitter, Facebook, Yahoo forum, PatientsLikeMe, TeenSMART, MySpace, Chatroom and other web forums) can be used in the surveillance of both infectious and non-infectious diseases [74]. They showed how the use of social media data can provide real-time surveillance of disease outbreaks, speed up management of such outbreaks, and can identify target populations in need of public health intervention.

Real-time modeling has also been found useful during past pandemics, as well as during the recent COVID-19 crisis, in providing public health agencies with outbreak analyses [72]. Andrews and Basu designed mathematical models to project cholera transmission in Haiti [75]. They further simulated future epidemic trajectories to estimate the effect of clean water, vaccination, and enhanced antibiotic distribution programs. Similarly, a simulation optimization approach was taken by Nsoesie et al. [76] for a real-time forecast of influenza epidemics. Mathematical simulation has also recently been used by Kissler et al. to model the spread of SARS-CoV-2 and predict its transmission through next few years [77].

19.3.3 Computational biology in pathogen identification and disease surveillance

Alongside applications in evolutionary studies, (improvement of livestock, metagenomics, epigenetics, forensic sciences and single cell genomics, etc.), DNA sequencing is gradually finding huge application in biomarker discovery, vaccinology, therapeutics and pharmacology [78–87]. DNA sequencing technologies have evolved substantially in the last 15 years. Advances in sequencing techniques, coupled with computer-aided faster analyses, have made it possible to identify and characterize several emerging respiratory viruses [88]. Methods have developed to make sequencing fast, accurate, yet inexpensive [89,90]. Since the introduction of next-generation sequencing (NGS) technology in 2005, a number of high-throughput sequencing platforms have emerged, such as MiSeq, GAIIx, MiSeqDx, NextSeq, NovaSeq, MiniSeq, and iSeq from Illumina, Ion Torrent/Ion S5 platform from Life Technologies, PacBioRS/RSII and Sequel from Pacific Biosciences (PacBio) as well as MinION, PromethION and GridION from Oxford Nanopore Technologies. These high-throughput sequencing platforms come with variety of capacities, costs and applications, Illumina MiSeq being the most commonly used platform for identification of pathogens and surveillance of infectious diseases [91–95]. Although considered error-prone, MinION is portable and can be easily transported to the site of an outbreak. It was successful in the real-time surveillance of Zika virus in Brazil, in the year 2016, and was also used previously in 2015 to monitor the Ebola outbreak in Guinea [96,97]. MinION showed the potential for rapid sequencing (as little as 15 min) and pathogen detection during an ongoing outbreak.

Metagenome sequencing has been found to be useful in cases of pathogenesis, where identification of the pathogen is carried out from samples like blood, stool, nasopharyngeal swab, urine or cerebrospinal fluid [98,99]. However, there remains a challenge of host genomic contamination, which needs to be managed logically while sequencing metagenome [100,101]. Enrichment methods such as negative selection, like the CRISPR-based approach used in the detection of meningitis and encephalitis-associated pathogens [102], or probe-based enrichment used for detection of viral genomes in Ebola virus outbreaks, Zika virus epidemics and respiratory virus surveillance [93,103,104] have proved useful in avoiding background noise in such metagenomic approaches. Positive enrichments, however, are more efficient as negative selection might lead to the loss of genomic content of the targeted pathogen itself [105].

Software used to process this vast amount of data has also evolved over the years to facilitate fast, accurate and easy detection of pathogens. Software readily used for genome construction and pathogen identification are Geneious, CLC Genomics Workbench, DNAStar, Mauve, Geneious, MEGA, Nextstrain, RAST, Taxonomer, EDGE, Pathosphere, etc. These are easy to use and often require less computer proficiency. On the other hand, tools like VirusSeeker, BWA, ngs_mapper, SOAPdenovo, Bowtie2, MetaRay, Trinity, Bandage, BEAST, Mr. Bayes, PhyML, RaxML, RDP4, GATK, iMetAMOS and ShortBRED are utilized by advanced users [105]. Tools like MEGA and Geneious have been found helpful for estimating the frequency of mutations responsible for conferring antiviral resistance to viruses like HIV or influenza [91,106]. Similarly, viral phylodynamics aid in the early identification of viral epidemics. For example, whole-genome sequence data was found to be critical in the early identification of HIV spread in North America as well as determining the spatial origins of the 2009 H1N1 pandemic, 2016 Zika virus outbreak in Florida and local H7N9 outbreaks in China [107,108].

Apart from genomic information, other "omics" approaches (e.g., proteomics, transcriptomics and metabolomics) have also added further to the repertoire of information associated with various pathological conditions. One such example is the identification of various tuberculosis biomarkers based on different omics approaches [109]. However, the major challenge that arises from such advanced techniques is post processing of the "big" data generated, often referred to as "omics data." Computational biology has thus become indispensable when it comes to storage, management and analysis of such data.

19.4 Need of biosensors for disease diagnosis

Contagious diseases continue to be a major health concern in Southeast Asia; in the developing world, countries such as China, Malaysia, Thailand, India, and Pakistan are in a transitional period, suffering from both communicable and noncommunicable diseases. With drastic changes in the environment and farm management practices, we must arm ourselves and prepare for developing and re-emerging animal diseases such as cancer, prion diseases, COVID-19, influenza, and others. Recent outbreaks of infectious diseases, due to rapid industrialization, population rise, urbanization and other such factors have underlined the importance of early and correct diagnosis of diseases in order to frame policies for accurate monitoring, proper treatment strategies, disease containment, and stopping the spread of disease. For clinical care of infectious and noninfectious animal diseases, rapid and precise diagnosis along with focused treatments are critical. Presently, centralized testing facilities are inconveniently located, and the need for patients to travel to these centers plays a critical role in promoting disease transmission. Current diagnostic approaches, on the other hand, are invasive/intrusive and create a significant bottleneck by necessitating specialized equipment and skilled workers. In addition, the effectiveness of biological sampling and prompt analysis are also essential components of successful diagnostic procedures.

The ability to treat and control emerging and reemerging diseases effectively requires an early diagnosis of the condition which is currently based on parameters such as basic biochemical, microbiological, parasitological, and serological testing of biological samples. Other methods include ELISA, PCR, DNA chips, and chromatographic procedures that require sophisticated instruments, trained operators, and expensive reagents. Thus, existing disease diagnosis methods need to be improved in terms of readability, reproducibility, reliability, specificity, sensitivity, and cost, so as to make them suitable for use as a point-of-care (POC) device, which the patient can use with ease. One of the most common examples for this are the present day glucose sensors which reads the glucose level in blood and informs the patient about their blood glucose level, indicating the presence or absence of diabetes. Another example is pregnancy detection kits that detect the presence or absence of HCG hormone in expecting mother's urine. Biosensor-based detection kits like these and many more, have made their presence in to most homes due to its ease of usability and readability.

There are other biosensors as well for which the protocol may be complex in some circumstances, and it must be automated and bundled into a simple device to be acceptable for POC use. Although we always require biological samples from patients which are collected in the form of blood, saliva, urine, oral swab, nasal swab, tissue biopsy, stool or genital swabs which are although noninvasive, may still cause difficulties for the patient and/or increase the chances of exposure to external pathogens.

19.4.1 What are biosensors?

Biosensors are miniaturized bio-analytical devices that provides a platform for a biochemical reaction to take place and report the changes (qualitative or quantitative) that occurred due to enzymatic reactions or another biochemical reaction. They can be successfully applied for detecting a wide variety of target biological moieties that include nucleic acids, proteins, enzyme specific substrates and many more. The detection is based on diverse biochemical changes like changes in temperature, light intensity, ion concentration, conductivity, etc.

19.4.2 Biosensors, internet of medical things and point-of-care testing devices

In the present era when industrialization is rapidly overtaking farm and forest land, thus promoting urbanization, people are bound to live in densely populated areas where hygiene and cleanliness is a major challenge, especially in developing world countries. Such conditions favor the origin and rapid spread of communicable diseases. In these conditions, quick diagnosis of any kind of spread of disease is very crucial for public health. With rapid digitalization, the internet is increasing its presence and reachability. This can be best utilized for the development of POC testing devices that can give on-site results which will help in rapid containment or spread of disease. This concept, referred to as Internet of Medical things (IoMT) allows POC devices to communicate with health experts and medical centers via wireless technology. It is not a rule of thumb, but observations have indicated that any disease, before reaching the pandemic stage, undergoes typically four stages. These include: (1) the first report of positive patients in a healthy community, (2) local transmission among the people who directly or indirectly came in contact with first known patient, (3) rapid spread at the community level called community transmission, and (4) reduced pathogen transmission due to acquired immunity and/or clinical interventions. Combined application of vaccines, drugs, diagnosis, and a therapeutic coordinated response helps in dealing with the danger of converting and epidemics into pandemics. Biosensors have a crucial role to play in the early disease detection system and control of disease spread.

19.4.3 Types of biosensors

The selection of a target for diagnosis can vary from an antigenic epitope to a pathogenic nucleic acid, a secondary metabolite of pathogen origin or a hormone. Subsequently, antibodies, enzymes, complementary sequences of nucleic acids, certain receptors for metabolites or even whole cells can be utilized as bioreceptors. Antibodies are more commonly used than nucleic acid probes in biosensors designed for disease detection. Out of three major components of a biosensor, the transducer, enables detection and communication of the biochemical changes that take place due to the interaction between the biological sample and the bioreceptor or sensor. Thus the choice and placement of the transducer is very important. With the introduction of immobilization technologies, it has become more convenient to use receptors and transducers of choice and obtain desirable combinations with needed sensitivity and specificity.

Depending on the method of transduction used, biosensors are classified as optical, electrochemical, piezoelectric, calorimetric, or mass-sensitive. The first three types of biosensors are currently the most developed for use in detecting infectious agents (Fig. 19.1).

19.4.3.1 Cellular components as biosensors

Glycoproteins, which are usually carbohydrates containing lipids with a sphingosine backbone, are found in practically every microorganism's cellular structure, including bacteria and viruses. They participate in a variety

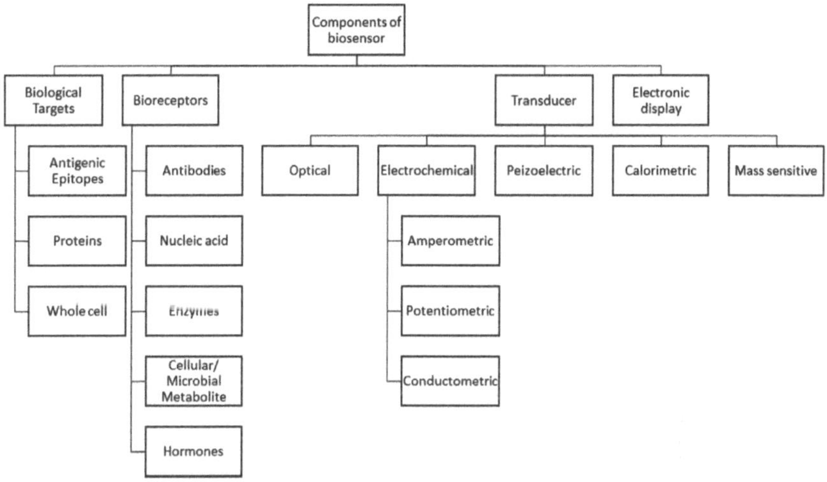

Figure 19.1 Components and classification of biosensor.

of biological interactions and hence can be used as signature molecules for specific proteins and glycan moieties. These glycoproteins, for example, lectins, mucines, etc., consist of oligosaccharides, attached to proteins via either *N*- or *O*-glycosidic bonds. These unique carbohydrate moieties that are associated with many surface proteins of pathogens can differentiate not only among pathogens, but also among subtle differences at the species and strain level. These include structures like capsular polysaccharides (CPS) and lipopolysaccharides (LPS), as well as relatively conserved structures with the polymer of amino sugars attached to a glycophospholipid like enterobacterial common antigen (ECA), or the diversity of O antigens found in Gram negative bacteria like *Vibrio cholerae* (> 190 different O-antigens), *Pseudomonas aeruginosa* (20 different O-antigens), *Salmonella enterica* (54 different O-antigens), and *Escherichia coli* serotypes (181 O-antigens).

Specific changes in N- and O-linked glycoproteins are highly linked to the development and progression of various diseases, so this can be used to tag proteins present on the outer surface of the cell as well as an identification marker for disease diagnosis and prognosis. Optimum utilization of such glycoproteins for development of diagnostic devices and their applications in clinical settings requires a thorough understanding of the biosynthetic pathways involved, as well as the roles that these molecules play in host–cell interactions (Fig. 19.1).

19.4.3.2 Glycane-based biosensors

The interaction between lectin carbohydrates and glycoproteins creates the basis for glycan biosensors, which is based on the fact that the presence of glycans on the outer surface of pathogens can be employed as signature molecules. In order to develop diagnostic devices or glycan-based biosensors for disease diagnosis, the whole-cell assay strategy is applied which involves the attachment of carbohydrates to the solid surface and the binding of glycoproteins of pathogen origin. One strategy to accomplish this is by conjugation methods, such as attaching biotin to streptavidin, a natural process that relies on the binding affinity of biotin for streptavidin. The other method is the production of self-assembled monolayers (SAMs) of alkanethiols on gold-coated substrate that exposes the functional groups thus facilitating noncovalent and covalent link formation on the surface. *N*-Acetyl glucosamine groups presented on the surface can be glycosylated by incubating with galactose and GalTase which can then be recognized via lectin binding to the newly formed disaccharide. The biotin−streptavidin-based carbohydrate presentation has the limitation that it can display a unique substrate on an individual surface and it requires each biomolecule to be coated separately in a 96 well plate while this is not so in SAMs method. The SAMs comes with key advantages like utilizing the synthesis of carbohydrate arrays which are easy to construct, and are more inert towards protein adsorption and have demonstrated utility in conjunction with cell culture. Thus SAMs shows higher potential to be utilized for development of biochips in disease diagnosis.

Recently, a glycan-based biosensor was developed for the detection of influenza haemagglutinins that worked on the principle of electrochemical impedance spectroscopy. The device is sensitive enough to detect the presence of influenza haemagglutinin molecules in concentrations as low as a single molecule. In addition, the biosensor had a controlled density of sialylated glycans on the surface via the covalent immobilization of amine-modified glycan on a mixed self-assembled monolayer of −COOH groups. The working concentration range of the biosensor was attomolar to nanomolar for both lectins and influenza haemagglutinins.

19.4.3.3 Lectin-based biosensors

The specificity in Lectin-carbohydrate binding has led to priority use of lectin protein for electrochemical and optical biosensor development. Commonly used biosensors for this principle include glucose- and mannose-specific lectin protein concanavalin A (Con A) lectin. The Con

A is widely used for development of diagnostic sensors for glucose, cancer cells, pathogenic bacteria, and hydrogen peroxide due to its capability to immobilize glucose oxidase (GOx) and horseradish peroxidase (HRP) enzymes.

Studies done in last few years suggest that by applying the concept of layer-by-layer deposition of Con A and enzymes on the surface of the electrode, dual enzyme-based sensors (GOx and HRP) were successfully prepared, and the electrochemical responses were recorded against aromatic amines, phenols, and sulfides. Presently, nanomaterials are also frequently employed in biosensor development to enhance the lectin-specific signal via amplification. Glycoproteins expressed on living cells are yet another emerging target of lectin-based biosensors.

19.4.3.4 Nanoparticle-based biosensors

Nanomaterials-based biosensors have recently gained a lot of attention because of their biological activities and unique sensing qualities, which allow them to detect analytes such as nucleic acid (DNA or RNA), aptamers, and proteins in clinical samples. Nanotechnology advancements also allow for the creation of smaller detection systems for POC biosensors, which could be a novel way to detect human viral illnesses. Nanoparticles are tiny structures or substances or materials ranging from 1 to 100 nm. Several unique properties of nanoparticles, such as enormously good chemical reactivity and high strength conductivity, have received a lot of interest for the use in electrochemical biosensor development. Carbon nanotubes, silica nanoparticles, quantum dots, polymeric nanoparticles, and metal nanoparticles (MNPs) are among the nanostructures that have been explored in the context of biological detection. Out of diverse MNPs, silver and gold nanoparticles are widely researched and used for this purpose due to their physicochemical, morphological and structural details available at the nanoscale. These characteristics allowed MNPs to be produced and modified with various chemical functional groups for conjugation with DNA probes, antibodies, and ligands, all of which are important in the development of biosensors.

19.5 Conclusion

It is critical to strengthen and promote basic and applied research on pathogens, hosts, and environmental factors by leveraging our understanding of these to improve our ability to forecast and avoid the conditions that lead to human diseases. The next stage is to develop new diagnostics, vaccines, therapeutics, and sequencing methods, as well as conduct post-genomics research on EID agents and animal vectors to uncover the genetic basis for microbial/vector development, adaptation, and pathogenicity. It is also vital to track better control efforts and develop national and worldwide scientific skills required to respond to future health concerns by supporting research and training programs.

Surveillance, research, and training, as well as enhanced diagnostic facilities and public health systems, must all be prioritized. At both the regional and national levels, readiness for disastrous circumstances must continue to increase. The regional stockpiles will be checked and updated on a regular basis. An electronic-mail system will be used to construct regional communicable disease information networks that will connect WHO collaborating centers, international agencies, and institutions. These networks will be linked to the WHO's worldwide communicable disease information system. The region's antimicrobial resistance surveillance will be bolstered by the addition of more monitoring laboratories in various places. The regional antibiotic resistance system (Network on Antimicrobial Resistance) for general bacterial illnesses will be integrated into global antimicrobial resistance surveillance.

References

[1] D.M. Morens, G.K. Folkers, A.S. Fauci, The challenge of emerging and re-emerging infectious diseases, Nature 430 (6996) (2004) 242–249.
[2] S. Binder, A.M. Levitt, J.J. Sacks, J.M. Hughes, Emerging infectious diseases: public health issues for the 21st century, Science 284 (5418) (1999) 1311–1313.
[3] P. Daszak, A.A. Cunningham, A.D. Hyatt, Emerging infectious diseases of wildlife—threats to biodiversity and human health, Science 287 (5452) (2000) 443–449.
[4] L.H. Taylor, S.M. Latham, M.E. Woolhouse, Risk factors for human disease emergence, Philos. Trans. R. Soc. London. Ser. B: Biol. Sci. 356 (1411) (2001) 983–989.
[5] J.A. Patz, et al., Unhealthy landscapes: policy recommendations on land use change and infectious disease emergence, Environ. Health Perspect. 112 (2004) 092–1098.
[6] R.A. Weiss, A.J. McMichael, Social and environmental risk factors in the emergence of infectious diseases, Nat. Med. 10 (2004) S70–S76.
[7] M.E.J. Woolhouse, S. Gowtage-Sequeria, Host range and emerging and reemerging pathogens, Emerging Infect. Dis. 11 (2005) 1842–1847.
[8] S.S. Morse, in: S.S. Morse (Ed.), Emerging Viruses, Oxford Univ. Press, New York, 1993, pp. 10–28.
[9] K.E. Jones, N.G. Patel, M.A. Levy, A. Storeygard, D. Balk, J.L. Gittleman, Global trends in emerging infectious diseases, Nature 451 (2008) 990–993.
[10] N.I. Nii-Trebi, Emerging and neglected infectious diseases: insights, advances, and challenges, BioMed Res. Int. 2017 (2017).

[11] C.J. Peters, Emergence of Rift Valley Fever pp253−264. Factors in the Emergence of Arbovirus Diseases, 1997.

[12] R. Daubney, J.R. Hudson, Enzootic hepatitis or Rift Valley fever. An un-described virus disease of sheep, cattle and man from East Africa, J. Pathol. Bacteriol. 34 (1931) 545−579.

[13] G.A. Eddy, C.J. Peters. The extended horizons of Rift Valley fever: current and projected immunogens. Army Medical Research Inst of Infectious Diseases Fort Detrick MD, 1980.

[14] T.A. Madani, Y.Y. Al-Mazrou, M.H. Al-Jeffri, A.A. Mishkhas, A.M. Al-Rabeah, A. M. Turkistani, et al., Rift Valley fever epidemic in Saudi Arabia: epidemiological, clinical, and laboratory characteristics, Clin. Infect. Dis. 37 (8) (2003) 1084−1092.

[15] Q. Mehmood, M. Irfan, I.O. Ogunkola, F. Jaguga, I. Ullah, Rift Valley fever and COVID-19 outbreak in Kenya: a double whammy, Ethics Med. Public Health 19 (2021) 100685.

[16] C.A. Whitehouse, Crimean−Congo hemorrhagic fever, Antiviral Res. 64 (3) (2004) 145−160.

[17] Ö. Ergönül, Crimean-Congo haemorrhagic fever, Lancet Infect. Dis. 6 (4) (2006) 203−214.

[18] D.A. Bente, N.L. Forrester, D.M. Watts, A.J. McAuley, C.A. Whitehouse, M. Bray, Crimean-Congo hemorrhagic fever: history, epidemiology, pathogenesis, clinical syndrome and genetic diversity, Antiviral Res. 100 (1) (2013) 159−189.

[19] N.Y. Pshenichnaya, I.S. Sydenko, E.P. Klinovaya, E.B. Romanova, A.S. Zhuravlev, Possible sexual transmission of Crimean-Congo hemorrhagic fever, Int. J. Infect. Dis. 45 (2016) 109−111.

[20] A. Papa, F. Weber, R. Hewson, M. Weidmann, I. Koksal, G. Korukluoglu, et al., Meeting report: first international conference on Crimean-Congo hemorrhagic fever, Antiviral Res. 120 (2015) 57−65.

[21] P. Fillâtre, M. Revest, P. Tattevin, Crimean-Congo hemorrhagic fever: An update, Med. Mal. Infect. 49 (8) (2019) 574−585.

[22] E.B. Hayes, Zika virus outside Africa, Emerg. Infect. Dis. 15 (9) (2009) 1347.

[23] A. Geser, B.E. Henderson, S. Christensen, A multipurpose serological survey in Kenya: 2. Results of arbovirus serological tests, Bull. World Health Organ. 43 (4) (1970) 539.

[24] R.S. Lanciotti, O.L. Kosoy, J.J. Laven, J.O. Velez, A.J. Lambert, A.J. Johnson, Propriedades genéticas e sorológicas do vírus Zika associado a uma epidemia, estado da Micronesia, 2007, Emerg. Infect. Dis. 14 (8) (2008) 1232−1239.

[25] G.W. Dick, S.F. Kitchen, A.J. Haddow, Zika virus (II). Pathogenicity and physical properties, Trans. R. Soc. Trop. Med. Hyg. 46 (5) (1952).

[26] G. Dick, S. Kitchen, A. Haddow, Zika virus. I. Isolations and serological specificity, Trans. R. Soc. Trop. Med. Hyg. 46 (5) (1952) 509−520.

[27] D. Musso, C. Roche, E. Robin, T. Nhan, A. Teissier, V.M. Cao-Lormeau, Potential sexual transmission of Zika virus, Emerg. Infect. Dis. 21 (2) (2015) 359.

[28] F.N. Macnamara, Zika virus: a report on three cases of human infection during an epidemic of jaundice in Nigeria, Trans. R. Soc. Trop. Med. Hyg. 48 (2) (1954) 139−145.

[29] J.P.T. Boorman, J.S. Porterfield, A simple technique for infection of mosquitoes with viruses. Transmission of Zika virus, Trans. R. Soc. Trop. Med. Hyg. 50 (3) (1956).

[30] D.Á. Moore, O.R. Causey, D.E. Carey, S. Reddy, A.R. Cooke, F.M. Akinkugbe, et al., Arthropod-borne viral infections of man in Nigeria, 1964−1970, Ann. Trop. Med. Parasitol. 69 (1) (1975) 49−64.

[31] A. Fagbami, Epidemiological investigations on arbovirus infections at Igbo-Ora, Nigeria, Trop. Geogr. Med. 29 (2) (1977) 187−191.

[32] Y. Robin, J. Mouchet, Serological and entomological study on yellow fever in Sierra Leone, Bull. Soc. Pathol. Exot. Filiales 68 (3) (1975) 249–258.
[33] A.H. Fagbami, Zika virus infections in Nigeria: virological and seroepidemiological investigations in Oyo State, Epidemiol. Infect. 83 (2) (1979) 213–219.
[34] J.G.i Olson, T.G. Ksiazek, Zika virus, a cause of fever in Central Java, Indonesia, . Trans. R. Soc. Trop. Med. Hyg. 75 (3) (1981) 389–393.
[35] C. Jan, G. Languillat, J. Renaudet, Y. Robin, A serological survey of arboviruses in Gabon, Bull Soc Pathol Exot Filiales 71 (2) (1978) 140–146.
[36] J.F. Saluzzo, J.P. Gonzalez, J.P. Herve, A.J. Georges, Serological survey for the prevalence of certain arboviruses in the human population of the south-east area of Central African Republic (author's transl), Bull. Soc. Pathol. Exot. Filiales 74 (5) (1981) 490–499.
[37] J.F. Saluzzo, B. Ivanoff, G. Languillat, A.J. Georges, Serological survey for arbovirus antibodies in the human and simian populations of the South-East of Gabon (author's transl), Bull. Soc. Pathol. Exot. Filiales 75 (3) (1982) 262–266.
[38] J. Liu, Y. Liu, C. Shan, B.T. Nunes, R. Yun, S.L. Haller, et al., Role of mutational reversions and fitness restoration in Zika virus spread to the Americas, Nat. Commun. 12 (1) (2021) 1–12.
[39] J.D. Frame, J.M. Baldwin Jr, D.J. Gocke, J.M. Troup, Lassa fever, a new virus disease of man from West Africa, Am. J. Trop. Med. Hyg. 19 (4) (1970) 670–676.
[40] S.M. Buckley, J. Casals, Lassa fever, a new virus disease of man from West Africa, Am. J. Trop. Med. Hyg. 19 (4) (1970) 680–691.
[41] S.M. Buckley, J. Casals, W.G. Downs, Isolation and antigenic characterization of Lassa virus, Nature 227 (5254) (1970) 174.
[42] N. Sarute, S.R. Ross, New World arenavirus biology, Annu. Rev. Virol. 4 (2017) 141–158.
[43] C. Huang, O.A. Kolokoltsova, N.E. Yun, A.V. Seregin, S. Ronca, T. Koma, et al., Highly pathogenic New World and Old World human arenaviruses induce distinct interferon responses in human cells, J. Virol. 89 (14) (2015) 7079–7088.
[44] T.P. Monath, V.F. Newhouse, G.E. Kemp, H.W. Setzer, A. Cacciapuoti, Lassa virus isolation from Mastomys natalensis rodents during an epidemic in Sierra Leone, Science 185 (4147) (1974) 263–265.
[45] A. Olayemi, D. Cadar, N.F. Magassouba, A. Obadare, F. Kourouma, A. Oyeyiola, et al., New hosts of the Lassa virus, Sci. Rep. 6 (1) (2016) 1–6.
[46] I.S. Lukashevich, S. Paessler, J.C. de la Torre, Lassa virus diversity and feasibility for universal prophylactic vaccine, F1000Research 8 (2019).
[47] D. Safronetz, C. Mire, K. Rosenke, F. Feldmann, E. Haddock, T. Geisbert, et al., A recombinant vesicular stomatitis virus-based Lassa fever vaccine protects guinea pigs and macaques against challenge with geographically and genetically distinct Lassa viruses, PLoS Negl. Trop. Dis. 9 (4) (2015) e0003736.
[48] K.M. Johnson, J.B. McCormick, P.A. Webb, E.S. Smith, L.H. Elliott, I.J. King, Clinical virology of Lassa fever in hospitalized patients, J. Infect. Dis. 155 (3) (1987) 456–464.
[49] A.N. Happi, C.T. Happi, R.J. Schoepp, Lassa fever diagnostics: past, present, and future, Curr. Opin. Virol. 37 (2019) 132–138.
[50] J. Purushotham, T. Lambe, S.C. Gilbert, Vaccine platforms for the prevention of Lassa fever, Immunol. Lett. 215 (2019) 1–11.
[51] V. Raabe, J. Koehler, Laboratory diagnosis of Lassa fever, J. Clin. Microbiol. 55 (6) (2017) 1629–1637.
[52] N. Baillet, S. Reynard, E. Perthame, J. Hortion, A. Journeaux, M. Mateo, ... S. Baize, Systemic viral spreading and defective host responses are associated with fatal Lassa fever in macaques, Commun. Biol. 4 (1) (2021) 1–19.
[53] E.M. Leroy, I. Labouba, G.D. Maganga, N. Berthet, Ebola in West Africa: the outbreak able to change many things, Clin. Microbiol. Infect. 20 (10) (2014) O597–O599.

[54] A.E. Thorson, G.F. Deen, K.T. Bernstein, W.J. Liu, F. Yamba, N. HabibSierra Leone Ebola Virus Persistence Study Group, Persistence of Ebola virus in semen among Ebola virus disease survivors in Sierra Leone: a cohort study of frequency, duration, and risk factors, PLoS Med. 18 (2) (2021) e1003273.
[55] M. Goeijenbier, J.J. Van Kampen, C.B. Reusken, M.P. Koopmans, E.C. Van Gorp, Ebola virus disease: a review on epidemiology, symptoms, treatment and pathogenesis, Neth. J. Med. 72 (9) (2014) 442–448.
[56] G. Rizzardini, T. Saporito, A. Visconti, What is new in infectious diseases: Nipah virus, MERS-CoV and the blueprint list of the World Health Organization, Infez. Med. 26 (3) (2018) 195–198.
[57] R.K. Singh, K. Dhama, Y.S. Malik, M.A. Ramakrishnan, K. Karthik, R. Tiwari, et al., Zika virus–emergence, evolution, pathology, diagnosis, and control: current global scenario and future perspectives—a comprehensive review, Vet. Q. 36 (3) (2016) 150–175
[58] R.K. Singh, K. Dhama, Y.S. Malik, M.A. Ramakrishnan, K. Karthik, R. Khandia, et al., Ebola virus–epidemiology, diagnosis, and control: threat to humans, lessons learnt, and preparedness plans—an update on its 40 year's journey, Vet. Q. 37 (1) (2017) 98–135.
[59] R.K. Plowright, D.J. Becker, D.E. Crowley, A.D. Washburne, T. Huang, P.O. Nameer, et al., Prioritizing surveillance of Nipah virus in India, PLoS Negl. Trop. Dis. 13 (6) (2019) e0007393.
[60] C.C. Broder, K. Xu, D.B. Nikolov, Z. Zhu, D.S. Dimitrov, D. Middleton, et al., A treatment for and vaccine against the deadly Hendra and Nipah viruses, Antiviral Res. 100 (1) (2013) 8–13.
[61] R.K. Singh, K. Dhama, S. Chakraborty, R. Tiwari, S. Natesan, R. Khandia, et al., Nipah virus: epidemiology, pathology, immunobiology and advances in diagnosis, vaccine designing and control strategies—a comprehensive review, Vet. Q. 39 (1) (2019) 26–55.
[62] S.B. Kasloff, A. Leung, B.S. Pickering, G. Smith, E. Moffat, B. Collignon, et al., Pathogenicity of Nipah henipavirus Bangladesh in a swine host, Sci. Rep. 9 (1) (2019) 1–11.
[63] World Health Organization. WHO Director-General's opening remarks at the media briefing on COVID-19-11 March 2020.
[64] Su, P. H., Gasic, M., Mrksic, N., Rojas-Barahona, L., Ultes, S., Vandyke, D., et al. Continuously learning neural dialogue management. *arXiv:1606.02689*, 2016.
[65] L. Zhu, Z.G. She, X. Cheng, J.J. Qin, X.J. Zhang, J. Cai, et al., Association of blood glucose control and outcomes in patients with COVID-19 and pre-existing type 2 diabetes, Cell Metab. 31 (6) (2020) 1068–1077.
[66] J. Cui, F. Li, Z.L. Shi, Origin and evolution of pathogenic coronaviruses, Nat. Rev. Microbiol. 17 (3) (2019) 181–192.
[67] R.A. Khailany, M. Safdar, M. Ozaslan, Genomic characterization of a novel SARS-CoV-2, Gene Rep. 19 (2020) 100682.
[68] M. Yüce, E. Filiztekin, K.G. Özkaya, COVID-19 diagnosis—a review of current methods, Biosens. Bioelectron. (2020) 112752.
[69] Y. Chen, S.L. Klein, B.T. Garibaldi, H. Li, C. Wu, N.M. Osevala, et al., Aging in COVID-19: vulnerability, immunity and intervention, Ageing Res. Rev. (2020) 101205.
[70] L. Dai, G.F. Gao, Viral targets for vaccines against COVID-19, Nat. Rev. Immunol. 21 (2) (2021) 73–82.
[71] J. Murray, A.L. Cohen, Infectious disease surveillance, Int. Encycl. Public Health. 2017 (2016) 222–229.
[72] S. Bhatia, B. Lassmann, E. Cohn, A.N. Desai, M. Carrion, M.U.G. Kraemer, et al., Using digital surveillance tools for near real-time mapping of the risk of infectious disease spread, NPJ Digit. Med. 4 (2021) 73.

[73] J. Ginsberg, M.H. Mohebbi, R.S. Patel, L. Brammer, M.S. Smolinski, L. Brilliant, Detecting influenza epidemics using search engine query data, Nature 457 (2009) 1012–1014.
[74] L.E. Charles-Smith, T.L. Reynolds, M.A. Cameron, M. Conway, E.H. Lau, J.M. Olsen, et al., Using social media for actionable disease surveillance and outbreak management: a systematic literature review, PLoS One 10 (2015) e0139701.
[75] J.R. Andrews, S. Basu, Transmission dynamics and control of cholera in Haiti: an epidemic model, Lancet 377 (2011) 1248–1255.
[76] E. Nsoesie, M. Mararthe, J. Brownstein, Forecasting peaks of seasonal influenza epidemics. PLoS Curr. 5, 2013: ecurrents.outbreaks.bb1e879a23137022ea79a8c508b030bc.
[77] S.M. Kissler, C. Tedijanto, E. Goldstein, Y.H. Grad, M. Lipsitch, Projecting the transmission dynamics of SARS-CoV-2 through the postpandemic period, Science 368 (2020) 860–868.
[78] B. Rabbani, H. Nakaoka, S. Akhondzadeh, M. Tekin, N. Mahdieh, Next generation sequencing: implications in personalized medicine and pharmacogenomics, BioSyst. 12 (2016) 1818–1830.
[79] P.M. Woollard, N.A. Mehta, J.J. Vamathevan, S. Van Horn, B.K. Bonde, D.J. Dow, The application of next-generation sequencing technologies to drug discovery and development, Drug Discov. Today 16 (2011) 512–519.
[80] W. Zhao, J.J. Chen, S. Foley, Y. Wang, S. Zhao, J. Basinger, et al., Biomarker identification from next-generation sequencing data for pathogen bacteria characterization and surveillance, Medicine 9 (2015) 1253–1264.
[81] Chee-Seng K, En Yun L, Yudi P, Kee-Seng C. Next generation sequencing technologies and their applications. eLS, 2010.
[82] A.N. Desai, A. Jere, Next-generation sequencing for cancer biomarker discovery. In Next generation sequencing in cancer research, Springer Int. Publ. 2 (2015) 103–125.
[83] L. Barzon, E. Lavezzo, V. Militello, S. Toppo, G. Palù, Applications of next-generation sequencing technologies to diagnostic virology, Int. J. Mol. Sci. 12 (2011) 7861–7884.
[84] H. Beltran, R. Yelensky, G.M. Frampton, K. Park, S.R. Downing, T.Y. MacDonald, et al., Targeted next-generation sequencing of advanced prostate cancer identifies potential therapeutic targets and disease heterogeneity, Eur. Urol. 63 (2013) 920–926.
[85] J.P. Lopez, C. Cruceanu, L.M. Fiori, S. Laboissiere, I. Guillet, J. Fontaine, et al., Biomarker discovery: quantification of microRNAs and other small non-coding RNAs using next generation sequencing, BMC Med. Genomics 8 (2015) 1.
[86] F. Luciani, R.A. Bull, A.R. Lloyd, Next generation deep sequencing and vaccine design: today and tomorrow, Trends Biotechnol. 30 (2012) 443–452.
[87] P. Mathonet, C. Ullman, The application of next generation sequencing to the understanding of antibody repertoires, Immunology 4 (2013) 265.
[88] T. Jartti, L. Jartti, O. Ruuskanen, M. Söderlund-Venermo, New respiratory viral infections, Curr. Opin. Pulm. Med. 18 (2012) 271–278.
[89] M.L. Metzker, Emerging technologies in DNA sequencing, Genome Res. 15 (2005) 1767–1776.
[90] M.L. Metzker, Sequencing technologies — the next generation, Nat. Rev. Genet. 11 (2010) 31–46.
[91] A.M. Wensing, D.A. van de Vijver, G. Angarano, et al., SPREAD Programme. Prevalence of drug-resistant HIV-1 variants in untreated individuals in Europe: implications for clinical management, J. Infect. Dis. 192 (2005) 958–966.
[92] M.R. Wilson, S.N. Naccache, E. Samayoa, et al., Actionable diagnosis of neuroleptospirosis by next-generation sequencing, N. Engl. J. Med. 370 (2014) 2408–2417.

[93] N.D. Grubaugh, J.T. Ladner, M.U.G. Kraemer, et al., Genomic epidemiology reveals multiple introductions of Zika virus into the United States, Nature 546 (2017) 401–405.
[94] J.C. Kwong, K. Mercoulia, T. Tomita, et al., Prospective wholegenome sequencing enhances national surveillance of Listeria monocytogenes, J. Clin. Microbiol. 54 (2016) 333–342.
[95] I. Maljkovic Berry, M.C. Melendrez, T. Li, et al., Frequency of influenza H3N2 intra-subtype reassortment: attributes and implications of reassortant spread, BMC Biol. 14 (2016) 117.
[96] J. Quick, N.J. Loman, S. Duraffour, et al., Real-time, portable genome sequencing for Ebola surveillance, Nature 530 (2016) 228–232.
[97] N.R. Faria, E.C. Sabino, M.R. Nunes, L.C. Alcantara, N.J. Loman, O.G. Pybus, Mobile real-time surveillance of Zika virus in Brazil, Genome Med. 8 (2016) 97.
[98] K. Schmidt, S. Mwaigwisya, L.C. Crossman, et al., Identification of bacterial pathogens and antimicrobial resistance directly from clinical urines by nanopore-based metagenomic sequencing, J. Antimicrob. Chemother. 72 (2017) 104–114.
[99] E.H. Graf, K.E. Simmon, K.D. Tardif, et al., Unbiased detection of respiratory viruses by use of RNA sequencing-based metagenomics: a systematic comparison to a commercial PCR panel, J. Clin. Microbiol. 54 (2016) 1000–1007.
[100] S.J. Salter, M.J. Cox, E.M. Turek, et al., Reagent and laboratory contamination can critically impact sequence-based microbiome analyses, BMC Biol. 12 (2014) 87.
[101] H.K. Lee, C.K. Lee, J.W. Tang, T.P. Loh, E.S. Koay, Contaminationcontrolled high-throughput whole genome sequencing for influenza A viruses using the MiSeq sequencer, Sci. Rep. 6 (2016) 33318.
[102] W. Gu, E.D. Crawford, B.D. O'Donovan, et al., Depletion of abundant sequences by hybridization (DASH): using Cas9 to remove unwanted high-abundance species in sequencing libraries and molecular counting applications, Genome Biol. 17 (2016) 41.
[103] Y. Yang, S. Walls, S.M. Gross, G.P. Schroth, R.G. Jarman, J. Hang, Targeted sequencing of respiratory viruses in clinical specimens for pathogen identification and genome-wide analysis, in: A. Moya, B.P. Brocal (Eds.), The Human Virome Methods and Protocols, Humana Press, New York, NY, 2018.
[104] S.E. Mate, J.R. Kugelman, T.G. Nyenswah, et al., Molecular evidence of sexual transmission of Ebola virus, N. Engl. J. Med. 373 (2015) 2448–2454.
[105] I.M. Berry, M.C. Melendrez, K.A. Bishop-Lilly, W. Rutvisuttinunt, S. Pollett, E. Talundzic, et al., Next generation sequencing and bioinformatics methodologies for infectious disease research and public health: approaches, applications, and considerations for development of laboratory capacity, J. Infect. Dis. 221 (2020) S292–S307.
[106] W. Toledo-Rueda, N.H. Rosas-Murrieta, J.E. Munoz-Medina, C.R. Gonzalez-Bonilla, J. Reyes-Leyva, G. Santos-Lopez, Antiviral resistance markers in influenza virus sequences in Mexico, 2000–2017, Infect Drug Resist 11 (2018) 1751–1756.
[107] J. Liu, J. Xu, L. Liu, et al., Sudden emergence of human infections with H7N9 avian influenza A virus in Hubei province, central China, Sci. Rep. 8 (2018) 2486.
[108] I. Mena, M.I. Nelson, F. Quezada-Monroy, et al., Origins of the 2009 H1N1 influenza pandemic in swine in Mexico, Elife 5 (2016). pii: e16777.
[109] Y.K. Yong, H.Y. Tan, A. Saeidi, W.F. Wong, R. Vignesh, V. Velu, et al., Immune biomarkers for diagnosis and treatment monitoring of tuberculosis: current developments and future prospects, Front. Microbiol. 10 (2019) 2789.

Index

Note: Page numbers followed by "*f*" and "*t*" refer to figures and tables, respectively.

P

Q

R

FSC
www.fsc.org
MIX
Papier | Fördert gute Waldnutzung
FSC® C083411

Kolibri 360 GmbH
Ferdinand-Jühlke-Straße 7
99095 Erfurt, Deutschland
produktsicherheit@kolibri360.de